Robert Barnes, Benjamin Frederich Dawson

Lectures on Obstetric Operations

Including the Treatment of Hæmorrhage

Robert Barnes, Benjamin Frederich Dawson

Lectures on Obstetric Operations
Including the Treatment of Hæmorrhage

ISBN/EAN: 9783337811860

Printed in Europe, USA, Canada, Australia, Japan

Cover: Foto ©berggeist007 / pixelio.de

More available books at **www.hansebooks.com**

LECTURES

ON

OBSTETRIC OPERATIONS

INCLUDING

THE TREATMENT OF HÆMORRHAGE

AND FORMING

A Guide to the Management of Difficult Labour

BY

ROBERT BARNES, M.D. Lond., F.R.C.P.

OBSTETRIC PHYSICIAN AND LECTURER ON OBSTETRICS AND THE DISEASES OF WOMEN AND CHILDREN TO
ST. GEORGE'S HOSPITAL; EXAMINER IN OBSTETRICS TO THE ROYAL COLLEGE OF PHYSICIANS AND THE
ROYAL COLLEGE OF SURGEONS; PRESIDENT OF THE METROPOLITAN BRANCH OF THE BRITISH
MEDICAL ASSOCIATION; LATE EXAMINER TO THE UNIVERSITY OF LONDON; FORMERLY
OBSTETRIC PHYSICIAN TO THE LONDON AND TO ST. THOMAS' HOSPITALS; AND LATE
PHYSICIAN TO THE EASTERN DIVISION OF THE ROYAL MATERNITY CHARITY

Third Edition, Revised and Extended

LONDON
J. & A. CHURCHILL, NEW BURLINGTON STREET
1876

TO PROFESSOR PAJOT.

———

DEAR PROFESSOR PAJOT,

You have thrown the Ægis of your great reputation over the French Editions of my works. You have thus given me a claim to place this third English Edition under the same protection.

In dedicating this book to you, who inherit the glorious traditions of the French school, and enrich that noble heritage, permit me to express my gratitude for all I owe to that school, to many of its teachers, and especially to him whose memory we alike revere, your illustrious predecessor, Dubois.

ROBERT BARNES.

December, 1875.

———

DURING the last thirty years French obstetric physicians have paid
but little attention to the advances made in this department by their
neighbours.

Two manuals of merit, though incomplete, by Naegelé and Scanzoni,
are almost the only foreign works which have been translated into our
language since the beginning of the latter half of this century.

This sad indifference has, under the influence of our public mis-
fortunes, now given way to an earnest desire to observe the progress
made by foreigners of distinction in the different branches of science and
art. Every effort, therefore, which is destined to spread in France the
knowledge acquired by other nations deserves encouragement; and
French physicians will surely be grateful to the industrious translator *
of these Lectures on Obstetric Operations, the most positive branch of
Medicine. The comparison of the foreign methods, modes, and instruments with our own may become the source of improvements advantageous to women and children, whose lives so often depend on our
means of action and on our skill.

To add to our French science that of other countries is not only to
know *more*, but to know *better*.

Without going quite so far as Dr. Barnes, when he says, " Even
more than surgery and medicine, obstetrics calls for promptitude in
judgment, courage under difficulties, and skill in execution," we will
say, to keep within the mark, *at least as much*, and those physicians
and surgeons who are ignorant of the difficulties in obstetrics will alone
venture to dispute this assertion.

The book of Dr. Barnes is not, properly speaking, a dogmatic treatise
on obstetric operations; it is a series of original lectures, comprising at
one and the same time a practical analysis of the serious accidents in

* Dr. Cordès, of Geneva.

parturition, the reasoned-out indications for, and the most judicious researches in, the manner of operating, the method to choose, the instrument to prefer, and the details of the manœuvres required to insure success. The clearness of the style is perfect. The order, without being altogether rigorous, is what it is able to be generally in a series of clinical lectures. But this translation will have, above all, the great advantage of making known in France the actual state of obstetric surgery in England.

It will be remarked, with a real feeling of satisfaction, that certain prejudices which were rife in the United Kingdom only forty years ago, at length show signs of being laid aside. For example, the use of our long forceps and of the cephalotribe, which were formerly absolutely discarded and almost ridiculed, now find among the leading accoucheurs of England more than one partisan and more than one staunch defender; nay, the author himself has not hesitated to assign an important place to the use of these instruments.

There is, however, as regards cephalotripsy, and I might say for my part, as regards all the operations, an opinion peculiar to the author, and one to which I should not readily give my assent. I believe it is dangerous, full of peril of all kinds, and only calculated to please a public, always an ignorant judge, always ready to misinterpret an operation not completed at one sitting.

Doing me the honour to quote my method of cephalotripsy, Dr. Barnes adds, "However, I cannot help thinking that it should be terminated at one sitting."

In 1853, having been witness more than once of the deplorable consequences of the obstinate tradition of wishing to finish " then and there " an impossible operation, I had already written against this pretended necessity of finishing obstetric operations at a single sitting, thus—" Whence this important precept, the more important because openly opposed to the accepted surgical rule: 'in labours one must know when to stop in time.'" The surgical precept, "always finish," would be disastrous in obstetrics. The best way to do for the mother and child that which science and our conscience bid us, "is to dare to stop in time," not to fear to give up for the moment a dangerous or impossible operation, the end of which will oftentimes be made more easy and less perilous by a few hours' patience."

Granted with reference to the forceps, these precepts seem to me equally applicable to cephalotripsy and to other operations, and the twenty years which have elapsed since the day when I enunciated them, have convinced me more and more of their utility and their prudence.

Non vi, sed arte, was the rule of a fellow-countryman of the author's. It is mine too.

Setting aside this difference of opinion, and a certain distrust on the subject of the possibility of the application of a new method of cephalotomy, I am happy to be able to bear witness to the numerous clinical truths sown over every page of this book.

The description of the instruments, the application of the forceps, cephalotripsy, embryotomy, Cæsarian section, the practical reflections on narrowing and malformation of the pelvis, ruptures of the uterus, placenta prævia, hæmorrhage, and, in fact, all the grand questions in obstetrics are treated with accurate good sense. At each instant by some remark or other is revealed a superior mind, ripened by having seen much and meditated much.

Certain judgments doubtless will be questioned, some operations contested; but these lectures will none the less remain interesting, instructive, useful; and will prove once more that the fatherland of so many celebrated men as Chamberlin, Smellie, Denman, Burns, Ramsbotham, Simpson, and many others, possesses to-day their worthy successors. Uniting their talents and their knowledge, they have founded the eminent Obstetrical Society of London, of which Dr. Barnes has had the well-merited honour of being a President.

PAJOT.

AUTHOR'S PREFACE TO THE THIRD EDITION.

WHEN I first ventured to describe this book as "a guide to the management of difficult labour," I did so with hesitation. Methods of practice which, by multiplied experiences and anxious thought, had assumed in my mind something of the precision of definite rules, might not commend themselves to the judgment of others, and I might fail to set them forth in such a manner as to make them intelligible. In the latter object, at any rate, success has been achieved. A large number of practitioners have assured me that by following the directions in this book they have acquired practical skill in obstetric operations; and many more have assured me that when called upon to act in the sudden and pressing emergencies of obstetric practice, they have found information and counsel which have brought safety to their patients, and comfort to themselves. The somewhat ambitious title, then, has been justified to a degree surpassing my expectation.

The success achieved increases responsibility and the correlative duty to make the work still more useful to those who honour me by their confidence. To this end I have done my best. Without materially modifying the doctrines set forth in former editions, I have sought to make them more clear by new illustrations and careful revision.

A short lecture on the signs of "Dystocia," and the general

indications for operative interference, has been added as an introduction.

In discussing "Inversion" and "Rupture"—calamities fraught with so much peril to the patients, and so much risk to the reputation of the surgeon—I have striven not only to explain more fully the conditions under which these injuries occur, with a view to their prevention, but also to illustrate the difficulties of diagnosis, and other cognate points, in such a manner as to aid in the appreciation of the medico-legal questions that arise in connection with these catastrophes.

I have felt doubly bound to introduce this edition by Professor Pajot's Preface to the French edition. This seemed to be the most fitting way in which to express my sense of the honour he has conferred upon me; and I am anxious to give prominence to his objection therein stated to what I have said upon the importance of completing delivery after cephalotomy at one sitting. One cannot differ from authority so great without misgiving. But I venture to plead that the difference is not so wide as it appears. It is not one of principle, but one of minor application. The principle of deliberation, to the extent even of interrupted action during an operation, is one of large application. I give it my cordial assent. The whole tenor of my teaching as regards the treatment of Placenta Prævia is in harmony with this principle. But in the case of cephalotomy it may be easily overstrained. As a general rule it is surely right to complete as quickly as can reasonably be done, what, in the interest of the .patient, it is recognized as necessary to do. Until within recent years, the fault, in this country at least, has not been on the side of precipitate resort to operation. The patient has often drifted dangerously near to the rock of irretrievable disaster before help was given. Help, then, should not be tardy or long drawn-out, lest the patient sink under the trial. But does not the question turn greatly on the method of operating? Granted that extraction after cephalotomy can be safely carried out at one sitting, should it not be done? Delivery is necessary to save the woman from the perils of protracted labour. Why leave to an exhausted woman the labour of moulding the crushed head, when, by the skilful use of instruments, you can accomplish the end? If cephalotripsy be supplemented by picking off the broken bones of the vault of the cranium, or by making sections by the wire-écraseur—operations bearing upon the fœtal head, and not adding materially to the suffering of the mother—then, it appears to me, " haud inexperto," that delay is more dangerous than action.

In revising the lectures on " Hæmorrhage," I have carefully weighed the theoretical objections and the authentic facts which have been urged against the intra-uterine application of styptics. This treatment no longer rests upon my individual authority. By wide experience of men entitled to credit, it is amply vindicated in those most critical cases where the uterus is dead to reflex or centric stimulus to contraction ; and where, but for the application of the new principle of local hæmostasis, the woman would too probably perish.

I cannot conclude without making my best acknowledgments to Dr. Cordès for the enterprise and care he has shown in translating this work into French ; to Dr. Conradi, of Christiania, for the like service in translating it into Norwegian; to Professor Lazaréwitch, of Kharkhof, for his proposal, I believe in course of execution, to direct a translation into Russian ; and, lastly, to my son, Dr. Fancourt Barnes, for useful aid in the irksome task of revision.

ROBERT BARNES.

31, Grosvenor Street, London,
    December, 1875.

CONTENTS.

INDEX TO ENGRAVINGS.

OBSTETRIC OPERATIONS.

LECTURE I.

BEFORE entering upon a description of the particular forms of Difficult Labour and of the modes of dealing with them, it will be useful to acquire a general idea of what Difficult Labour is.

There are two ways of studying medicine: the abstract or theoretical, and the clinical or analytical.

The latter is the method which the practitioner must of necessity follow. At the bedside he finds his patient suffering from a combination of symptoms, the significance of which is the problem he must try to solve. In this analytical task he summons to his aid the theoretical knowledge drawn from his own experience and the teaching of others.

The problem before us is the estimation and management of Difficult Labour. If we try to define this term, we encounter at the outset the difficulties that surround all attempts at definition. A simple object or idea may be defined or described with tolerable precision. But a complex state, the outcome of a number of concurrent or conflicting conditions, will always evade rigorous definition. This is eminently the case with Difficult Labour. In dealing with this subject, therefore, we are compelled to start with the most general idea; and then to seek for

B

precision by a process of disintegration, breaking up the general summary into parts, which admit of more particular examination and description.

The term "Dystocia" comes down to us from Hippocrates. Its counterpart is "Eutocia." These terms are convenient from their conciseness. It only remains to endeavour to attach to them a definite meaning. We may sufficiently define *"Eutocia" as labour proceeding smoothly and terminating favourably under the natural forces.* Accepting this, we should be driven to define "Dystocia" as labour in which the converse conditions occur. It would not be easy to devise a better positive definition than that of Harvey, a man who would have been revered as the most illustrious of obstetricians if he had not blinded us to all his other merits by the dazzling splendour of the grandest discovery in medicine. *"Fit partus,"* he says, *"difficilis et laboriosus, quod nec modo neque ordine debito res peragatur aut pravis aliquibus symptomatibus impediatur."*

Many authors invest the subject at the very threshold with needless difficulty by confounding definition with classification, by drawing descriptions based upon knowledge which can only be completed when the labour is over. That is, too late to be of any use to the practitioner at the bedside watching for indications when and how to act. Of what clinical value, for example, is a definition based upon the time expended in the labour? We can only know that an unduly long labour was a case of dystocia when it is all over. A similar objection applies to other definitions, especially to those which invoke a knowledge of the *causes* of the difficulty, such as contracted pelvis, mal-position of the child, and so forth. "Felix qui potuit rerum cognoscere causas !" The discovery of the causes is a subsequent and distinct problem.

We want to know, as early as possible, when in the presence of a woman in labour, whether her case is going on smoothly or not, whether or no she is drifting into danger, and what are the symptoms that dictate the necessity of interfering.

We must then begin by observing the course of the labour. The points for observation are: 1, the time spent in the process ; 2, the character of the pains ; 3, the effects on the patient as revealed by the study of the subjective and objective signs. Observation of these points will tell if the labour is difficult.

Examination internally may reveal the cause of the difficulty, and supply the indication for treatment.

Labour is a problem in dynamics. Three factors are concerned in the solution : 1st, there is the fœtus, the body to be expelled ; 2ndly, there is the channel, made up of the bony pelvis and soft parts, through which the body must be propelled ; these two together constitute the resisting force, the obstacle to be overcome ; 3rdly, there is the expelling power, the uterus and auxiliary forces. This last factor presupposes a healthy condition of the vascular, respiratory, and nervous systems. All these factors must be harmoniously balanced to produce a healthy labour. Labour may come to a stand, or be disturbed from error in any one of these factors, or from disorder in their correlation. What are the signs of disorder in the correlation of the forces of labour ?

Natural or easy labour, propitious both for mother and child, is a continuous process going steadily on towards the accomplishment of its object. It is divided into three stages, the conclusion of each of which is marked by a distinct event. Thus, the first ends with the expansion of the cervix uteri and the rupture of the membranes ; the second, with the extrusion of the child ; and the third, with the casting-off and removal of the placenta. These three stages not uncommonly overlap each other, or are so far continuous that two of the events named, that is, the bursting of the membranes and the expulsion of the child, or the expulsion of the child and the casting-off of the placenta, are accomplished by the same effort. Still these stages are marked by distinctive characters, and include distinct physiological acts. The great feature pervading all three is uterine contraction. It is by this that each step of the process and the final safety against bleeding, when the vessels which united the placenta to the uterus are severed, are accomplished. But the characters of the contraction change with the stage of labour ; and especially do they change if undue resistance or other disturbing influence intervene.

By a well-known figure of speech, we use the word " pains " to describe the " expulsive forces." That is, we put the effect for the cause. It is too true that expulsive effort and pain commonly go together. But they are essentially distinct ; they have no necessary correlation. In the most propitious labours, although the contractile power is active and vigorous, there may be little or no pain. It may even be stated as a general truth that the greater the pain the wider is the departure from propitious labour. Where pain predominates, more or less disturbance of the typical harmony between the factors of labour is certainly entailed. The nervous energy, instead of being concentrated upon the duty of supplying the uterus and auxiliary muscles with contractile power, is in great measure wasted in other directions. Two evil consequences result: first, the labour itself fails to proceed as it ought; secondly, the system at large is injuriously affected. It is in the observation of these two conditions that we find the proofs of dystocia. These proofs are thus naturally divided into two orders: 1, the local; 2, the general. Our attention is, in the first place, necessarily drawn to the general signs. The woman shows manifest evidence of suffering, of distress.

A function which is going on naturally to the accomplishment of its object may, indeed, sometimes be attended with some degree of pain; but that pain exerts no injurious or depressing effect upon the system.

On the contrary a sense of ease, even of satisfaction, is felt. Labour is no exception to this law. If pain arise, it is because the resistance is in excess of the expelling force, or because the subject is abnormally sensitive. In either case pain, an abnormal element, disturbs the equable course. The moral, or rather the emotional element is also disordered. Instead of the healthy invigorating influence of satisfaction and confidence which attends a smoothly advancing labour, mental depression and loss of nervous energy ensue.

The characters that mark propitious labour then are:—A cheerful contented state of mind, absence of bodily restlessness; ability to walk about, to sit, or lie down without distress; regularly recurring uterine contractions, not inducing acute pain; perfect remission of pain during the intervals between the uterine contractions; recovery of strength between the contractions; a sense of something gained by the expulsive efforts; a moderate increase in the frequency of the pulse during the uterine contraction, and return to healthy frequency and rhythm on the subsidence of the contraction; preservation of normal temperature; easy excretion of pale urine.

The manifestations of the contractions vary with the stage of labour. In the first stage the contractions are strictly uterine; they are periodical, but the intervals are longer than in the later stages; they are sometimes painless, but the hand laid on the abdomen feels a wavy (peristaltic) roll, the uterus harden and slightly change its position behind the abdominal wall. These contractions are involuntary, but they are under the influence of emotions just as are the movements of the heart, bladder, or rectum. A little later these contractions generally evoke pain; this is when they become stronger and when resistance is encountered. The contractions observe a regular course: there is first a stage of increment; 2, of acme; 3, of decrement. These combined form the systole. Then follows diastole, a stage of repose, which should be painless.

These contractions cause the membranes to bulge through the os uteri and move the child. When fairly begun they cannot be arrested. The resistance to the contractions during the first stage of labour is mainly in the cervix uteri. This is the first obstacle to be overcome.

At this time when the cervix is expanding under the contact and pressure of the membranes and head, shivering, even vomiting, not seldom attend the uterine contractions. This is simply a healthy reflex act. It is even observed that the process of dilatation often goes on more easily under its influence. It is no more dangerous than is the ordinary vomiting of early pregnancy.

At the commencement of the second stage the membranes should burst, and a great part at least of the liquor amnii should escape. Unless this take place the uterus cannot act at full advantage. Part of the expulsive force is lost in hydrostatic compression. The labour then would linger, and the "pains" would exhibit an abortive character. But in the normal event, the uterine contractions, acting directly on the body of the fœtus, drive it onward through the cervix and os externum uteri, acquiring power as the head descends and rotates in the pelvis. The "pains" become expulsive or "bearing-down;" evoking powerful reflex action they call forth the stored-up nervous energy, and excite the semi-voluntary, semi-reflex contractions of the diaphragm, abdominal and other expiratory muscles.

The uterus must be supported by the diaphragm and abdominal muscles. The diaphragm is not a direct agent; it acts by supplying a *point d'appui* to the abdominal muscles; the chest expands, the lungs fill, the glottis closes. The diaphragm then contracting gives to the base of the thorax, supported by the lungs distended with air, the solidity required for the effective action of the muscular forces. The uterus in its turn, supported and compressed by the diaphragm and abdominal muscles, is excited to further exertion. The body is arched forward, the thighs are flexed upon the belly, the pelvis is tilted up so as to bring its axis more nearly into coincidence with the axis of the trunk. Everything conspires to aid extrusion. The uterine contractions become more frequent; and when the head comes down upon the perinæum, pressing upon the anal and vulvar sphincters, reflex action is irresistibly exerted, and new vigour is added to the expulsive efforts. At this critical moment the nervous, vascular and muscular systems are at the highest pitch of tension. Unless the various forces are exactly balanced, or unless provision is at hand to moderate the resisting force on the one hand, or the driving force on the other, danger is imminent. Thus the vascular tension may issue in apoplexy, or the muscular tension may issue in bursting of the uterus or laceration of the perinæum. These evils are averted by the yielding of the vulva before the advancing head, and often by the opening of the glottis under the sense of pain; when, the *point d'appui* being lost, the abdominal muscles quickly relax, and the dangerous strain is taken off.

The conditions ascertained on examination are: the regular succession and progress of dilatation of the cervix uteri, the rupture of the membranes, discharge of liquor amnii, and propulsion of the fœtus.

Now, if any cause of obstruction intervene, the progress or order of the labour is interrupted, and the characters of the " pains " change.

The characters that mark an unpropitious labour are: irritability of mind, anxiety, bodily restlessness; continuous pain, exacerbated on return of the uterine contractions; tenderness on pressure upon the uterus; the uterine contractions assuming a peculiar abortive or ineffectual character, that is, having a wearying, irritating effect upon the system, and leaving a sense of having been of no use in advancing the labour: this is expressed in the utterances of the woman in a way easily interpreted by the skilled observer. If, instead of aiding the " pains " with a cheerful will, she dread their return, and is careful not to add force to an effort which she feels will be useless, and which exhausts her strength ; if the pulse rise to 100 or above, and maintain this during the intervals of uterine contraction ; if the temperature rise to 100° F. or above ; if there be excessive perspiration, or a hot dry skin ; scanty secretion of high-coloured urine ; or vomiting, which is now of ominous import, being no longer the result of healthy reflex excitation, but of prostration and perverted nervous action, there is obstructed labour ; danger is at hand.

The conditions made out by examination are : occasionally, great tenderness and heat of the vaginal passage ; tumid or unyielding state of the os and cervix uteri ; stationary position of the fœtus ; increasing tumefaction of the scalp (caput succedaneum) if the child is alive.

When these conditions are manifested there is dystocia ; and the physician is called upon to act. His first duty plainly is to discover the cause of the dystocia ; the second is to study and to apply the appropriate remedy.

LECTURE II.

Two things have to be considered when attempting to describe
the operations in midwifery—

1. What are the emergencies which call upon the practitioner to operate?

2. What are the means, the instruments at his disposal?

If each accident or difficulty in labour were uniform and
constant in all its conditions, it might be possible to apply to
its relief the same operation or the same instrument. Could we
accept Levret's definition that "Labour is a natural operation,
truly mechanical, susceptible of geometrical demonstration," the
history of operative midwifery might be told in an orderly
series of simple mechanical formulæ. But how different is the
case in practice! How infinite is Nature in her phases and
combinations! The dream of Levret will never be realized.
In proportion as observation unfolds these combinations, ingenuity is ready to multiply the resources of art. To describe
these combinations, and the means of meeting them, is a task of
ever-growing difficulty. Partial success only is possible.

The multitudinous array of instruments exhibited at the
Obstetrical *conversazione* in 1866, vast as it was, gave but a
feeble idea of the luxuriant variety that have been devised. If
all these had their individual merits and uses, endless would be
the labour of appreciation; the task of describing the operations
of midwifery would be hopeless. It is, indeed, true that every
instrument, even every modification of an instrument, represents
an idea, although sometimes this idea is not easy to understand.

Fortunately, it is not always important that the idea should be understood. Many of these instruments are suggested by imperfect observation, by ill-digested experience ; many are insignificant variations upon an idea which, in its original expression, was of little value. Huge heaps, then, of instruments may, without loss to science, and to the great comfort of womankind, be cast into the furnace ; the ideas of their inventors melted out of them. All that is necessary in relation to them is, to preserve examples in museums, where they may serve as historical records marking the course of obstetric science in its ebb and flow ; for, strange to say, obstetric science has its fluctuations of loss as well as of gain, of going back as well as of going forward. These historical specimens will also serve the useful office of warning against the repetition of exploded errors, and of saving men the trouble and vexation of re-inventing.

When Science finds herself in the presence of complicated and disordered facts and ideas, her resource is to classify—that is, to seize a few leading ideas under which the subordinate ones may be grouped. In the first instance, the minor or subsidiary ideas—the epigenetic ideas they may be called—are disregarded. The grand or governing ideas only are studied. Then the process of analysis, the descent to details, to particulars, begins ; and again, unless we keep a steady eye upon the governing principles, we are in danger of losing ourselves in the infinitely little, of falling into chaos, of running astray from the parent or guiding truth, in fruitless chase of the multitudinous splinters into which it has been subdivided.

What, then, have we to do ? Knowing that, we will see how we can do it. Nature, although always requiring skilful watching, in the majority of cases does not want active assistance. But the cases are many in which pain, agony, may be averted ; in which positive danger has to be encountered and thrust aside ; in which action must be prompt and skilful.

To produce a healthy labour its three factors must be harmoniously balanced. Labour may come to a stand from error in any one of these factors, or from disturbance of correlation. The permutations are almost infinite in kind and degree. There are many ways in which disturbance may arise. There are not so many ways in which compensation or correction may be made—that is, treatment is more simple than are the causes of disturbance. Let us take one factor, *the expelling power*. *This may be deficient*, the other factors preserving their due relations. This power is a *vis à tergo*. The want of it may be made good in one of two ways. We may, in some cases, spur the uterus and its auxiliary muscles to act. The power may be dormant only; it exists potentially, capable of being roused by appropriate stimulus. This is the case for oxytocics, such as ergot, cinnamon, borax, or cinchona. But the power may not be there, or, if there, it may not be wise to provoke it to action.

An interesting question arises here :—Can we, without resorting to oxytocic medicines, arouse or impart a *vis à tergo ?* Can we apply direct mechanical force to push the fœtus out of the uterus, instead of dragging it out ? Now, in some cases this is possible. Von Ritgen,* in a memoir on " Delivery by Pressure instead of Extraction," adverting to the fact that the *natural* mode is by pushing out, said that the *artificial* mode is by dragging out; and asked very pertinently, " Why do we always drag and never push out the fœtus ?" Dr. Hoenning (*Scanzoni's Beiträge*, 1870) says this was known to Celsus. It was strongly urged by Benjamin Pugh. The principle is seen rudely applied amongst some savage races, in the practice of sitting upon the belly of the woman in labour. Dr. Kristeller† has carried the idea into practice. By means of a dynamometric forceps he has shown that a force of five, six, or eight pounds only is often sufficient to extract a head that has lain for hours unmoved; so that the force to be administered in the form of pressure need not be very great. Poppel ‡ estimates, from experiments made to determine the power necessary to burst the membranes, that the force necessary to effect an easy labour does not much exceed four pounds. The driving-power exerted in labour must necessarily be equal to the weight of the child. This has to be propelled horizontally, retarded to a variable extent by friction. Taking the average weight at seven or eight pounds, and adding as much for friction, we shall arrive at a fair approximate estimate, probably not less trustworthy than estimates based upon experiments in which some factor or other is generally overlooked. It is needless to premise that the presentation and the relations of fœtus and pelvis must be normal.

* " Monatsschr. f. Geburtsk," 1856.   † Ibid., 1867.   ‡ Ibid., 1863.

The method is as follows:—The patient lying on her back, the operator places his hands spread on the fundus and sides of the uterus, and combining downward pressure with the palms on the fundus with lateral pressure by means of the fingers, the uterus being brought into correct relation with the pelvic axis, its contents are forced down into the cavity. The pressure is so ordered as to resemble the course and periodicity of the natural contractions. Of course, the pressure will often excite uterine contraction to aid or even supplant the operator. But it seems that pressure alone is sometimes sufficient. As an adjuvant to extraction, pressure is, I know, of great value. I never use the forceps or any extracting means without getting an assistant to compress the uterus firmly, to maintain it in its proper relation to the axis of the brim, and to help in the extrusion of the fœtus. This resource, then, should not be lost sight of. In certain cases it may obviate the necessity of using the forceps ; or it may stand you in good stead when instruments are not at hand. When a *vis à tergo* cannot be had, we have the alternative of supplying power by importing a *vis à fronte*. In the case we are supposing, the means of doing this reside chiefly in two instruments : the forceps when the head presents; the hand when other parts present.

In the second order of cases *there is a want of correlation between the body to be expelled and the channel which the body must traverse.* There are many varieties of this kind of disturbance. The progress of the head may be opposed by rigidity of the soft parts, especially of the cervix uteri. Patience is one great remedy for this. A dose of opium and a few hours' sleep will sometimes accomplish all that is desired. But patience may be carried too far. If the pulse and temperature rise, and the patient show signs of distress, it is proper to help. I have no faith in belladonna. To excite vomiting by tartar emetic is to add to the distress of the patient without the certainty of relieving her. To bleed is also to indulge in a speculation that will certainly cost the patient strength she will need, and it promises only a doubtful gain. We have two mechanical resources to meet this strictly mechanical difficulty. There is the hydrostatic dilator, which I have contrived for the express purpose of expanding the cervix. In the case of a cervix free from disease, dilatation will commonly proceed rapidly and smoothly under the eccentric pressure of these dilating water-bags, which closely imitate in their action the hydrostatic pressure of the liquor amnii. In the case of rigidity from morbid tissue, as from hypertrophy or cicatrices, something more may be necessary. The timely use of the knife will save from rupture, from exhaustion, or from sloughing. I have contrived a very convenient bistoury for this purpose. It is carried by the finger into the os uteri; multiple small nicks are made in its circumference; and by alternate distension with the water-bags the cervix may be safely and sufficiently dilated.

The fœtus and the channel may be duly proportioned, but *the position of the child is unpropitious.* In this case all there is to do is to restore the lost relation of position. The hand, the lever, and the forceps are the instruments.

There is disproportion. This may be of various kinds and degrees. The varieties will be more conveniently unfolded hereafter. It is sufficient to say here that all resolve themselves, in practice, into three classes—

1. Disproportion that can be overcome without injury to the mother, and with probable safety to the child.

2. Disproportion that can be overcome without injury to the mother, but with necessary sacrifice of the child.

3. Disproportion that can be overcome with possible or probable safety to both mother and child.

The first class of cases may be relieved by the hands, or by the forceps. The second by reducing the bulk of the child to such dimensions as will permit it to pass through the contracted channel. The perforator, the crotchet, the craniotomy-forceps or cranioclast, the cephalotribe, the forceps-saw, and the wire-écraseur are the principal instruments for bringing the bulk of the child down to the capacity of the pelvis. In the third class of cases we cannot insure the mother's safety by sacrificing her child. We therefore seek her *probable* safety by an operation—the Cæsarian section—which evades the difficulty of restoring the relation of bulk and capacity between fœtus and pelvis, by extracting the fœtus through an artificial opening in the mother's abdomen. The instruments required for this purpose are not specially obstetrical. But a bistoury, scissors, needles, and sutures, silk or silver, take but little room, and as they may at any unforeseen moment be wanted, they should always be found in the obstetric bag. And we shall have to put into it a few other instruments and accessories in order to be prepared for all emergencies. Let us enumerate all in order.

OBSTETRIC INSTRUMENTS—THE OBSTETRIC BAG.

To save the child.

1. A lever. (This may be dispensed with.)
2. A long double-curved forceps.
3. Roberton's apparatus for returning the prolapsed funis.
4. Richardson's bellows to restore from asphyxia.

To reduce the bulk of the child.

5. A craniotome or perforator.
6. A crotchet.
7. A craniotomy-forceps.
8. A cephalotribe.
9. A wire-écraseur and an embryotomy-scissors.
10. Ramsbotham's decapitating hook.
11. A blunt hook, slightly flexible.

To induce or accelerate labour.

12. A blunt-ended straight bistoury, with a cutting edge of three-quarters of an inch, to incise the os uteri in cases of extreme contraction or cicatrization.

13. A Higginson's syringe, fitted with a flexible uterine tube nine inches long, which serves for the injection of iced water or iron styptic, and which also serves to expand

14. A set of my caoutchouc hydrostatic uterine dilators.

15. Three or four elastic male bougies (No. 8 or 9).

16. A porcupine quill to rupture the membranes.

17. A flexible male catheter. The short silver female catheter is often useless, and is generally less convenient.

18. A pair of scissors and thread.

For the Cæsarian section, and to repair a torn perinæum.

19. A bistoury, forceps, director, sutures, silk, and silver needles.

To restore the mother.

20. A transfusion apparatus.

MEDICINES.

1. Chloroform or ether and inhaler.
2. Chloral.
3. Laudanum.
4. Hofmann's anodyne.
5. Ergot of rye.
6. Perchloride, persulphate, or chloroxyde of iron.
7. Permanganate of potash.

The iron and permanganate may be conveniently stowed in a separate case with the Higginson's syringe, the dilators and uterine tube. Both should be in the solid form, so as to occupy less room, and to avoid risk of injuring instruments by leakage. ʒij. of the perchloride or persulphate dissolved in ʒx. of water makes the "*styptic injection*." ʒj. of the permanganate in ʒx. of water makes the "*antiseptic injection*," to be injected through the uterus after labour to counteract septicæmia.

The most convenient mode of packing these things is to adapt a travelling leather bag. There is always spare room for any-thing likely to be wanted, besides its ordinary furniture, or for bringing away a pathological specimen; and by turning out the obstetric furniture you have a travelling bag again.

I will now say a few words in explanation of the instruments recommended.

1. *The Lever.*—The form adopted is that of Dr. Uvedale West. The blade or fenestra is moderately arched, making it easy to introduce; the fenestra is wide to grasp a large segment of the head; and there is a joint in the shank, enabling the instrument to be doubled up for greater convenience of carrying.

2. *The Forceps.*—There are several excellent models. I am not bigoted in favour of my own. Amongst the best are Simp-son's and Roberton's. The essential conditions to be contended for are :—That the blades have a moderate pelvic curve ; a head-curve also moderate ; an extreme divergence between the fenes-træ of three inches; the length of the arc of the cranial bow of about seven inches, to adapt it to the elongation of the foetal head during protracted labour. There should be between the springing of the bows and the lock a straight shank to lie parallel with its fellow, to carry the lock clear of the vulva and to save the perinæum. In my forceps the shank is further

lengthened by a semi-circular bow, which forms a ring with its fellow when locked. The use of this is to give a hold for the finger of one hand whilst the other grasps the handles. In Simpson's instrument there is a hollowed shoulder at the head of each handle which answers a similar purpose, and perhaps better. The lock should be easy—a little loose. The English lock is not, I think, surpassed for convenience; but the French lock is a good one. The handles should not be less than five inches long. They should afford a good grasp. Unless they are strong and of fair length, they cannot exert any compressive force for want of leverage, for the fulcrum is at the lock. I think all forceps that have very short handles, especially if not provided with some means, such as the ring or projecting shoulders, which will enable the operator to use both hands, ought to be rejected. A two-handed instrument can be worked with the utmost nicety and economy of muscular force. A single-handed instrument is necessarily a weak one. The absurd dread of possessing powerful instruments has long been the bugbear of English Midwifery. It has been sought to make an instrument safe by making it weak. There can be no greater fallacy. In the first place, a weak instrument is, by the mere fact of its weakness, restricted to a very limited class of cases. In the second place, if the instrument is weak, it calls for more muscular force on the part of the operator. Now, it is sometimes necessary to keep up a considerable degree of force for some time, and not seldom in a constrained position. Fatigue follows; the operator's muscles become unsteady; the hand loses its delicacy of diagnostic touch, and that exactly balanced control over its movements which it is all-important to preserve. Under these circumstances he is apt to come to a premature conclusion that he has used all the force that is justifiable, that the case is not fitted for the forceps, and he takes up the horrid perforator; or he runs the risk of doing that mischief to avoid which his forceps was made weak. The faculty of accurate graduation of power depends upon having a reserve of power. Violence is the result of struggling feebleness, not of conscious power. Moderation must emanate from the will of the operator; it must not be looked for in the imperfection of his instruments. The true use of a two-handed forceps is to enable one hand to assist, to relieve, to steady the other. By alternate action the hands get rest, the muscles preserve their tone, and the accurate sense of resistance which tells him the minimum degree of force that is necessary, and which warns him when to desist. A similar reasoning applies to the perforator and the craniotomy-forceps.

Roberton's Tube for Prolapsed Funis.—There are many contrivances for returning the prolapsed cord. Braun's is an excellent one ; one by Hyernaux, of Brussels, is also very ingenious and useful ; but Roberton's appears to me the most simple. By the knee-elbow posture, indeed, all instruments may occasionally be dispensed with ; but still it is well to be provided with this very simple apparatus.

The Perforator.—The instruments designed to open the skull are classed in the Obstetrical Society's Catalogue under four types :—1. The wedge-shaped scissors, having blades cutting on the outer sides. 2. The spear-head. 3. The conical screw. 4. The trepan. Mr. Roberton uses a spear-head. The form most in use in this country is some modification of Smellie's wedge-shaped scissors. But many of these instruments are very clumsy and inefficient. It requires sometimes considerable force to penetrate the cranium. A weak instrument here is especially dangerous : it is apt to slip, to glide off the globe of the head at a tangent, and to tear the uterus. The conditions of efficiency are these :—The perforating blades must be strong and *straight*. The curve sometimes given is of no use whatever, as it throws the force out of the perpendicular. The shanks must be long, eight inches at least, so as to reach the pelvic brim without interfering with the working of the handles. There should be a broad rest for the hand to give a powerful and steady hold. Almost all the instruments in use fail in this point. The best of all those I have tried and seen is the modification of Holmes' and Naegele's by Dr. Oldham ; it fulfils every indication. On the Continent, especially in Germany, the trepan, first introduced by Assalini, and variously modified, is greatly used. To use a trepan, the crown of which can hardly be less than an inch in diameter, you must have at least an equal amount of surface of the cranium accessible, and the crown must be applied quite perpendicularly to the cranium. Now, these conditions are not always present. I have been much pleased in some cases with the trepan of Professor Ed.

Martin, of Berlin. But in others, where the pelvic deformity was great, and especially where it was necessary to perforate after the body was born, there was no room for the passage or application of the instrument. I found no difficulty with Dr. Oldham's perforator; it will run up through the merest fissure wherever the finger will go to guide it, and will readily penetrate any part of the skull. This, then, is the perforator to be preferred. Various guards have been adapted to the scissor or spear-head perforators. These consist of sliding sheaths, which cover the sharp edges of the instrument during introduction, and which can be drawn back when the adjustment for operation is made. They appear to me to constitute unnecessary complications; they rather hinder than aid the working of the instrument. The most trustworthy guard is the finger of the operator.

The Crotchet.—The design of the crotchet was to seize and extract, by taking a hold inside the cranium, after perforation. For this purpose the best crotchet is the one used in the Dublin Lying-in Hospital. It has a curve in the shank, which is set in a transverse bar of wood for a handle. This gives an excellent hold for traction, that does not fatigue or cramp the operator. The crotchet, however, as an extracting instrument, has been greatly displaced by the craniotomy-forceps and the cephalotribe. The use to which I now almost restrict the crotchet is to break up the brain and tentoria, so as to facilitate the evacuation and collapse of the skull.

The Craniotomy-Forceps.— The use of this instrument is twofold. It should be able to break up and pick away the bones of the cranial vault, and to grasp firmly the skull to serve as an extractor. In the majority of cases the latter action alone is necessary. For extraction, the essential condition is to have the blades so made that when grasping they shall be perfectly parallel. Unless this be obtained, the blades will only pinch at one point, and the effect will be to break through the bone, to tear through the scalp, and to come away. Each time the attempt is renewed, if ever so little traction is necessary, you are exposed to the same mishap, until you may find no place left that will afford a hold. To remedy this defect, many instruments are armed with horrent teeth and spikes, which only add to the evil. Whereas, if the blades are parallel, they

c

gripe firmly over a wide surface, and do not break away. The
hold is obtained by compression, by accurate apposition, not by
teeth or spikes. To secure the grasp without fatiguing the
hands by compressing the handles, I have adapted a screw to
bind the handles together. It is also important that the blades
should be distinct, so as to admit of being introduced separately,
like the ordinary forceps. These principles are fairly carried
out in my craniotomy-forceps, and others as well as myself are
well satisfied with this instrument. In selecting the instrument, it is necessary to see that these essentials have been
preserved by the maker. I have been shown many specimens
as mine which I should utterly condemn.

The Cephalotribe.—In cases of great, but not of extreme,
distortion of the pelvis, the cephalotribe is an instrument capable
of materially accelerating delivery. After perforation, the
powerful blades applied to the head crush in and flatten it, so
that it can be drawn through a comparatively narrow passage.
The unwieldy bulk and formidable appearance of most of the
Continental cephalotribes, requiring, as they do, an assistant in
their use, must preclude their extensive adoption. Almost
every objection is removed in Sir James Simpson's instrument.
Dr. Braxton Hicks's modification of Simpson's cephalotribe is, in
my experience, a most effective and convenient instrument,
although a little too short in the blades. Dr. Kidd, of Dublin,
has also contrived an excellent instrument. All these can be
worked by the operator unaided; and can be carried in the bag
with the other pieces of the obstetric armamentarium. Sufficient
power to crush down the roof of the cranium upon the base
after perforation is combined with a minimum of size and
weight.

The Wire-écraseur.—This instrument I propose to add in
order to execute the new method of Embryotomy I have
designed, to effect delivery in extreme cases of pelvic deformity.
Its use is to make sections of the head *in utero*, so as to reduce
it to a bulk easy for extraction. The instrument must be a
powerful one; and it should have an endless or Archimedean-
screw movement on Weiss's principle, so as to work a loop of
wire large enough to take the head in its equator. Other instrument makers make excellent écraseurs, which can be used for this
and other surgical purposes. In connection with this operation

it is also necessary to be provided with a pair of *Embryotomy-scissors* to cut up the trunk. The wire can also be used as a decapitator, instead of Ramsbotham's hook. It further forms an excellent snare to catch a foot when too remote to be seized by a loop of tape carried by the fingers.

The Decapitating Hook.—This instrument will be rarely required, but when the occasion arises, the service it renders is very great. In a protracted transverse presentation, when the child is dead from compression, the uterus spasmodically and closely contracted upon the child, turning cannot be accomplished without subjecting the mother to much suffering and some danger. In such a case it is obviously preferable—the child being past help—to spare the mother to the utmost. This hook can be carried over the child's neck, and by a movement of sawing and traction, the head can be severed in a few seconds. Then the body is extracted by pulling on the prolapsed arm. The head, remaining alone *in utero*, can be easily extracted by the craniotomy-forceps. Thus delivery can be effected, with little cost to the mother, in a few minutes. The same object can be obtained by a pair of strong scissors (Dubois) which is made to divide the cervical spine. An excellent decapitator was exhibited* by Jacquemier; in general form it resembles Ramsbotham's, but it has a concealed or sheathed decapitator, the cutting being effected by movable blades and saw-links. Pajot, again, decapitates by carrying a strong cord round the neck. The operation may also be effected by the wire-écraseur.

The Blunt Hook.—A blunt hook at the end of a slightly flexible stem, and eighteen inches long, is a useful instrument in seizing the child's limbs in certain cases of turning, and in canting the base of the skull after craniotomy.

The Syringe, Uterine Tube, and Caoutchouc Dilators.—These are perhaps the most frequently useful of all the instruments enumerated. A Higginson's syringe is fitted with a mount, to which the flexible uterine tube or any one of my dilators can be adapted. Three sizes of the dilators are sufficient.

The *elastic male bougies* are useful as the best means of inducing labour, that is, of provoking labour.

The *porcupine quill* is a most convenient instrument for

* See Obstetrical Society's Catalogue, p. 47.

piercing the membranes; and although a common quill or steel pen will answer the purpose, these are not always at hand. The special instruments, as stilets, &c., are really superfluous.

The remaining instruments will be described with the operations in which they are used.

And, lastly, let me add a few words concerning the obstetric hand, as the master instrument of all, not only as guiding all the rest, but as performing many most important operations unarmed. In ordinary labour it is the only instrument required. It is also the only instrument called for in many of the greatest difficulties. In malpresentations, in placenta prævia, in many cases of contracted pelvis, in not a few cases where, after perforation, the crotchet and craniotomy-forceps have failed to deliver, the bare hand affords a safe and ready extrication. One cannot help seeing that practice is often determined by the accidental perfection of, or familiarity with, particular instruments. Thus, a man who has only reached that stage of obstetric development which is content with a short or single-curved forceps, will be armed with a good perforator and crotchet. He cannot fail to acquire skill and confidence in embryotomy, and greatly to restrict the application of the forceps. Again, the preference generally given on the Continent to cephalotripsy over craniotomy and extraction by the crotchet or craniotomy-forceps, is the result of the great study directed to the perfecting of the cephalotribe. At the present day we may boast of having good and effective instruments of all kinds, each capable of doing excellent work in its own peculiar sphere, and moreover endowed with a certain capacity for supplanting its rival instruments. For example, the long double-curved forceps is adapted to supplant craniotomy in a certain range of cases of minor disproportion. Hence it follows that it is of more importance to have a good forceps which can save life than it is to have a good perforator and crotchet which destroy life. At the same time, it is eminently desirable to possess the most perfect means of bringing a fœtus through a very narrow pelvis, in order to exclude or to minimise the necessity of resorting to the Cæsarian section. Our aim should then be to get the most out of all our instruments, to make each one as good of its kind as possible. And admirable is the perseverance, marvellous and fertile the ingenuity, that have been brought to this task. I will not say that it has all been misdirected; but certainly the cultivation of the hand, the study of what it can do in the way of displacing cold iron, has been much neglected. It would be not less instructive than curious to carry our minds back to the days when the forceps and other instruments now in use were unknown, and to confront the problem which our predecessors, Ambroise Paré, Guillemeau, and others, had to solve—namely, how to deliver a woman with deformed pelvis without instruments. That they did successfully accomplish in many instances with the unarmed hand what we now do with the aid of various weapons, there can be no doubt. If this implies greater poverty of resources on their part, it not the less implies also greater manual skill. I am confident that the possession of instruments, especially the craniotomy instruments, has led, within the present century, to a neglect of the proper uses of the hands, which is much to be deplored. We are only now recovering some of the lost skill of our ancestors.

Obstetric Surgery has this peculiarity: its operations are carried on in the dark, our only guide being the information conveyed by the sense of touch. The mind's eye travels to the fingers' ends. The hand thus possesses an inestimable superiority over all other instruments. Its every movement is regulated by consciousness. It is right, then, to ascend a little the stream of knowledge, and to endeavour to recover from the experience of our forefathers their secret of *chirurgery*; to regain, to extend, our power over that great instrument from which the Surgeon derives his name.

LECTURE III.

THE MECHANISM OF HEAD-LABOUR—THE POWERS OF THE FORCEPS
—THE FORCE BY WHICH IT HOLDS THE HEAD—THE COM-
PRESSIBILITY OF THE CHILD'S HEAD—THE LEVER—DEMON-
STRATION THAT THE LEVER IS A LEVER, NOT A TRACTOR;
ALSO THAT THE FORCEPS IS A LEVER.

THE description of the forceps may fitly be preceded by a sketch of the mechanism of head-labour. This sketch will also be useful to refer to subsequently when studying the mechanism of shoulder-presentation.

In the first head-position, the occiput is directed to the left cotyloid foramen, the face looks to the right sacro-iliac joint, the vertex points downwards to the os uteri, whilst the long axis or trunk corresponds with the long axis of the uterus, which is coincident, or nearly so, with the axis of the pelvic brim. The head in its progress to birth undergoes five successive movements.*

1. *A Movement of Flexion.*—The posterior fontanelle placed opposite the left cotyloid cavity descends and approaches the

---

* The student is earnestly advised to follow and execute the ensuing description with a fœtal skull and a female pelvis.

It will here be useful to define the meaning we attach to the terms, "Vertex," "Position," and "Presentation." 1. By *Vertex*, we mean the highest point of the head in the erect posture. This is the space between the two fontanelles and the parietal protuberances. Although the central point of this space rarely is exactly the lowest in labour, yet in natural head-labour the presenting part is always some point within this space. 2. By *Position*, we mean the relation of the head or other presenting part to the *plane* of the pelvic brim, that is, to the diameters of the brim. 3. By *Presentation*, we mean the part of the fœtus which is most prominent in the line of the *axis* of the pelvis whether at brim, cavity, or outlet.

centre of the brim, the chin is strongly pressed upon the chest, the back of the neck comes to bear upon the cotyloid wall, whilst the forehead rises on the right, and the anterior fontanelle is applied to the right sacro-iliac joint. By this movement the head fixes itself upon the trunk, and presents its smaller diameters to the greatest or oblique diameters of the pelvic brim. This movement is an exaggeration of the natural flexion which exists before labour begins. In several cases of women dying undelivered at the end of gestation I have had the opportunity of observing in the dead body the position of the child : the chin was flexed upon the chest, and the presentation was oblique.

2. *A Movement of Descent or Progression.*—This begins commonly with the escape of the head from the mouth of the uterus, the clearing of the brim, and ends with the total extrusion of the child. It continues in the same line—that is, in the axis of the brim—until the head reaches the floor of the pelvis.

3. *A Movement of Rotation.*—This takes place chiefly in the lower part of the pelvic cavity. The forehead and the anterior part of the region of the vertex resting on the right sacro-iliac ligament, or on the right posterior wall of the pelvic cavity, follow the incline backwards and downwards, turning towards the sacral cavity, whilst the back of the neck slides behind the left foramen ovale, or the left anterior wall of the pelvis, and, following the incline forwards, and a little upwards, turns towards the upper part of the pubic arch.

4. *A Movement in a Circle.*—The back of the neck is arrested under the symphysis pubis ; the posterior fontanelle is nearly in the centre of the pelvic outlet ; the occiput, the vertex, the forehead, the face, and lastly, the chin, roll successively over the posterior commissure of the vulva, traversing the concavity of the lower part of the sacrum and the distended perinæum. This movement in a circle does not begin until the head has reached the floor of the pelvis, nor until the movement of rotation (3) is completed, or nearly so, that is, when the occiput becomes engaged under the pubic arch.

5. *A Movement of Restitution.*—As soon as the head is freed from the pelvis, the occiput turns quickly to the left, and the face and forehead to the right. This last movement of the head is the effect of the first of a series of movements similar to those described by the head, and which are now described by the

trunk. The shoulders entering the brim in the left oblique diameter, turn the head, now freed from all restraint, bringing the face forwards to the right.

The movements undergone by the trunk are three :—

1. *A Movement of Descent or Continuous Progression.*—The right shoulder is forward to the right, the left is behind to the left : the child's back is directed forwards to the left.

2. *A Movement of Rotation.*—The shoulders and the upper part of the trunk having descended into the excavation, the right shoulder turns towards the apex of the pubic arch, and the left rotates towards the concavity of the sacrum. The child's back, after the rotation, is turned to the left.

3. *A Movement in a Circle.*—The right shoulder remaining fixed beneath the pubic arch, the left shoulder, followed by the corresponding side of the trunk and the left hip, describes the arc of a circle; and gradually the right shoulder rises over the mons Veneris, whilst the parts placed behind traverse the sacro-perinæal concavity.

These movements are governed by the form of the pelvis.

To arrive at a just idea as to the application of instruments in difficult midwifery, it is first of all necessary to study carefully what these instruments can do. What, for example, are the powers of the forceps, the lever, of the crotchet and craniotomy-forceps, and of the cephalotribe? When we know these, and have formed a correct idea of the nature of the labour—that is, of the difficulty to be overcome—we shall know which instrument to select, and how to use it. The powers of an instrument must obviously depend upon its construction; but this is true to an extent not often thoroughly appreciated. Take, for example, the noblest of all, the forceps. It is difficult to exaggerate the importance of developing to the fullest extent the powers of this instrument. The more perfect we make it, the more lives we shall save, and the more we throw back into reserve those terrible weapons which only rescue the mother at the sacrifice of her offspring.

Three distinct powers or forces can be developed in the forceps. First, by simply grasping the head and drawing upon the handles, it is a *tractor*, supplementing a *vis à fronte* for the defective *vis à tergo*. Secondly, the forceps consisting of two blades having a common fulcrum at the joint or lock, we can by a certain manipulation use it as a *double lever*. Thirdly, if the blades and handles are long enough and strong enough, and otherwise duly shaped, the forceps becomes a *compressive power* capable of diminishing certain diameters of the child's head, so as to overcome minor degrees of disproportion.

Now, all these powers may be brought into use, and all may be in great measure lost, according to our choice of a good or a bad model. Thus, if we rest satisfied with the short forceps of Denman, we shall only have a feeble tractor, a feeble lever, and an instrument having almost absolutely no compressive force. It is obvious that such a forceps can have but a restricted application. It can only serve to deliver the child when the head is in the pelvis, when a leverage or very little tractile power is required. Ask yourselves what this means. What is the consequence in practice? Simply this : you are driven in a multitude of cases to perforate, to destroy the child. Such an alternative may well make us reflect whether we cannot extend the powers and the application of the forceps. By simply lengthening the blades and shanks and giving the blades an additional curve adapted to the curved sacrum, we can reach the head detained on the brim of the pelvis. By moderately lengthening the handles and making the instrument stronger, we increase the leverage and tractile power, and we gain a moderate compressive power. Thus we bring within the saving help of the forceps a further number of children that must otherwise be given up to the perforator, or run the risk of turning. You ask, Why hesitate to endow the forceps with this great privilege? Why has the feeble forceps of Denman so long held its sway in this country? The reason is that there are limits beyond which we cannot push the saving powers of the forceps. If we pass beyond these limits, we run into danger of injuring the mother and of losing the child. Now, the great contest in all matters of strife is about boundary lines ; and it is concerning these limits that authorities have differed. Some men are afraid of giving power, lest it should be abused. They are so terrified at the possible mischief which great power may work, that they would rather abandon the good which great power is equally capable of working. They tremble lest we should be unable to acquire the skill and the discretion necessary to direct that greater power. Such men virtually say, You shall not apply the

forceps where the head has not descended into the pelvic cavity
—an arbitrary limit dictated by fear, and fixed by ignorance
that the forceps is just as capable of safely delivering a child
whose head is arrested at the brim. For here, as is continually
the case in Medicine, experience, arbitrarily limited, excludes
progress in knowledge and bars improvement in practice. For
example, how can a man acquire a just knowledge of the power
of the forceps to deliver a head delayed, by slight disproportion
at the brim, if he always delivers under this difficulty by perforating ? Clearly, he bars himself from acquiring that knowledge ; and, giving up his intelligence to the delusive dictates of
his wilfully limited experience, he refuses even to accept the
evidence of those whose experience is greater, because it is
directed by a freer spirit of research, by greater confidence in
the resources of art.

Let us, then, go back to the study of the powers of the forceps,
unshackled by any preconceived opinions as to what the instrument can do or can be permitted to do. First, as to its *tractile*
power. In order to draw, the instrument must take hold.
How does it take hold ? You may at first sight suppose that
this is accomplished by grasping the handles. But in the case
of the ordinary forceps, especially the short-handled forceps,
there is little or no compressive power, so that the hold cannot
be due to the handles. The hold is really due to the curvature
of the blades, which fit more or less accurately upon the globular
head, and to the compression of the bows of the blades against
the soft parts of the mother, supported by the bony ring of the
pelvis. This may be made clear by a simple experiment. Take
an india-rubber ball, slightly larger in diameter than a solid
ring ; place the ball upon the ring. Then seize the ball through
the ring by the forceps. The blades will be opened out by the
ball. Then drawing upon the handles, even without squeezing
them together, you will see the blades pressed firmly upon the ball
by gradual wedging, as the greatest diameter or equator of the
ball comes down into the ring. Just so is it with the child's
head and the pelvic brim and canal. The blades are held in
close apposition to the head by the soft parts and the pelvis of
the mother. The effect of pressure upon the bows of the blades
in maintaining the hold is again proved by the readiness with
which the blades slip off as soon as the equator of the head has

cleared the outlet. In many cases, this outward pressure upon the bows of the blades is enough to serve for traction. It is not necessary to tie the handles of the forceps. You may even do without handles altogether. Thus, one of the earliest attempts, stimulated by the desire to realize the concealed discovery of the Chamberlens—that of Palfyn—consisted in applying two opposed levers, which did not cross, and therefore could not exert any compressive force. Assalini's forceps is constructed on this principle. It is essentially a tractor with slight leverage power. Professor Lazarewitch, of Charkoff, brought to the Obstetrical Exhibition a beautiful forceps constructed on Assalini's principle. This instrument I tested in two cases on the same day. It held admirably ; but all its holding power is due to the great curvature of the blades, and hence to the pressure exerted by the mother's parts upon the blades. Dr. Inglis, of Aberdeen, proposed a forceps in which the handles are done away with altogether, there being nothing but a short curve of the shank, representing the shoulders on the handles of Simpson's forceps, to serve for traction. I think this sacrifice of all compressive and leverage power, reducing the instrument to a weak tractor, is a retrograde movement. But it proves the proposition that the hold upon the child's head is the result of the adaptation of the curved blades and the outward wedge-pressure of the mother's parts upon the bows of the blades. Now, the strength of the hold depends mainly upon the degree of curvature of the blades and the width of the fenestræ. If the curve is one of large radius, so that the two blades, when in opposition, approach parallelism, and especially if the fenestræ be narrow, the hold will be feeble, and moderate traction will cause the forceps to slip, and this in spite of any compression you can exert upon short handles. But increase the curve so that the blades in opposition form nearly a circle, and the instrument will not slip. This increased head-curve is one feature of the French or Continental forceps. The hold is further strengthened by making the points approach nearer together. In the English patterns the points are generally distant from each other an inch or more. In the foreign forceps the distance is often much less than an inch. There is some danger from this proximity of pinching or abrading the skin of the face. So much for the grip and traction.

Let us now study the *compressive power*. This is inconsiderable in almost all the English forceps, but is an important feature in most of the foreign long forceps. The essential conditions for compression are, indeed, present in English and foreign. These consist in the crossing of the blades, and in the greatest divergence of the blades, when the handles are brought together, being less than the greatest tranverse diameter of the child's head. This diameter is normally from 3¾ to 4 in.; the greatest divergence of the blades is rarely more than 3 in. Therefore, when the blades are sitting loosely on the head, the handles diverge. Practically, the head is rarely grasped exactly in its transverse diameter, but generally in one more or less oblique—something between the transverse and the longitudinal diameter. This, of course, is even longer than the transverse. Now, if we are to exert any direct compression upon the head, we can only do it by squeezing the handles together. For this

Fig. 1.

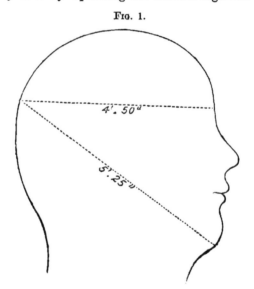

THE ORIGINAL FORM OF THE HEAD BEFORE BEING AFFECTED BY LABOUR.

purpose the handles must be long and strong on one side of the lock, and the blades must be strong, but not much longer, on the other side of the lock, than are the handles.

It would be useless to provide this compressing power if the head were not compressible. That the head is compressible—that is, that we may diminish some of its diameters by lengthening others—is easily proved.

Firstly. It is known that in normal pelves the head in passing, if the labour be protracted, undergoes elongation ; from spherical it becomes conical ; the greatest transverse diameter—the inter-parietal—becomes reduced to approximation to the lesser or interauricular, whilst the longitudinal diameters are correspondingly increased.

These changes I have demonstrated by actual measurements and outlines.* Diagrams 1 and 2 may be taken as types of the normal head and of the form impressed in protracted labour.

Thus, just as the pressure of the soft parts and the pelvis is a main agent in fixing the forceps upon the head, so it is in moulding the head to allow of its passing, Indeed, I think this pressure almost entirely accounts for the alteration of form the

Fig. 2.

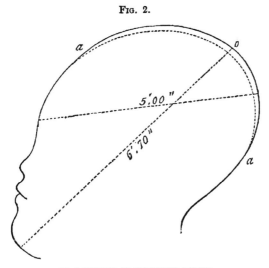

HEAD MOULDED BY PROTRACTED LABOUR.

head undergoes when the English forceps is applied. I can show outlines of heads as strongly altered under the natural forces of labour as they often are under forceps delivery.

Secondly. Numerous experiments have been made with strong forceps upon dead children to determine this point. Baudelocque found that he could lessen the transverse diameter by a quarter to a third of an inch. Siebold gained half an inch. Osiander

* " Obstetrical Transactions," vol. vii.

and Velpeau claim quite as much. More conclusive are the observations of M. Joulin and of M. Chassagny. These gentlemen, in experiments designed to demonstrate the utility of continuous compression and traction by powerful forceps upon the head in difficult labour, have completely proved that a degree of moulding may be effected much beyond that commonly observed. This moulding consists in the elongation of the head, the elongation being gained by the lessening of the equatorial diameters. The process resembles that of reducing wire by drawing it through holes in an iron plate.

Now, another question arises : the head is indeed compressible, but to what extent is it compressible without sacrificing the child's life ? For if the maximum of plasticity compatible with life be represented by that degree which is common in severe first labours, then we ought to give the mother all the ease in our power by lessening the diameter of the child's head by perforating. It is very difficult to fix this limit with accuracy. Baudelocque thought compression to the extent of a quarter or a third of an inch was compatible with the safety of the child. The important fact is, that in many cases the child survives, although its head has undergone very great compression and moulding. Experiments on the dead fœtus are of doubtful value. The conditions are so different, that application of these experiments to clinical use might lead to serious error. The degree of compressibility compatible with life is no doubt a variable quantity. The following conditions influence the result :—The degree of development of the head as to size and ossification ; and the mode in which the compressing force is applied. If this force be applied *gradually* and *continuously*, a much greater extent of moulding with less injury to the child may be obtained than what Baudelocque thought possible.

At one time it was the practice—more probably with the view of securing the hold than of compressing the head—to tie the handles together ; and even now that tying is generally abandoned and condemned, the old custom asserts itself in the preservation of the grooves near the extremities of the handles made to receive the ligature. The objection to tying is this— the continuous compression is opposed to the course of Nature, which intermits the expulsive act, giving periods of rest during which it is presumed that the brain may better adapt itself, and its circulation be maintained. Hence the law that we ought in forceps labours, and, generally in all operative labours, to imitate this intermitting action by interposing intervals of rest, endeavouring so to time our efforts as to be simultaneous with, and in aid of, the natural expulsive efforts. The argument is good, both in logic and in physiology. It is not wise to disregard it. But experience proves that there are cases where the moulding of the head can be accomplished more quickly, and without endangering the child, by continuous pressure. Some practitioners, therefore, have recurred to the old practice. Delore,* who has made many dynamometric observations, concludes that pressure exerted either by the forceps or by the genital organs, may be harmless to the head if spread over a large surface. It is limited and angular pressure that is dangerous. He has also shown that *the greater the traction the greater is the pressure.* The pressure is equal to about half the traction. Thus, if you exert a traction force of fifty pounds, the pressure upon the head is about twenty-five pounds.

To economize traction, then, is to economize pressure. How do we economize traction?

There are three principal rules.

First. Take sufficient time to allow the head to mould.

Secondly. Take care to draw in the axis of the brim—that is, traction must be perpendicular to the plane of the brim. If this is neglected, additional force is required, increasing with every degree of angular difference.

Thirdly. To use slight movements of laterality or oscillation.

This uncertainty and inconstancy in the degree to which compression may be carried with safety to the child, is a justification for tentative or experimental efforts with the forceps. It is the reason why in doubtful cases, where the disproportion in size between the pelvis and head is not very decided, we are called upon to make a reasonable trial of the forceps before resorting to craniotomy. It appears to me quite certain that in this country we are yet far from having utilized the powers of the forceps to the highest legitimate extent. I might go further, and say that during Denman's time, and until quite recently, we had actually lost ground in this respect, and had reverted to the use of instruments scarcely better than the

* "Gazette Hebdomadaire," 1865.

original rude forceps of the Chamberlens. More than one hundred years ago Smellie contrived and used the long forceps. Perfect used it, and it seems that in his time the long forceps was better known in England than it was during the first half of the present century.

It is no valid objection to the forceps to urge that in extreme cases it will probably deliver the child dead from excessive compression. The child has had the best chance of life. The alternative, craniotomy, gives none. We may now study the various cases in which the instrument may be used, and the modes of applying it. It is well to begin with the simplest case. This occurs when the head, presenting in the first position, has descended into the cavity of a well-formed pelvis, and is arrested on the perinæum from want of expelling power. In such a case very moderate leverage and tractile power—a force of a few pounds, perhaps—is all that is required. Often the lever or tractor will be quite sufficient. The moulding or diminution of the equatorial diameters will be effected by the sole compression of the mother's structures. The occiput lying behind the left foramen ovale, the lever may be slipped over it, and the head drawn down towards the pubic arch, using your fingers as a fulcrum. This may be enough, for often, when the head is once started, expulsive action returns. If not, then the lever may be shifted to the opposite side, so as to lie over the child's face and chin in the hollow of the sacrum. Then drawing down, you give the extension movement to the head, and the cranium soon emerges through the outlet. Several skilful practitioners, who frequently resort to this instrument, contend that it is a true tractor, and point, in confirmation of this view, to the great curve of the blade. But I think reflection will show that it is essentially a lever. It does not directly draw down the head, but by pressing upon one side or point of the head-globe, it causes the globe to revolve upon its opposite point or pole as a centre, its axis representing another lever. If the point opposite to that seized by the lever be moveable, of course, when leverage is applied, the head will roll up on one side as it comes down on the other; but if the opposite point be more or less fixed, as the occiput generally is, against the foramen ovale or left ramus of the pubes, then leverage on the face and chin will effect rotation on that fixed point as a centre, and the bulk of the head will have descended.

The following series of diagrams will illustrate the action of the lever bringing the head down by alternate flexion and extension. The lever is supposed to be applied alternately

FIG. 3.

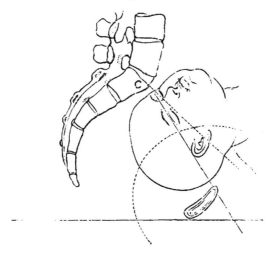

over the occiput and the face. In Fig. 3, c represents the

FIG. 4.

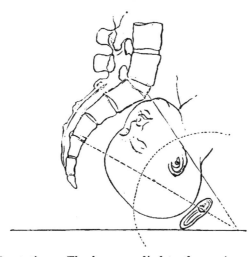

centre of rotation. The lever applied to the occiput will bring

D

down the pubic hemisphere of the head-globe, the forehead remaining nearly fixed against the sacrum at C. Flexion is preserved.

FIG 5.

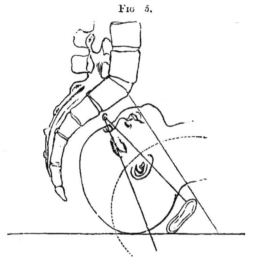

In Fig. 4 the lever is reversed. The centre, C, is at the

FIG. 6.

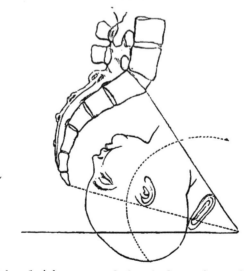

pubes; the facial or sacral hemisphere descends with extension.

In Fig. 5 the lever is shifted back to the occiput, which is made to descend by flexion, the face resting in the sacrum, but at a lower point than in Fig. 3.

In Fig. 6 the lever shifts to the face. The centre, c, is again at the pubic arch. The facial hemisphere is now made to sweep down over the perinæum, performing the extension-movement of delivery.

Fig. 7.

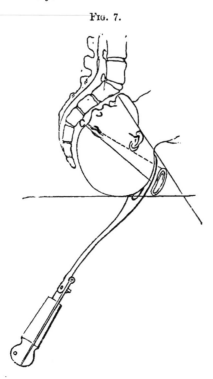

Figs. 7 and 8 further illustrate the same points. In Fig. 7 the lever is seen applied to the occiput, bringing down the pubic hemisphere, whilst the opposite point is fixed in flexion at c in the sacrum.

In Fig. 8 the lever is applied over the face, which is brought down in extension, the occiput resting against the pubes.

An instrument which claims to be considered with the lever is the *whalebone fillet or loop.* Its action is entirely similar. If the loop of the fillet be supposed to be substituted in Figs. 7

and 8 for the lever, the demonstration will equally apply. The
fillet has lost its place in scientific works; but it is, I believe,
largely used by some practitioners, and with great success. The
instrument deserves to be remembered for this reason: it can
be extemporized. Under exceptional circumstances, when no
instruments are to be had, a bit of whalebone can be bent and
applied over the occiput or face, in a case of arrest of the head

Fig. 8.

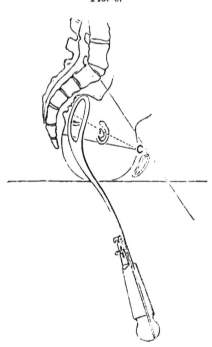

in the pelvis, thus enabling the practitioner to rescue his patient
and himself from a trying situation. Fig. 9 represents an
excellent form of this instrument. It is that of Dr. Westmacott.
I have not used the fillet, having been accustomed to rely
upon the forceps, which is undoubtedly a superior instrument.
It will do all that the fillet or lever will do, and more.

A similar principle of leverage may be applied by the two
blades of the short forceps. But in this case the leverage is
applied more to the transverse diameter of the head. The lever

can in like manner be applied to the side of the head if necessary. When the blades are crossed and locked, the common fulcrum is at the lock. Then by gently bearing upon either handle
alternately, swaying the instrument backwards and forwards,

FIG. 9.

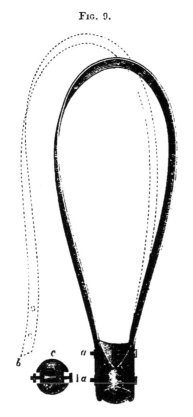

DR. WESTMACOTT'S FILLET.

a a, screw-pins and nuts to fix b. By removing the nuts, the end b of loop is released;
c, top of handle.

avoiding all pressure against the pelvic walls, you cause the
head-globe to rotate to a small extent alternately in opposite
directions upon its own centre. At each partial rotation a little
descent is gained, owing to the point opposite to the lever in
action being partially fixed by the other blade; and by gentle
traction upon the handles. In very many cases this gentle

double leverage is enough to effect delivery. Traction is hardly called for at all. The alternate action of the forceps is illustrated in Figs. 10 and 11. In Fig. 10, the head grasped transversely, the handles are first carried to the left. The right or pubic hemisphere descends.

In Fig. 11 the handles are carried across to the right. The left or sacral hemisphere descends.

Fig. 10.

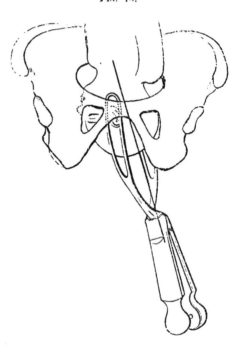

SHOWING LEVERAGE ACTION OF FORCEPS.

It is easy to demonstrate this simple leverage action on the phantom. Thus, if I take each blade of the forceps alternately, unlocked, and use it as a lever, the head advances by a series of alternate side-movements, until it is actually extracted by this power alone. Is it reasonable to throw away a power by means of which we can safely economize the more hazardous traction-force? It is, however, disapproved of by some authorities, who enjoin traction alone. But I believe that pure traction is almost impossible, and I am equally certain that a gentle and careful leverage will enable you to deliver with a great economy of

FIG. 11.

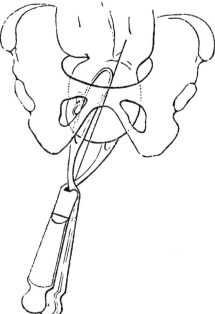

SHOWING LEVERAGE ACTION OF FORCEPS.

force and time, which means, of course, greater safety to the mother.

LECTURE IV.

THE APPLICATION OF THE SHORT FORCEPS—HEAD IN FIRST POSITION—HEAD IN SECOND POSITION—OBJECTIONS TO SHORT FORCEPS—THE APPLICATION OF THE LONG OR DOUBLE-CURVED FORCEPS—INTRODUCTION OF THE BLADES—USE OF LONG FORCEPS—LOCKING—CAUSES OF FAILURE IN LOCKING—EXTRACTION—HOW TO MEASURE THE ADVANCE OF THE HEAD—THE MANŒUVRE OF "SHELLING OUT" THE HEAD DELAYED AT THE OUTLET—RELOCKING—THE HEAD IS SEIZED OBLIQUELY BY THE FORCEPS—TIME REQUIRED FOR EXTRACTION.

WE now come to the mode of applying the short forceps. The head we assume to be in the pelvis, lying in the right oblique diameter, occiput forwards. The child's right ear will be a little to the right of, and above, the symphysis pubis. We have first to consider certain conditions, some of which are necessary to the proper use of the forceps; some which are not necessary, but favourable. 1. The membranes must be ruptured, 2. The cervix uteri must be fairly dilated. 3. The bladder should be empty. 4. The patient must be in a convenient position. Abroad, the patient is usually placed in lithotomy position, on the edge of the bed. With us the pelvis is simply drawn to the edge of the bed, the patient lying on her left side. I think it needless to enter into controversy upon the relative advantages of the two positions. We shall probably adhere to custom. The English method involves much less disturbance of the patient; it involves no exposure; it requires no second assistant ; and is in many respects most convenient in home practice. Certainly in the last stage of extraction, when the handles have to be carried round the symphysis, the dorsal position gives more power and precision. And in cases of convulsions, where the patient is unconscious or unmanageable, it is at times necessary to apply the forceps in the dorsal position. If we use the long French

FIG. 12.

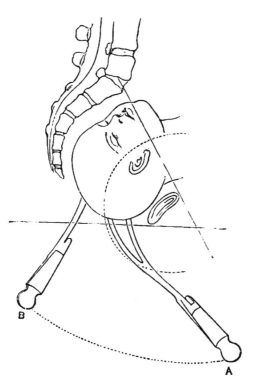

SHOWING APPLICATION OF FIRST OR SACRAL BLADE OF THE SHORT FORCEPS.

A, first stage, blade being guided on to the head. The handle A is then carried slightly downwards and backwards, to get the point of the blade round the head and up into side of the pelvis in the line A B. At B the blade is *in situ*.

forceps, there is, indeed, little choice. The patient must be in lithotomy position, or if on her side, the pelvis must overhang the edge of the bed to an inconvenient extent. The conditions rendering the dorsal position preferable will be pointed out as the occasions arise.

The operation may be divided into four stages or acts.

1. Introduction of the blades; 2. Locking; 3. Traction, sometimes compression, and leverage; 4. Removal of the instrument.

1. *Which blade do you pass first?*—In the case of the single-curved forceps, both blades being alike, you cannot take up the wrong one. Seizing, then, either blade, you have to pass it between the head and the sacrum, and feeling the pubic ear, you know the sacral ear is exactly opposite. This blade becomes the posterior or sacral blade. Holding the blade lightly in the right hand, the handle raised and directed forwards, so that the blade shall cross the mother's right thigh obliquely, the point will be guided over the perinæum by two fingers of the left hand, which are passed up carefully between the child's head and the cervix uteri. The all-essential point is to make out clearly the edge of the os uteri, to pass your fingers inside this edge, and to touch the head itself; then slipping the point of the blade along the inside of your fingers, the os uteri resting on the outside of your fingers, the blade will strike the head. This done, you have to adapt the blade to the convexity of the head. The point, therefore, must follow this convexity. This is done by lowering the handle and drawing it backwards, the point being still guided by the fingers of the left hand. When the convexity is well grasped, the handle is further pushed well back against the perinæum, to give room for the manipulation of the second or pubic blade.

Introduction of the second blade.—The fingers of the left hand are shifted forward, so as to raise the os uteri from the pubic side of the head. The handle is held very low, and slightly forwards, crossing the mother's left thigh obliquely. Running the point along the palmar aspect of the fingers behind the right pubic ramus, when the point strikes the head, the handle is raised and carried backwards, so as to take the blade over the convexity of the head. Here you must proceed with the utmost gentleness. It is not by force that you will succeed in passing the blade. Force is quite out of place. You may take this as an axiom: If you are met by resistance that only force can overcome, you are going wrong ; and, *vice versâ*, if the blades are slipping in easily, the probability is that you are going right. The rule, then, is this—hold the blade lightly; let it feel its way, as it were; let it insinuate itself into position. It will be sure to slide into the space where there is most room—that is, one blade will go nearly opposite the sacro-iliac synchondrosis, the other will go opposite the foramen ovale.*

FIG. 13.

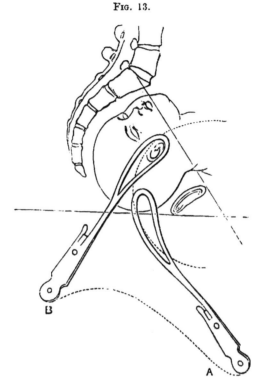

INTRODUCTION OF THE SHORT FORCEPS—THE SECOND OR PUBIC BLADE.

A, the first stage. As the blade passes up the pelvis and round the head, the handle travels in the direction of the line A B. At B the blade is *in sitú*. The two blades of Figs. 12 and 13, therefore, correspond at B, and will lock.

The blades introduced, the left hand is withdrawn from the vagina, and the *second* act, or *locking*, is to be done. You seize lightly a handle with each hand, draw them into opposition, and if they have been correctly introduced they will readily lock. A smooth lock is generally an indication that the head is

* The adherents of the short forceps generally recommend to pass *the upper or anterior blade first*. It would not be easy to prove any advantage in this method. I believe the most skilful practitioners in London, Dublin and Edinburgh now follow the method recommended in the text—namely, of passing the lower or sacral blade first.

properly grasped. During locking be careful to pass your finger round the lock, in order to remove any hair or skin that might otherwise get pinched. This is especially necessary in using the single-curved forceps.

Then come, thirdly, *traction, compression* and *leverage.* Traction must be exerted in the direction of the pelvic axis. Thus at first the traction will be backwards in a line drawn from the umbilicus to the coccyx. Gradually, as the head descends, the handles will come more forwards, and the face turning a little backwards into the hollow of the sacrum, the handles will also rotate, so that the instrument will approach the transverse diameter of the pelvis. As the head emerges, the vertex appearing under the pubic arch in the genital fissure, the handles, following the extension movement of the head, will describe a circle round the symphysis as a centre, and will therefore at the moment of exit be applied nearly to the mother's abdomen. At this moment, and even earlier if active uterine action have set in, the fourth act—*removal* of the instrument—must be effected. This often requires some smartness. You abandon the grasp of the handles, so as to unlock, seize the handle of the pubic blade, draw it downwards and backwards off the head; then, taking the sacral blade, draw it upwards and a little backwards.

Head in the Second Position, or occiput to right foramen ovale. In this case you still feel for the pubic ear, which will guide you to the other ear opposite the right sacro-iliac synchondrosis. As the rule is to apply the short forceps over the ears, the introduction of the blades must be governed by the position of the head. You must first then determine the position of the head. So say most, if not all, our systematic authors. So many positions of the head, so many varying modes of applying the forceps! Now listen to the voice of Experience—Experience that so often sets at nought the refinements of theory, and clears out for herself a straight and simple path through the intricacies woven in the closet. Dr. Ramsbotham says:* " In employing the short forceps I lay it down as a rule that the blades should be passed over the ears: the head is more under command when embraced laterally, and there is less danger of injuring the soft parts during extraction. *But I confess that I have for many years been accustomed, however low the head may be, to introduce the*

* " Medical Times and Gazette," 1862.

blades within each ilium, because they usually pass up more easily in that direction." I think I am not wrong in believing that many others do the same thing, some not knowing it, and even imagining that they are following the ancient rule. It is a habit of mine to examine the head in every case after delivery. I have thus many times seen the stamp of the fenestræ on the brow and side of the occiput. This is as clear to read as the impression of a seal on wax. It says, unmistakably, that the blades found their way into the sides of the pelvis with at most a slight deviation towards an oblique diameter.

All this suggests the question, whether it be really so necessary to "feel the ear" before applying the forceps as has been imagined. If the blades *will* find their way to the sides of the pelvis, clearly it is not necessary to know where the ears are. To feel an ear must in most cases put the patient to much suffering. Roberton, an operator like Ramsbotham, trained in the field of obstetric difficulties, says : " There is some haziness of mind about position of the fœtal head—about the necessity of feeling for the ear." The terms " sacral " and " pubal " are misleading. He, too, applies the blades in the sides of the pelvis.

There is one case in which the short forceps is of especial value. It is when the head descends into the pelvis, its long diameter keeping nearly in the transverse diameter of the pelvis, until it is arrested on the shelf formed by the sacro-sciatic ligaments. At this point, from want of propelling power, the head does not take its screw-movement of rotation on its axis so as to bring the occiput forwards. If the short forceps be now applied in the transverse diameter of the head, by a slight rotatory movement, the axial turn is given, the occiput comes forward, the face goes to the sacrum, and the head is released. In two cases of this kind I thus easily succeeded in delivering after failing with my long forceps, the blades of which lying in the transverse diameter of the pelvis, grasped the head in its long or fronto-occipital diameter. These are the only two cases in which I have ever found the short forceps preferable to the long. And the simple lever would, perhaps, have answered as well.

There are *objections* to the single-curved forceps, short or long :—

1. One objection is in the introduction ; others in the injuries

likely to be inflicted on mother or child. To introduce the
second or upper blade, the handle must be much depressed,
nearly at right angles with the mother's left thigh, which is
flexed upon her abdomen. Now, to do this the patient's nates

FIG. 14.

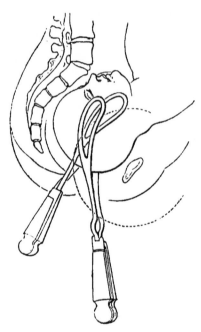

SHOWING THE SINGLE AND DOUBLE-CURVED FORCEPS IN SITŮ.

The single-curved forceps presses back upon the perinæum, putting this structure on the
stretch. The shanks of the double-curved forceps keep clear of the perinæum, the whole
instrument approximating to Carus' curve.

must be dragged over the edge of the bed. To procure and to
maintain this position is often a matter of great difficulty and
inconvenience.

You may facilitate the introduction of the second blade by
introducing a joint into the shank, so as to allow the handle to
be doubled up out of the way. Dr. Giles showed at the Obstetrical Exhibition an instrument so modified.

2. In extraction, the handles, nearly to the last moment,
must be directed more backwards than is necessary with the
double-curved forceps, and owing to the bows springing directly

from the lock, the perinæum is wedged open, and not seldom unavoidably torn. In some cases, this injury may be avoided by taking off the blades before the greatest diameter of the head passes. But then the work is not always done, and you may have to put them on again. I may perhaps be told that to suffer the short forceps to tear the perinæum implies want of skill. I reply that men of the highest skill and the largest experience with this instrument have confessed to me that this objection is a real one.

The best single-curved forceps is that of Dr. Beatty, of Dublin. I used it for some time, but gave it up because of these two faults, and of its inadequacy to cope with a large range of cases which come within the power of the long forceps.

3. The posterior or sacral blade is extremely apt to bruise by one of its edges the sciatic nerve. The effect is the crushing of some fibres and more or less protracted paralysis of the leg.

4. If the blades be applied as usually taught—*i.e.*, nearly in the transverse diameter of the head—an edge is very likely to press upon the portio dura as it emerges from the temporal bone. The result is paralysis of the facial muscles to which the branches are distributed. The child cannot shut the eye; it cannot suck. I have known a child die of starvation from this cause.

5. If applied, as usually inculcated, the anterior blade will often go directly behind the pubes. The edge will bruise the urethra. Hence vesico-vaginal fistula.

From all these objections the long forceps I recommend is nearly altogether free. I have described the short single-curved forceps in deference to a still common prejudice, and because many men possess only this instrument; but this description may be conveniently passed over by those who are ready to adopt the double-curved forceps.

THE LONG OR DOUBLE-CURVED FORCEPS.

The application of this instrument is governed by a different law from that which governs the use of the short forceps. The short forceps, according to the recognized rule, must be applied with the blades quite or *nearly* over the transverse diameter of the head. The head determines the manner of applying it. But with the long forceps it is the pelvis that rules the application. The position of the head may be practically disregarded. The pelvic curve of the blades indicates that these must be adapted to the curve of the sacrum in order to reach the brim. They must therefore be passed as nearly as may be in the transverse diameter of the pelvis. One blade will be in each ilium, and the head, whatever its position in relation to the pelvic diameter, will be grasped between them. The universal force of this rule much simplifies and facilitates the use of the instrument. Not only does it apply to the position of the head in relation to the pelvic diameters, but also to all stages of progress of the head, from that where it lies above the brim down to its arrest at the outlet.

It has been contended that the short forceps should be preferred in cases where the head is arrested in the cavity ; and as a corollary it is urged that in cases of arrest at the brim, where the head has been brought into the cavity by the long forceps, this instrument, after serving so far, should be discarded and replaced by the short forceps. I do not concur in this view. I doubt whether any one who has had any considerable practice with the long forceps has found it worth while to change instruments in the course of delivery. The long forceps possesses a more scientific adaptation to the pelvis throughout the whole canal than the short forceps. And if the long forceps is found in practice capable of taking the head through the pelvis from brim to outlet, it follows that, since the whole contains the parts, the long forceps is qualified to take up the head at any point below the brim.

The pelvis has been compared to a screw. I think a better idea may be formed of its mechanical properties by comparing it to a rifled gun, and the child's head to a conical bullet. But even then the comparison is not complete, for the pelvis, unlike a gun, is a bent tube. Now, just as the head must traverse the pelvis in a helicine course, determined by the relation of form between pelvis and head, so is it natural that an instrument designed to grasp the head should be so modelled as to be fitted to follow this helicine course during introduction and extraction. This indication a well-modelled long forceps fulfils ; no single-curved forceps can fulfil it.

First, as to the application when the head is delayed at the brim.

Mode of applying.—Position of the patient.—The patient should lie on her left side, the knees drawn up towards the abdomen; the head should be only slightly raised. She should lie across the bed, with the nates near the right edge, about midway between the head and foot. This will facilitate the introduction of the blades, and give room for the sweep of the handles round the pubes at the end of the operation. I do not find it necessary to bring the nates to hang over the edge of the bed. Exposure of the patient is unnecessary. The whole operation is conducted under the bed-clothes or a sheet. You trust to touch, not to sight.

FIRST ACT—INTRODUCTION OF THE BLADES.

Selection of the blades.—Dip them in warm water, wipe dry, and lubricate them. Join them, and, holding the instrument with

FIG. 15.

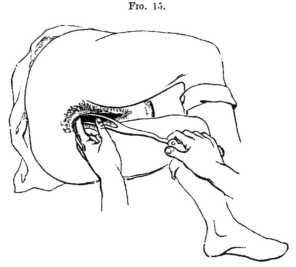

SHOWING THE FIRST STAGE OF INTRODUCTION OF THE FIRST BLADE.

the concavity of the pelvic curve forwards, and the blades in the position which they are to occupy in the pelvis, you take that one first which is to lie in the left or lower side.

E

First stage.—One or two fingers of the left hand are passed
in at the perinæum and between the cervix uteri and the head.
Then, bearing in mind the relative forms of the instrument, the
head, and the pelvic canal, the point of the blade is passed
along the palmar aspect of the fingers at first nearly directly
backwards towards the hollow of the sacrum (*see* Fig. 15, p. 49).

Second stage.—The handle is now raised so as to throw the
point downwards upon the left side of the head.   As the point

FIG. 16.

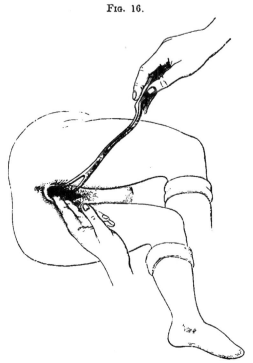

SHOWING SECOND STAGE OF INTRODUCTION OF FIRST BLADE.

of the blade must describe a double or compound curve—a
segment of a helix—in order to travel round the head-globe,
and at the same time to ascend forwards in the direction of
Carus' curve so as to reach the brim of the pelvis, the handle
rises, goes backwards, and partly rotates on its axis (*see* Fig. 16).

Third stage.—The handle is now carried backwards and
downwards to complete the course of the point around the

head-globe and into the left ilium. Slight pressure upon the handle ought to suffice. This will impart *movement* to the blade; the *right direction* will be given by the relation of the sacrum and head. The blade is now *in situ;* the shank is to

Fig. 17.

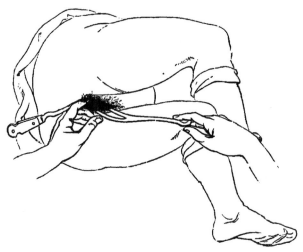

SHOWING LAST STAGE OF INTRODUCTION OF THE FIRST BLADE AND THE CROSSING THE SHANK OF THE FIRST BLADE BY THE SECOND BLADE IN THE FIRST STAGE OF ITS INTRODUCTION.

be pressed against the coccyx by the back of the operator's left hand whilst he is introducing the second blade. Its weight aids in maintaining it *in situ* (*see* Fig. 17).

INTRODUCTION OF THE SECOND BLADE.

First stage.—Two fingers of the left hand, the back of which is supporting the first blade against the perinæum, are passed into the pelvis between the os uteri and the side of the head which lies nearest to the right ilium. The instrument held in the right hand lies nearly parallel with the mother's left thigh, or crossing it with only a slight angle. The point of the blade is slipped along the palmar aspect of the fingers in the vagina, across the shank of the first blade *in situ*, inside the perinæum towards the hollow of the sacrum (*see* Fig. 17).

Second stage.—As the point has to describe a helicine curve to get round the head-globe and forwards in the direction of Carus' curve, the handle is now depressed and carried back-

E 2

wards until the blade lies in the right ilium. When it has reached this position the handle will be found near the coccyx, nearly in opposition to the first blade.

The application of the long forceps is further illustrated in the following diagrams.

FIG. 18.

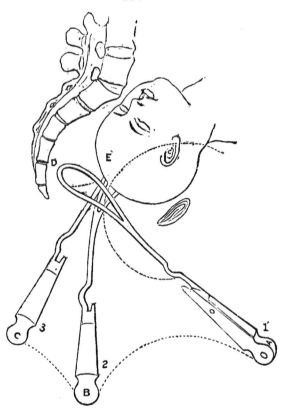

INTRODUCTION OF THE FIRST OR LEFT BLADE OF THE LONG FORCEPS.

1. First stage, or introduction of point of blade in the hollow of the sacrum : A, the handle, is then raised, and at the same time carried across, rotating partly on its axis to B, so that the point C, turning round in the hollow of the sacrum to E, strikes the head, and rises towards the left side of the pelvis. 2. The second stage, or advance of the blade round the head and up in the left ilium. 3. Third stage: the handle B has travelled in the direction B C, still rotating slightly, until at C it is at rest *in situ*, the shank near the coccyx, where it is held by the back of the operator's left hand, whilst the point of the second blade is passed over and across it inside the perinæum, as seen in the next figure.

The locking.—This is effected by a slight movement of adaptation. A handle is seized in each hand. The handle of the first blade is brought a little forwards over the handle of the second blade. If one blade is a little deeper in the pelvis than the other, it is either brought out, or the other is carried in until the lock is adjusted. This is commonly facilitated by pressing both handles backwards against the coccyx. This movement, by throwing the blades well into the ilia, where

Fig. 19.

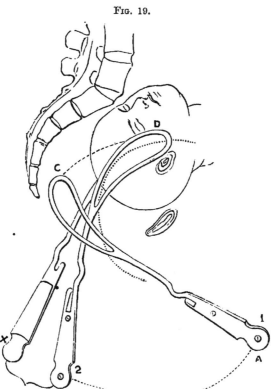

INTRODUCTION OF THE SECOND OR RIGHT BLADE OF THE LONG FORCEPS.

1. First stage of second blade; 2. Second stage of second blade; x, the first blade *in sitú*. A, the handle at the moment of passing c, the point, inside the perinæum into the hollow of the sacrum, across x, the first blade; the handle then drops and goes backwards to B, the point c travelling round the head, and advancing into the right ilium in the direction of the axis of the brim to D; when it has reached this position, it will be found nearly opposed to the first blade, x; the locking is effected by bringing the handle x over the handle D.

there is room, allows the handles to be rotated a little, so as to fall into accurate relation (*see* Fig. 20).

Accurate locking is generally evidence that the blades are properly adjusted to the head, and that the pelvis admits of the successful use of the instrument. On the other hand, their not locking is proof of their not being properly introduced, or *of the pelvis not admitting of their application.* In the first case, that of improper introduction, the failure is generally due to neglect in passing the blades exactly in the same diameter of the pelvis— that is, in passing the second blade exactly opposite to the first, so that if the first blade is applied in the left ilium, opposite one end of the transverse diameter, the right does not lie at the opposite end of that diameter. To remedy this error, the blade must be partly or wholly withdrawn and readjusted.

In the second case, that of pelvic unfitness, the locking is prevented by the projecting promontory or other deformity so

FIG. 20.

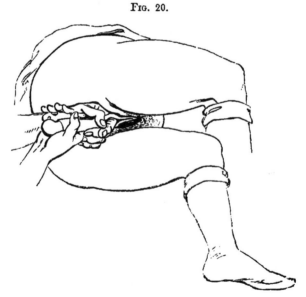

SHOWING THE LONG FORCEPS LOCKED, AND GRASPED BY THE TWO HANDS.
The head being at the brim, traction is backwards.

distorting the pelvic diameters that the two blades cannot find room to lie in the same diameter opposite to each other. It will commonly be found that the blades will pass one on each side of the promontory, the inside of the blade not looking towards its fellow, but towards the opposite foramen ovale, where you cannot get a blade to lie. When you find this happen you must give up the attempt to use the forceps. Pass the hand into the pelvis, if necessary; explore its dimensions and form carefully; and determine between turning and craniotomy. A correlative proposition may here be stated:— *Wherever the long forceps will lock without force, it may be reasonably concluded that the case is a fit one for the trial of this instrument; and a reasonable attempt should be made to deliver by its aid before passing on to turning or perforation.*

The Extraction.—Get the nurse to press upon the right hip and support the back. Grasp the handles with one hand, and

FIG. 21.

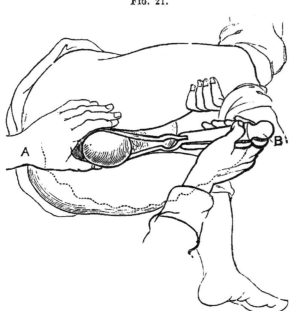

REPRESENTING THE LAST STAGE OF EXTRACTION.

The handles have travelled from A to B, so as at last to touch the abdomen. The dotted line shows the course of the handles, and the slight oscillations practised during the descent of the head.

apply the fingers of the other hand to the ring or shoulders at the lock. Draw at first backwards in the axis of the brim, during the pains if any be present, and at intervals of a minute or so if there be none. Concurrently with traction, alternate slight leverage movements may be executed by swaying the handles gently from side to side, always taking care not to press the shanks against the pelvic walls. Each blade is the fulcrum to its fellow. The finger which is used in the ring from time to time gauges the advance of the head (*see* Fig. 21).

It is further extremely useful, if you have a competent assistant, to get him to support and press upon the fundus during extraction. This helps to keep the axis of the uterus and of the child in proper relation to the axis of the pelvis; and in proportion to the aid thus given *à tergo*, you lessen the amount of extracting force required. Placing the woman in the dorsal position also facilitates extraction.

The advance of the head is measured by the following standards:—First, you feel if the occiput approaches the pubic arch by passing a finger below and behind the pubic bones. Secondly, you sweep your finger round the circumference of the brim, and thus feel if the equator of the head-globe is pressing lower down through the brim. Thirdly, by feeling the direction of the sagittal suture. If you find that it is approaching parallelism with the conjugate diameter, you may be certain that the head is descending. Further evidence is found in the rotation of the forceps. As the head can hardly turn upon its cervico-vertical axis without at the same time descending in the pelvis, if the handles of the forceps are observed to rotate, this rotation being imparted by the head, is evidence of advance. Again, as the head descends, of course more and more of the shanks and blades will become visible. This, indeed, is open to a fallacy. Allowance must be made for some degree of slipping, which takes place with all the English instruments whose blades have only a moderate bow. And further, when the head is fairly in the pelvic cavity, the blades lose something of that external support which, as explained in Lecture III., is the principal force in maintaining the grasp upon the head. This is still more marked when the head has partly emerged from the vulva. At this time the blades will be apt to slip away altogether, and it will be necessary to increase the compression on the handles in order to keep your hold. Fourthly, by two or more fingers you measure the space or degree of tightness between the vertex and the floor of the pelvis. At first the fingers find free space;

gradually the vertex leaves no room for the fingers. Then the soft floor of the pelvis, the perinæum, is distended by the advancing vertex ; it bulges out ; it puts the perinæum tightly on the stretch. The anus is protruded. Fæces are often squeezed out. Indeed, the pressure upon the sphincter ani at this stage sets up reflex action. The call to strain or bear down to expel the pelvic contents, whether uterine or rectal, is uncontrollable. Turbulent expulsive action, then, and defecation, constitute certain signs that the head is advancing. To some extent the increasing scalp-swelling or caput succedaneum may give a false impression that the cranium itself is descending. But a little practice and attention will correct this error. When the vertex has reached the floor of the pelvis, the handles of the forceps are found to have turned a little upon their axis, to lie more nearly in the transverse diameter of the pelvis. This is the result and the indication of the screw-rotation of the head. You have no hand in producing it. It is effected by the descending head adapting itself to the cavity of the pelvis. The handles may now be directed more forwards during traction. The shanks thus avoid stretching the perinæum, and the traction is in the axis of the outlet. An assistant is now useful in holding up the right knee, so as to leave room for the operator to carry the handles well round the pubes in Carus' curve. Here it is often convenient to push the handles forwards rather than to pull. This action is seen in Fig. 21.

In the last stage of extraction it is often useful to turn the patient on her back. In this position you get the aid of gravity ; your assistant can support the uterus better in the axis of the pelvis ; and the handles of the forceps travel more easily round the symphysis.

The forward direction of the handles, or the revolution in Carus' orbit, must not begin until the occiput is well under the pubic arch.

During extraction it occasionally happens that the blades will lose their hold, that the handles will twist in opposite directions, and thus unlock. This is generally owing to the operator carrying the handles forward too early. The effect of this is to throw the blades off the head-globe over the face. It is another illustration of the law that the position of the forceps is determined by the relation of the head to the pelvis, and that if you reverse the order by attempting to make the forceps alter this relation you are immediately at fault. The remedy is to carry each handle well back again towards the perinæum, when they will re-lock.

If the head is in the genital fissure, and there is sufficient uterine energy, you may proceed to the

4th Act. The Removal of the Blades.—If the head should not be propelled, you may often assist it by a manœuvre which it is well to understand. You apply the palms of both hands one on either side and behind to the perinæum distended by the head; and bearing upon this structure so as to press it a little backwards, whilst the head is pushed forwards towards the pubic arch, the head is, as it were, shelled out by being made to complete its movement of extension. Steady pressure by the hands of an assistant or by a binder upon the fundus uteri will much assist the extension of the head. By this combination of perinæal squeezing and of compression of the fundus and body of the uterus I once extricated myself and my patient from an awkward predicament. I had been summoned into the country without knowing the nature of the case, and had no instruments. I found a lady who had been many hours in labour, the head on the perinæum, and no pains. The lever or the forceps would have delivered her in a minute. Neither was to be had. But the manœuvre I have described perfectly succeeded, and put an end to a state of extreme anxiety, and even danger.

Another manœuvre is occasionally serviceable. This is to pass a finger into the rectum, so as to get a point of pressure upon the forehead. In this way it is sometimes possible to bring the face downwards, to start the extension movement, and thus to extricate the head delayed at the outlet. And if at the same time firm downward pressure be made upon the breech through the fundus, as described in Lecture II., the force propagated through the spine will aid materially in giving the extension movement. This combination of the principles of "pushing," of leverage, and of "shelling-out," may in certain cases enable you to deliver without resorting to the forceps or lever.

When the blades are adjusted, they will not lie exactly in the transverse diameter of the pelvis. The head, lying between the transverse and right oblique diameters, will tend to throw off the blades towards the opposite or left oblique diameter. The head then will be seized obliquely, one blade grasping the right brow, the other the left occiput. This is clearly demonstrated by the impressions of the fenestræ left on the scalp. The blades naturally find their way into this position if they are introduced gently. One tendency of this oblique seizure is to assist the head in its axial rotation, face sacrumwards, as it descends into the pelvis. It is also an answer to an objection urged against the use of the long forceps at the brim—namely, that by seizing the head in its long or fronto-occipital diameter, compression in this direction makes the opposite or bi-parietal diameter bulge out, thus increasing the difficulty of passing the small or conjugate diameter of the pelvis. In most cases the objection is theoretical only—it is mainly based upon experiments made on the dead fœtus on the table.

Elongation or moulding of the head, we have seen, is the result of gradual compression of the equatorial zone. Now the pelvis and the forceps together constitute the compressing ring. Pressure, then, upon the transverse diameter of the head by the opposing points of the sacrum and pubes, simultaneously with pressure upon the longitudinal diameter between the blades of the forceps, tends to *diminish both diameters* by lengthening out the head. Of course it must be understood that the pelvic contraction is of moderate degree only—in short, that the case is a proper one for the forceps. If the conjugate diameter be less than 3·25 inches, the prospect of effecting the desired elongation within a reasonable time is greatly diminished.

I have said that the head is very rarely seized exactly in its longitudinal diameter. An exception occurs in the case of the very flat pelvis, in which there is conjugate contraction with very little projection of the promontory. In this case the head will lie very nearly in the transverse diameter. If, in presumed contraction of the brim, the marks of the blades are on the brow and side of the occiput, the projection of the promontory is not great.

The time required for extraction.—If the head be delayed in the cavity of the pelvis for want of expulsive action, or because it rests upon the ischia, maintaining a too near approach to

the transverse diameter, and there is no marked hindrance on
the part of either the anterior or posterior valve, it is generally
sufficient to use slight traction and oscillation for a few minutes.
As soon as the head is started by the forceps, the uterus takes
up its work, helps the operator, and the labour is quickly over.

If the uterine and perinæal valves obstruct the passage of
the head, a little more time and caution are required. (*See*
Lecture V., Figs. 22, 23.)

If the head has to be seized at the brim on account of delay
from want of uterine action, time may often be saved by placing
the patient on her back, and supporting the uterus against the
spine by the hands of an assistant or a binder. This proceeding, by adjusting the axis of the uterus to that of the brim,
and getting the aid of gravitation, will greatly facilitate the
entry of the head and encourage the action of the uterus. If
there is no obstacle from narrowing of the pelvis or want of
dilatation of the soft parts, gentle traction and oscillation during
ten minutes will generally complete the labour.

In the event, however, of arrest from pelvic contraction or
from want of dilatability of the soft parts, time is a necessary
element. The process of moulding, of elongation of the head,
can only be effected gradually. Here oscillation or leverage must
be used with great care. What is wanted is steady compression
and traction extended, with moderate intervals of rest, over
thirty minutes, or even an hour. Should the head be found to
make no advance in entering the brim in that time, and if the
handles of the forceps maintain a marked degree of divergence,
the question whether the forceps must not be laid aside for
turning or perforation will have to be considered.

LECTURE V.

A VERY large proportion of cases that call for the forceps are *first labours*. It is therefore well to take a survey of the conditions which lead to this necessity. Disproportion as a cause of arrest we will put aside for the present. In the great majority of first labours the difficulty does not arise from disproportion. The frequency of an easy second labour proves this. The diffi-culty, then, lies in the soft parts of the parturient canal, or in faulty position of the head. And this may be either from want of contractile energy of the uterus or from excessive resistance of the os uteri, vagina, or vulva. I will endeavour to explain the nature of these cases. First, the suspension of uterine and other muscular force. This may be the result of exhaustion from fatigue, or of the discharge of the *vis nervosa* in other directions—metastatic labour, as Dr. Power calls it. Emotion, fear, the shrinking before pain, will frequently cause such a deprivation of nerve-force that all labour is suspended. It is in such cases that chloroform finds one of its happiest offices. By removing the sense of pain and of fear, the emotional disturbance is eliminated, the nerve-force responds to the natural call, and labour is frequently resumed and carried out to a successful termination. It is not a figure of speech to say that here chloroform acts like a charm. It may even save the necessity of resorting to instruments.

Arrest or delay in the second stage of labour, from whatever cause, is a source of danger to the mother and child. The procrastinating school insists upon the dogma that we should wait four, six, or more hours from the commencement, or after an ear is felt, before interfering. This course is full of peril. Whilst we are waiting the woman is suffering—suffering needlessly, her nervous energy is being used up, she is drifting into exhaustion. Often aid is imperiously indicated long before the os is fully dilated, or an ear can be felt. If there is conjugate contraction the head cannot descend enough to stretch the cervix ; and in primiparæ the anterior or uterine valve remains capping the head down to the very floor of the pelvis. (*See* Fig. 22.) It comes out very clearly from the histories of the cases in the Dublin Lying-in Hospital, that the maternal mortality has decreased since the forceps has been more frequently, that is, earlier, resorted to. Upon this point, the evidence of Dr. Johnston, the Master, is very decided. Dr. G. Hamilton* also says if the patient remains undelivered much more than two hours after the os uteri is dilated, and the head has entered the pelvis so that an ear can be felt, the danger to the child becomes imminent. Delay in the second stage is eminently a case for help as soon as it is clearly ascertained that the natural powers are not efficient. It is often for want of driving-power that the head rests on the floor of the pelvis in a transverse position, so that a double cause of delay arises. This is almost instantly relieved by the forceps.

In no respect has modern midwifery given more satisfactory evidence of progress than in the extending practice of applying the forceps to obviate delay in the second stage of labour.

Not seldom, combined with more or less emotional disturbance, the expelling force gives way before a real mechanical obstacle. It is this :—In primiparæ the cervix dilates slowly. The vertex partly enters the pelvis, capped by the cervix. The anterior portion of the cervix, especially, is carried down before the head, much below the brim. It even gets jammed between the head and the symphysis, and becomes, perhaps, more unyielding from œdema. Now this anterior segment of the uterus

* "Edinb. Med. Journ.," 1861.

forms a valve or plane which guides the head backwards into the sacral hollow in the direction of the axis of the brim. So far it fulfils a useful function; but, having done this, it ought to retire. In pluriparæ it commonly does so early. When this anterior valve has retreated, the head encounters the second valve formed by the perinæum, which is exactly opposed to the first or uterine valve. The function of this is to guide the head forwards under the pubic arch in the direction of the outlet. Now it frequently happens in primiparæ that these valves maintain their resistance too long. The uterine valve may still cap the head when it is propelled to the very floor of the pelvis. In this case the head is prevented from receiving the full impact from the inclined planes of the ischia; it is impeded in its half-

FIG. 22.

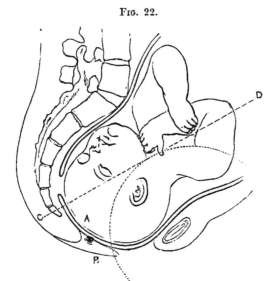

SHOWING THE HEAD ARRESTED IN THE PELVIS BY THE ANTERIOR OR UTERINE VALVE A, WHICH IS CARRIED DOWN INTO CONTACT WITH THE POSTERIOR OR PERINÆAL VALVE B.

The uterine valve A helps to guide the head into the pelvis, in the axis of the inlet C D.

quarter axial turn, occiput forwards, and also in its movement of extension. Hence a double difficulty: there is the opposing valve, there is malposition. Clearly the valve must be got out of the way. How to do it? Sometimes patience will do it; but as patience on the part of the physician may involve agony and danger to the woman, this should not be overstrained. Sometimes one or two fingers may be insinuated between the valve and the head in the intervals of pains, and then the valve may be held back so that the equator of the head may pass it. But you must be careful lest by over-meddling you cause more swelling and rigidity. You may pass up the lever or one blade of the forceps, and bearing upon the occiput, just as you use a shoehorn, the valve, like the heel of the shoe, is held back whilst the head descends upon the inclined plane of the instrument. And here you often get another beneficial result. The head-globe has been lying closely fitting to the ring of the cervix uteri like a ball-valve, ponding up the liquor amnii behind, and impeding the full action of the uterus by over-distending it.

Fig. 23.

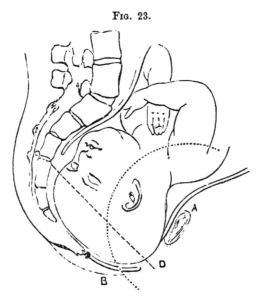

SHOWS THE HEAD ARRESTED AT THE OUTLET BY THE POSTERIOR OR PERINÆAL VALVE B; THE ANTERIOR OR UTERINE VALVE A HAS SLIPPED UP ABOVE THE EQUATOR OF THE HEAD.

The posterior valve guides the head out of the pelvis in the axis of the outlet c d.

The lever or forceps opens a channel for the escape of the pent-up fluid. The uterus then acts immediately, and the labour proceeds. I have often used the forceps successfully for no other purpose than this. One blade will sometimes suffice.

Well, we have now disposed of the uterine valve. The perinæal valve and the vulva oppose another barrier, all the more troublesome because it has to be encountered by diminished force. Arrest on the floor of the pelvis, nothing but this valve obstructing, is very common. The lever applied alternately over the occiput and face, or over the sides of the head, may answer well in this case. But many will prefer the forceps.

The delay at the vulva is often further increased by intense emotional and sensational nervous disturbance. The uterus seems instinctively to hesitate to contract, lest, by forcing the head upon the acutely sensitive structures of the vulva, it cause intolerable pain. The consequence of this protracted shrinking before pain is twofold:—There is, first, exhaustion of nerve-force; there is, secondly, a condition which I can best describe as a kind of shock, producing prostration, if not collapse, which supervenes whenever an urgent function is suspended or remains unfulfilled.

Another cause of delay may of course reside in the mechanical condition of the resisting structures; rigidity of the cervix uteri, of the perinæum or vulva, may be added to other unfavourable conditions. Rigidity may be due to thickening, œdema, hypertrophy of the tissues, or there may be rigidity, without discoverable alteration of texture; and there are the more serious cases of partial or complete occlusion from cicatricial tissue, the result of previous injury or disease. Obstruction from these causes demands special treatment, which will be discussed hereafter.

Most authors describe the application of the *forceps to the after-coming head*—that is, when the head is delayed after the birth of the trunk in breech, footling, or turning labours. The position of the child with its head delayed at the brim, probably compressing the cord, is indeed perilous. Prompt delivery alone can rescue it from asphyxia. How shall we best reconcile the two conditions of promptitude and the minimum of force? It is a point of extreme interest to know what is the greatest time a child can endure being cut off from placental and aërial respiration, and yet recover; for within that time the head must generally be extricated in order to save life. The time is certainly very brief. Here it may truly be said that "horæ momento cita mors venit, aut victoria læta." The data

F

are necessarily wanting in precision. Hugh Carmichael[*] in two cases removed the fœtus from the uterus within fifteen minutes from the death of the mother. In both cases the fœtus was quite dead, although on the mother's evidence, it was living just before her dissolution. A similar case has occurred to me. Dr. Ireland was called to a woman who had died suddenly from a blow received from her husband. The Cæsarian section was performed, and a live child was extracted. The interval was here estimated at eight or ten minutes. The following case occurred at St. Thomas's[†] :—A woman in her ninth month was run over in St. Thomas-street, at 7.35, and carried to the Hospital. She died at 7:55. Mr. Green opened the abdomen as 8.8, and the child was withdrawn by Dr. Blundell asphyxiated. Its lungs were inflated, and it survived thirty-four hours.[‡] Here, then, we have an instance of partial recovery at the end of thirteen minutes from the mother's death. Dr. Pingler relates a case in which a live child was extracted by Cæsarian section about fifteen minutes after the mother's death; and another case in which it was established that the mother had been dead twenty-three minutes.[§] Cases of extraction of live children within ten minutes are not very rare. But perhaps examples of this kind are not exactly in point. They are not quite analogous to the case of compression of the cord during labour. Numerous observations lead me to conclude that the child will be asphyxiated beyond recovery if aërial respiration do not begin within three, or at most five, minutes after the stoppage of the placental respiration. I think it must be accepted as a general law that if the head compress the cord, the child should be extracted within three minutes. Even if this be done, there will commonly be considerable asphyxia and cerebral congestion, and restorative means will be required.

Now the practical question arises—What is the readiest way of delivering the after-coming head? We can extract by the

[*] "Dublin Journal of Medicine," vol. xiv.

[†] "Med. Chir. Trans.," 1822.

[‡] It is worthy of remark that in the history of this case Mr. Green especially calls attention to the depressing effect of the warm-bath—a point since enforced by Milne-Edwards and Marshall Hall.

[§] "Monatsschr. f. Geburtsk," 1869.

hands or by the forceps. Which is to be preferred? In many cases undoubtedly the hands are the best instrument. Where the cervix is fully expanded, and the brim of the pelvis is roomy, well-directed manipulation will deliver in a few seconds. And again, if there be any marked contraction of the conjugate diameter, the forceps will probably fail, whereas the hands may extricate the head very quickly. But still some cases may occur in which the forceps will be useful. How to apply it? In the first place, draw down the cord gently, so as to take off any dragging upon the umbilicus, and lay the part which traverses the brim in that side in which the face is found ; there is most room for it there. The head is engaged with its long axis more or less nearly in the transverse diameter of the brim. The blades should grasp it in an oblique diameter approaching the antero-posterior. To be able to effect this, the trunk must be carried well forwards over the symphysis in the direction of Carus' curve, and held there by an assistant, so as to leave the outlet clear for manipulation. Then passing your left hand into the vagina, you carry the fingers to the left side of the pelvis, between the cervix uteri and the head. The blade is slipped up along the palmar aspect of the fingers to its place. The like proceeding is then repeated on the right side of the pelvis, and the blades are locked. The assistant supporting the child's body, you then draw the head into the pelvis in the axis of the brim. As soon as this is cleared, you may take off the blades and finish the extraction by the hands. This is done by hooking two fingers of the right hand over the back of the neck, on the shoulders, whilst the left hand seizes the feet above the ankles, a napkin interposed. You then draw in the axis of the outlet. It is the work of a few seconds. Some will prefer completing the extraction with the forceps. If you select this mode, you find the face turned towards the sacral hollow when the head has cleared the brim ; the forceps, following this, rotates a little in your hands. When the occiput is appearing under the pubic arch, carry the handles well forwards, so as to bring the face over the perinæum with the least possible strain upon this structure. The face and forehead sweep the perinæum, describing a curve round the occiput resting upon the pubes. The use of the forceps in this case was strongly inculcated by Busch, of Berlin, who attributes to this practice the extraordinary success of turning in his hands. Of forty-four cases of turning, only three children are stated to have been lost from the effects of the operation. The late Dr. E. Rigby and Dr. Meigs insist also upon the advantage of the practice. The traction-force being applied to the head itself, you avoid the dangerous traction upon the neck incidental to delivery by the hands pulling upon the shoulders and trunk.

FIG. 24.

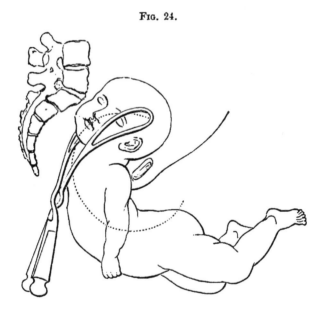

ILLUSTRATES THE APPLICATION OF THE LONG FORCEPS TO THE AFTER-COMING HEAD.

Such, then, is the story of the long forceps applied to the head in the first position at the brim. If the head be in the second position, the blades will seize it—one on the left brow, the other on the right occiput. The occiput will emerge under the right pubic ramus. Here special care is necessary. When the head emerges occiput to the right, if the shoulders are so large as to demand extraction in aid of expulsion, be very careful to direct the face downwards—*i.e.*, to the mother's left thigh —for if through inadvertence you turn the face upwards to the right thigh, you may give a fatal twist to the child's neck, or impede the turn of the shoulders into the antero-posterior diameter of the outlet.

In the case of the third and fourth positions of Naegele, the head will still be seized obliquely; and as it enters the pelvic cavity, it will generally make a quarter axial rotation, face backwards, so as to bring the occiput under a pubic ramus. In the case of fronto-pubic position, the head will be grasped more nearly in its transverse diameter. As it descends into the pelvis, this position may be preserved; and it becomes a question whether delivery should be completed with the forehead forwards, or an attempt made to turn it back into the hollow of the sacrum.

The cause of arrest of labour, of difficulty, when the position is occipito-posterior, is, I believe, this : The head imprisoned in the pelvis is not able to take its normal extension movement. In occipito-anterior positions, the propelling force propagated through the spinal column causes the head to roll up from the floor of the pelvis out by the open space under the pubic arch. But in occipito-posterior positions, the propelling force acts against the escape of the head by driving it against the floor of the pelvis, the occiput naturally rolling back into the hollow under the promontory. If extension-movement then takes place, this, by throwing the occiput against the back, rather increases the difficulty. Release can only be obtained by a movement of flexion. Now, flexion may be useful under two circumstances—first, as already explained, by supplying the essential condition for the spontaneous turn of the face into the sacrum ; secondly, by taking the symphysis as the centre of rotation, and the point against which the root of the nose or the forehead is fixed, whilst the vault of the cranium is made to roll over the floor of the pelvis and through the outlet.

The first question that arises in the presence of an occipito-posterior position is, whether we can hope for the change, spontaneously or by art, to an occipito-anterior position ?

Smellie accomplished this change by the forceps. Clarke and Burns said the rectification could be made by the fingers. Dr. R. U. West[*] has proved the practicability of procuring the rotation face backwards by artificial means. He applied his fingers to the frontal bones, turning this part backwards, and at the same time raising it up until he felt the posterior

[*] " Glasgow Medical Journal," 1856.

fontanelle come down. In another case he brought the occiput down by the lever. As soon as the occiput came down, the rotation seems to have been effected by Nature. This, indeed, is the essential thing to do—to get the occiput down, to restore flexion.

On the other hand, I am persuaded that the head often turns of its own accord, when we think we are helping it. The evidence of Dr. Millar is quite to the purpose. "I met," he says, "a good many cases of occipito-posterior positions in which anterior rotation was effected; but the efficiency, I believed, belonged to me, and not to Nature, because I laboured assiduously to promote it after the manner recommended by Baudelocque and Dewees. . . . I have since experimentally allowed Nature to take her course in a considerable number of such cases, and *I find that the desired mutation is generally accomplished about as well without as with my assistance.*"

Dr. Leishman says :* "We may succeed in amending the position in two classes of cases. In the first, the head is free at the brim, or at least has not encountered any serious pelvic resistance ; and here rotation may be effected by the forceps. . . . In the second, the head has reached the floor of the pelvis, where we have natural rotatory forces operating in our aid ; but no attempt, while the head is in a position intermediate between these two, is likely to be attended with success. . . . In the second class of cases the forceps is quite inapplicable. Simple rotation must not be attempted. We must employ our whole efforts in promoting the preliminary flexion of the head. This is done most effectively by bringing two fingers to bear upon it, and pressing in the direction indicated during a pain. This, in the first instance, will probably have little effect in displacing the forehead, but if we can only succeed in preventing its further descent we thus transfer the propulsive force to the occiput, and encourage the essential movement of flexion." Should this fail, the occiput may be pulled down by the vectis as R. U. West describes.

I have found that the occiput must be brought down below the edge of the sacro-sciatic ligament in order to permit of the rotation face backwards.

* "System of Midwifery," 1873.

It is judicious, I think, to make a reasonable attempt, after
the methods of Leishman and Dr. R. U. West, to bring the
occiput down and forwards. But it will be seen that Leishman's manœuvre postulates a driving force. This is often
wanting. The "pains" commonly flag when the head is
arrested on the floor of the pelvis.

The leverage may be applied by the forceps. The head
being grasped in its transverse diameter, or with only moderate
obliquity, a movement of rotation of the instrument on its axis
will turn the face backwards into the sacrum. But the forceps
cannot at the same time so well bring down the occiput as the
lever combined with the fingers can do. Professor Elliot, in
his admirable practical work (*Obstetric Clinic, New York*,
1868), gives cases in which he rotated the occiput forwards
by the forceps with success.

But I cannot give more than a qualified assent to the
propriety of attempting to rectify the position. It is only
exceptionally useful; still more rarely is it necessary, and it
is not free from danger. The head can be born very well
preserving the occipito-posterior position throughout. Indeed
I think this occurs more frequently than Naegele represents.
Nor does the case call for any amount of force. By aid of
the forceps the delivery is nearly as easy as when this instrument is applied to an occipito-anterior position. In the
event of delay, I therefore advise resort to the long forceps.

The blades should be applied in the sides of the pelvis;
they will be guided by the head into the most suitable position. *Extraction*, then, *simply*, without troubling yourselves
about rotation, is all that is necessary. If Nature prefer or
insist upon rotation, your business is to consent. As the head
advances, the occiput may come forwards, and you will feel
the handles of the forceps turn upon their axis. But in a
large proportion of cases Nature will not insist upon bringing the occiput forwards; and here again your part is simply
that of a minister of Nature. The forehead will emerge
under the pubes; the cranium will sweep the sacrum and
perinæum.

As the blades of the forceps preserve their original relation,
the handles will turn with the head. It is labour lost—it is
encumbering Nature with superfluous help—it is a sin against

that most excellent maxim, "*ne quid nimis*," to attempt to promote this turn by twisting with the forceps.

In this latter case there are two things to be observed:
first, the perinæum is put more upon the stretch, and therefore requires more care; second, if the handles of the forceps
are carried forward towards the mother's abdomen too soon, the
blades will be apt to slip off. The superiority of the long
forceps in saving the perinæum is very marked.

The propriety of not attempting to turn the face backwards
is even more decided in those more marked cases of fronto-

FIG. 25.

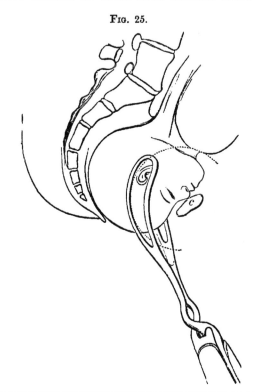

SHOWING THE APPLICATION OF THE FORCEPS TO THE HEAD IN FRONTO-ANTERIOR POSITION.

The promontory of sacrum shows very little projection. The head is seized nearly in its
transverse diameter. The symphysis, c, is the centre of rotation. The vertex and occiput
sweep the perinæum, producing a movement of flexion.

anterior positions in which *the forehead looks nearly directly
forwards*. It appears to me that this position is sometimes due
to unusual flatness of the promontory—a very slight projection

of this part. A pronounced projection of the promontory will scarcely permit the head to occupy the antero-posterior diameter; it will throw the occiput to one or the other iliac hollow, so that the moment the head dips into the pelvis the anterior pole is turned into the hollow of the sacrum, or to one side of it. The connection between a particular form of pelvis and occipito-posterior positions seems proved by the frequent recurrence of these positions in the same subjects. Thus, I have delivered the same lady three times successively by forceps on this account.

Upon this point I am glad to quote the authority of Dr. Ramsbotham,—" I prefer extracting it, if possible, with the face under the arch of the pubes, because, as the rotation is made over only one quarter of the half-pelvis, there is less chance of injuring the soft parts. Besides, should the child's body be strongly embraced by the uterine parietes while we are acting, and should it not follow the turn which we are forcing the head to take, we should twist the child's neck, perhaps fatally." In truth, there is no very serious difficulty in extracting with the face forwards. In this case the fillet or lever seizing the occiput would find its most scientific application.

When the head is delivered, if the child is of large size, there may still remain considerable difficulty in extracting the shoulders and trunk. And during the delay the child may die of asphyxia. The shoulders will be lying nearly in the trans-verse diameter just above the outlet, breast forwards, an unfavourable position since the shoulders will hitch on the ischial tuberosities. To liberate the child it may be necessary to rotate the shoulders so as to bring the dorsum forwards, or at least to bring one shoulder under the pubic arch. This may be done by a bi-polar manœuvre, pressing backwards one shoulder with one hand, and the other shoulder forward with the other hand. As soon as a little movement is effected in the desired direction, uterine action, or in its default firm compression of the uterus to impart onward movement, will complete the necessary rotation. The delivery is then easy. The manœuvre is in strict accordance with the law that dictates the conversion of abdomino-anterior into dorso-anterior positions.

The case is, however, more severe if it is a complete *face-*

presentation. You can hardly, by aid of the forceps, so far modify the position of the head as to render its course through the pelvis easy; and when you have succeeded in dragging it into the cavity, you may find yourself left with no alternative but to perforate. It is very true that a large proportion of face-labours end happily without assistance. It is equally true that face-presentations supply some of the most difficult cases in practice.

It is convenient in this place to examine *how brow-presentations and face-presentations are produced.* Dr. Carl Hecker attributes face-presentations to an original dolico-cephalic condition. He supposes that the increased power of the occipital arm of the lever throws the face down. I have taken photographs and outline-tracings of heads born after face and other varieties of labour, and think I am entitled to affirm that Dr. Hecker's hypothesis is drawn from an erroneous interpretation of facts. It is true that in face-labours the head is found elongated; but it is easy to trace this altered form to the compression undergone during labour. There is no evidence to show that it exists before labour. Fig. 1 represents the common globular shape of the head before it enters the pelvis; in Fig. 29 the head, taken from a photograph of an actual case, represents the dolico-cephalic condition *produced* during its compression in the pelvis.

Brow-presentations may be regarded as transitional between vertex and face-presentations; and by analyzing the mode in which brow- and face-presentations arise, we shall have the best indications for prevention and treatment. Consider the head as a lever of the third order, the power acting about the middle. The fronto-occipital diameter or axis represents the lever; the atlanto-occipital articulation is the seat of the power. Riding upon this point, the head moves in seesaw backwards and forwards. A force which is generally unnoticed in obstetrics is *friction;* and if friction were uniform at all points of the circumference of the head, it would be unimportant, from a purely dynamic point of view, to regard it. But it is not always so. Friction at one point of the head may be so much greater than elsewhere, that the head at the point of greatest resistance is retarded, whilst at the opposite point the head will advance to a greater extent; or resistance at one point may quite arrest the head at that point. In either case the head must change its position in relation to the pelvis.

Let us, then, take the case where excess of friction bears upon the occiput directed to the left foramen ovale. This point

FIG. 26.

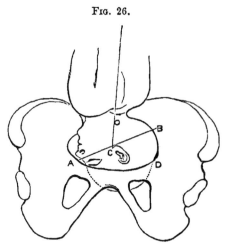

REPRESENTS A CHANGE IN PROGRESS FROM AN ORIGINAL VERTEX PRESENTATION TO A FOREHEAD.

o is the atlanto-occipital joint, or point where the force propagated through the spine e o impinges upon the lever A B C. D is the point of greatest resistance. Therefore the arm A of the lever descends. c o, the force, forms an obtuse angle with the arm A.

will be more or less fixed, whilst the opposite point or forehead, receiving the full impact of the force propagated through the spine to the atlanto-occipital hinge, will descend—that is, the forehead will take the place of the vertex, and be the presenting part. If this process be continued, the head rotating back more and more upon its transverse axis, the face succeeds to the forehead. A condition that singularly favours this excess of occipital friction, and consequent rotation in extension, is *lateral obliquity of the uterus*. The want of coincidence between the axis of the uterus and child and of the axis of the pelvis, disturbs the equilibrium of resistance and friction.

Now, if we can transpose the greatest friction or resistance to the forehead, and still maintain the propelling force, it is clear that the occiput must descend, and that the normal condition may be restored. In practice this is actually done. When at an early stage of labour we find the forehead presenting, we can, by applying the tips of two fingers to the forehead, during a pain, retard its descent, and the occiput comes down. This effected, the rest will probably go on naturally, because, the atlanto-occipital joint being somewhat nearer the occipital than the frontal end of the lever, the shorter or occipital arm of the lever will keep lowest. But if there should still be excess of resistance at the occipital end, we have only to add so much resistance to the frontal end as

FIG. 27.

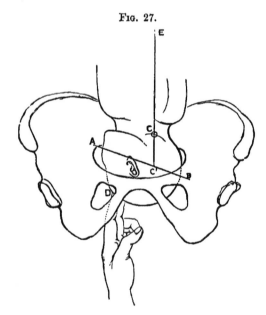

C

THE FINGERS APPLIED TO THE FOREHEAD AT D TRANSPOSE THE GREATEST RESISTANCE TO THIS POINT.

The force propagated from E to c will therefore drive down B, the occipital or shorter arm. The force E c c' will form an acute angle with the long arm A, and the tendency will thus be greater to keep the occipital arm B lowest in the pelvis. Or we may help to overcome the resistance at the occipital end of the lever by applying the palm of the right hand externally and pressing the occiput downwards.

will maintain the lever in equilibrium. This manœuvre is illustrated in diagrams 26, 27. It is at the same time desirable to restore the due relations between the axis of the uterus and the axis of the pelvis. This is best done by placing the patient on her back, and supporting the uterus in the median line by the hands.

The mechanism of face-presentation must detain us a few

minutes. As in the case of the cranial presentation, there are
four positions; and these may be described as conversions or
accidental departures from the respective cranial positions.
Thus, taking the first cranial position at the brim, in which, as
we know, the occiput is directed to the left foramen ovale.
Imagine that the occiput instead of descending into the pelvic
cavity hitches against the edge of the brim, or is kept up from
some other cause, whilst, the driving-force continuing, the head
will rotate on its transverse axis and the forehead and face will
successively come lowermost. Thus is formed the *first face-
position*. The second face-position is formed in a precisely

FIG. 28.

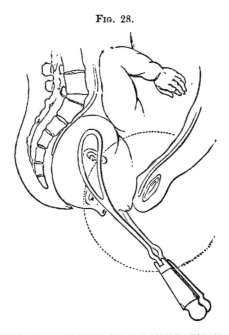

SHOWS THE LONG FORCEPS APPLIED TO THE HEAD IN FACE-PRESENTATION DELAYED IN THE PELVIS.
The curve of Carus—the dotted circle—indicates the direction of traction, restoring flexion.

similar manner, the forehead coming to the right foramen ovale,
and the fronto-mental line of the face lying in the second
oblique diameter of the pelvis. The third and fourth face-
positions may, in like manner, be traced to the third and fourth
cranial positions. In all the cases the forehead represents the
occiput. And in this position the head might very well proceed

through the pelvic cavity and outlet, following corresponding rotations to those observed in cranial positions, *if the head only had to be transmitted.* But the child's trunk must follow; and it is this which causes the difficulty, and which *compels* a change. If you look at diagram 29, you will see that if the head get into the cavity of the pelvis, face-presenting, the occiput will be jammed between the shoulders, which, entering the pelvis along with it, make a wedge whose base is too large for the space. Hence impaction, unless rotation is effected. This is usually accomplished. The chin comes forward, getting under the nearest pubic ramus, as in diagram 28. The pelvis here being shallow, the chin emerges, and, the whole head rotating upon its transverse axis in the curve of Carus, the occiput quits the shoulders; the wedge is decomposed; the difficulty is overcome; the trunk follows. The sum and object of all this mechanism is to restore flexion.

But the face may enter the pelvis, take its turn forwards, and then be arrested, just as the head in cranial presentation may be arrested. In such a case the forceps may be useful. The application is as follows. Assume that it is the first face-position; remember that the object to be accomplished is to make the vault of the cranium and the occiput roll over the floor of the pelvis around the symphysis as a centre, so as to restore flexion. The blades should seize the head nearly in its transverse diameter. Now, the face presents some degree of obliquity in relation to the pelvis. The first or sacral blade, therefore, must pass up the left side of the pelvis, somewhere between the sacro-iliac joint and the left extremity of the transverse diameter. The second or pubic blade will pass in the opposite point of the pelvis, that is, between the foramen ovale and the right extremity of the transverse diameter. When locked, traction is at first directed downwards, to get the chin fairly under the pubic arch. Then the traction is directed gradually more and more forwards and upwards, so as to bring the vault of the cranium out of the pelvis. The posterior part of the head puts the perinæum greatly on the stretch. It requires great care to extract. Give time for the perinæum to dilate. Carry the forceps well forwards, so that the shanks are out of the way; but not too soon, lest the blades slip off. Extract gently.

But we shall not always be so fortunate even as this. Several of the most difficult cases in which my assistance has been sought have been face-presentations. In some, *the face will not enter the brim.* This is the first order. What shall we do here? If we apply the forceps, one blade is likely to seize beyond the jaw and compress the neck, bruising the trachea. If the attempt be made to seize the head by applying the blades in the oblique diameter, they must be passed very high, and even then may slip. If firmly grasped and traction be made, the faulty extension of the head is increased; the compression of the vessels of the neck and the danger of apoplexy are augmented; and, after all, extraction may have to be completed by perforation. Turning can be effected with infinitely less trouble, and with a better prospect for the child. In the second order of cases, the *face has descended into the cavity.* The birth of a full-grown living or recently dead child, with the forehead maintaining its direction forwards, is almost impossible. The extension of the neck is extreme, the head being doubled back upon the nucha. The chin represents the apex of a wedge, A B C, the base of which is formed by the forehead, the entire length of the head, and the thickness of the neck and chest (*see* Fig. 29). This must be equal to at least seven inches. The bregma and occiput become flattened in, it is true, but much is not to be expected from moulding. Compression, bearing upon the neck, if great and long-continued, is almost necessarily fatal to the child. Hence arrest or impaction. The turn of the chin forwards under the pubic arch, so as to release the head by permitting flexion round the symphysis, cannot take place. Aid becomes necessary. We have to consider the following points :—

1. Can the head be rotated on its transverse axis, restoring flexion, and so to bring the cranium down? This may be accomplished whilst the head is above the brim, but scarcely when it is squeezed into the cavity.

2. Can the turn of the chin forwards be effected by the hand, the lever, or the forceps? This is sometimes possible, and should be tried. It is thus described by Smellie: "After applying the short or long-curved forceps along the ears, push the head as high up in the pelvis as is possible, after which the chin

is to be turned from the os sacrum to either os ischium, and afterwards brought down to the inferior part of the last-mentioned bone. This done, the operator must pull the forceps with one hand, whilst two fingers of the other are fixed on the lower part of the chin or under-jaw to keep the face in the middle and prevent the chin from being detained at the os ischium as it comes along, and in this manner move the chin round with the forceps and the above fingers till brought under the pubes, which done, the head will easily be extracted."

3. Can the head be brought down by the forceps without turning the chin forwards ? This is a practice against Nature. If the forceps grasp, and it will generally slip, it will bring more of the base of the wedge into the brim. The head must be small, or the pelvis large, to admit of success by this mode.

4. Shall we extricate the head by perforating ? The wedge may be lessened, but even after this, delivery is not always easy unless part of the cranial vault be removed, so as to allow of the flattening in of the head.

5. Can we turn simply ? It is the best course, but if the head is low it may be difficult to accomplish.

6. The chin will sometimes turn forwards at the very last moment, when the face is quite on the floor of the pelvis. If not, it may be possible to hitch the chin over the perinæum by drawing the chin forwards by forceps, and pulling the perinæum backwards (see Fig. 29). The chin thus outside, the forceps or lever may be applied to draw the occiput down under the pubes and backwards, so as to make the head revolve on its transverse axis, thus restoring flexion. You are in fact decomposing the base of the wedge. You deliver by a process the reverse of that of ordinary occipito-anterior labour. In this, the occiput escapes by a process of extension. In the mento-sacral position you deliver by promoting flexion. Or, to take our illustration from the mechanism of face-labour, you obtain flexion by causing the chin to turn over the coccyx or sacro-sciatic ligament as a centre, instead of over the symphysis. The latter is the natural mode, but it may be that the first alone is possible. This is a case in which incision,

bilateral, of the perinæum, here acting as an obstructing posterior valve, may be performed in order to facilitate the release of the chin.

Fig. 29.

AN EXTREME CASE OF FACE-PRESENTATION IN WHICH THE ROTATION CHIN-FORWARDS WILL NOT TAKE PLACE.

The face, much swollen by pressure, is near vulva; the cranium, although flattened on vertex, cannot descend, because the occiput, rolled back upon the nucha between the shoulders, forms, with the thickness of the chest, the base of a wedge A B C much longer than the conjugate or oblique diameters of the pelvis. If traction in the axis of the outlet be made by forceps this will only increase the stretching of the neck, and jam the head and chest more strongly in the brim. It will be seen that, if the head can be grasped by the forceps, and drawn down in the curve D E round the coccyx as a centre, the chin riding over the perinæum, the wedge A B C will be decomposed, and flexion being restored, the head may be delivered.

LECTURE VI.

Now we have to consider what the forceps can do in cases of
disproportion; for instance, where the brim is too small to
allow the head to pass by the unaided powers of the uterus.
This brings up the problem of the compressibility of the head
under the forceps, and the comparison of the advantages of
the forceps with those of turning. The degrees of contraction
of the brim may be classified approximately in the following
manner :—

Scheme of Relation of Operations to Pelvic Contractions, Labour at Term.

	Conjugate diameter reduced to	Operations practicable.
The first degree	4 to 3¼ in.,	admits the forceps, opposed to the bi-parietal diameter of 3½ to 4 in.
The second degree..	3¾ to 3 in.,	„ of turning, opposed to the bi-mastoid diameter of 3 in.
The third degree	3¼ to 1¾ in.,	„ of craniotomy and cephalotripsy.
The fourth degree..	below 1¾ in.,	„ of Cæsarian section.

If you have the advantage of bringing on labour at seven
months, then you may eliminate the Cæsarian section, and
slide down the scale of operations, so that craniotomy shall

correspond with the fourth degree, turning with the third, and the forceps with the second, whilst the first degree, being reduced to the conditions of natural labour, may require no operation at all.

Scheme of Relation of Operations to Degrees of Pelvic Contraction under Labour at Seven Months.

	Conjugate diameter reduced to	Operations practicable.
First degree ..	.. 4 to $3\frac{1}{4}$ in., admits spontaneous labour.
Second degree ..	.. $3\frac{3}{4}$ to 3 in., ,, of forceps.
Third degree ..	.. $3\frac{1}{4}$ to 2 in., ,, of turning.
Fourth degree ..	.. below $1\frac{3}{4}$ in., ,, of craniotomy.

Cæsarian section is eliminated.

The range of application of the forceps is, I believe, not great. The head cannot be compressed by it quickly. The

FIG. 30.

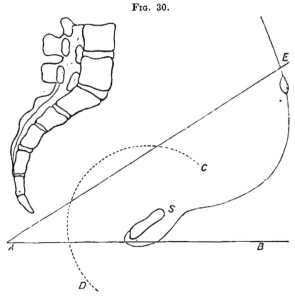

NORMAL PELVIS.

s, the symphysis pubis, the centre of Carus' curve C D; A E, the axis of the brim, forming an acute angle, not less than 30°, with the datum-line A B. The uterus and the child's body nearly corresponding with the axis of the pelvic brim, the head enters its natural orbit, represented by Carus' curve, at once.

proper use of it is to aid that natural process of moulding which always takes place in protracted labour. Now, this is a gradual, even a slow process. The head is seized by the long

forceps in the way already described. The handles are firmly
grasped with both hands, and especial care is required to
extract well backwards in the axis of the brim, so as to make
the head revolve round and under the projecting, overhanging
promontory as a centre. Here I may pause to show that, in
labour with conjugate contraction from rickets, the promontory
possesses a like importance at the brim or entry of the pelvis
to that which the symphysis pubis possesses at the outlet.
The promontory is a turning-point—a centre of revolution of
the head, just like the symphysis. The curve round the pubes,
which Carus described, has its counterpart in a curve round
the promontory. In ordinary labour, with a well-constructed
pelvis, the head enters the pelvis, and reaches nearly to the

Fig. 31.

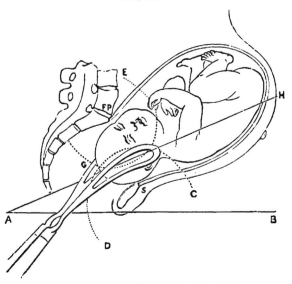

PELVIS CONTRACTED BY RICKETS TO SHOW THE CURVE OF THE FALSE PROMONTORY.

s, the symphysis, the centre of Carus' curve c d ; f p, the false promontory, the centre of the
false curve e g ; g, the point of intersection of the two curves where the head passes from the
false to the true orbit; a h, the axis of the brim, forming a very acute angle, varying from
30° to 20° or less, with the datum-line a b. The head is thrown over the pubic symphysis by
the projecting promontory. The forceps draws backwards in the line a h to bring the head
under the promontory in the orbit of the false promontory before it can enter at g the true or
Carus' orbit c d.

This diagram should be contrasted with Fig. 30. It will be seen how remarkably the axis of
the pelvic brim a h, is altered in relation to the datum-line a b. It is brought nearer to the
horizon, forming a more acute angle. It thus throws the uterus more forwards, and becomes
the chief cause of the pendulous belly, so marked a feature in cases of contracted pelvis.

floor, without deviating much from the straight line which represents the axis of the brim. Thus it enters its orbit, the circle of Carus, at once.

But a projecting promontory, involving, as it commonly does, a scooped-out sacrum below, disturbs this course. The promontory must be doubled. The head must move round this before it can strike into its natural orbit. I propose to call this first curve, *the curve of the false promontory.*

This curve is the chart by which to steer in turning on account of contracted pelvis. Bearing this in mind, and assuming, that the head is seized nearly in its transverse diameter, the blade corresponding to the anterior or pubic side of the head must describe a large circle, whilst the sacral side of the head, and the blade in relation with it, move but little until the promontory is rounded, and the head has entered the pelvis. When this point is reached, the direction of traction is that of Carus' curve. The head which was compelled to traverse the brim nearly in the transverse diameter, will quickly rotate, face to sacrum. This turn, imparted to the handles of the forceps, and sudden transition from resistance to ease, a sort of jerk, mark the completion of the first circuit and the beginning of the second. The rest falls within the laws of natural labour.

It happens, however, in these cases of contracted conjugate diameter, that the head commonly presents at the brim with its long diameter very nearly, if not quite, in correspondence with the transverse diameter of the pelvis. The blades of the forceps, also finding most room in this diameter, will grasp the head in its longitudinal diameter. In extraction, therefore, both blades will move equally around the false promontory, so that when the head has doubled the promontory, and entered the cavity, it will be useful to remove the forceps, and to readjust the blades on the sides of the head, should traction-force still be necessary.

What is the extreme degree of narrowing that will admit of the useful application of the forceps? I have stated it in the table at three and a quarter inches, but it cannot be defined absolutely. A head slightly below the normal size, and less firmly ossified than usual, may be brought through a conjugate diameter of only three inches. And as we cannot know with

sufficient precision what the properties of the head still above
the brim are, we are justified in making tentative, experimental
efforts with the forceps before resorting to turning, which is,

FIG. 32.

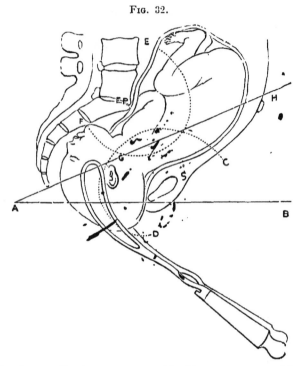

A B, datum-line; C D, Carus' curve; E F, curve of the false promontory; F P, false promontory; G, point of intersection of the two curves where the head passes from the false to the
true orbit. The forceps now draws the head in the direction of the outlet in Carus' curve.

perhaps, more hazardous to the child, or to craniotomy, which is
certainly destructive to it.   This uncertainty, or want of fixity,
in the relations between the head and the pelvis, compels us to
leave a range or border-land of debatable territory between the
more clearly recognized or conventional limits assigned to the
several operations.   This debatable territory is further liable to
invasion from either side, according to the relative skill with
which the competing operations are carried out.   And herein
lies the source of the great controversies in obstetric practice.
Thus one operator, possessing a good long forceps, and confident
in his skill in handling it, will use this instrument with success

where the contraction is three and a quarter inches, whilst another possessing only a single-curved forceps or a bad double-curved one *must* either turn or perforate. So again, the region between the second and third degrees of contraction, the region assigned to turning, may be invaded on the one side by the forceps, on the other by the perforator, and become the subject of a partition-treaty, which shall dispossess turning, the rightful power, altogether. It unfortunately happens that perforation, being an easy operation, is apt to carry its inroads further than the forceps ; and thus the child falls under a wide, arbitrary, and fatal proscription.

There is a condition causing dystocia called the *pendulous abdomen*. It is most frequent—First, in women who have borne many children, and in whom the abdominal walls are much relaxed. Second, where the pelvis is contracted. Third, when there is separation of the recti muscles, constituting hernia of the gravid uterus. When it exists, the uterus hanging down in front of the pubes is out of the axis of the brim, and, if it contracted, would only direct the child over the brim, backwards against the promontory. This may sometimes be remedied by putting the patient on her back, and making up for the want of support from the abdominal muscles by applying a broad binder so as to lift the fundus of the uterus upwards and back-wards. This will restore the relation between the axis of the uterus and that of the pelvic brim. But if contractile energy be still insufficient, the long forceps will come into use. And this is a case where the dorsal decubitus will much assist the delivery. If the patient continue on her side, the uterus not only hangs forwards, but swags downwards to the dependent side, constituting a further deviation, and increasing the obstacle to parturition.

How to determine the choice between forceps and turning ? There are two cases. First, the liquor amnii has drained away, and the head is pressing into the brim: the forceps is strongly indicated here. Secondly, the head is mobile above the brim, and not easy to grasp in the blades: here turning may be preferable. I have several times rescued a living child by turning under these circumstances.

The second case may sometimes be reduced to the first, and thus brought within the more desirable dominion of the forceps. One result of the pendulous abdomen and uterus is to form a kind of reservoir in which the liquor amnii is dammed up. Hence an added impediment to contraction of the uterus. Now, the waters can be drained off by lifting the fundus uteri up to its normal position against the spine, laying the patient on her back, and making a channel past the head to the uterine reservoir by introducing one blade of the forceps. Having accomplished this, the uterus, under the combined advantages of restoration to its natural axis, and of steady pressure by the hands or a belt, may recover its power, and expel the child. If not, the forceps supplies an easy remedy.

In the following diagrams (Figs. 33, 34), the mechanism of labour obstructed by this form of malposition of the uterus is illustrated.

Until the uterus is brought back to its normal position, it is clear that two causes concur to render labour difficult. First, the uterus being thrown forwards, its fundus is carried away from the diaphragm and upper part of the abdominal walls. It loses, therefore, the aid which the expiratory muscles, acting powerfully when the glottis is closed and the chest is fixed, usually give. This expellent power of the expiratory muscles is so great, that it appears to be of itself sufficient in some case to complete labour, the uterus remaining quite passive. When the uterus is thrown forwards across the pubes, any force propagated downwards from the diaphragm will strike the posterior wall of the uterus at a right angle with the body of the uterus, and of the long axis of the foetus. It will, in short, drive the uterus and its contents down upon the symphysis, or even more forwards still, since the body of the child, which lies in front of the symphysis, forms the longer arm of a lever, and the force is expended upon it.

Secondly, the uterus itself, if not paralyzed, acts in a wrong direction. It loses the stimulus to action which the normal pressure and support of the diaphragm and abdominal walls supply, and therefore acts languidly. Its independent power is also weakened by another circumstance. It is a law, of which the patient observer will not fail to discover many proofs in the progress of difficult labour, that, whensoever a mechanical obstacle is encountered, before long, the uterus, conscious, as it

were, of the futility of its efforts, intermits its action, takes a
rest, lies dormant, until the time shall arrive when it can act

Fig. 33.

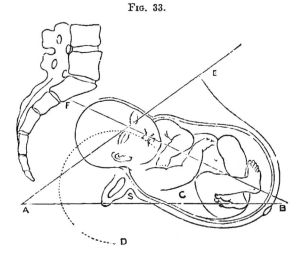

A B, datum-line; A E, axis of pelvic brim and normal axis of uterus; F B, axis of uterus
pointing to sacrum; C D, Carus' curve.

with advantage. This provision protects, for a long time, against
exhaustion from protracted labour. Indeed, what appears to be
protracted labour is often simply suspended labour; and suspended labour may even pass into what Dr. Oldham has so aptly
called "missed labour."

The remedy is obviously to restore the uterus to its normal
position. In Fig. 34, the uterus and child are represented
lying across the symphysis pubis, like a sack across a saddle.
H I is the line in which the proper uterine force would be
exerted; F G is the line of force of the expiratory muscles
striking the long axis of the uterus behind and nearly at a right
angle. These two forces, which ought to coincide, thus cross
each other, and the error is but imperfectly compensated by the
resultant force obtained between the two. But raise the uterus
to its normal position, as indicated by the dotted outline, and,
immediately, the expiratory force and the uterine force coincide
with the axis of the child and of the pelvic brim, and both
conspire to expel the contents of the uterus. Not even the
forceps will act efficiently until this restoration is made.

In the case of pendulous belly from contracted pelvic brim the restoration of the axis of the uterus to its proper relation with that of the pelvis can rarely be effected until the head is

FIG. 34.

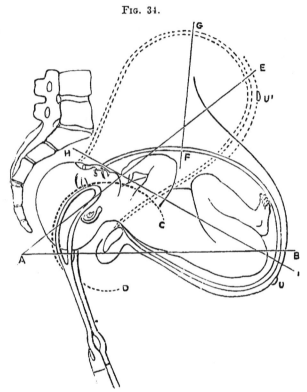

SHOWING THE MECHANISM OF LABOUR IN PENDULOUS BELLY.

A B, datum-line determining position of the pelvis; C D, Carus' curve; H I, axis of uterus and of child directed towards the promontory; F G, line of force of expiratory muscles cutting the axis of the uterus and of child; A E, normal axis of pelvis; U U', the umbilicus.

brought into the pelvic cavity; and this has to be done after craniotomy. Then the dorsal decubitus, and upward support of the abdomen and uterus come in effectively.

LECTURE VII.

*Dystocia from faulty condition of the soft parts of the parturient
canal* is only incidentally and occasionally related to the history
of the forceps. This incidental relation, however, makes it convenient to discuss the question in this place.

The cervix uteri, the vagina, or the vulva, including the
perinæum, may refuse to yield a passage to the child, and to
permit the application of the forceps. The conditions which
lead to this result are various. First, as to the cervix uteri,
including the os externum uteri, the following causes of obstruction may be observed:—1. Spastic annular contraction.
2. Thickening from œdema. 3. The cervix may have an
abnormal direction and position. The cervix is not in the
line of the axis of extrusion. Perhaps it is bent down at a
more or less acute angle upon the body. A more frequent
condition is the pointing of the os uteri backwards towards

the promontory, and very high up, so that it is difficult to reach, perhaps impossible, without passing the hand into the vagina. In such a case, the head bears unduly upon the anterior segment of the lower part of the uterus. This is often the result of slight narrowing of the pelvic brim, which throws the head upon the anterior wall of the pelvis. This may persist so long that the tissues become worn and their texture softened, so that when the head is driven down into the pelvis the damaged cervix rends. 4. From contraction of the brim, or from the presentation of some part of the child—as the arm, face, or feet —not adapted to descend easily and fairly upon the cervix, there is insufficient dilatation. 5. The cervix may be organically diseased. The most marked causes of this kind are hypertrophy, occlusion from false membrane described by Naegele, fibroid tumour, or cancer. In one case to which I was called, although the lips of the os could be distinguished, a thick membrane was continuous with the circumference of the os, completely occluding it. In this case I arrived at the opinion that the occluding membrane was formed by amnion and chorion which had become closely adherent to the lower zone of the uterus. 6. In another case, that of a Maternity patient of St. Thomas's, the os was completely closed by cicatricial tissue contracting concentrically. Severe labour pains had no effect in opening it. The pulse rose to 120. I pushed a hernia-knife through the centre, and nicked freely all round. The cervix expanded with wonderful rapidity as soon as the unhealthy tissue was divided. Mother and child did well.

There is another condition, which, although not in itself abnormal, will be properly considered in connection with the above. The os and cervix may be met with closed or only imperfectly dilated under circumstances which render speedy delivery eminently desirable. In such a case, the cervix must be treated as one that is rigid or otherwise diseased. It obstructs labour, and, just as in the cases where the closure of the os is the primary cause of obstruction, it must be opened.

The first of the conditions enumerated includes what is commonly understood as *rigidity*. It is really much more rare than is supposed. Most frequently when the os will not dilate, it is because the presenting part of the child cannot come down upon it. But if the membranes are ruptured prematurely, and the presenting part comes to press upon the os before this is at all dilated, then it often acts as a source of irritation, and produces this spastic annular contraction.

Before discussing the second cause of rigidity, it will be useful to examine *what are the forces that dilate the cervix.*

This study will throw light on the causes and pathology of rigidity, and furnish useful indications in treatment. By some it is held that the opening of the cervix is the direct result of the active contractions of the longitudinal uterine muscles, ⅄ which, pulling the os towards the fundus, thus draw it open. No doubt this is an important factor; but it is ineffectual alone.

It is a matter of observation that the os uteri does not expand in any marked degree until either the bag of membranes or the child's head comes to bear upon it. These distend the cervix and os as a direct mechanical force; they are, in fact, wedges, themselves inert, but propelled by the contractions of the uterus and the abdominal muscles. Under this distending force, the circular fibres of the cervix yield, just as the sphincter ani or the sphincter vesicæ yields under ⅄ the pressure from above. The yielding of the cervix uteri is indeed a question of the preponderance of the *vis à tergo* exercised by the body of the uterus and the expiratory muscles over the resistance offered by the cervix. Sometimes the normal harmony between this preponderance and the resistance is disturbed; the active force or the passive resistance is in excess; or the resistance may become active, and the force may be reduced to inefficiency. There is, in fact, a translation or metastasis of the nervous energy from the body of the uterus to the neck. This disturbance most frequently arises from an inversion in time, in the order of succession of the parturient phenomena. Thus, if the liquor amnii escape prematurely, the presenting part of the child will bear too early upon the cervix, and excite it to irregular action. This, by diverting and disordering the nervous supply to the body of the uterus, disables this part of the organ; and concurrently the cervix itself, becoming congested and thickened by undue pressure, irritation, and action, loses its natural capacity for dilatation.

A very instructive illustration of the theory, that the dilatation of the cervix uteri is essentially dependent upon the eccentric pressure exerted by the liquor amnii and fœtus driven into it, is found in the equivalent action of my hydrostatic cervical dilator. This instrument is inserted in a collapsed state within the cervix, and then gradually distended with water, as seen in Fig. 35. This very nearly represents the normal action of the liquor amnii distending the sac of the amnion. Under this pressure, the cervix yields smoothly and gradually, just as in natural labour; the speed, however, being very much within the discretion of the operator. In this instrument we possess a power in midwifery, at once safe and efficient, that brings the cervix, and therefore the course of labour, completely within the control of skill.

How shall we restore the due relation between the expulsive and the resisting forces? How, in other words, shall we overcome the rigidity of the cervix uteri? This may be done in one of two principal ways. We may increase the power of the body of the uterus, so as to restore its preponderance over the cervix; or we may apply direct means to the cervix to dilate it, doing ourselves the work that the uterus cannot do. Great judgment is necessary in selecting between these two courses. Before deciding in favour of the first, we must be satisfied that the resistance opposed by the cervix is of such kind and degree that it may be overcome by moderate force. We must also be satisfied that there is potential energy enough in the system and in the uterus to respond to the stimulus, to the whip we propose to administer. To give ergot, for example, when the frame and the uterus are exhausted, is to equal the folly of the heavy rider who drives his spurs into his jaded horse when he ought to dismount and lead him.

It will almost always be proper, as the first step—that is, before seeking to rouse the uterus to increased action—to secure a more favourable condition of the cervix. How is this to be done? Let us take the case of spastic rigidity, the cause of which we have just glanced at. The first indication is to soothe, to subdue nervous irritability. Belladonna in the form of extract has been smeared upon the part. One sees the action of this drug in expanding the pupil. I have never felt it on the os uteri. The analogy is probably defective in theory. I believe it is not in the least degree to be relied upon in practice. It is at best an expedient for passing ⤴ time.

Chloroform is often of signal service. It acts by annulling the sense of pain and the fear of pain, and by restoring the equilibrium of the nervous system by removing disturbing causes that divert the nerve-force from its appropriate distribution; the sphincteric spasm relaxes, the body of the uterus contracts as it ought to do, and the labour proceeds.

A remedy sometimes of equal value to chloroform is *opium*. Thirty drops of the tincture or twenty of the sedative liquor combined with thirty drops of *compound sulphuric ether* will assuage pain, procure rest, and restore the harmony of the distribution of nerve-force; and if not in itself sufficient, it will aid in the carrying out of other measures.

Chloral is sometimes superior to opium. It produces unconsciousness, without stopping uterine contraction. Mr. Lambert (*Edinburgh Medical Journal*, 1870) recommends it to be given in gr. xv. doses every fifteen minutes, until its action is observed. It is especially useful in the first stage of labour.

Tartar emetic in small doses to provoke nausea has been recommended. In some cases I have proved its use, but I am not now disposed to resort to it, at the cost of postponing means at once more prompt and less distressing in their action.

Bleeding has been much extolled. In certain cases, as of convulsions, apoplexy, or such states of system as threaten these catastrophes, this proceeding may be adopted. But, apart from a decided special indication of this kind, it is not wise to bleed a woman in labour. Nor can bleeding be depended upon, as may be frequently seen in cases of placenta prævia, where even flooding *ad deliquium* will sometimes fail to relax the rigid or spastic cervix.

Warm baths have been much praised, and no doubt have a certain degree of power in inducing relaxation of tissue. But a warm bath is rarely at hand, and, if it were, the inconvenience of putting a woman in labour into it must often be insurmountable.

The most valuable of all preparatory measures is the *irrigation of the cervix and vagina* with a stream of tepid water. ⤴

We know that this is even efficacious in the induction of labour. And it is obviously useful to apply our knowledge of the agents that are effective in the solution of the major problem to the minor one : how to facilitate, to accelerate labour that has begun. In many cases this irrigation will be enough. Presently the cervix softens and yields, spasm is subdued, and abnormal action of the cervix is turned into normal activity in the body of the uterus. The mode of proceeding is simple. Introduce the vaginal tube connected with Higginson's syringe into the vagina, guided by the fingers of the left hand to the os uteri—*not into the os uteri*, there is danger in that—so that the stream of water shall play upon the cervix and fundus of the vagina. This may be continued for ten or fifteen minutes at a time, and repeated after an equal interval.

To accelerate labour-pains, Valenta (*Wien. Med. Presse*, 1870) extols the plan of placing an elastic bougie in the uterus, without rupturing the membranes, — another application of the labour-provoking agents to the task of accelerating labour.

When the cervix has become disposed to yield, it may not yield. The dilating force has still to be found. You may now, perhaps, give *ergot*. But when you have given ergot you are likely to be in the position of Frankenstein. You have evoked a power which you cannot control. Ergotism, like strychnism, will run its course. If it act too long or too intensely, you cannot help it. You may try "epichonto-cics," as chloral or nitrite of amyl, or possibly physostigma. But these may fail. The ergotic contraction of the uterus, when characteristically developed, resembles tetanus. Then, woe to the mother if the cervix does not yield, if the pelvis is narrowed, if, in short, any obstacle should delay the passage of the child ! And woe to the child itself if it be not quickly born ! I very much prefer to use weapons that obey me, that will do as much, or even less, than I wish. I fear to use weapons that will do more.

The cervix may be dilated *by the hand*. Two or three fingers are insinuated within the os, one after another, so as to form a conical wedge. This wedge is gently and gradually pushed forward into the cervix, and, widening as it goes, the cervix gives way. This wedge has the advantage of being

a sentient force.   It tells you what it is doing.   But what
it will tell you is sometimes this : it is that the fingers, with
their hard joints, form a rather painful and irritating wedge.
As it proceeds it is apt to renew the spasmodic contractions
you have taken such pains to allay.   If the head is pressing
upon the cervix, you may, as has been already mentioned,
help the dilatation by hooking down the anterior lip with
one or two fingers, holding the os open, as it were, to allow
the head to engage in it.   But this application is limited ;

Fig. 35.

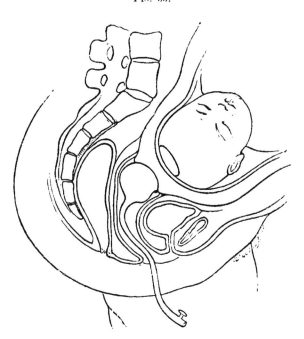

THIS FIGURE SHOWS THE HYDROSTATIC DILATOR DISTENDED IN SITÛ WITHIN THE CERVIX UTERI.

and, I think, what is called manual dilatation of the spasmodic
cervix should be abandoned, except in the case of spasmodic
contraction after the expulsion of the child—as, for example,
when the placenta is retained, or clots are filling and irritating the uterus.   In such a case, the steady onward pressure of the hand-wedge will in a few minutes wear out the

H

spasm and effect a passage, enabling you to clear out the cavity.

Water pressure is the most natural, the most safe, and the most effective. An os uteri that will admit one finger will admit No. 2 dilator in the collapsed state. The introduction is effected in this way : Insert the point of the uterine sound, of a male catheter, or any convenient stem, into the little pouch at the end of the bag ; roll the bag round the stem, anoint it with lard or soap, then pass it into the cervix, guided by the forefinger of the left hand, which is kept on the os uteri; or sometimes it is easier to roll up the bag, and to seize it by a long speculum-forceps. When the bag is passed so far that *the narrow or middle part is fairly embraced by the cervical ring*, withdraw the sound or forceps, keeping the guiding finger on the os to insure the preservation of the bag *in situ*. Then pump in water gradually. Continue distending the bag until you feel it is tightly nipped by the os. When this is done, wait a while ; close the stop-cock, and give time for the distending eccentric force to wear out the resistance of the cervix. No muscle can long resist a continuous elastic force. From time to time inject a little more water, so as to maintain and improve the gain. But be careful not to distend the bag beyond its strength. There is of course a limit to the distensibility, even of india-rubber ; and I have been told of cases where the bag has burst. I think this accident ought to be avoided. It has never happened to me, and I think it need not happen if the bags are well made. When you have got all the dilatation out of No. 2 that it is capable of giving, remove it, and introduce No. 3, which is larger and more powerful. The dilatation No. 3 will give is commonly enough to afford room for the forceps or the hand. The time required for this amount of dilatation will range from half an hour to two hours. But, not to lose time, it is desirable to keep your finger on the edge of the os, so as to be sure that the bag does not slip forward into the uterus altogether, or is not driven down into the vagina by uterine action. If it slips wholly into the uterus, it may displace the head. When you have gained your end, open the stop-cock, the water is ejected in a stream, and the bag is easily withdrawn. The cervical dilator serves yet another purpose. Taking the place of the liquor amnii, it does duty for the bag of membranes. It not only directly expands the cervix, but, setting up a quasi-normal reflex excitation, it evokes the regular action of the body of the uterus.

The proceeding I have described will succeed in the great majority of instances, especially where the closure of the cervix is due to spasmodic action ; or where, the tissue of the cervix being normal, it cannot expand for want of an eccentric expanding force, as when the bag of membranes or the child does not bear upon it. But in certain cases where there is rigidity from alteration of tissue, as œdema, hypertrophy, cicatrix, something more is required; and that is found in the *knife.* There is nothing new in this use of the knife. It is an old resource too much neglected. Coutouly, Velpeau, Hohl, Scanzoni, indeed all the most eminent Continental practitioners, advocate it. Judiciously employed, the knife can do no harm. It will save life when nothing else can.

You are sometimes in the presence of this alternative : exhaustion, sloughing, or rupture of the uterus, on the one hand, or the timely use of the bistoury on the other. It would be as absurd to hesitate as it would be to refuse to perform the Cæsarian section to give birth to a child which cannot be delivered by the natural passages. Indeed, it would be far more absurd, for the Cæsarian section is a most dangerous operation, whilst vaginal hysterotomy of the kind under discussion is comparatively free from danger.

There are various cases in which *vaginal hysterotomy,* or *dilatation of the cervix by incisions,* is necessary.

First, no os uteri is to be found. Of course, at the time of conception there was an os uteri ; it may have been subsequently closed by a false membrane or by cicatricial contraction. You will rarely fail to feel a nipple or depression where the os ought to be. It is generally very high up and far backwards, near the promontory. Pressure with a sound or the finger will mostly break down a false membrane and offer a sufficient opening to admit a hernia-knife or the special knife described in the first lecture. This is long and straight, probe-blunted at the end, having a cutting edge of about three-quarters of an inch near the end. The forefinger of the left hand is kept on or in the os uteri, as a guide. (*See* Fig. 36.) The knife is then slipped up, lying flat upon this finger, until its cutting edge is within the os. This edge is then turned up, the back supported by the guiding-finger, which takes cognizance of what is to be done and of what is done; and an incision of about a quarter of an inch deep, a slight nick rather, is made in the sharp ridge of the os. The knife is then carried round to another part of the ring of the os, and another nick is made. In this way four or five nicks are effected. Each gives perhaps little; but the aggregate gain of these minute multiple incisions

FIG. 36.

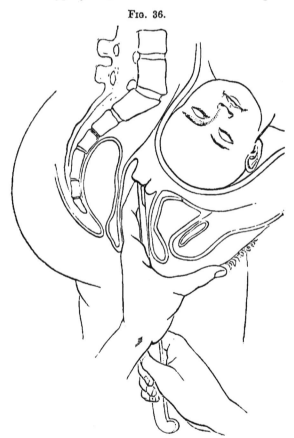

THIS FIGURE SHOWS THE OPERATION OF DILATING THE RIGID OR HYPERTROPHIED CERVIX UTERI BY INCISION.

is considerable. I do not think it matters much at what particular points of the circumference of the os these are made; perhaps the two sides are to be preferred.

Before extending or repeating these incisions, it is proper to observe the effect of uterine action in continuing the dilatation. And if nothing is gained in this way, introduce the hydrostatic dilator; distend this gently, carefully testing by the finger its action. This plan of *combining the water-dilator with incisions* is especially valuable in cases of rigidity from hypertrophy of the cervix, or of atresia of the os or vagina from cicatrices.

When the forceps will pass—and it is quite possible to apply it when the os will allow the three fingers to pass as far as the knuckles—this instrument may serve to dilate further. The head being grasped, you may draw steadily down ; and by keeping up gentle traction, the wedge formed by the blades and the head will gradually dilate the os, perhaps enough to allow the head to pass, and thus to save the child's life. But this must be done with great caution, and not be persevered in unless the cervix yield readily. If this resist long, the continuous pressure upon the child's head is very likely to prove fatal. Moreover, this mode of dilating the cervix is not free from danger to the mother.

But it will occasionally happen that, neither by incision, water-pressure, the hand, nor the forceps, will you obtain an opening sufficient without danger of laceration or other mischief. In such a case, you are justified in reducing the head to the capacity of the cervix by perforation.

Narrowing and rigidity may exist in the vagina in consequence of similar conditions, and may be treated on the same principle. The small rigid vagina of a primipara is best dilated by irrigation and the hydrostatic dilator. This will often singularly shorten labour. Atresia from cicatrices presents a more formidable obstacle. I have found the passage constricted by dense cartilaginous tissues, so as to permit no more than a probe to pass. In such a case, a careful process of incisions, multiplied in all points of the circumference, alternating with water-pressure, is necessary ; and it is, after all, probable that you will have to meet the difficulty halfway by perforating the head.

Lastly, obstruction may occur at the vulva and perinæum. In primiparæ, especially, the vulva may form a small rigid oblong ring scarcely permitting the scalp of the presenting head to show through it. The expulsive pains cause the perinæum behind this ring to protrude; but the ring itself will not open; in fact, the perinæum will yield first. It bulges more and more, and may give way in the raphe, just behind the commissure, this part remaining, for a time at least, intact. A central rent is thus made, through which the child has occasionally been expelled, instead of through the vulva. Or if the perinæum does not yield, something else must. The uterus will cease to act; or, struggling in vain, it may burst itself or the vagina. These dangers you may avoid by incisions. The fore-finger is passed between the head and the edge of the vulva, and two or three small nicks are made on either side, nearer to the posterior commissure than to the anterior. The relief sometimes gained in this way is surprising. Spasm, irritation, pain subside; the vulva dilates, and labour is soon happily at an end. The bleeding is insignificant; and the minute wounds left when the parts have contracted quickly heal.

Sometimes the vulva, including the labia majora, is so greatly swollen by *serous infiltration*, as to offer a serious obstacle to labour. This condition is commonly associated with albuminuria and convulsions. And out of this association contending difficulties arise. The convulsions urge to the acceleration of labour; the state of the soft parts forbid active interference. If the head comes down through tissues thus distended by fluid, not only laceration, but subsequent sloughing or gangrene may result. The obstacle to labour, and the local mischief, may be avoided by pricking the skin and mucous membrane in numerous points, so as to let the serum drain off. The operation is performed by a lancet held by the blade between finger and thumb, at a distance of a quarter of an inch from the point, so that the stabs made shall not exceed that depth.

Any point of the parturient canal may be swollen, so as to impede the descent of the child, by a sub-mucous infiltration of blood—the so-called *thrombus*. I have seen a large tumour formed in this way on the os uteri; but the more common seat is the labia of the vulva. If the obstacle be so great that the head, in passing, threatens to burst and rend the tumour, it is better to open it with a lancet. As soon as the child is born, the part should be carefully examined to see if it bleeds; and pressure upon it by compresses soaked in perchloride of iron should be applied. The nature and treatment of thrombus will be further discussed under " Hæmorrhage."

In cases of formidable obstacle from *cancerous or fibrous growth*, recourse to the *ultima ratio*—the Cæsarian section—may be indicated.

LECTURE VIII.

Turning.

IF we were restricted to one operation in midwifery as our sole resource, I think the choice must fall upon turning. Probably no other operation is capable of extricating patient and practitioner from so many and so various difficulties. In almost every kind of difficult labour with a pelvis whose conjugate diameter exceeds three inches, it would be possible to deliver by turning with a reasonable prospect of safety to the mother, and in many instances with probable safety to the child. We might very greatly restrict craniotomy. We might dispense with the forceps ; but neither forceps nor craniotomy will serve as a substitute for turning in its special applications. It is difficult, therefore, to exaggerate the importance of carrying to the utmost limit the perfection of this operation. Yet the text-books exhibit a very inadequate appreciation of the subject. Turning by the feet was once said, not inaptly, to be the master-stroke of the obstetric practitioner. And still the operation was very imperfectly developed.

I propose to describe and illustrate with some fulness the

conditions upon which mobility of the fœtus *in utero* depend; the various modes by which the fœtus may be made to change its position; and the applications of this knowledge to the practice of turning, embodying the teaching of Wigand, D'Outrepont, Radford, Simpson, Esterlé, Lazzati, Braxton Hicks, myself, and others.

Having regard to the various allied operations which it is convenient to class under a general description, I would define *Turning as including all those proceedings by which the position of the child is changed in order to produce one more favourable to - delivery.*

There are three things which it is very desirable to know as much about as possible before proceeding to the study of turning as an obstetric operation—

1. What are the conditions which determine the normal position of the fœtus *in utero*?

2. What are the conditions which produce the frequent changes from the ordinary position?

3. What are the powers of Nature, or rather the methods employed by Nature, in dealing with unfavourable positions of the fœtus?

1.—*The Conditions that determine the Normal Position of the Fœtus in Utero.*

It would be idle to do more than glance at the fanciful ideas upon this subject that have obtained currency at various times, although most have an element of truth in them. Ambroise Paré believed that the head presented owing to the efforts made by the child to escape from the uterus. Even Harvey believed that the fœtus made its way into the world by its own independent exertions. Dubois endeavoured in a long argument to show that the fœtus has *instinctive power*, which determines it to take the head-position. Simpson, rightly concluding that the maintenance of normal position depends very much upon the life of the fœtus, observes that it has no power of motion except muscular motion, and infers that the fœtus adapts itself to the uterus by *reflex muscular movements* excited by impressions—as by contact with the uterus—upon its surface. Thus we come down by a curious scale of theories, in which the philosopher may trace the influence of contemporary physiological doctrines. First, the fœtus is endowed with the high faculty of *volition ;* then it falls to the lower faculty of *instinct ;* and, lastly, it is degraded to the lowest nervous function—that of *reflex* motion. I am disposed to estimate at a still lower point the influence of the fœtus as an active agent in maintaining its position during pregnancy or labour. It is incontrovertibly true that the normal position of the fœtus and the course of labour are intimately dependent upon the life of the fœtus. But I think I am enabled to affirm from very close observation that the fœtus, if full grown and *only recently dead*— that is, for a few hours—may be nearly as well able to maintain its position, and to conduce to a healthy labour as one that is alive. How is this ? It depends simply upon the preservation of sufficient tone and resiliency in the spinal column and limbs to maintain the form and posture of the fœtus. Whilst alive, or only recently dead, the spine is firmly supported in a slight curve, the limbs are flexed upon the trunk, the whole fœtus is packed into the shape of an egg, which is very nearly the shape of the cavity of the uterus. It has a long axis, represented by its spine. This long axis, being endowed with sufficient solidity, resembles a rod, rigid or only slightly elastic. It is a lever. Touched at either pole, the force is propagated to the opposite pole. If the head impinge upon one side of the uterus, the breech will be driven into contact with the opposite point of the uterus; head and breech will move simultaneously in opposite directions. In labour, when the uterus is open to admit of the passage of the fœtus, the propelling power applied to the breech is propagated throughout the entire length of the spine or long axis, so that the head, the end furthest from the direct force, is pushed along in the direction of least resistance, turning at those points where it receives the guiding impact of the walls of the canal.

When the fœtus has been sometime dead, the elasticity and firmness of its spine are lost; flaccidity succeeds to tonicity. Force applied to one extremity is not propagated to the other extremity—or, at least, it is very imperfectly so ; the long axis bends, doubles up like a rod of gutta-percha softened by heat. If, the fœtus *in utero* being in this state, pressure be applied to one side of the head, the head will simply move towards the opposite side of the uterus. And if labour be in progress, the propelling force applied to the breech will not be duly transmitted to the head, but will tend to double up the trunk, to make it settle down in a squash in the lower segment of the uterus or in the pelvis. The head—the cervical spine having lost its resiliency—will not take the rotation and extension turns. It will run into the pelvis like plaster into a mould. Or, at an earlier stage, the limbs, especially the arms, having lost their tonicity, drop or roll in any direction under the influence of gravity or of pressure; and hence may fall into the brim of the pelvis, constituting what are called transverse presentations. The influence of this law is clearly seen in the course of that process called "spontaneous expulsion," by which a dead child is expelled, a shoulder presenting. A dead child is so efficient a cause of dystocia that it may even lead to rupture of the uterus.*

Another property of the dead child which it is useful to bear in mind, is its extreme *ductility*. If any degree of decomposition have taken place, it easily elongates under traction, and may be drawn through a very narrow pelvis.

Still another property to remember is *fragility*. According to the degree of change the body has undergone, will be the facility with which its parts give way and separate under traction. Thus, the head being arrested on the brim, and traction-force applied to the trunk, the amount of force required to separate trunk from head is much less in the dead child than in the living.

Other factors besides the child have to be considered. Scanzoni correctly observes that the frequency of head-presentation is dependent on the operation of various causes. 1. There is the force of gravitation; 2. The form of the uterine cavity; 3. The form of the foetus (to which must be added the properties I have described due to life or death); 4. The quantity of amniotic fluid; 5. The contractions of the uterus during pregnancy and the first stage of labour; 6. The coincidence of the uterine and pelvic axes. In the early stages of pregnancy the embryo is so small relatively to the cavity containing it

---

* *See* "A Memoir on Pelvic Hæmatocele," by the Author, in St. Thomas's Hospital Reports, 1870.

that it floats suspended in the liquor amnii. But about the middle of pregnancy the fœtus grows rapidly ; it acquires form ; and, at the same time, the uterus grows more in its longitudinal than in its transverse diameter. As soon, therefore, as the fœtus —an ovoid body—attains a size that approaches that of the capacity of the uterus, the walls of the uterus will impose upon the fœtus a vertical position. The fœtus has become too long to find room for its long diameter in the transverse diameter of the uterus. Mutual adaptation requires that the long diameters of fœtus and uterus shall coincide.

A condition not, to my knowledge, hitherto noticed, which has a powerful influence upon the determination of the child's position *in utero*, is the normal flattening of the uterus in the antero-posterior direction. In the non-pregnant uterus, the cavity of the body—the true and only gestation-cavity—is a flat triangular space, the angles of which are the orifices of the Fallopian tubes and the os internum uteri. A similar triangular superficies is marked out on each half of the uterus, anterior and posterior. The anterior superficies lies flat against the posterior superficies, touching it as if the two were squeezed together. When pregnancy supervenes, these surfaces are necessarily separated to form a cavity for the growth of the ovum. But the original form is never entirely lost. The cavity is always more contracted from before backwards than from side to side. This is proved by direct observation if the fingers are introduced after abortion, or the hand after labour at term. The uterine cavity is closed by the flattening of the anterior and posterior walls together. This takes place the moment the uterus contracts. If the finger or hand be in the uterus at the time, this is plainly felt. Now, this flattened form of the uterus is the reason why the fœtus takes a position with either its back or belly directed forwards. The fœtus is broader across the shoulders than from back to front, and therefore its transverse diameter is fitted to the transverse diameter of the uterus. There is a physiological design that dictates the downward position of the head. The fundus is the part designed for the implantation of the placenta, where it can grow undisturbed, and continue its function during the expulsion of the child. The lower part of the cavity is therefore left free for the development of the embryo. Why the back is commonly directed forwards to the mother's belly is this : The child's back is firm and convex; its head is also firm and convex behind. The anterior aspect of the child's body is plastic and concave, and therefore fits itself better to the firm convexity of the mother's spine. It is clear that the two solid convex spines of mother and child would naturally repel each other; and the child being movable, it is its back that recedes, turning forwards.

2.—*The Conditions which produce the frequent Changes in the Child's Position.*

Any considerable disturbance of the correlation of the factors which keep the fœtus in its due position, of course favours malposition. The principal disturbing conditions may be stated as follows :—An *excess of liquor amnii* acts in two ways : first, it favours increased mobility of the fœtus ; secondly, it tends to destroy the elliptical form of the uterus. The transverse diameter, increasing in greater proportion than the longitudinal, the cavity becomes rounder. Hence the fœtus is no longer kept in a vertical position, for want of the proper relation between its form and size and those of the uterus.

Obliquity of the uterus was considered by Deventer to be a main cause of malposition. It is now very much discredited, but I am disposed to believe that it has, not seldom, a real influence. Dubois and Pajot showed that in one hundred women the uterus in seventy-six exhibited a marked lateral obliquity to the right, in four to the left, and in twenty an anterior obliquity. Wigand had shown that deviations of the uterus to the right and forwards were far the most frequent. The normal direction of the non-pregnant uterus is nearly that of the axis of the pelvic brim. As the uterus grows during pregnancy, rising above the brim, the projecting sacro-vertebral angle and the curve of the lumbar column deflect its fundus to one or other side ; and, if the abdominal walls be very thin and flaccid, the fundus will fall forwards. The tendency of these obliquities, if carried beyond ordinary measure, is to throw the axis of the uterus out of the axis of the pelvic brim, and to bring some other part than the vertex of the fœtus to present. The probability of this will be increased by the irregular contractions of the uterus likely to be excited by parts of the fœtus pressing unequally upon its walls. For example, in extreme lateral obliquity the breech may press strongly upon one side of the fundus; contraction taking place here, will drive the head further off the brim on to the edge, where, if it find a *point d'appui*, it will rotate on its transverse axis, producing forehead or face-presentation, and favouring the descent of the shoulder. Wigand explains how a too loose and shifting relation of the uterus to the pelvis disposes to cross-birth. In this condition it is observed that the head is fixed now in one place, now in another, and now not felt at all.

He further* says that any obliquity of the uterus exceeding an angle of 25° is unfavourable ; and that even a lesser obliquity, with excess of liquor amnii or a small child, is likely to cause the presenting head to be displaced, and to bring a shoulder into the brim, especially if strong pains or bearing-down efforts be made *early* in labour.

He explained that the os uteri might be brought over the centre of the brim by internal drawing upon the os, combined with external pressure upon the fundus in the opposite direction, thus putting in practice the principle of acting simultaneously upon the two poles of the uterus.

Deformity of the pelvis or lumbar vertebræ is often a powerful factor. The comparative frequency of transverse presentations in cases of deformed pelvis is certainly greater than where the pelvis is well formed. I think, however, that *slight* deformity has more influence in causing malposition than extreme degrees. In these latter, malpositions are rarely observed. In marked deformity, the head cannot enter the pelvic brim, but floats free above it, and therefore is not influenced by hitching on the edge.

The attachment of the placenta to the lower segment of the uterus is, as Levret has clearly shown, a cause of malposition, by forming a cushion or inclined plane, which tends to throw the fœtal head out of the pelvic axis across the brim. Hence the frequency of cross-birth and of funis-presentation in cases of partial placenta prævia. But there are numerous cases in which the placenta dips into the lower zone, growing downwards from

* " Die Geburt des Menschen," Berlin, 1820, vol. ii. p. 137.

the posterior and lateral walls of the uterus, without leading to
hæmorrhage, and thus not suspected to be cases of placenta
prævia, which, nevertheless, form an inclined plane behind or on
one side, and produce malposition.

Then there is the *influence of external forces*, as of pressure
applied to the uterus through the abdominal walls. The dress
of a woman at the end of pregnancy is a matter of no small
moment. The pressure of a rigid busk of wood or steel upon
the fundus of the uterus, modified by the various movements
and postures of the body, may flatten in the fundus, thus
reducing the longitudinal diameter of the uterus, or it will
push the fundus to one side, causing obliquity. It will, at the
same time, press directly upon the breech, and thus tend to give
the fœtus an oblique position, throwing the head out of the
pelvic axis. Pluriparæ should do the reverse of this. They
should wear an abdominal belt, which supports the fundus of the
uterus from below upwards.

Want of tone in the uterus, which implies inability to preserve
its elliptical form, and a tendency to fall into rotundity, a form
which evidently favours malposition. Scanzoni says laxity of
uterus is a chief cause. As soon as contraction begins, the
uterus tends to resume its ovoid form.

Excess of liquor amnii is an efficient factor: first, by impairing the tone of the uterus; secondly, by permitting the fœtus
to float too freely. A very slight force applied to either pole
will cause the head to deviate from the axis of the brim; and
if, at such a moment, the membranes burst, the fœtus may
be fixed in its unfavourable position by the contracting uterus.

Hœning (*Scanzoni's Beiträge*, 1870) calls special attention to
the *development of the fœtus* in the later months as a factor. A
large fœtus cannot so easily change position. The cranial presentations have the greatest stability. A 1st cranial changes
to a 2nd, and *vice versâ*, but the cranial rarely changes to
a breech. Cranial and breech presentations are most stable
in primiparæ; oblique presentations in pluriparæ.

Irregular or partial contractions of the uterus cause malposition.
Naegele insisted upon this. He found that in some cases malposition was averted by allaying spasm. Heyerdahl says
contractions of the uterus are a chief factor, and these are often
caused by palpations. This, no doubt, accounts for a large

proportion of the changes of position encountered by the too industrious German observers. They produced the changes they observed.

Credé, Hecker, Valenta,[*] Gassner, Heyerdahl, Schultze, have made repeated observations, and found change of position even more frequent than did other observers. They establish the fact that the fœtus changes its position with remarkable frequency. Valenta examined 363 multiparæ and 325 primiparæ in the latter months of pregnancy. He found that a change of position took place in 42 per cent. Change was more frequent in multiparæ, and in these in proportion to the number of previous pregnancies. Narrow pelves very frequently cause change of position. Circumvolutions of the cord, so often observed, are produced by changes of position, and hence bear evidence to the correctness of the proposition. It is interesting to observe that the general tendency of changes of position is towards those which are most propitious. Thus, cranial positions are least liable to change. Oblique positions are especially liable to change. These mostly pass into the long axis by spontaneous evolution. *Self-evolution is a very frequent resort of nature.* In some cases several changes of position have been observed in the same patient. The presentations are made out by external manipulations. Valenta thus describes his method of ascertaining a breech-position during pregnancy :— He lays his right hand flat on the fundus uteri, and then strikes the tips of the fingers as suddenly as possible towards the cavity of the uterus, against the part of the child lying at the fundus. By this manœuvre he has always succeeded in recognizing the head, if lying at the fundus, by its peculiar hardness and evenness. He detects the head in oblique and cross positions in the same manner. P. Müller[†] relates a case in which, within five days, a complete version of the fœtus was effected six times.

Esterlé [‡] gives abundant evidence to the same effect.

Yet the fact of the " spontaneous evolution " of a living child, as described by Denman from actual observation, has been doubted !

We have now to study—

* "Monatsschr. f. Geburtsk," 1866.   † Ibid., 1865.
‡ "Sul rivolgimento esterno, Annali Universali di Medicina," 1859.

3.—*The powers of Nature, or rather the methods employed by
Nature, in dealing with unfavourable positions of the fœtus.*

I will do no more at present than glance at those minor
deviations from the natural position in which the long axis of
the child's body stlll maintains its coincidence with the axis of
the pelvic brim.   With some additional difficulty, Nature is in
most of these cases able to effect delivery without materially
modifying the position.   Forehead and face positions have,
indeed, already been described in some detail.   Difficult breech
positions will be specially considered at a later period of the
description of turning.

From the time of Hippocrates, who compared the child
in utero to an olive in a narrow-mouthed bottle, it has been
known that the child could hardly be born if its long axis lay
across the pelvis.   But before the time of Denman it was not
clearly explained that a correction of the position, or a restitution of the child's long axis to coincidence with the axis of the
pelvic brim, could be brought about by the spontaneous operations of Nature.   And observations of this most deeply interesting of natural phenomena are so rare, that many men,
even at the present day, do not hesitate to deny the accuracy
of Denman's description.   I would, with all deference, suggest
for the consideration of these sceptics, whether they do not
carry too far their regard for the maxim, "*Nulla jurare in verba
magistri.*"   In matters of deduction, of theory, that maxim
can hardly be too rigorously applied.   But to reject as false
or impossible matters of fact, observed and recorded by men
of signal ability and conscientiousness like Denman, is to push
scepticism to an irrational degree.   There are subjects, and this
is one, which are not questions of opinion, but of evidence.
Shall we reject the testimony of Denman ?   Whose shall we,
then, accept in contradiction ?   Shall it be the testimony of
those who deny that Denman saw what he says he saw,
because they themselves have never seen it ?   This is simply
to give the preference to negative over positive evidence, to
say nothing of the relative weight or authority of the witnesses.   There is no man whose experience is so great that
nothing is left for him to learn from the experience of others.

I

Let us first call Denman into the box. He says: "In some cases . . . . the shoulder is so far advanced into the pelvis, and the action of the uterus is at the same time so strong, that it is impossible to raise or move the child. . . . This impossibility of turning the child had, to the apprehension of writers and practitioners, left the woman without any hope of relief. But in a case of this kind which occurred to me about twenty years ago, I was so fortunate as to observe, though it was not in my power to pass my hand into the uterus . . . . that, by the mere effect of the action of the uterus, an evolution took place, and the child was expelled by the breech. . . . . The cases in which this has happened are now become so numerous, and supported not only by many examples in my own practice, but established by such unexceptionable authority in the practice of others, that there is no longer any room to doubt of the probability of its happening, more than there is of the most acknowledged fact in midwifery. As to the manner in which this evolution takes place, I presume that, after the long-continued action of the uterus, the body of the child is brought into such a compacted state as to receive the full force of every returning pain. The body, in its doubled state, being too large to pass through the pelvis, and the uterus pressing upon its inferior extremities, which are the only parts capable of being moved, they are gradually forced lower, making room as they are pressed down for the reception of some other part into the cavity of the uterus which they have evacuated, till the body turning, as it were, upon its own axis, the breech of the child is expelled, as in an original presentation of that part. I believe that a child of the common size, living, or but lately dead, in such a state as to possess some degree of resilition, is the best calculated for expulsion in this manner. Premature, or very small children, have often been expelled in a doubled state, whatever might be the original presentation; but this is a different case from that we are now describing."

Denman cited, in confirmation, the evidence of Dr. Garthshore, Consulting Physician of the British Lying-in Hospital, who related to him a case of the kind, in which the child was living, and the not less trustworthy evidence of Martineau of Norwich. But, before Denman's time, similar cases had been observed, although not understood. Thus, Perfect: " The arm presented ; and after endeavours were ineffectually made to get at the feet to turn the child, the patient was thereupon left to herself, and delivered, in a few hours, of a live child, without any assistance whatever."

D'Outrepont* cites Sachtleben, Löffler, Christoph von Siebold, Wilhelm Schmitt, Wiedemann, Vogler, Saccombe, Ficker, Simons, Elias von Siebold, Hagen, Wigand, as all having witnessed self-turning, chiefly, indeed, by the head. He says he himself has frequently witnessed it.

Since Denman's time, evidence has accumulated. Professor Boer, of Vienna, a name of the first rank of the illustrious in Medicine, described, in 1801, a case of arm-presentation, the fingers having been seen at the vulva. He was preparing to turn, when he found the hand higher than when he had examined before. As the pains continued, Boer rested with his hand in the pelvis. The arm distinctly moved up. At this time the whole cavity of the pelvis was filled with the breech of the child. The body and head of a fresh living child were expelled. Velpeau, a man remarkable for the precision of his observations, is equally decided in corroboration.

What observations can be more positive, exact ? Who can give evidence more carefully ? Who is more worthy of belief? Upon what grounds is evidence so distinct impeached ? There are two grounds. In the first place, there is the observation of a fact, of a different method of spontaneous or unaided delivery under arm-presentation from that which Denman described. In the second place, there is the assumption that this different method is the only true one.

Now, let us admit the accuracy of the observation, which we may do unreservedly: does it follow that the assumption which excludes the possibility of the occurrence of any other mode of unaided delivery is to be received ? Denman, more logical and more philosophical than his opponents, is not so ready to impose limits upon the resources of Nature. He not only observed the " spontaneous evolution " or version of living children, and described this as one resource ; but he also observed the " spontaneous expulsion " of dead or premature children by doubling-up, and described this as a second and

* " Abhandlungen und Beiträge." Würzburg. 1822.

different resource. Not only, therefore, did Douglas fail to correct or to displace the explanation of Denman, but Denman had actually left nothing for Douglas to discover. In two papers published in the *London Medical Journal* in 1784, Denman relates several cases of spontaneous birth with arm-presentation—some observed by himself, some communicated to him. In these the child was born dead, and the shoulder remained fixed at the pubes. They are clearly described, and certainly anticipate the description given by Dr. Douglas in 1811. But it is fair to add that at this time Denman had not arrived at that sharp distinction which he afterwards drew (1805, see fifth edition *Introduction to Midwifery*) between spontaneous version and expulsion.

These are the facts, the evidence. The assumption to which I have referred is further rebutted by abundant collateral testimony. The observation of Denman, so far from being incredible or improbable, is in entire harmony with the phenomena of gestation and labour.

We will now endeavour to trace, with more precision, the modes employed by Nature in dealing with shoulder-presentations. The mechanism of labour with shoulder-presentation is strictly analogous to that of ordinary labour. It is, therefore, desirable to set before our minds the picture of an ordinary head-labour. For this purpose reference is made back to Lecture III.

Labour with shoulder-presentation must obey the same laws as labour with head-presentation.

Shoulder-presentations may be *primitive* or *secondary*. The *primitive* exist before labour has set in, and are almost necessarily associated with obliquity of the uterus. The *secondary* are formed during the initiatory stage of labour, under conditions which lead to the deflection of the head from the pelvic brim when it is made to move under the influence of force applied to the breech or trunk.

In Nature we observe two chief shoulder positions, and each of these has two varieties. In the *first position*, the head lies in the left sacro-iliac hollow. In the *second position*, the head lies in the right sacro-iliac hollow. Now, in either position, either the right or the left shoulder may present. Thus, if the head is in the left ilium, the right shoulder will descend when the child's back is directed forwards; and the left shoulder will descend when the child's belly is directed forwards. In the case of the *second* or right cephalo-iliac position, the right shoulder will descend when the child's belly is turned forwards, and the left shoulder when the child's back is turned forwards.*

* FIG. 37.

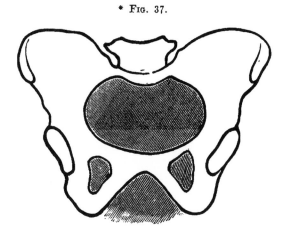

An easy method of realizing a description of some positions of the fœtus is to follow them with a pelvis and a small lay-figure, such as is used by artists. It is not even necessary to have a pelvis. A partial equivalent may be made by tracing a drawing of a pelvis on a piece of cardboard, and cutting out the oval which represents the brim and the space beneath the symphysis pubis. The above figure will serve as a model. Cut out the dark parts. A small lay-figure of corresponding size must be procured.

LECTURE IX.

IT is especially necessary, before we proceed, to define with
precision the significance that attaches to the terms employed,
the more especially that I find it desirable to use some terms
in a different sense from that current in this country. Dr.
Denman used the term "spontaneous evolution" to express
the natural action by which the pelvis or head was substituted
for the originally presenting shoulder. The term "spontaneous
expulsion" has been applied to the process of unaided delivery
described by Douglas, in which the child is driven through the
pelvis doubled up. Neither of these terms is free from objection. The first especially is inaccurate, and has given rise to
much misapprehension. The process described by Denman is
a true *version* or *turning*. The position of the child is changed.
All French, German, Italian, and Dutch authors apply to this
process the term " spontaneous version"—"*versio spontanea.*"
It might be called *natural version*, to distinguish it from artificial version effected by the hand of the obstetrician. All
Continental authors likewise call Douglas's process by the

name, "spontaneous evolution," the process being one of un-folding, as it were, of the doubled-up fœtus. It is of great consequence to bring our nomenclature into harmony with that of our brethren abroad, and it is of still greater consequence to bring our nomenclature into harmony with Nature. It is clear, therefore, that the change in terms should be made by us. I shall use the terms "version" and "evolution" in the correct sense.

There are *two varieties of spontaneous version*—one in which the head is substituted for the shoulder, the other in which

FIG. 38.

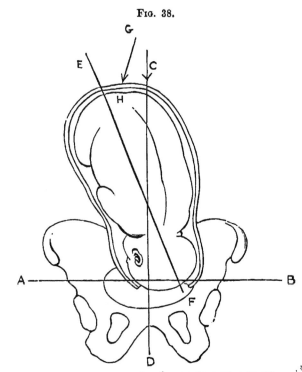

THE BEGINNING OF SHOULDER-PRESENTATION, DORSO-ANTERIOR.

A B is the horizontal line of the pelvic brim; C D the axis of pelvic brim; E F the axis of fœtus and uterus inclined obliquely, so that the head projects a little over the pelvic edge; G H, the direction of the force sent by diaphragm and abdominal muscles striking fundus uteri on one side.

the pelvis is substituted for the shoulder. These varieties of spontaneous version correspond with two similar varieties of artificial turning.

There are likewise *two varieties of spontaneous as well as of artificial evolution.* The head or the trunk may be evolved or extracted first.

These processes I will describe successively, beginning with the spontaneous or natural operations, since these are conducted in obedience to mechanical laws which must be respected in the execution of the artificial operations.

In Fig. 38 I have endeavoured to represent the very earliest

FIG. 39.

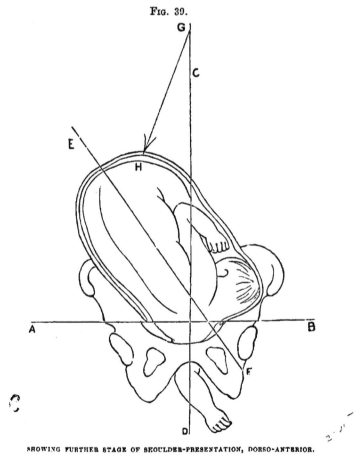

SHOWING FURTHER STAGE OF SHOULDER-PRESENTATION, DORSO-ANTERIOR.

The letters have the same significance as in Fig. 38. The force o n bearing more still on the side of the uterus throws it and the fœtus more nearly into transverse position.

stage or condition of things in shoulder-presentation. The long axis of the child, and of the uterus, stands obliquely to the plane of the pelvic brim. It is not, indeed, very distant from the perpendicular. It is a very serious error to regard these presentations as entirely cross or transverse. It is only in the advanced stages of labour with shoulder-presentation, when the liquor amnii has been long drained off, when the uterus has been contracting forcibly, driving the shoulder deeply into the pelvis, that the child can truly be said to lie across the pelvis. Diagrams copied from text-book into text-book seem to have fixed this false idea firmly in the obstetric mind. Wigand insists that transverse positions are rare. Esterlé and Lazzati say the same, and maintain that the oblique position is favourable to spontaneous version.* I venture to say that, except in cases of dead, monstrous, or small children, or with loss of form of the uterus through excess of liquor amnii, a true cross-birth, such as is commonly pictured and generally accepted, does not exist at the commencement of labour. It would be better, because certainly true as a fact, and because it does not commit us to any theory, to call these presentations *shoulder-presentations,* and to discard the terms "cross-birth" and "transverse presentation" altogether. In shoulder-presentation, an oblique position of the child *becomes* transverse in the course of labour; but the presentation is not transverse *ab initio.*

The neglect of this fact has been a main cause of the errors that prevail in the doctrine and practice of turning.

In the diagram (Fig. 38), the child and the uterus, E F, stand obliquely, at an angle of about 15° or 20° to a perpendicular C D drawn upon the plane A B of the pelvic brim. The child's head is nearly in a straight line with its spine. It stands partly over the brim, and partly projecting beyond into the left iliac fossa. That is the *first act.* This act may pass into natural head-labour. Wigand, Jörg, and D'Outrepont say this position is common, and that the effect of the first uterine contractions is usually to bring the long axis of the uterus and of the child into due relation with the pelvic brim. This phenomenon is, in fact, a form of self-turning or natural rectification.

* It has been my habit, when making notes of cases coming under my observation, to record the position of the child by means of sketches. It is from these graphic memoranda that most of the illustrations in these lectures of the phenomena of shoulder-presentation and turning will be taken.

If this attempt at rectification fail, then we have the transition into shoulder-presentation. The shoulder or arm cannot come down into the pelvis until the *second act*, a movement of flexion of the head upon the trunk, takes place.

This happens in the following manner :—The muscles of the fundus uteri contracting, aided or not by the downward pressure of the abdominal muscles and diaphragm, bring a force acting primarily upon the breech which lies at the fundus. This force will strike with greatest effect upon the left or uppermost side of the breech, at an angle with the long axis of uterus and child. The line G H represents the direction of this force. The result is that the breech descends. And now mark what follows :—If the cavity of the uterus were as broad as long—that is, a flattened sphere or short cylinder like a tambourine—the child's long axis, formed by spine and head, might preserve its rectilinear character ; and as the breech descended, the head would simply rise on the opposite side until it came round to the spot abandoned by the breech, performing, in fact, complete version. But the uterus, we know, is narrower from side to side than from top to bottom. The head will find great difficulty in rising ; it therefore bends upon the neck. The shoulder, pertaining to the trunk, is kept at the lowest point in a line with it. The head is thrown more into the iliac fossa, where it rests for a while. Fig. 39 represents this second position of the fœtus. A B is the plane of the brim ; C D the perpendicular to the plane, representing the axis of entry to the pelvis; E F is the axis of the child's trunk; and G II shows the direction of the downward force, which now strikes the uterus and breech at a greater angle with the perpendicular.

Now, the arm will commonly be driven down, and the hand may appear externally. The observation of the hand will tell the position of the child. The back of the hand looks forwards, the palm backwards, the thumb to the left. All this tells plainly that the head is in the left iliac fossa, and that the child's back is turned forwards to the mother's abdomen. The right scapula will lie close behind the symphysis pubis; the acromion and right side of the neck will rest upon the left edge of the pelvic brim ; and the right axilla and right side of the chest will rest upon the right edge of the pelvic brim ;

whilst the belly and legs of the child, turned towards the mother's spine, will occupy the posterior part of the uterus.

At this stage, even after the liquor amnii has been drained off, spontaneous or natural version may still be effected. The process described as the second act still continuing, the breech is driven lower down ; the trunk bends upon its side ; the curve thus assumed by the long axis carries on the propelling force in a direction across the pelvic brim ; the head tends to rise still higher into the left iliac fossa ; the presenting shoulder and prolapsed arm are drawn upwards a little out of the pelvis. The *third act*, one of increased lateral flexion of the child's body, and of movement across the pelvic brim, is represented in Fig. 40.

FIG. 40.

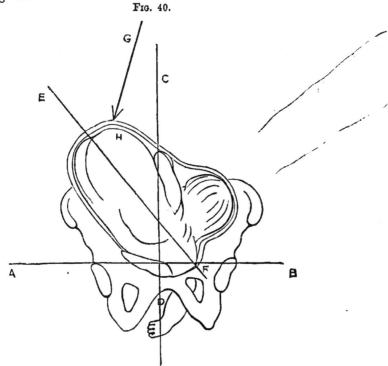

THIRD STAGE OF SHOULDER-PRESENTATION.
The head is high in left iliac fossa. The body is undergoing flexion upon its side.

If spontaneous version is to be completed, the *fourth act* succeeds. The breech being the most movable part, and the

trunk being capable of bending upon itself, partly on its side,
partly on its abdomen, is driven lower and lower, the right
shoulder being forced well over to the left side of the pelvic
brim, and the head being fairly lodged in the upper part of
the iliac fossa, the brim is comparatively free for the reception
of the trunk.   This enters in the following manner :—The
right hip comes first into the brim ; it is forced lower, and is
followed by the breech.   As soon as the breech enters the pelvis
—that is, as soon as it gets below the sacral promontory—a
movement of rotation takes place analogous to the rotation which
the head takes in head-labour.   There is most room in the
sacral hollow, and there the breech will turn.   This turn of the

FIG. 41.

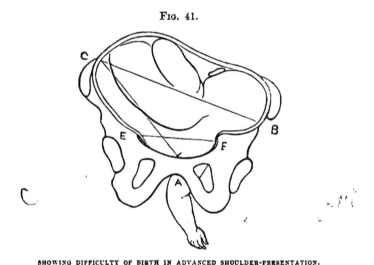

SHOWING DIFFICULTY OF BIRTH IN ADVANCED SHOULDER-PRESENTATION.

E F represents transverse diameter of pelvic brim ; A.B C triangle formed of child's body,
whose apex is A, and whose base B C much exceeds E F.

trunk brings the body from the transverse position it occupied
above the brim to one approaching the antero-posterior ; and
commonly the head yields somewhat to the altered direction
of the spine by coming more forward.*
    When this rotation-movement is effected, or rather simul-
taneously with it, a *movement of descent or progress* in an arc of
a circle round the pubic centre goes on.   The flexion of the

    * This part of the mechanism of spontaneous version will be illustrated
further on.

spine is now reversed. Above the brim the trunk was concave on its left side, as seen in Figs. 40 and 41. When the breech has dipped into the pelvis, the trunk becomes concave on its right side. The breech descends first. The right ischium presents at the vulva. Then the whole breech sweeps the sacral concavity and perinæum. The trunk follows. The right arm, which has not always completely risen out of the way, comes next; then the left arm ; and lastly the head, taking its rotation-movement, and its movement in a circle.

The cause of the difficulty that opposes delivery in shoulder-presentation is obvious. The pelvic canal is too narrow to permit the child to pass freely when its long axis lies across the entry. On looking at the diagram (Fig. 41), we see the shoulder driven into the pelvis, forming the apex A of a triangle whose base B C is considerably longer than E F, the transverse diameter of the pelvic brim. To overcome this difficulty, Nature struggles to shorten the base B C. To a certain extent she generally succeeds, and occasionally she succeeds completely.

The uterus contracts concentrically, tending to shorten all its diameters, especially its transverse diameter. The axis formed by the trunk and head of the child, which go to make up the resisting base of the triangle, is flexible; therefore B and C admit of being brought nearer to each other. But, when the utmost approximation has been obtained in this manner, we still have the entire thickness of the head, equal to four inches, and only very slightly compressible, plus the thickness of the body, which, after all possible gain by compression is effected, is equal to at least two inches more, being an inch or more in excess of the available space in the pelvic brim. As a general rule, it may be stated that no part of the child, except a leg or an arm, can traverse the pelvis along with the head, the head alone being quite large enough to fill the pelvis.

One result of the great compression exerted by the concentric contraction of the uterus is to cause such pressure upon the chest and abdomen of the child, and so to compress the placenta and cord, that the child is asphyxiated and killed. The death of the child, leading to the loss of resiliency, will, after sufficient time, admit of a much further degree of com-

pression, and then, possibly, the child may be so doubled up
and moulded that it may enter the pelvis.

*The condition, therefore, of spontaneous evolution is the death of
the child.* If not already dead at the commencement, the child
will almost certainly, if of medium size or larger, be killed
in the course of the process. Herein lies a great distinction
between spontaneous evolution and spontaneous version. A
living child is favourable to version, a dead one to evolution.

Spontaneous evolution from the first position proceeds as
follows :—At *first*, we have the oblique position of fœtus and
uterus represented in Figs. 38 and 39. *Secondly*, strong flexion
of the head upon the trunk, and descent of the shoulder into
the pelvis (*see* Figs. 40 and 41); the head is in one iliac fossa,
the trunk and breech in the other. At this stage, commonly,

Fig. 42.

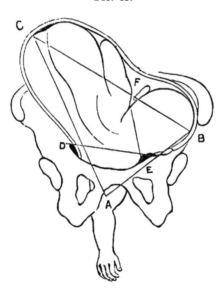

SHOWS THE POSITION OF THE CHILD AFTER THE ESCAPE OF LIQUOR AMNII.

The head is strongly flexed upon the trunk, forming together the base of a wedge A B C, too
large to enter the brim. The line E F represents the line of decapitation, by which proceeding
the base of the wedge is decomposed. The head thus being put aside, the axis of the trunk
will easily be brought into coincidence with the axis of the brim, permitting delivery.

the membranes burst, and the arm falls into the vagina, the
hand appearing externally. *Thirdly*, increased descent of the
shoulder and protrusion of the forearm, doubling with com-

pression of the body, so that the breech is driven into the pelvis; as soon as this takes places, a movement of rotation succeeds (*see* Fig. 43). The inclined planes of the ischia direct the breech backwards into the sacral hollow; this backward movement of the trunk throws the head forwards over the symphysis pubis; from transverse, as the child was above the brim, it now approaches the fore-and-aft direction; the right side of the head, near its base, is forcibly jammed against the symphysis; the side of the neck corresponding to the pre-

Fig. 43.

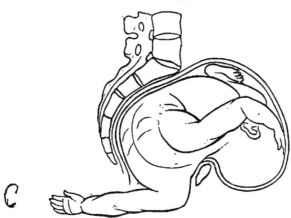

RIGHT SHOULDER; FIRST POSITION AFTER ROTATION.

senting shoulder is fixed behind the symphysis pubis, and the shoulder itself is situated under the pubic arch. *Fourthly*, the expulsive force continuing, can only act upon the breech and trunk, the shoulder being absolutely fixed; the trunk bends more and more upon its side, the presenting chest-wall bulges out, and makes its appearance under the pubic arch. Then, *lastly*, the movement in a circle of the body round the fixed shoulder is executed. The side of the trunk and of the breech sweep the perinæum and concavity of the sacrum; the legs follow. When the whole trunk is born, the movement of restitution is effected, the back turning forwards, the belly backwards. The uterus rights itself. The head escapes from its fixed position above the symphysis, the chin turns downwards, the occiput looks upwards to the fundus uteri, the nucha is turned to the right foramen ovale. It enters in the left oblique diameter, it takes the rotation-movement in the pelvis, the occiput coming under the pubic arch; then the movement in a circle is executed. The chin first appears, followed by the mouth, nose, and forehead, which successively sweep the perinæum. The occiput, which had been applied to the symphysis, comes last. So strict is the subjection throughout this process

FIG. 44.

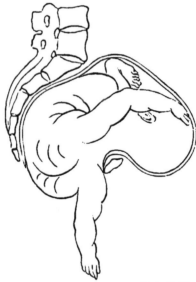

RIGHT SHOULDER, FIRST POSITION; DURING MOVEMENT IN CIRCLE AROUND SYMPHYSIS.
BULGING OF RIGHT SIDE OF CHEST.

to the laws which govern the mechanism of ordinary labour, that Lazzati * does not hesitate to describe spontaneous evolution as the natural delivery by the shoulder.

The case we have just described is the most common form of spontaneous evolution. It is the type of the rest. Keeping its mechanism well in mind, there will be little difficulty in tracing the course of spontaneous evolution when the child presents in any other position.

If the head lies in the right iliac fossa, constituting the second shoulder-position, as in the first position, the child's back may be directed forwards or backwards. In the first case, we have exactly the counterpart of the process described.

* "Del parto per la Spalla," 1867.

It would be superfluous to repeat the description, when all is told by simply substituting the words "right" for "left" and "left" for "right." It is, however, useful to trace the course of a labour in which the child's belly is directed forwards. Let us take the second position—head in right iliac fossa. This will involve the presentation of the *right* shoulder. (*See* Fig. 45.) A represents the presenting shoulder forming the apex of the triangle, whose base B C is formed by the long axis of the child's body. The expulsive force and the concentric contraction of the uterus draw the head towards the breech, shortening the opposing base by bending the head upon the chest and the trunk upon itself. This is the movement of flexion. This movement continuing is combined with

Fig. 45.

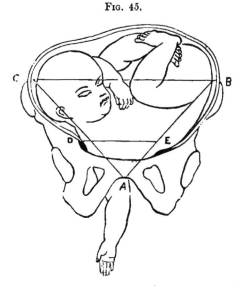

SECOND POSITION OF SECOND SHOULDER-PRESENTATION (ABDOMINO-ANTERIOR) ABOVE THE BRIM; STAGE OF FLEXION.

A, apex of triangle wedged into pelvis; B C, base of triangle opposing entry into D E, brim of pelvis.

movement of descent. The right side of the chest is driven more deeply into the pelvis, and is followed by the breech. Then rotation takes place. (*See* Fig. 46.) The head comes forward over the symphysis; the breech rolls into the sacral hollow. The right side of the chest emerges through the vulva;

K

the trunk and breech sweep the perinæum; the left arm follows, and lastly the head, the occiput taking up its fixed point at the pubic arch, forming the centre of rotation.

The presentation of the left shoulder in the *first position* offers no essential difference in its course from that pursued in the case of right shoulder with dorso-anterior position. The fœtal head is in the left iliac fossa; the sternum is directed forwards; the thumb of the prolapsed arm is turned to the left, the back of the hand looks backwards, the palm towards the pubes. The lateral flexion of head upon trunk and of trunk upon itself taking place, the left shoulder with the correspond-

Fig. 46.

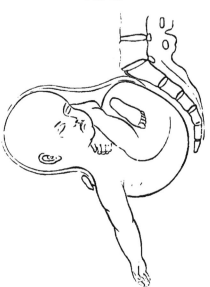

RIGHT SHOULDER ABDOMINO-ANTERIOR.—SECOND POSITION AFTER ROTATION.

ing side of the chest descending into the pelvic cavity, the rotation-movement takes place, and carries the head over the symphysis pubis. (*See* Fig. 47.) The basilar part of the left temporal region will be applied to the anterior part of the brim; the sternum will turn to the right, the dorsum to the left and backwards. Then the movements of descent and in a circle follow. The side of the chest, trunk, and breech sweep the sacrum and perinæum. The body having escaped, the movement of restitution is performed—the back will be directed to the left and forwards. The head will be above the brim, with the nucha turned to the left and forwards behind the left foramen ovale, the face looking to the right sacro-iliac joint. Thus it will be born according to the mechanism observed in breech-labour.

In the case of the dorso-anterior position, with the head in the right ilium, we have, as has been stated, simply the reverse of the dorso-anterior position with the head in the left ilium—the left shoulder becomes wedged in the brim, the left

FIG. 47.

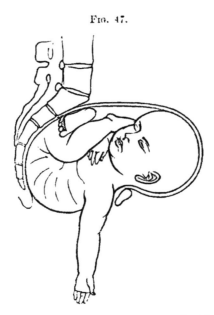

LEFT SHOULDER—FIRST POSITION AFTER ROTATION.

side of the head gets fixed upon the symphysis, the left side of the chest bulges out of the vulva. (*See* Fig. 48.)

Such, in brief, is the description of spontaneous evolution. The process is the normal type of labour in shoulder-presentation. Were it more often justifiable to wait and watch the efforts of Nature, we should probably not seldom enjoy opportunities of observing it; but the well-grounded fear lest Nature should break down disastrously impels us to bear assistance. To be useful in the highest degree, that assistance must be

K 2

applied in faithful obedience to the plans of Nature. In seeking to help, we must take care not to defeat her objects by crossing the manœuvres by which she attains them. Whenever we lose sight of this duty, whenever we try to overcome a difficulty by arbitrary operations, greater force, running into violence, is required, and the risk of failure and of danger is increased.

It has been already said that spontaneous evolution may be effected by the head traversing the pelvis first. The case

FIG. 48.

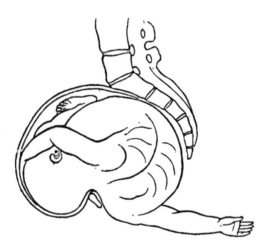

LEFT SHOULDER—SECOND POSITION AFTER ROTATION.

is indeed rare, but the process and the conditions under which it occurs deserve attention. The essential idea of spontaneous evolution is that the presenting shoulder remain fixed, or, at least, should not rise up out of the pelvis into the uterus. Therefore, if the head comes down, it must do so along with the prolapsed arm. This simultaneous passage of the head, arm, and chest can hardly take place unless the child is small. If the child is very small, the difficulty is not great. If the child be moderately large, it will be far more likely to be born according to the mechanism already described and figured, in which the side of the chest corresponding to the presenting shoulder emerges first, and the head last. But some cases of head-first deliveries have been observed. Pezerat* relates a case that seems free from ambiguity. The child was large, the shoulder presenting. Pezerat tried to push it up, but could not. A violent pain drove the head down. Fichet de Flichy† gives two cases. In both the midwife had pulled upon the arm. Balocchi relates a case.‡ It was an eight-months' child. He says the case is unique rather than rare, but still regards it as a natural mode of delivery in shoulder-presentation. Lazzati thinks the descent of the head in these cases is always the result of traction upon the presenting arm. As the expelling power is exerted mainly upon the breech, tending to drive the head away from the brim, it is indeed not easy to understand how spontaneous action can restore the head, if the shoulder is forced low down in the pelvis. Monteggia§ held the same

Fig. 49.

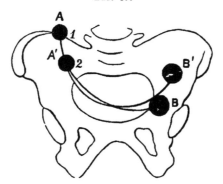

DIAGRAM TO REPRESENT SPONTANEOUS VERSION.

A B, a flexible elastic rod as is spine of live child. Force at A 1 bends the rod, B rises to B' A descending to A'.

opinion. He relates two cases, in both of which tractions had been made. I myself have seen an instance of the kind.

Fielding Ould (1742) relates the following: He was called to assist a midwife, who had been pulling at the child's arm, which came along with the head, yet she could not bring it forth. The head was so far advanced that it could not be put

---

* " Journ. Complementaire," tome **xxix.**
† "Observ. Med. Chir."
‡ "Manuale Completo di Ostetricia." Milano, 1859.
§ "Traduzione dell' Arte Ostetricia di Stein," 1796.

back, in order to come at the feet. However, after an hour of excessive toil, he brought forth a living child, with a depression of the parietal and temporal bones proportional to the thickness of the arm. Next morning the bones had recovered. Child and mother did well.

It is so important, as a guide to the artificial means of extricating a patient from the dangers of shoulder-presentation, to possess accurate ideas of the mechanism of spontaneous version and evolution, that I am led to present a further illustration of these processes.

To make the mechanism of spontaneous version clearer, let us represent the child's body by a rod, flexible and elastic, as the spine really is. In Diagram 49, A B 1 is the rod fixed at B by a sort of crutch, formed by the head and neck against an edge of the pelvis. A, the breech, being movable, receives the impulse of the force, and is driven downwards. This rod, or spine, therefore bends. But the rod, being elastic,

Fig. 50.

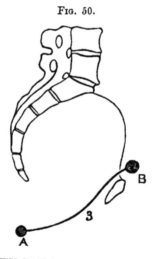

REPRESENTS FURTHER STAGE OF SPONTANEOUS VERSION AFTER ROTATION.

B, head has risen from symphysis, the rod straightening, A breech escapes.

constantly tends to straighten itself. This effort will, if the head is not immovably fixed, lift the head off the edge of the pelvis, and carry it higher into the iliac fossa. The force continuing to press upon A, as in 2, will drive it still lower, and the rod still bending, and tending to recover its straightness, the head will rise further from the edge of the pelvis. At last (*see* Fig. 50) there will be room for the end A to enter the pelvis, and the rod springing into straightness by the escape of A from the pelvis, the whole may emerge, B coming last. For this process to take place, it is obvious that the rod must be endowed with elasticity or spring; and, therefore, as Denman said, a live child is best adapted to undergo spontaneous version.

The mechanism of spontaneous evolution may also be illustrated in like manner. Let us represent the child's body by a rod, flexible, but almost without elasticity. In Diagram 51,

Fig. 51.

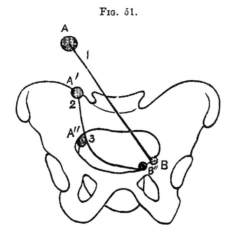

ILLUSTRATING MECHANISM OF SPONTANEOUS EVOLUTION (OR EXPULSION).

A B, flexible inelastic rod, as is body of dead child, bending so that B rests on edge of pelvis, whilst A is driven into pelvis.

one end of the rod, B, is fixed against the edge of the pelvis; the other end, A, being movable, receives the impulse of the downward force, and is driven first to 2; the rod continuing to bend, A falls to 3, and, as B is fixed, the rod forms a strong curve, with its convexity downwards, in the cavity of the pelvis. This convexity will be the first part of the rod to emerge. The force urging on the end A, more and more of the convex rod will emerge, until A itself escapes. Then, and not till then, can the rod recover its straightness, and the end B will follow. (*See* Diagram 52.)

In the case of spontaneous version, as well as in that of spontaneous evolution, it is necessary to exhibit first a pelvis seen from the front, then a section as seen from the side; because in the earlier stages the movement is across the pelvis, and in the later stages the head comes forward above the symphysis, and the movement in a circle around this centre is from behind forwards.

Now, we may ask, What are the conditions required for the execution of spontaneous version, or natural turning? Some of them, probably, are not understood. Certain it is that we are hardly yet in a position to predicate in any given case of shoulder-presentation, seen at an early stage, that spontaneous version will take place, as we might be, if all those conditions were known and recognizable. They would be more familiar if the law to turn were not laid down in such imperative terms —if the dread of evil as the consequence of neglect of that

FIG. 52.

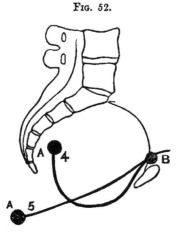

ILLUSTRATING END OF SPONTANEOUS EVOLUTION, ROTATION COMPLETED.

B, the head is fixed above symphysis; A escapes from pelvis after doubling of rod.

law were not so overwhelming. But if Nature be always superseded, if the physician always resort to artificial turning as soon as he detects a shoulder presenting, how can we obtain sufficient opportunities for discovering the resources of Nature, and how she acts in turning them to account? The principal conditions seem, however, to be these: 1. *A live child*, or one so recently dead that the tone or resiliency of its spine is still perfect. 2. A certain degree of *mobility of the child* in utero. 3. Strong action of the uterus and auxiliary muscles. A roomy pelvis does not appear to be always necessary.

Spontaneous version is not likely to take place when the shoulder has been driven down in a point with a part of the chest-wall low in the pelvis, and the uterus is strongly grasping the fœtus in every part, bending its long axis by approximating the head and breech. It is not likely to take place when the head has advanced towards a position above the symphysis pubis, that is, when the movement of rotation has commenced.

But the practical question will arise, Is spontaneous version ever so likely to occur, that we shall be justified in trusting to Nature? Ample experience justifies an answer in the affirmative. But it appears to me that the great lesson taught by the observation of the phenomena of spontaneous version is this : If Nature can by her unaided powers accomplish this most desirable end, we may by careful study and appropriate manipulation assist her in the task. We shall be the better ministers to Nature in her difficulties as we are the better and humbler interpreters of her ways. *Natura enim non nisi parendo vincitur.*

It has been already stated that spontaneous version may take place either by the head or by the pelvis. Examples of spontaneous version by the pelvis have been given in Lecture VIII. I will give two examples of spontaneous turning by the head.

Velpeau * relates the following case of cephalic version. A woman was in labour at the École de Médecine (1825). The os was little dilated. The left shoulder was recognized. The waters escaped five hours after this examination. Four students recognized the shoulder. The pains were neither strong nor frequent: and "*being not without confidence in Denman,*" Velpeau did not search for the feet. In five hours later, the shoulder was sensibly thrown to the left iliac fossa. The pains increased, and the head occupied the pelvic brim. The vertex came down, and the labour ended naturally.

Dr. E. Copeman, of Norwich, records the following case : †

---

* " Traité complet de l'Art des Accouchements," 1835.
† J. G. Crosse's " Cases in Midwifery," 1851.

Some time after the waters had escaped in great quantity, the child was found lying across the pelvis, with the back presenting: neither shoulders nor hips could be felt. At a later period, preparing to turn, Dr. C. was surprised to find the pelvis filled. He endeavoured to pass his hand over the right side of the child towards the pubes, but in so doing he felt the child recede, and therefore confined himself to raising the child's pelvis with his flat hand and fingers; whilst the pains forced down the occiput, the head descended, and delivery was quickly completed. Dr. C. thinks, if he had waited a little longer, spontaneous evolution would have occurred, and the child would have been born even without manual interference. The child was a full-grown male, lively and vigorous.

LECTURE X.

FROM the observation of the spontaneous or accidental changes of position of the fœtus *in utero*, the transition is natural to the account of those changes which can be effected by art. The observations already referred to prove that the fœtus *in utero* may, under certain conditions, change its position with remarkable facility. It follows that the judicious application of very moderate forces may, under favourable circumstances, effect similar changes.

We have seen that spontaneous version may be effected by the substitution of the head for the shoulder, and of the pelvic extremity for the shoulder; also that spontaneous evolution may be effected by the descent of the head with the presenting shoulder and arm, or by the descent of the chest and trunk with the presenting shoulder and arm. Now, each of these natural or spontaneous operations for liberating the child may be successfully imitated by art. Let us study the conditions which guide us in the selection of the natural operation we should imitate, and the methods of carrying out our imitations.

A successful imitation of natural version by the head or by the inferior extremity demands the concerted use of both hands. You must act simultaneously upon both poles of the fœtal ovoid. This combined action may be exerted altogether externally—*i.e.*, through the walls of the abdomen—or one hand may work externally, whilst the other works internally through the os uteri. The first method—that practised by Wigand, D'Outrepont, Esterlé, and others—has been called the bi-manual proper. The second, which was first clearly taught by Dr. Braxton Hicks, has been called by him combined "internal and external version." But the same principle governs both. As I have already said, you must act at the same time upon both poles of the long axis of the fœtus. It would be more correct to describe them both as forms of *the bi-polar method of turning;* and by this designation, proposed by me in the first edition of these lectures, they are now generally known. It is an accident, not a fundamental difference, if, in one case, it is more convenient to employ the two hands outside; and in another, to employ one hand outside, and the other inside. Each form has its own field of application. We should be greatly crippled, deprived of most useful power, if we were restricted to either form. At the same time, I am of opinion that the combined internal and external bi-polar method has the more extensive applications to practice.

I have found the bi-polar method serviceable, adjuvant in every kind of labour in which it is necessary to change the position of the child. It is true that a rather free mobility of the fœtus *in utero* is most favourable to success; it is true that the external bi-polar method can hardly avail unless at least a moderate quantity of liquor amnii be still present; it is true that the internal and external bi-polar method requires, in its special uses, if not the presence of liquor amnii, at any rate a uterus not yet closely contracted upon the fœtus. But I am in a position to state that amongst more than 200 cases of turning of which I have notes, there was scarcely one in which I did not turn the bi-polar principle to more or less advantage; and in not a few cases of extreme difficulty from spasmodic concentric contraction of the uterus upon the fœtus, with jamming of the shoulder into the pelvis, where other practitioners had been foiled, I have, by the judicious application of this principle, turned and delivered safely.

The history of the bi-polar method of version, the steps by which this, the greatest improvement in the operation, has been brought to its actual perfection, deserve to be carefully recorded.

From what has been already said it is clear that Wigand, D'Outrepont, and others who took up Wigand's views, had acquired an accurate perception of the theory of bi-polar turning, and had, moreover, successfully applied that theory in practice. They had applied it to the purpose of altering the position of the child before labour, chiefly by bringing the head over the centre of the pelvis, restoring at the same time the uterus and fœtus from an oblique to a right inclination. This they did generally by external manipulation; but not exclusively, for sometimes one or two fingers introduced into the os uteri served to drag the lower segment or pole of the uterus to a central position, whilst the hand outside acted in the opposite direction upon the upper pole. Here the application seems to have stopped short. At least, I am not aware of any distinct description of the application of the bi-polar principle to produce version.

In one form, indeed, the bi-polar principle of turning by the feet has been in use for a long time. It not uncommonly happens, when turning is attempted after the waters have escaped, and when the uterus has contracted rather closely upon the child, that, even when one or both legs have been seized and brought down, the head will not recede or rise from the pelvis—that is, version does not follow. It then becomes obvious that by some means you must push up the head out of the way. The operation by which this is effected—an exceedingly important one—will be fully explained hereafter. It is enough to say in the present place that it consists in holding down the leg that has been seized, whilst a hand or a crutch introduced into the pelvis pushes up the head and chest. In this operation it will be observed that both hands work below the pubes, whilst in the true bi-polar method one hand works below and inside, and the other above and outside.

In several obstetric works (Moreau, Caseaux, Churchill, &c.), diagrams illustrating the operation of turning are given repre-

senting one hand applied to the fundus uteri outside, and the
other seizing the feet inside. But it would be an error to infer
that these indicate an appreciation of the principle of bi-polar
turning. They simply indicate the principle of *supporting the
uterus*, so as to prevent the risk of laceration of the cervix whilst
pushing the hand through the uterus and up towards the fundus.
The true bi-polar method does not involve passing the hand
through the cervix at all.

The following passage from the late Dr. Edward Rigby
(*Library of Medicine*, Midwifery, 1844), may be taken as a
description of the diagrams referred to :—" In passing the os
uteri . . . we must at the same time fix the uterus itself
with the other hand, and rather press the fundus downwards
against the hand which is now advancing through the os uteri.
In every case of turning we should bear in mind the necessity
of duly supporting the uterus with the other hand, for we thus
not only enable the hand to pass the os uteri with greater ease,
but we prevent in great measure the liability there must be to
laceration of the vagina from the uterus in all cases where the
turning is at all difficult."

The same precept is even more earnestly enforced by Professor Simpson*:—" Use both your hands," he says, " for the
operation of turning. In making this observation, I mean that
whilst we have one hand *internally* in the uterus, we derive the
greatest possible aid in most cases from manipulating the uterus
and infant with the other hand placed externally on the surface
of the abdomen. Each hand assists the other to a degree which
it would not be easy to appreciate except you yourselves were
actually performing the operation. It would be extremely
difficult, if not impossible, in some cases to effect the operation
with the single introduced hand ; and in all cases it greatly
facilitates the operation. The external hand fixes the uterus and
fœtus during the introduction of the internal one ; it holds the
fœtus *in situ* while we attempt to seize the necessary limbs, *or it
assists in moving those parts where required towards the introduced
hand*, and it often aids us vastly in promoting the version after
we have seized the part which we search for. Indeed, this
power of assisting one hand with the other in different steps of

the operation of turning forms the principal reason for introducing the left as the operating hand."

Here the consentaneous use of the two hands is well described. But the bi-polar principle is at best but dimly foreshadowed.

Dr. Robert Lee, in his *Clinical Midwifery*, relates several cases in which he succeeded in converting a head or shoulder presentation into a pelvic one by introducing one or two fingers only through the os uteri, when, indeed, this part was so little expanded that to introduce the *hand* would have been impossible. These cases were mostly cases of placenta prævia, the fœtus being premature and small. He managed this by gradually pushing the presenting part towards one side of the pelvis until the feet came over the os uteri. Then he seized the feet and delivered. But there is no mention of the simultaneous or concerted use of the other hand outside, so as to aid the version by pressing the lower extremity of the child over the os, or to carry it within reach of the hand inside. His is a manœuvre of limited application. It differs in principle from the bi-polar method, which requires the consentaneous use of both hands, and which enjoys a far wider application.

A process of synthetical reasoning, especially if informed by the light of experience in practice, might construct out of the elements thus contributed by Wigand and his followers, by Rigby, Simpson, and Robert Lee, a complete theory and practice of bi-polar turning in all its applications to podalic as well as to cephalic version. I am conscious myself of having in this manner evolved that theory, and applied it in practice. Dr. Rigby's was the work I had adopted as my guide from the commencement of my career; and my attention was especially directed by Dr. Tyler Smith to the admirable lecture of Professor Simpson, from which the passage above quoted is drawn, at the time of its appearance. Since then I have turned more than two hundred times. In no case have I failed to observe the precept of using both hands; and gradually I found out that the external hand often did more than the internal one—so much so, indeed, that the introduction of one or two fingers through the os uteri to seize the knee pressed down upon the os by the outside hand was all that was necessary. I feel that

I am entitled to say this much; and not a few of my professional brethren who have honoured me by seeking my assistance can bear witness to the fact that it was by the application of the bi-polar method that I have been enabled to complete deliveries where others had failed.

But, in saying this, I should be sorry indeed if it were interpreted as a desire on my part to detract in any degree from the merit of my colleague, Dr. Braxton Hicks. His claim to originality in working out and expounding the application of the external and internal bi-polar method of podalic version is indisputable. I know of few recent contributions to the practice of obstetrics that possess greater interest or value than his memoirs on *Combined External and Internal Version*, published in the *Lancet* in 1860, in the *Obstetrical Transactions*, 1863, and in a special work in 1864.

If the proposition which I have already urged with reference to the forceps be true — namely, that the carrying to the greatest possible perfection of an instrument that saves both mother and child is an object of the highest interest—it is scarcely less true of turning, also a saving operation. I cherish a fervent hope that the exposition of the principles and methods of turning which will be made in the following lectures, will, in conjunction with those on the forceps, be the means of materially enlarging the field of application of the two great saving operations, and, as a necessary result, of supplanting, in a corresponding degree, the resort to the revolting operation of craniotomy.

As head-presentation is the type of natural labour, it follows that to obtain a head-presentation is the great end to be contemplated by art. It seems enough to state this proposition to command immediate assent. But in practice it is all but universally contemned. No one will dispute that the chance of a child's life is far better if birth takes place by the head than if by the breech or feet. Yet delivery by the feet is almost invariably practised when turning, or the substitution of a favourable for an unfavourable presentation, has to be accomplished. Why is this?

The answer is not entirely satisfactory. It rests chiefly upon the undoubted fact that in the great majority of instances, at the time when a mal-presentation comes before us, demanding skilled assistance, turning by the feet is the only mode of turning which is practicable. Frequent experience of one order of events is apt so to fill the mind as to exclude the recognition of events that are observed but rarely. Many truths in Medicine escape recognition because the mind is preoccupied by dogmas and narrowed by an arbitrary and enslaving empiricism. Many things are not observed because they are not sought for with an intelligent and instructed eye. And then, reasoning in a vicious circle, some men will boldly deny the existence of that which their untrained faculties cannot perceive. They go further: by doggedly and consistently following a practice which arrests Nature in her course, substituting a violent proceeding of their own, they never give Nature a chance of vindicating her powers, and they consequently never give themselves a chance of learning what those powers are, or of realizing the imperfection of their own knowledge. They close the shutters at noon-day, and say the sun does not shine.

In the seventeenth and in the beginning of the eighteenth centuries, Velpeau remarks, cephalic turning was hardly ever mentioned unless to be condemned. But if the practice of podalic turning was then so general, it was justified because the forceps was not known. In many cases it is not enough to correct the position; it is also necessary to extract. Without the forceps our predecessors could only extract by the legs. But now, if the head is brought to the brim, the forceps affords a ready means of extraction.

Flamant appears to have been amongst the first to revive the practice of turning by the head; he did it by external manipulation. Osiander and Wigand (1807) investigated the subject with remarkable sagacity and skill. D'Outrepont pursued it; and many other names might be cited. The researches of Wigand, however, contain the germ of all the subsequent inquiries.

Flamant strenuously contended that head-turning was best. In two cases of arm-presentation, he raised the breech towards the fundus uteri. The head thus made to descend was seized by the hand. The liquor amnii had long escaped. He worked in these cases entirely by *internal* manipulation. Wigand accomplished the same object by *external* manipulation, saving

L

the children. D'Outrepont had a case of a woman who had lost three children by foot-turning. In her fourth labour she had a shoulder-presentation. There was slight conjugate contraction. The head lay to the right, the feet to the left; the back of the chest was above the brim. He seized the child by the back, placed his right thumb and the right side of four fingers on its left side ; then he pushed it to the left and upwards ; then he released the back, and seized the neck, whilst he pressed upon the shoulder with his thumb, and the palm and four fingers on the back. The head came over the brim, and the child was safely delivered. In a second case, the breast was on the brim, the head to the left ; he pushed up the chest and brought down the head, which entered by the face, and was so delivered. Strong pains prevented his reducing the face to a cranial position. In a third case he was equally successful. D'Outrepont afterwards practised with success Wigand's method of head-turning by external manipulation.

Here is a case of bi-manual and bi-polar head-turning by D'Outrepont : The head lay in the right side. He placed the patient on her left side raised. During each pain he imparted gentle pressure on the side where the head lay, directing it towards the brim; and *at the same time* he pressed with his other hand in the opposite direction upon the fundus where the breech lay. In the intervals of pain, he planted a pillow in the side where the head lay. The head was brought into the pelvis, and a large living child was born.

Professor E. Martin * has carefully described the operation, and practised it with great success.

Hohl† says turning by the head is much less esteemed than it ought to be, and it would be more esteemed if more pains were taken to instruct pupils how to do it on the phantom.

Head-turning, or simple rectification of the presentation, may be indicated under the following circumstances :—

A. *Before the Accession of Labour.*—When the uterus and fœtus are placed obliquely in relation to the pelvic brim; and in some cases where the shoulder is actually presenting.

B. *When Labour has Begun.*—1. When the uterus and fœtus

---

* " Froriep's Notizen," 1850.
† " Lehrbuch der Geburtshülfe," 1862.

are placed obliquely in relation to the pelvic brim, which obliquity may be preparatory to the complete substitution of the shoulder for the head.

2. In some cases of shoulder-presentation, the membranes being still intact.

3. In some cases of shoulder-presentation, the membranes having burst, but considerable mobility of the child being still preserved.

4. The forehead or face presenting.

5. Descent of the hand by the side of the head.

6. Prolapse of the umbilical cord by the side of the head.

A. *Head-turning, or Rectification before Labour.*—This has been often practised by Wigand, D'Outrepont, and others. I will describe the operation after Esterlé. It was the observation of the frequent occurrence of spontaneous version in the eighth and ninth months of gestation that led this eminent obstetrician to practise external bi-polar version.* He observed that a large number of shoulder-presentations in the last two months, if left to themselves, were converted into natural presentations, either on the approach of labour or after the beginning of labour. He had further remarked that spontaneous version had occurred after the escape of the liquor amnii, and the shoulder was sensibly down. The most efficient cause of spontaneous version, he says, is the combined action of the movements of the fœtus and of its gravity, the centre of gravity not being far from the head. The extension of the feet must drive the breech away from the uterine wall as the feet strike it, and so the head is brought nearer to the brim. His method was as follows :—

The patient must be placed in such a posture as to produce the greatest possible muscular relaxation. Bearing in mind the conditions which take part in spontaneous version, it is necessary to imitate them as much as possible. Amongst these is the lateral and partial contraction of the uterus, which diminishes the transverse diameter, and which exerts a convenient pressure upon the ovoid extremities of the fœtus ; and the movements of the fœtus, the re-percussion of its head, and its descent when the centre of gravity of the fœtal body favours its fall. To imitate this, the lateral contractions must be

* "Sul Rivolgimento Esterno." Annali Universali di Medicina, 1859.

L 2

replaced by lateral pressure. This is applied towards the
fundus or the neck, according to the situation of the part
which it is sought to raise or to depress. This pressure is
assisted greatly by gentle strokes or succussions made by the
palm of the hand alternately towards either ovoid extremity.
These strokes are then made, in rapid succession, simultaneously upon the two extremities, one giving a movement
of ascent, the other a movement of descent; or we may act
upon the head alone, whilst the other hand makes a steady
pressure on the contrary side, the more to diminish the transverse diameter. The desired position being effected, it is
necessary to maintain it. This may be done by the adaptation of a suitable bandage.

Lazzati operated in a similar manner. He maintained the
uterus and fœtus in due position by the adaptation of cushions
or of pads to the sides of the opposite poles of the fœtal
ovoid.

B. 1, 2, and 3.—Head-turning or correction of the presentation may be attempted in cases of moderate obliquity, where
the liquor amnii is still present or has only recently escaped.
It is also necessary that the action of the uterus be moderate.
Correction, as we have seen, consists in restoring the head,
which has passed across the brim of the pelvis into the ilium,
back to its due relation to the brim. This operation involves
the rectification of the uterus, as well as of the child. It may
in certain cases be effected entirely by external manipulation.
Supposing the case be one in which the head is deviated to the
left ilium, and the fundus, with the breech, are directed to the
right of the mother's spine, the first step is to place the patient
in a favourable position. Now, by laying her on her left side,
the fundus of the uterus, loaded with the breech, and being
movable, will tend to fall towards the depending side. This
will act as a lever upon the uterine ovoid, and raise the lower
or head end of the uterus, so as to facilitate its return to the
brim. In such cases Wigand recommends that the posture
should be repeatedly changed, so as to ascertain which is the
best to maintain the head in the central line of the pelvis.
When this is found, the sooner the membranes are ruptured
the better. The patient must thenceforward be kept carefully
in the same posture, the uterus being supported in due relation

by the hands externally. But I believe that in many cases
the dorsal position will lend the greatest facility.

We must apply pressure to the uterus towards the median
line of the mother, both at its fundus and at the lower part,
which contains the head. The head will thus be pushed by
one hand to the right, whilst the fundus uteri is pushed by
the other hand to the left. When the head has been thus
brought over the brim, the difficulty is to secure it there.
If the correcting pressure be removed, the uterus tends to
resume its obliquity.

If labour has begun, we may combine internal with external
manipulation. We may press upon the fundus with one hand,
whilst with a finger in the os uteri we pull this over the centre
of the brim (Wigand). External pressure by a cushion or
pillow laid in the hollow of the ilium in which the head lay
will aid this manœuvre. Then, having got the head into
proper position, and whilst it is kept so by aid of an assistant,
rupture the membranes. *The contraction of the uterus tends to
restore its natural ovoid shape.* And this will tend to keep the
child's long axis in due relation. If by this contraction the
head should happily become fixed in the brim, the manœuvre
has succeeded; the labour has become natural. Velpeau and
Meigs relate instances of the successful application of this
practice. But if the head still show a disposition to recede,
grasp it at once with the long double-curved forceps, and hold
it in the brim until it is sufficiently engaged to be safe.

4. The mode in which *forehead- and face-presentations* arise
out of excess of friction or resistance encountered by the occiput
has been described in Lecture V.

Sometimes correction of these presentations may be effected
by restoring the equilibrium of resistance to the anterior part
of the head. Sometimes this is effected by simply keeping the
tips of the fingers upon the forehead, trusting to the expulsive
efforts propagated through the child's spine to cause the head
to rotate upon its transverse axis, and bring down the occiput.
Sometimes further aid is necessary. The tips of two fingers of
the left hand are applied internally upon the forehead, and at
the same time the occiput must be pressed down by the fingers
of the right hand applied externally in the iliac fossa.

In some cases a rougher method has been pursued. The

hand introduced into the uterus has seized the head by the occiput, and brought it down. This manœuvre is by no means easy, and, if the child is mature, will rarely succeed.

Wigand, when the head was not too low in the pelvis, first pushed the face upwards, so as to convert the face into a forehead-presentation, if not into a cranial; then he applied the forceps.

Smellie had already deliberately put in practice the restoration of a lost head-position.* In one case, feeling the face presenting through the membranes, he raised the forehead; then letting the waters escape, the head was fixed in its proper position, and the labour terminated successfully. In the second case, a hand presented. Smellie grasped the head and brought it into the brim, having pushed up the shoulder. In this position the head was fixed by the escape of the liquor amnii and bearing-down pains. The child was delivered naturally. In a third case, in which the breast presented, he was equally successful in bringing down the head.

5. *Descent of the hand by the side of the head.* When this accident occurs, it is apt to proceed to shoulder-presentation, the hand and arm slipping down and wedging the head off the brim to one or other iliac fossa. Hence the importance of correcting this presentation as early as possible. Whilst the parts are still movable, it is commonly possible to push up the presenting hand by means of your left fingers in the vagina: and at the same time, by pressing down the head by the external hand towards the brim, you make the head fill the space until the double-curved forceps is applied. Then, drawing the head into the brim, the hand cannot again descend.

6. *Prolapse of the umbilical cord by the side of the head* may sometimes be managed successfully in a similar manner, having first replaced the cord above the presenting head. In the next lecture we may conveniently examine the mode of dealing with *prolapse of the umbilical cord,* and discuss the subject of asphyxia generally.

* "Cases and Observations," vol. ii., 1754.

PROLAPSE OF UMBILICAL CORD—HOW PRODUCED—HOW TREATED
—THE CORD ROUND THE NECK—SIGNS OF DEATH OF CHILD
—ASPHYXIA OF CHILD, PARALYTIC AND SIMPLE—INFLUENCE
OF PLACENTAL AND AERIAL RESPIRATION COMPARED—TREAT-
MENT OF ASPHYXIA : BY REFLEX IRRITATION ; BY ARTIFICIAL
RESPIRATION ; MARSHALL HALL, SYLVESTER, PACINI, BAIN,
RICHARDSON.

MOST of the causes which favour the production of malposition of the child, such as pelvic distortion, placenta prævia,
excess of liquor amnii, also favour prolapse of the umbilical
cord. All causes that hinder the complete filling of the lower
segment of the uterus and the pelvic brim by the presenting
part of the child, of course leave room for the falling through
of a loop of cord. And the cord is especially apt to be hooked
down if it be over-long, if it take its origin near the orificial
zone of the uterus from the lower margin of the placenta, and
if there be a copious or sudden rush of liquor amnii when the
membranes burst. In healthy labour, the lower segment of
the uterus, which contains the presenting head, sinks down at
the very earliest stage into the pelvic brim. Thus, when the
cervix dilates and the membranes burst, the head instantly
descends into contact with the uterine orifice, and closes it like
a ball-valve. So perfect is this action, that usually a portion
of liquor amnii is even retained until the child's body is born.
Whenever the cord falls through, something has prevented this
sinking of the lower segment of the uterus into the pelvis.
One thing deserves attention. Prolapse of the funis has frequently happened when the patient has been sitting or standing

up at the time the membranes have burst. Hence it is desirable to keep the patient on her bed when this event is expected;
and then, should the funis come down, you will have the further
advantage of detecting the accident at the earliest moment, a
point of paramount importance.

There are *two periods of prolapse of the cord ;* and the
management must be modified accordingly.

The first case is when the cord is felt below the presenting
part of the child whilst the membranes are still entire. The
liquor amnii being of course preserved, the risk of pressure upon
the cord is but small. It is rarely necessary to interfere acting
before the membranes burst. But when the cord is felt
through the membranes, we must be prepared for the coming
difficulty. The moment the membranes burst the cord will
probably be carried down by the torrent of liquor amnii and
the expulsive force. If the cervix uteri is freely open at the
moment of bursting of the membranes, the condition will be so
far favourable for immediate efforts to replace the cord or to
deliver. But should the cervix be only moderately open, the
risk is greater. In this case we should do well to dilate the
cervix, before the membranes burst, by the water-bags, so as to
ensure the desirable freedom for manipulation. We must be
ready to act the moment the membranes burst, according to the
indications of danger to the child and the presentation.

Since, during the passage of the child, the prolapsed cord
must undergo protracted compression, the child's life is in
imminent danger. Rescue from this danger will depend upon
one of two conditions: first, upon the re-position of the prolapsed cord above the presenting part of the child, and its
retention there; or secondly, upon the speedy delivery of the
child, before fatal asphyxia has taken place. Sometimes we
have a choice in the method of proceeding; sometimes the
course of action is imperatively dictated by circumstances. For
example, if there is a shoulder-presentation, there is obviously a
major reason for turning. If there is placenta prævia with
profuse hæmorrhage, to secure the safety of the mother must
be our first object; for this, turning may be indicated: indeed
that which is best for the mother is often here the best chance
for the child also. In the case where the cord springs from the
margin of the placenta, and this margin descends to near the

orifice of the womb—and the two things commonly go together —it is clear that there is not much hope of keeping up the cord even if you succeed in putting it back into the uterine cavity. The next pain will expel it again. The prospect of saving the child depends upon prompt delivery. If the head present, put on the forceps at once. If the os is not sufficiently dilated, use the water-bags. Or if you think delivery can be effected more quickly by turning, adopt this operation.

If there be contraction of the conjugate diameter of the pelvis so that the cord slips down on one side of the promontory where the brim cannot be occluded by the head, you may try to replace the cord before proceeding to expedite delivery. The same course should be adopted when there is no deformity and when the cord springs from the higher zones of the uterus. *As to the mode of re-position of the umbilical cord.* Much ingenuity has been brought to bear upon this problem. You may occasionally—rarely, it is true—take the prolapsed loop in the fingers of your left hand, and in the absence of uterine contraction, carry it up into the uterus above the presenting head, and even hang it over the child's knee or foot. Unless you succeed in thus finding a peg for it, it will almost certainly follow your hand down again as you withdraw it. And whenever you have succeeded in carrying the cord into the uterus, in whatever way, you must immediately try to fill up the pelvic brim with the child's head so as to obviate a recurrence of the prolapse. This may be done, sometimes by external pressure by the hands, or a bandage so applied as to fix the head in the brim, or better still by the forceps. Where the cord comes down in feet-presentation, you may do as Wigand did: carry up the cord with your hand, and immediately bring down a leg into the os uteri.

Another mode of replacing the funis is by help of the so-called *knee-elbow position.* By placing the woman on her elbows and knees, the pelvis is necessarily raised above the level of the fundus of the uterus. Gravity, therefore, favours the descent of the child from the pelvic brim, thus giving room for the cord to fall back too, and for the obstetric hand to help to replace it. This is an excellent plan. But even here, you must not neglect to bring the head into the brim either before or immediately after allowing the patient to resume the horizontal posture. The resort to the knee-elbow position is not altogether new. I was taught it by the late Dr. Bloxam nearly thirty years ago. But it has lately been enforced, and successfully applied by Dr. Thomas, of New York, under the name of the " Postural Method " (*Transactions of the New York Academy of Medicine*, 1858); W. Theopold (*Deutsche Klinik*, 1860); by Dr. Wilson (*Glasgow Medical Journal*, 1867); and others. You may, as these practitioners have done, succeed

FIG. 53.

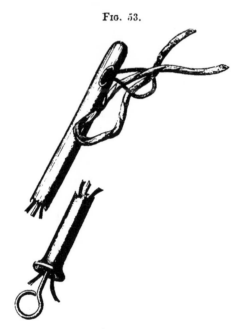

ROBERTON'S FUNIS-REPLACER.

by the knee-elbow position, aided by your hand alone. But it may be useful to employ a special instrument to replace the cord. A multitude of instruments for this purpose have been invented. None surpass in simplicity and effectiveness the contrivance of Dr. Roberton. This is nothing more than a large catheter of vulcanized india-rubber mounted on a stilet. It should be fifteen inches long, of large calibre, with a large eyelet-hole near the blind end. Figure 53 represents it prepared for use. A loop of coarse, soft silk, or worsted, is carried through the tube and brought out at the eyelet; or you may

simply tie a bit of worsted in form of a loop round the end of
the catheter. The prolapsed cord is caught in this loop. The
tube is then carried by means of the stilet past the presenting
part of the child until the cord is fairly lodged in the uterus.
Then, applying a finger of one hand to the lower end of the
tube, you withdraw the stilet with the other hand, leaving the

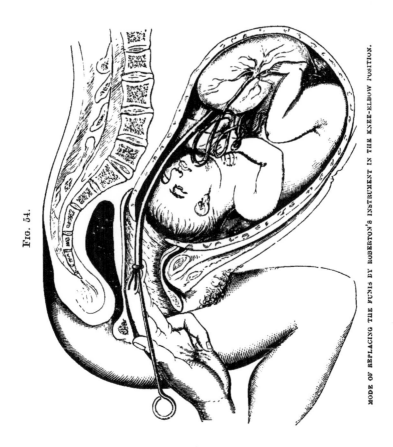

FIG. 54.

MODE OF REPLACING THE FUNIS BY ROBERTON'S INSTRUMENT IN THE KNEE-ELBOW POSITION.

tube *in situ*. Being perfectly smooth and flexible, it gives rise
to no sort of inconvenience. It comes away with the placenta
when the child is born.

Efficient as this little contrivance often is, I however advise
you to combine the advantages offered by the knee-elbow position, as represented in Figure 54.

It is well to keep the tube always ready armed with the silk-loop. If not at hand, you will find an ordinary elastic catheter an excellent extemporaneous substitute.

If the cord is quite flaccid and pulseless, and especially if a loop outside the vulva is also cold; and if tickling of the child's feet excite no reflex movement, it may be *presumed* that the child is dead. If known to be dead, the obvious indication is to turn our attention exclusively to the interest of the mother. But mere absence of pulsation in the cord must not hastily be taken as certain evidence of the child's death. It is proper in doubtful cases to auscultate the abdomen for the fœtal heart; and if the hand is passed into the uterus we should take the opportunity of *feeling* the child's chest for the heart-beat.

Observations on the signs of impending asphyxia will be made when describing breech-labour.

Allied to prolapsus is the *entanglement of the cord round the child's neck*. In this case, the cord being perhaps of excessive length, and the child's head being prevented from filling the lower segment of the uterus, a loop of cord lodges there, so that when the cervix opens, the head passes through the loop. I have no doubt this is the common way in which the accident arises. But there is reason to believe that it is sometimes due to versions of the child during or before labour. Occasionally, the cord forms two coils, or even three, round the neck. The effect of the accident is that, as the child descends through the pelvis, the cord tightens round the neck, tending to strangle it, and at the same time the circulation through the cord is interrupted. And there is, moreover, likely to be arrest of the labour from the child being held back by the cord dragging upon the uterus. It is obvious that the child is in imminent peril. It is, therefore, a good rule, the moment the head is born to pass your finger round the neck to feel if the cord encircles it. If it be found loose, and the body is advancing, so that you have no time to bring the loop of cord over the head, open the loop and let the child's body pass through it. But if the head is not advancing; if you observe that after every pain the head seems to be retracted; and especially if you see the face becoming congested, and you feel the cord tight and pulseless round the neck, lose not a moment in passing a finger under the cord, and sever it. Dr. Haake (*Zeitschr. für Med. Chir. u. Geburtsk*, 1865) suggests that this complication may be detected by examining by the rectum. Thus the finger can be easily, he says, carried above the head, so as to feel the umbilical cord and its pulsation. This may give timely indication of the child's position, and tell when it is necessary to accelerate labour. Then accelerate the birth of the body; and tie the cord at the usual distance from the navel. If you fear that the child may bleed you may secure both ends of the cord by tying, before the body is expelled or extracted. But this is not really necessary.

Asphyxia of the new-born child. In cases of labour with prolapsus of the cord, and when the cord has tightly embraced the neck, we should always be prepared with means of exciting respiration. Asphyxia in the new-born is a large and interesting subject, which I cannot fully discuss in this place. Asphyxia from prolapse of the cord is one of the most simple cases. It is generally entirely due to pressure upon the cord interrupting the circulation between the fœtus and the pelvis. There is not so often as in other causes of asphyxia, for example, in head-last labour, contraction of the brim, difficult labour, or forceps delivery, complication with compression of the child's brain. Where this occurs there is often injury to the brain-tissues, meningeal effusion of blood, or at least congestion of the pons Varolii and medulla oblongata. These conditions may be so severe as to be incompatible with recovery. The *child cannot breathe*, the lesion of the nervous centres renders this act impossible. This we may call *paralytic asphyxia*.

But in simple asphyxia there is hope. The great indication is obviously to establish aërial respiration to compensate for the loss of the placental circulation. How is this best done? In minor degrees of asphyxia where the excitability of the medulla oblongata remains, you may work upon the reflex function. Irritation of the respiratory nerves will often succeed. Cold aspersion on the face and chest, gentle friction of the chest, dipping the child for a moment in warm water, may be enough to excite respiration. But if the nervous centres will not respond to peripheral irritation, what next? Here comes the question as to the absolute and relative merits of the various methods of artificial respiration. The object of all is to get some air into the lungs. The oldest method is by direct insufflation by mouth to mouth, or by a tracheal tube. An objection to it is that it is difficult to avoid rupturing the delicate air-cells.

In discussing the principles of dealing with asphyxia in the new-born infant, it is necessary to consider how it is produced and the phenomena which attend it. We may start with the undisputed proposition that aërial or lung-respiration is the necessary substitute for placental respiration. If the child is to live, the lung-respiration must quickly take the place of arrested placental respiration. Usually, this substition takes place immediately the child is born. What is the cause of the first inspiration? The following events may be observed. The child commonly breathes the moment its chest emerges from the mother's pelvis. Here, two conditions conspire to set up breathing: first, the elasticity of the chest-walls causes them to expand when external pressure is taken off; secondly, the impression of cold air, or the cold produced by the evaporation of the fluid on the surface of the face and chest, acts as an excitor of the respiratory nerves, which, acting on the medulla oblongata, excites reflex action of the inspiratory muscles.

Another event is not uncommon. Breathing begins as soon as the head is born, and *before* the chest is expelled. In this case the elasticity of the chest does not enter as a factor. The impression on the respiratory nerves of the face may do so, but the researches of Hecker, Krahmer, Schwartz, and others prove that an *attempt at inspiration may take place even before the face comes into the external world.* We must, therefore, seek further back for the essential cause of aërial respiration. We find evidences of this attempt whenever there has been interruption of the placental circulation. These evidences are: the presence of liquor amnii, foetal epithelial scales, meconium in the air-passages, and ecchymoses or puncta of blood under the pleura and pericardium. Now, the fluid in which the foetus floats can only be drawn into its trachea by an inspiratory effort. This effort is the result of the accumulation of carbonic acid in the child's blood from the interruption of the placental circulation. There is an immediate want of oxygen, and the inspiratory act is excited to supply it. There can be no doubt generally, that complete cutting-off of the placental communication with the child is a sufficient provocative of the act of inspiration. This complete breaking off commonly takes place with the powerful contractions which drive the head into the world. But it has been further contended that every uterine contraction during the second stage of labour, since it must partially interrupt the placental circulation, by constricting the uterine placental vessels, must also provoke an effort to respire. If this proposition were true, we ought constantly to find extraneous fluids in the child's air-passages, and the risk of asphyxia would be much higher than it is actually found to be. I cannot therefore admit, that in ordinary labours the partial and periodical interruptions to the placental circulation provoke inspiratory efforts.

Nor must it be concluded that even complete placental interruption is necessarily provocative of an inspiratory effort. In cases where the brain has been long compressed, and probably from other causes, the medulla oblongata may be insensible to the stimulus—there is *paralytic asphyxia.*

A circumstance observed in asphyxia is that the heart continues for a time to pulsate. I do not know, indeed, that recovery has ever been established when the heart has ceased to beat. To maintain the pulsation of the heart, purification of the blood circulating in the vessels must be effected. This must be done either by means of the placenta or of the lungs. If both these organs are idle, the heart's action will gradually flag, and soon cease. The experiments of Brodie long ago demonstrated that artificial respiration would keep up the action of the heart. But I am not aware of any observations quite so apposite to the present question as some made by myself. (See *London Hospital Reports*, vol. i.) I induced labour, at the seventh month, in a woman who had a contracted pelvis, chiefly by galvanism. The cord came down, and by holding it I had accurate information of the variations of the fœtal pulse. In the absence of uterine action, the pulsations in the cord numbered eighty in the minute. Galvanism being applied during contractions, these were sensibly increased in power, and the pulsations became intermitting, and occasionally stopped. As the contractions relaxed, the pulsations resumed their strength and regularity. I also initiated contractions by galvanism, and then, too, intermittences of the fœtal pulse ensued. The same effect upon the pulse was also observed under the influence of normal contractions. There was never any pressure upon the cord.

In the course of a case of placenta prævia I had an excellent opportunity of observing and comparing the respective influences of placental and aërial respiration upon the fœtal circulation. The gestation was between six and seven months. The feet presented. By stethoscope I heard the fœtal heart beating ninety times in the minute. The mother's pulse was eighty. During uterine contractions the fœtal pulse fell to sixty. I presently seized the feet to aid expulsion. The child when born gave a feeble gasp, followed by an *emissive* cry. Heart and cord both pulsating. I did not sever the cord, but endeavoured to excite more perfect respiration by stimulating the respiratory nerves. The gasp and emissive cry recurred every minute. The pulsations were sixty in the minute only; but they increased to ninety immediately after every inspiration, and then sank again to sixty. The same phenomena were observed for some time after the cord was severed. We here see that placental and aërial respirations were of equivalent value in maintaining the fœtal pulse at ninety.

We have it then clearly established: 1. That the uterine contractions of labour, whether occurring spontaneously, or artificially produced, lower the heart's action ; 2. That placental or aërial respiration raises and sustains the heart's action, and 3. That unless aërial respiration be kept up in the new-born child the heart's action will speedily fail altogether. Here, then, we find the scientific basis for resorting to artificial respiration when natural respiration cannot be provoked.

Is it worth while to defer cutting the cord when a child is born asphyxiated ? Can we depend upon the placental circulation fulfilling, for a time, the function of breathing until the lungs will take up the work ? It is not easy to answer this question absolutely. Tying the cord sometimes seems to be an immediate stimulus to respiration. On the other hand, as it is possible the heart's action may be sustained, to a moderate extent, by the limited placental circulation, it may be wise to avail ourselves of this help. But generally it is better to sever the cord at once.

The course recommended is this :—

1. Place the child on a flannel on its back in a position favourable for respiration.

2. See that its mouth, nostrils, and trachea are clear. If obstructed by mucus or indrawn fluids, remove them before attempting to set up respiration. B. S. Schultze* describes a method of rejecting indrawn fluids, which embraces a method of producing artificial respiration. He seizes the child by the shoulders in such fashion that his thumbs are applied on either side to the anterior surface of the thorax, his index fingers in the axillæ, and the other fingers to the child's back; the head is supported by the ulnar edges of his palms. He then swings the depending child upwards about 45° above the horizon, so that the lower end of the child and its weight rests on his thumbs applied to the thorax. The chest in this way is compressed upwards by the abdominal cavity, and the indrawn fluids are driven out. This being effected, the body is allowed to fall back to its original position, when the chest being relieved from pressure, it expands by its own elasticity and draws in air. This may be repeated several times. I cite this method because Schultze speaks highly of its efficacy. But I should use it no further than to get rid of the tracheal obstruction, not as a method of inducing artificial inspiration. It is too violent, and certainly inferior to other methods of artificial respiration. Dr. J. G. Wilson, of Glasgow, has contrived a convenient apparatus, consisting of a tracheal tube attached to an elastic ball, by which, he says, mucus is readily sucked up and removed. His instrument is designed for insufflation; it is very simple, and is certainly preferable to the ordinary tracheal tube. By working *one ball only* of Richardson's bellows, that one unconnected with the secondary ampulla, B (Fig. 55, p. 164), you have an aspirator which will suck mucus out of the air-passages.

3. Try if the medulla oblongata will respond to reflex excitation. Blow upon the face; dash cold water on the face and chest; sprinkle a little brandy on the chest, and use gentle friction on the chest; flick the chest or buttocks with the

* " Jenaische Ztschr. für Medicin., iii. Band."

corner of a wet towel; dip the child for a moment in warm water, say at 90° F.

4. Reflex irritation failing, turn to artificial respiration. This may be practised by one of two principal methods—first, by direct insufflation into the lungs; or secondly, by expanding the chest-walls, creating a vacuum which draws air into the lungs. The first method is the most ancient. It was first practised by blowing from mouth to mouth, with or without the intervention of a piece of linen. It was found that the air often passed over the trachea and went into the stomach. A tube was therefore passed into the trachea, to serve for insufflation. It has been objected to this plan, first, that the air from the surgeon's lungs must itself be charged with carbonic acid, and, secondly, that it is difficult to moderate the force of the insufflation so as to avoid rupturing the tender air-vesicles and causing emphysema. The first objection is not of much force. If, before operating, you fill and empty your lungs fully several times in succession, and then having well filled them, blow into the tube gently the air with which your mouth is filled, very little carbonic acid will go into the child's lungs. The second objection is more valid. With a tracheal tube it must be difficult to regulate very nicely the force of the current of air.

This objection, and the fame of Marshall Hall, led in this country to the extensive use of the " Ready Method." There are *four methods of practising artificial respiration by manipulation of the body.* The *first, Marshall Hall's,* is effected in this way: —First lay the child on its back, the head slightly raised; then roll the trunk over a little more than a quarter turn, on its side, until the chest looks a little downwards; then roll the trunk back to its original position; repeat this movement twelve to sixteen times a minute. The weight of the trunk upon the chest compresses it, and the elasticity of the chest when the weight is taken off opens it, and air is drawn in.

2. The same object is effected by *Sylvester's method.* To practise this, place the child on its back, the head slightly elevated; seize the hands or fore-arms, one in each of your hands, and extending them outwards so as to bring the child's hands above the level of its head; then bring the arms down again to the sides; repeat this manœuvre twelve to sixteen times

in the minute. The chest is thus pulled open by the attachments of the pectoral muscles, and it collapses again when the arms come down.

3. *Pacini's method.*—Place the child on its back, and standing behind the patient's head, insert your hands in the axillæ on the dorsal aspect, then pull the shoulders towards you with an upward movement ; then let them fall again. This manœuvre is repeated twelve or sixteen times in the minute.

4. *Bain's method* is a modification of Sylvester's and Pacini's. The child laying on its back, place your fingers in the axillæ in their front aspect, with your thumbs over the outer ends of the clavicles, and draw the shoulders towards you. On relaxing your hold, the shoulders and chest return to their former position.

In appreciating the relative value of these methods, I can appeal to my own experience of each. Under each, asphyxiated children have recovered. I prefer Sylvester's and Bain's, because they are more easily practised on the infant. The rolling about of the head under Marshall Hall's method is awkward. In using Sylvester's, it is convenient to get the child's feet steadied so as to counteract the dragging on the arms. Dr. Bain's method has been admitted by a committee of the Medical and Chirurgical Society, on which Dr. Burdon-Sanderson and Mr. Savory served, to draw in more air than the other methods; it certainly is more easy to carry out, involving less disturbance of the body; and Dr. Bain informs me that he succeeded in restoring a child by it, after Hall's and Sylvester's plans had been tried ineffectually.

A signal advantage, common to these four methods, is that they are independent of all apparatus, and can be applied under all circumstances without loss of time. A drawback, also common to all, but applying in least degree to Bain's method, is the protracted disturbance of the patient. I have seen reason to conclude, in some cases, that the extensive movements imparted did more harm than good ; that, in short, they helped to extinguish the latent scintillula which required the most gentle treatment to fan into life.

I, therefore, to a great extent, abandoned these methods, and where means of exciting reflex action failed, I thought it

best to wrap the infant in a warm flannel and get it nursed
before the fire. In this way, by inaction—deliberate, if not
masterly—the flickering spark will sometimes gather strength,
and respiration be established. But lookers-on are rarely
competent to appreciate inaction. Inaction may be the best
practice, but it bears the outward likeness of neglect. Many a
man has earned high credit for strenuous exertions which were
really injurious. Of course, this should not lead us to do what

FIG. 55.

DR. RICHARDSON'S DOUBLE-BELLOWS FOR ARTIFICIAL RESPIRATION.

A takes air from without, and, on compression, propels it along the tube into the lungs.
B draws back from the lungs, and discharges externally.

we ought not to do. But where inaction fails, we should
anxiously seek for means of acting beneficially. Dr. B. W.
Richardson has lately returned to the old idea of insufflation, and
has devised an apparatus at once scientific and practical for
applying it. It is represented in Fig. 55. It consists of a
double bellows of india-rubber. The nozzle is introduced into a
nostril. Both bellows are grasped in the hand, and compressed
twenty times in the minute. Since one of the balls acts for

insufflation and the other for expiration, it is scarcely necessary to compress the chest after insufflation, as is done in insufflation by an ordinary tracheal tube. This supplies the quantity of air that is required. To make sure that the air goes into the lungs you may close the other nostril. If it be feared that the force of insufflation may injure the air-vesicles, leave the nostril open. This will act as a safety-valve, the air returning by it from the lungs, avoiding undue tension. But I think the instrument is so constructed that the force of the air-current is not likely to do injury. Dr. Richardson insists that warm dry air be used ; moist air frustrates attempts at artificial respiration. Let the child, then, be held in a blanket before the fire during the proceeding.

There would be no difficulty in using this instrument in cases of head-last labour, where the head is delayed, and the child is in danger from compression of the cord. The nozzle can be slipped into a nostril ; and, if the chest be tolerably free from pressure, respiration can be thus kept up until the head is born.

There is a class of cases of *asphyxia in which the face is livid, indicating congestion of the brain*. This is especially apt to occur when the cord is round the child's neck, or where the head being born, the neck is gripped by the vulva. Under these conditions, the face is seen to rapidly become purple and bloated from compression of the veins in the neck. In such cases, it is generally advised, I think rightly, to let a teaspoonful or so of blood escape from the cut cord before tying it. It is one of the most remarkable and gratifying illustrations of the phenomena of respiration and circulation, to observe how quickly the bloated cyanosed aspect of the face vanishes, giving place to a healthy hue, when a good respiration has taken place.

LECTURE XII.

TURNING (*continued*): THE MANAGEMENT OF CERTAIN DIFFICULT BREECH-PRESENTATIONS — THE MANAGEMENT OF TWIN-LABOUR—LOCKED TWINS—DORSAL DISPLACEMENT OF THE ARM—DOUBLE MONSTERS—SINGLE MONSTERS.

BEFORE proceeding to the discussion of podalic turning, strictly so called, it will be both convenient and useful to deal with certain cases of difficult breech-presentation. It will be remembered that I defined "*turning as including all those proceedings by which the position of the child is changed, in order to produce one more favourable to delivery.*"

Now, the cases of breech-presentation to which I refer cannot be brought to a satisfactory conclusion unless the position of the child, or at least of some of its parts, be changed. They, therefore, fall within our definition. But since the breech or podalic extremity of the child is already presenting, a great part of the end contemplated in podalic turning is already accomplished. The problem, so far, then, is simpler than that of effecting complete version, and may therefore logically precede the latter in the order of discussion. The simplicity is, indeed, more apparent than real, more theoretical than practical. The task of delivering a breech case such as I shall presently describe, vies in difficulty with that which has to be encountered in the most severe forms of shoulder-presentation.

In a considerable proportion of breech cases, the labour is premature. In these generally there is no difficulty. Indeed, I have commonly observed in these premature breech-labours a remarkably active, even stormy character in the uterine contractions, driving the child through with unexpected rapidity.

But when the child is mature and well developed, a breech-labour is by no means easy.

Although breech-presentations fall within the accepted definition of natural, inasmuch as the long axis of the child's body coincides with the long axis of the uterus and with the axis of the pelvic brim, they are less favourable to mother and child than head-presentations. The risk to the child is serious if there is any delay in the transit of the trunk and head through the pelvis. The chief cause of danger is compression of the funis, or interruption of the placental circulation in some other way. Danger may arise before as well as during the transit of the child through the pelvis. The breech plugging the os uteri imperfectly lets all or nearly all the liquor amnii drain off. Thus the child, placenta, and cord are liable to be deprived of the usual protection from pressure during the contractions of the uterus. This is dangerous in several ways: 1st, the cord may be compressed between the child's body and the uterine wall; 2nd, the placenta may be squeezed against the child's head, so that its vascular system is obstructed; 3rd, the more persistent and closer contraction of the uterus may so contract the uterine arteries and sinuses as to involve want of supply and stagnation of blood in the placenta.

When the trunk and head are traversing the pelvic canal, the danger of direct compression of the cord is increased. Perhaps the moment of greatest peril is when the head is arrested at the os uteri. The os uteri is probably imperfectly dilated, so that the head does not pass readily. The os uteri then encircles the neck like a ring, compressing the cord. Or the cord may be jammed against the edge of the brim. If the cord be found or be placed on one side of the sacral promontory, it may lie in the recess comparatively protected.

If delivery be hurried by traction, the cord may be retained by friction so that it is put on the stretch pulling upon the umbilicus. To obviate this, it is a good plan in every case, as soon as the way is cleared by the birth of the breech and legs, to pass the hand up along the child's abdominal surface to the umbilicus, and gently to draw down a sufficient loop of funis to allow of the progress of the trunk without pulling upon the umbilicus. We may at the same time take the opportunity to direct the funis into a sacro-iliac recess.

The probability of this dragging upon the umbilicus, and the not less serious one of the running up of the arms by the sides of the head, warns against precipitate delivery by traction. The reasons for traction are : 1st, impending danger to the child from interruption of placental circulation; 2nd, arrest of the labour from inertia ; 3rd, arrest of labour from impaction. The two last conditions involve danger to the mother as well.

How, then, do we know—1st, when the child is in danger ? If the placental circulation be materially interrupted, an attempt is excited to substitute aërial respiration. The child

Fig. 56.

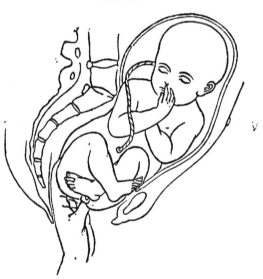

makes an effort to expand the chest. In correspondence with this reflex act the legs often twitch convulsively ; the sphincters relax, and meconium is often voided, and recognized by its green slimy character; the pulsation in the cord is felt to flag and cease. These three signs together indicate extreme peril ; but the escape of meconium, I know from frequent observation, is not in itself of much importance. The child's vitality may be experimentally tested by tickling its feet, thus producing reflex retraction of the legs ; by feeling the beating of the heart by direct application of your hand to the child's chest, which can generally be reached; and, especially, by tracing the cord up to its insertion in the navel, and feeling if there is any pulsation at this point. If the heart is still beating its impulse is generally felt here, even long after it has ceased to be transmitted to the cord.

There are two principal conditions of breech-presentation under which labour may become arrested or difficult. Whether

FIG. 57.

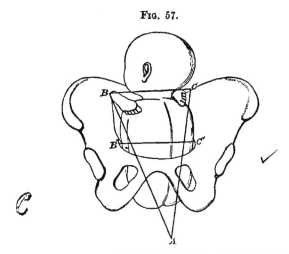

THE POSITION OF THE FŒTUS WITH THE LEGS EXTENDED, AS SEEN FROM A FRONT VIEW.

The breech has descended into the pelvis. The fœtus forms a wedge, of which the apex A is turning forwards under the pubic arch. The base B C formed by the head and legs is wider than B' C', the transverse diameter of the pelvis.

the position of the fœtus be dorso-anterior or abdomino-anterior, the legs may be disposed in one of two ways. First, and it is the most common case, the legs may be placed upon the thighs so that the heels are near the nates, and, what is very important to recollect, therefore not far from the os uteri. (Fig. 56.) Secondly, the legs may be extended so that the toes are pointed close to the face. (Fig. 57.)

In this second case several causes concur in obstructing delivery. The breech is not nearly so well adapted as the head to traverse the pelvis. Instead of taking a movement analogous to the extension of the head forwards under the pubic arch, the breech tends to bend backwards in the hollow of the sacrum.

The spine, tending to curve in a sigmoid form (Fig. 58), is not so well fitted to transmit the expulsive force applied to the head by the fundus uteri. Then there is the wedge formed by the legs doubled up on the abdomen, which does not easily allow of more than the apex, represented by the breech, descending into, or traversing, the pelvis.

Now the apex of this wedge, represented by the breech and

Fig. 58.

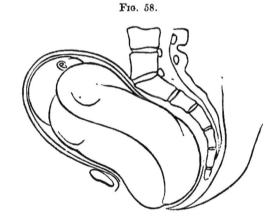

REPRESENTS A SIDE VIEW OF A BREECH-PRESENTATION, IN WHICH THE BREECH HAS ENTERED THE PELVIS.

It shows the sigmoid form imparted to the trunk in its effort to traverse the pelvis.

the thighs bent on the abdomen, can enter the pelvis very well. But then comes the widest part or base of the wedge, formed by the chest, shoulders, arms, head, and legs. This often exceeds the capacity of the brim in mere bulk. But, in addition, there is an impediment to rotation of the child on its long axis, which rotation is necessary to easy descent.

There is yet another obstacle. It arises out of the condition of the uterus. The cervix opens just in proportion to the dimensions of the body which traverses it. The breech, being of less bulk than the head and other parts constituting the base of the wedge, does not open the cervix widely enough to allow this base to descend. The uterus is apt to contract firmly upon the parts still retained in its cavity; and, the cervix encircling the wedge about its middle, a state of spastic rigidity ensues, which tends to lock up the head and chest and to impede descent and rotation. In Fig. 57 I have endeavoured to depict some of the conditions described.

Sometimes the cause of arrest is simple inertia: a little *vis à fronte* to compensate for defective *vis à tergo* may be all that is necessary. It is in the hope of extricating the child by this means that traction in various forms is resorted to. If this is unsuccessful, the case is rather worse than it would have been if left alone. The apex is dragged down a little more, the mother's pelvis is more tightly filled, and the uterus has become more irritable. I have on this account arrived at the conclusion that it is better not to resort to direct traction upon the breech in any case where there is arrest. The proper course is, I believe, to bring down a foot in the first instance. Then, traction, if still indicated, can be exerted by aid of the leg with safety and with increased power, and under the most favourable conditions for the descent and rotation of the child.

I have seen fruitless and injurious attempts made to extract by fingers, hooks, and forceps. I believe that all the best authors—that is, of those who have encountered and have had to overcome this difficulty, for it is little considered in our text-books—condemn the use of hooks and forceps. Chiari, Braun, and Spaeth,* Ramsbotham, H. F. Naegele, are decided in their reprobation; Hohl † says the forceps is neither necessary nor effectual. The breech is already in the pelvis. To apply the blades safely, the hand must be passed into the vagina, and having done this, it may as well do the right thing at once—that is, bring down a foot. Special forceps made to seize the breech are also superfluous.

I have always succeeded in delivering these cases by the simple use of the unarmed hand; and since the cases in which I did so succeed were the most difficult that can be encountered, it follows that the unarmed hand is sufficient to overcome the cases of minor difficulty of the same kind. To determine us to reject hooks and forceps, it should be enough to remember that the child is probably alive, and that, under proper skill, it may be born alive. Now hooks and forceps will, in all likelihood, either destroy the child or involve its death through the delay arising out of their inefficiency, or they may seriously

* " Klinik der Geburtshülfe," 1855.
† " Lehrbuch der Geburtshülfe," 1862.

injure the child.   The blunt hook may fracture the femur,
contuse the femoral vessels, or at least inflict severe bruises on
the soft parts.   The forceps may injuriously press upon the
abdominal viscera.

The difficulty is seldom manifest until the breech has entered
the pelvis, and this is the great cause of the obstacles opposing
operative measures.   To traverse the pelvis, the child's body
must take a sinuous course, represented in Fig. 58.

The clear indication is to break up or decompose the obstructing wedge.   (Fig. 59.)   This is done by bringing down one

FIG. 59.

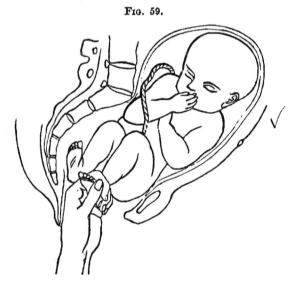

REPRESENTS A BREECH-PRESENTATION WITH THE LEGS FLEXED UPON THE THIGHS, AND THE
MODE OF SEIZING A FOOT.

foot and leg.   For this purpose, pass your hand through the os
uteri in front of the breech where the feet lie; seize one by
the ankle with two fingers; draw it down, and generally the
breech will soon descend.   It is better to leave the other leg
on the abdomen as long as possible, as it preserves greater
rotundity of the breech, and helps to protect the cord from
pressure.   It will escape readily enough when the breech comes
through the outlet.

The first thing to do is to determine the position of the
breech in its relation to the pelvis, in order that you may know

where to direct your hand to the feet. The breech simulates the face more than any other part, and so it is from the face that the breech has chiefly to be distinguished. There are four principal diagnostic points in the breech : the sacrum and anus behind, the genitals in front, an ischiatic protuberance on either side. The sacrum is distinguished by its uneven spinous processes from anything felt in a face presentation; and this is the most trustworthy characteristic, for the malar bones may pass for the ischia, and the mouth for the anus. In all cases of doubtful diagnosis it is well to pass the fingers, or hand, if necessary, well into the pelvis, so as to reach the higher presenting parts. In a breech case you will thus reach the trochanters, and above them the groins, where a finger will pass between the child's body and the thigh flexed upon it. Then in front will be the fissure between the thighs themselves ; and here, if the legs are flexed upon the thighs, will be the feet to remove all doubt. These are what you are in search of. But you only want one. It is much more easy to bring down one foot than both ; and it is, moreover, more scientific. The question now comes, Which foot to bring down ? I believe the one nearest to the pubic arch is the proper one to take. To seize it, pass the index finger over the instep; then grasp the ankle with the thumb, and draw down backwards to clear the symphysis pubis. When the leg is extended outside the vulva, it will be found that traction upon it will cause the half-breech to descend, and the child's sacrum to rotate forwards. The further progress of the case falls within the ordinary laws of breech-labour.

The second case—that in which the feet lie at the fundus of the uterus close to the face—is far more difficult. The wedge formed by the extended legs and the upper part of the trunk must, in some instances at least, be decomposed before delivery can be effected. The cause of the difficulty will be understood on looking at the diagram, Fig. 57 ; and on reflecting that the breech or wedge may in great part be driven low down into the pelvis, leaving but little space for the operator's hand to pass; further, that the hand must pass to the very fundus of the uterus to reach a foot. No ordinary case of turning involves passing the arm so far.

The mode of proceeding is as follows :—Place the patient on

her left side. (*See* Figs. 60, 61.) Produce anæsthesia to the surgical degree; support the fundus of the uterus with your right hand on the abdomen; pass your left hand into the uterus, insinuating it gently past the breech at the brim, the palm being directed towards the child's abdomen, until you reach a foot—the anterior foot is still the best to take—a finger is then hooked over the instep, and drawn down so as to flex the leg upon the

FIG. 60.

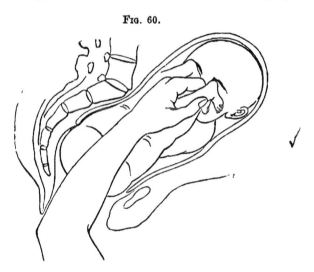

thigh. Maintaining your hold upon the foot, you then draw it down out of the uterus, and thus break up the wedge. The main obstacle is thus removed, and you have the leg to exert traction upon, if more assistance is necessary. One caution is necessary in performing this operation. It is this: the finger *must* be applied to the instep. It is of no use to attempt to bend the leg by acting upon the thigh or knee. You must therefore carry your finger nearly to the fundus of the uterus. This, and the filling up of the brim, and even of a part of the pelvic cavity sometimes, by the breech, render the operation one of considerable difficulty, demanding great steadiness and gentleness. I have brought a live child into the world by this proceeding on several occasions, where forceps, hooks, and various other means had been tried in vain for many hours. The *reason* of the operation, you will see, is analogous to that which indicates turning in arm-presentation. The further management of podalic or feet-first labours will be described under " Turning."

Fielding Ould (1742) seems to have clearly understood these cases. When the feet, he says, are near the outlet, seize them ; and at the same time that they are drawn forwards, the buttocks must be proportionately thrust into the womb by the fingers of the left hand ; for want of this precaution the thigh-bone of many an infant has been broken. " Both legs and

FIG. 61.

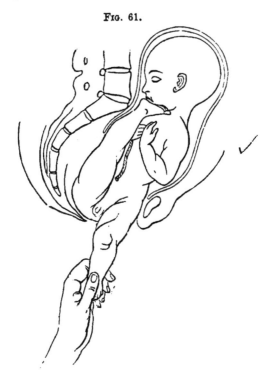

thighs may be extended along the child's body, so as to have a foot over each shoulder, which much increases the difficulty. In this case each leg must be taken separately, and the knee bent." I must observe, that it is mostly superfluous and injurious to take each leg ; one is enough, and better. It is quite excusable, before proceeding to so difficult an encounter, to try some other method. The child may be small and the pelvis large, and so a moderate degree of tractile force may be enough

to bring the wedge through without decomposing it. Various
manœuvres have been adopted. You may hook one finger in
a groin and draw down; or, what I have found better, you
may with the forefinger *hook down each groin alternately.* (*See*
Fig. 62.) In this way the breech will sometimes move. Or

FIG. 62.

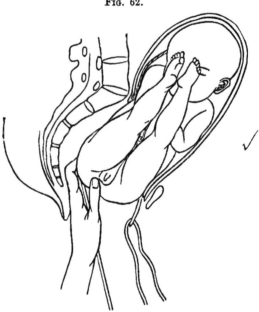

you may pass a piece of tape or other soft cord over the
groins, as Giffard did in a case quoted by Perfect. The left
buttock presented. Giffard, not being able with the forefinger of each hand, placed on each side of the thigh near the
groin, to draw out the feet, succeeded by putting a soft string
over the end of his finger; and getting that up on one side
over the thigh and a finger on the other side, he drew the
string out, and fixing it close up to the hips, he took hold of
the ends that hung out, and thus extracted, being aided by
the pains. An apparatus, having a curved flexible spring, might
be used to carry the string over the hips, or the object might be
accomplished by a catheter first carried across, and then, having
tied the string to the end it could be drawn through—the proceeding resembling that adopted to plug the posterior nares for
epistaxis. (*See* Fig. 62.)

Dr. Ramsbotham recommends the slipping a silk handkerchief over the groins. But it is possible that these and like measures may fail, and that you have nothing left but to break up the wedge by separating its component parts ; and this, I repeat, is the proper thing to do in the first instance.

Whenever traction is performed, and especially when rotation of the child on its axis is made by the operator, there is great risk of the arms hitching on the edge of the pelvic brim, and thus running up by the sides of the head. The way of remedying this difficulty will be explained in a succeeding lecture. (*See* Figs. 72, 73.) When the breech and trunk are delivered, the arms and head may follow with the aid of slight guiding force ; and it is important in the interest of the child that the slightest possible traction-force be used. If, however, we feel no pulsation in the cord, and there be convulsive twitching of the legs, with spasmodic heaving of the chest, there is no time to lose. The delivery must be accelerated. This may be done in one of two ways :—1st, we may hook two fingers of one hand over the shoulders, whilst the other hand holding the legs, we exert traction in the direction of the axis of the pelvis ; but if this manœuvre do not succeed readily, it is better, 2nd, to apply the forceps in the manner described in Lecture V., on the use of the forceps to the after-coming head. The traction exerted by the forceps obviates the danger of pulling upon the neck.

Twins.—Amongst the most puzzling and difficult cases requiring operative interference are *certain cases of twins.* Commonly, when twins are found, the embryos are lodged each in its own bag of amnion and chorion, apart from each other, and so packed in the uterus that when labour occurs, one presents at a time in the brim, and traverses the pelvis before the bag of liquor amnii of the other is ruptured. Under such circumstances, labour is apt to be tedious, because the uterus, being over-distended, acts at a great disadvantage.

There is another cause for this lingering, imperfect action, which I have not seen noticed : it is, that uterine force is lost because it can only be transmitted to the first child through the fluid contained in the membranes of the second. But in this case there is no mechanical interference of the children with each other. All that may be necessary is to supply a little *vis à*

N

fronte by means of the forceps, to make up for the defective or wasted *vis à tergo*. The question arises, Shall we expedite the delivery of the second child, or leave it for the natural powers to expel? When expectation has been the course adopted, hours, even days, have elapsed before the second child was born. The conditions of the case commonly declare that want of power is the cause of delay. And want of power is a cogent reason for giving help in labour. It is not wise to leave an inert uterus and an exhausted system to struggle alone. Dr. David Davis says he frequently saw flooding ensue in the practice of those who waited after the birth of the first child. The judicious course, then, is to allow a moderate time, say half an hour, for the system to rally from the first labour, and then to help the second labour. Immediately after the expulsion of the first child, apply the binder firmly to support the uterus. If the membranes bulge in the os uteri, rupture them ; whilst, at the same time, you increase the pressure on the fundus uteri so as to compensate for the escape of liquor amnii. If effective uterine action arise, let the uterus do its work, aiding only by external support. But, if deficient, apply the forceps if the head present. Be careful to follow down the child by external pressure, squeezing the child out as it were. It is especially necessary in twin-labour to help the uterus to the utmost. That the risk of hæmorrhage is far greater after twin-labour than after single labour, is well known. Not only is the uterus weakened by excessive distension, but there is a greatly-increased area whence blood may flow. I have ascertained that the superficies of an ordinary single placenta is about sixty square inches. If the uterine walls are relaxed, an equivalent surface may pour out blood. In twin-pregnancy, this bleeding area may equal one hundred square inches or more. Thus there is a vastly larger bleeding surface, and less contractile energy to close it.

When the embryos are both lodged in the same sac, that is, in one common chorion and amnion, an awkward complication may arise. Before or during labour, the limbs and heads may become so entangled or locked as to form virtually one body, which is too large to pass through the pelvis. The embryos may perform the most remarkable evolutions. Cases have been observed where one embryo has dived through a loop in the

other's umbilical cord; and knots have been formed involving
the two cords.

In the more ordinary and favourable course of twin-labour,
the first child presents by the head, and is entirely expelled

FIG. 63.

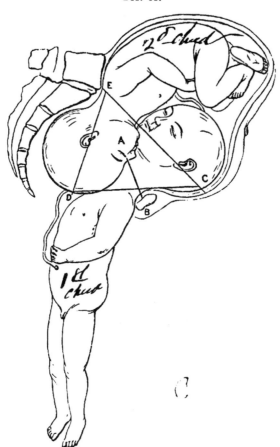

SHOWS HEAD-LOCKING, FIRST CHILD COMING FEET FIRST; IMPACTION OF HEADS FROM WEDGING
IN BRIM.

D apex of wedge, E C base of wedge which cannot enter brim, A B line of decapitation to
decompose wedge, and enable trunk of first child and head of second to pass.

before any part of the second becomes engaged in the pelvis.
Indeed, the membranes of the second do not burst until after

the first is wholly born. This second child may present either by the feet, or breech, or by the head.

The most common form of locking occurs through the hitching of one head under the chin of the other; and this may happen whether both children present head first, or one by the breech, the other by the head. The latter case appears to be the more frequent. A child presents by the feet or breech; and when born as far as the trunk or arms, it is found that the labour does not proceed, and on making traction to accelerate the delivery, unexpected resistance is encountered. You pull, but the child sticks fast in the pelvis. The first surmise is, probably, that the head is too large, from hydro-cephalus, or that the arms have run up by the sides of the head, wedging it in the brim. You liberate the arms, and pull again, and still the head refuses to move. And now you must explore fully. You may get information in two ways. First, produce anæsthesia, and pass your left hand into the cavity of the pelvis, so as to reach above the child's breast, feel-ing for its chin or mouth. Instead of feeling this first, you may be surprised at meeting a hard, rounded mass (see Fig. 63), jammed in the neck and chest of the presenting child, which can hardly be anything else than the head of another child, which has got in the way of the first. Secondly, by external palpation, you may succeed in making out through the abdominal walls the head of a child above the symphysis pubis, inclined to one or other side, in a position which its relation to the trunk partly born, and to the head you have felt whilst exploring the interior of the pelvis, will satisfy you is the head of the first child.

If the children are small, they may, with more or less difficulty, come through the pelvis together in this fashion.

Sometimes it has been possible to seize the second head by the forceps, and to extract it and the embryo to which it be-longs without disturbing the first child. But if the children be at all large, this proceeding is not likely to save them. The pressure to which both must be subjected is too hazardous. Even with children of the average size, the head of the second child, resting on the neck and chest of the first, form a wedge too large to clear the brim. You get the state of things represented in Fig. 63. D is the apex of the wedge driven into the pelvis; E C is the base, too large to enter; A B is the point at which the wedge may be decomposed, by detruncating the first child at the neck.

The apex of the wedge formed by the trunk of one child has traversed the pelvis; the base, formed by the head of one child pressed against the neck of the other, is too large to enter the pelvis; and if traction is exerted on the apex, the only effect is to jam the head tighter against the neck, hooking the two heads more firmly together. The problem is, how to extricate one head from the other, so as to allow one child to

FIG. 64.

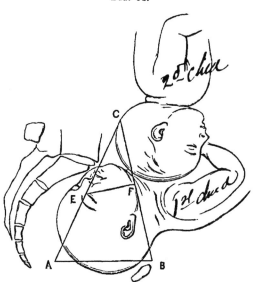

SHOWING HEAD-LOCKING, BOTH PRESENTING HEAD FIRST.

pass at a time. There are several methods of accomplishing this. But, before deciding upon one, it is well to study how the children are affected by the complication. Is one child in greater jeopardy than the other? If so, which? If we find that the situation involves extreme peril or death to one child, we shall of course not hesitate to mutilate this one, if, by so doing, we can promote the safety of the other.

The first thing you will try to accomplish is, to disentangle the heads without mutilating either child. It is still possible that both may be born alive. The patient being rendered insensible, you press back the trunk into the pelvis as much as possible, so as to lift the heads off the brim, and so to weaken the lock. Then by external manipulation, aided by a hand in the pelvis, you try to push the heads apart in opposite directions. If you succeed in unlocking them, support the head of the second child out of the way, whilst you or an assistant draw down the body of the first child and engage its head in the pelvis. If you can manage this, the difficulty is over.

Now, experience and reflection concur in showing that the first child whose trunk is partly born encounters by far the greatest danger. Its umbilical cord is likely to be compressed; its neck and chest are forcibly squeezed. On the other hand, the umbilical cord of the second child is comparatively safe, and the pressure upon its neck is less severe. You may, moreover, find, by feeling the cord of the first child, that it is pulseless and flaccid, and that tickling its feet excites no reflex action. Having thus determined that there is no hope for the first child, you turn to the best means of rescuing the second. You may decompose the wedge formed by the two heads by detaching the head of the first child. This is done by drawing the body of the child well backwards, so as to bring its neck within reach. Being held in this position by an assistant, you pass the fingers of the left hand into the pelvis, so as to hook them over the neck, and serve as a guide to Ramsbotham's or Braun's decapitator, or the wire-écraseur. If these are not at hand, the task can be accomplished by strong scissors, or even by a penknife.* (*See* Fig. 63; A B represents the line of decapitation.)

As soon as the neck is severed, the trunk will be extracted easily enough. The head of the first child will then slip up or on one side, or you may make it do so by passing your hand inside the uterus. If the head of the second child do not descend by the spontaneous action of the uterus, you may either seize it by the long double-curved forceps, or seize a leg and turn. The first head will follow last of all. If it offer any difficulty, it may be dealt with as described in Lecture XVI.

If there be reason to conclude that the second child is dead, it would be justifiable to perforate its head, and lessen its bulk

---

* *See* a case by H. Raynes, "Obst. Trans.," 1863.

by help of the crotchet. This is another mode of breaking up the base of the wedge. The head will then flatten in, and permit the trunk and head of the second child to be delivered.

In the other case, where the head of the first child presents, and gets locked by the head of the second, as in Fig. 64, a similar rule of action will apply. You may disentangle the heads by external and internal manipulation. Failing this you may seize the foremost head by the forceps; and whilst an assistant pushes away the second head, you can extract the first child. A good case, in which this plan succeeded, is related by Dr. Graham Weir.* Hohl recommends to apply forceps to the second head.

Dr. Elliot (Obstetric Clinic) relates an interesting case of twins with contracted pelvis, in which imminent locking of the heads was prevented by placing his hand on the abdominal wall over that head which was superior and to the left, and forcing it into the pelvis in advance of the other.

Another rule should be observed. When the first child is born, do not pull upon the cord, or you may do mischief in two ways. It is possible that the cord may be entangled round the neck or limb of the child still *in utero*, or the cord of the child *in utero* may be entangled in it, so that by pulling on the first cord you may strangle the second child, or arrest the circulation in its cord; or you may detach the placenta prematurely, thus giving rise to hæmorrhage, and, in the probable event of the placenta being united or common, again imperilling the child *in utero*. It would be better if the cord become tight, so as to drag on the umbilicus to divide it, without attempting to tie the cut ends.

Dorsal Displacement of the Arm.

There is a curious cause of dystocia resulting from the locking of an arm behind the neck, to which special attention has been drawn by Sir James Simpson† and Caseaux. As the difficulty is to be met by altering the position of the displaced limb, it comes under our definition of Turning.

* " Edin. Journ. of Med.," 1860. † " Obstetric Works," vol. i.

Fig. 65 represents this position of the arm. This displacement more commonly occurs in cases of podalic or breech-labour, as after turning; and I am very much disposed to think it is then most frequently produced by unskilful manipulation. I have already insisted upon the importance of avoiding the error of rotating the child upon its axis during extraction. If you commit this fault, this is what is likely to happen: the trunk revolving under your manipulation, the arm is caught against the wall of the uterus, and does not move round with the trunk, but comes to be applied to the nape of the neck. Dugès and Caseaux explain that this may happen in two different ways. First, the arm may cross behind the nucha after having been raised above the head; the crossing then takes place from above downwards and from before backwards relatively to the fœtus. Secondly, it may take place from below upwards, the arm rising on the back of the fœtus, and stopping below the occiput. This last method requires a little explanation. The arms are habitually placed by the sides of the chest. In rotating, the attempt is made to carry the abdominal aspect of the child towards the loins of the mother; the trunk only moves; the arm, therefore, remains behind; the operator, in performing extraction, draws the trunk down; the arm is caught by the symphysis pubis, where it is detained until the nucha comes down to clench it. It is obvious, therefore, that this dorsal displacement of the arm is manufactured by too great diligence on the part of the obstetrician. Those who anticipate Nature, thwarting her operations, must be prepared for the penalty. I am entitled to say, that by observing the rules I have laid down for version, this accident will not occur.

But to extract light from our errors is true wisdom. By reflecting on the mode in which the displacement is produced, we shall see how to remedy it. You must retrace your steps. By rotating the child back in the contrary direction, so as to restore the original position, you may possibly liberate the arm. At any rate, you will render the further proceeding that may be necessary more easy. You carry the trunk well backwards, so as to give room to pass your forefinger in between the symphysis pubis and the child's shoulder; then hooking on the elbow, draw this downwards, and then forwards.

It may be useful, as a preliminary step, to gain room by first liberating the other arm.

If the arm cannot be liberated, you may be driven to perform craniotomy.

In Sir J. Simpson's case, the head presented. It is not easy to imagine how the arm of a living child can get behind the neck when the head presents. Simpson throws out the suggestion that this occurs more frequently than is suspected; and that

Fig. 65.

SHOWING NUCHAL HITCHING OF THE ARM.

it accounts for many cases of arrest of the head where there is no disproportion, and which resist even traction by the forceps. Simpson recommended to bring down the hand and arm forward over the side of the head, converting the case into one of simple presentation of the head and arm. Or recourse might be had to turning, as was done successfully in a case related by J. Jardine Murray.*

* "Medical Times and Gazette," 1861.

The Delivery of Monsters.

We may here consider the mode of delivery of monsters. Double monsters may give rise to the same kinds of difficulty that sometimes occur in the case of twins. In many cases, Nature is able to deal with them. They are most frequently born dead, or possessed of little vitality—a circumstance not usually much regretted. The death is often the result of the mode of birth, one part of the monster pressing injuriously upon another. If, in any present case of obstructed delivery, it could be with certainty diagnosed that the cause of obstruction was a monstrous embryo, the indication would be clear to do our best to spare the mother, even at the cost of mutilating the embryo. But this cannot always be known in time to influence our choice of proceeding. As in the case of locked twins, we are therefore led to postpone mutilation, in the hope that the delivery may be accomplished without. Dr. W. Playfair has discussed this subject in an excellent memoir.[*] He divides monsters according to their obstetric relations into four classes.

A. Two nearly separate bodies are united in front to a varying extent by the thorax or abdomen.

In this case, the feet or heads may present. The most favourable presentation appears to be the feet. The trunks come down nearly parallel. The arms can be liberated without much trouble. It is when the heads come to the brim that the difficulty arises. The object then is to get one head at a time to engage in the brim. This has been done successfully, after turning, by Drs. Brie[†] and Molas,[‡] in the following manner. When the shoulders were born, the bodies were carried strongly forwards over the mother's abdomen. This manœuvre has the effect of placing the two heads on a different level, bringing the posterior head lower than the anterior one, which for the time is fixed above the symphysis. When the posterior head is in the pelvis, traction then will bring it through, and the second head will follow. If not, either the

[*] " Obstetrical Transactions," vol. viii.
[†] " Bulletin de la Faculté de Med., vol. iv.
[‡] " Mémoires de l'Acad.," vol. i.

first or second head can be detruncated by Ramsbotham's hook, or by scissors or knife.

The command thus obtained over the course of labour in podalic presentations renders it desirable to turn if the heads present.

But sometimes, when the heads present, a mutual adaptation takes place, which permits them to pass without mutilation. As in a case reported by Mr. Hanks,* one head got packed between the shoulder and head of the other body, so that both passed without great difficulty. One head is born first, either by aid of forceps or spontaneously, and the corresponding body may be expelled by a process of doubling up, or spontaneous evolution. If this does not proceed with sufficient readiness, decapitation or craniotomy of either the first or second child must be practised.

B. Two nearly separate bodies are united nearly back to back by the sacrum, or lower part of the spinal column.

In this case the mode of delivery is essentially the same as in A.

C. Dicephalous monsters, the bodies being fused together.

One head will come down first. The body follows, by doubling or spontaneous evolution. If this does not take place, decapitate the first head, and bring down the feet.

D. The bodies are separate below, but the heads are partially united.

Whether the head or feet present, if there be obstruction, perforate the head.

In the case of *dystocia from excessive size of the abdomen of the fœtus*, as happens sometimes from dropsy, from enlarged kidneys or liver perforation and evisceration may be indicated. In some of these cases the abdomen may burst, or be originally defective, so that the intestines will prolapse. Such a condition, if not produced deliberately by the surgeon, may puzzle extremely. The procident intestines may be taken to be those of the mother, and lead to the conclusion that there is rupture of the uterus. Or, on the other hand, where there has actually been rupture of the uterus or vagina, with protrusion of intestines, these may, under circumstances disturbing the surgeon's judgment, be assumed to belong to the child, and cut away.

* "Obstetrical Transactions," vol. iii.

Labour with *single monsters*, especially the *acephalous* monsters, is not only at times puzzling, but is apt to be protracted. Mr. Curtis, of Melbourne (*Australian Medical Gazette*, 1870), appears to have had unusual experience in cases of this kind. The diagnosis, of course, is apt to perplex. We are not accustomed to feel a head divested of cranial bones. He points out that labour with an acephalous fœtus is always protracted, and that there is more hæmorrhage. He attributes the protraction to the absence of the fully-developed skull. As the uterus presses upon the child, the child's body doubles upon itself, the back of the neck becoming the presenting part. He had to deliver all his cases by the forceps; and says resort to this instrument or turning is necessary. In fact, an acephalous fœtus acts like a dead fœtus in retarding labour.

A full description of the varieties of monsters, and of the difficulties into which they may lead the obstetric practitioner, would fill a considerable volume. They may simulate almost every complication. Tumours as big as the head itself may be attached to the head or to the sacrum. The limbs may be double or truncated. Such anomalies may defy diagnosis before delivery.

Hernia, inguinal or femoral, complicating pregnancy, may not seriously obstruct the progress of labour, but may prove a source of danger to the mother. A coil of intestine might get compressed against the pelvis by the descending child. A similar danger might also arise from vaginal hernia. Reduction should always be effected during pregnancy. The same observations apply to umbilical hernia, generally the result of separation of the recti muscles in a former labour. A well-adjusted abdominal belt should be worn during pregnancy and during labour to compensate for the deficiency in the normal support and action of the abdominal muscles, as well as to guard against the protrusion and injury of the intestines.

LECTURE XIII.

PODALIC BI-POLAR TURNING—THE CONDITIONS INDICATING ARTI-
FICIAL TURNING IN IMITATION OF SPONTANEOUS PODALIC
VERSION, AND ARTIFICIAL EVOLUTION IN IMITATION OF
SPONTANEOUS PODALIC EVOLUTION—THE SEVERAL ACTS IN
TURNING AND IN EXTRACTION—THE USE OF ANÆSTHESIA
IN TURNING—PREPARATIONS FOR TURNING—THE STATE OF
CERVIX UTERI NECESSARY—THE POSITION OF THE PATIENT
—THE USES OF THE TWO HANDS—THE THREE ACTS OF
BI-POLAR PODALIC VERSION.

THE conditions indicating podalic turning are :—

1. Generally, those which are not suited for head-turning, or
for the imitation of spontaneous evolution.

2. And more especially, shoulder-presentations of living
children, in which the knees or feet are nearer to the os uteri
than is the head.

3. Cases in which the shoulder has entered the brim of the
pelvis, and especially those in which the arm is prolapsed.

4. Most cases in which the cord has prolapsed with the arm
or hand, and some cases where the cord alone is prolapsed, and
cannot be returned or maintained above the presenting part of
the child.

5. Cases of shoulder-presentation in which the liquor amnii
has drained off, and in which the uterus has contracted so
much as to impede the mobility of the fœtus.

6. Certain cases in which it is desirable to expedite labour on
account of dangerous complications, present or threatening—as
hæmorrhage, accidental or from placenta prævia; convulsions.
In these cases it is indifferent what the presentation may be.

But the forceps would be preferred if promising equal expedition.

7. Some cases of inertia, the head presenting, as in pendulous belly and uterus, where the head cannot well be grasped by the forceps.

8. Certain cases of face-presentation. (*See* Lecture IV.)

9. Certain cases of minor contraction of the pelvis, or of the second degree (*see* Lecture V.), which are beyond the power of the forceps, and which ought not to be given over to craniotomy.

10. Certain cases of morbid contraction of the soft parts.

11. As a part of the operation for the induction of premature labour in certain cases in which the pelvis is contracted, or other circumstances do not permit the spontaneous transit of the fœtus with sufficient ease and quickness to secure a live birth.

12. Some cases after craniotomy, as the readiest mode of extracting the fœtus.

13. Certain cases of rupture of the uterus, the child being still in the uterine cavity.

14. Certain cases of monstrosity in the fœtus. (*See* Lecture XII.)

15. Certain cases of dystocia from tumours encroaching on the pelvis. (*See* Lecture XIX.)

16. Certain cases of death of the mother during labour, in the hope of rescuing the child, when Cæsarian section cannot be performed.

We will now discuss *what are the conditions necessary or favourable to turning in imitation of the spontaneous podalic version ?*

These are: 1st. The pelvis must be capacious enough to permit the passage of the fœtus without mutilation. 2nd. The vulvo-uterine canal must be dilated, or sufficiently dilatable to permit of the necessary manipulations and of the passage of the fœtus. 3rd. The presenting part must not be deeply engaged in the pelvic cavity. 4th. The uterus must not be contracted to such an extent that the fœtus has been in great part expelled from its cavity, which is hence so diminished that the presenting shoulder or head cannot be safely pushed on one side into the iliac fossa. If the *shoulder is free* above the brim, the hand not descended, it will be easy to push it across to the nearest iliac fossa. If the *shoulder is moveable*, even if the hand has fallen into the vagina, the operation is practicable, often not even difficult.

If, on the other hand, the *shoulder has been driven low down* into the pelvis, near the perinæum, the body being firmly compressed into a ball by the spasmodic contraction of the uterus, the child is almost certainly dead, and turning may be difficult or impossible without extreme danger to the mother. This is the indication for imitation of the natural spontaneous evolution.

Let us, first of all, take the more simple order of cases where turning is resorted to on account of symptoms indicating danger to the mother, as hæmorrhage from placenta prævia, the head presenting, and the cervix uteri sufficiently dilated. I take such a case first because it requires *complete turning*, and therefore best illustrates the mechanism of the bi-polar method.

It is important, at the outset, to bear in mind that turning —that is, the changing the position of the child in order to produce one more favourable to delivery—is one thing, and that extraction, or forced delivery, is another thing. Sometimes turning alone is enough, Nature then taking up the case and completing delivery. Sometimes extraction, or artificial delivery, must follow, and complement turning.

It will, however, give a more complete exposition of the subject to describe the two operations of turning and extraction continuously, assuming a case in which both are necessary.

Each operation again admits of useful division into stages or acts. *The several acts in turning are these :—*

1st Act.—The removal of the presenting part of the child from the os uteri, and the simultaneous placement there of the knees.

2nd Act.—The seizure of a knee.

3rd Act.—The completion of version by the simultaneous drawing down of the knee, and the elevation of the head and trunk.

These three acts complete turning.

The several acts in extraction are :—

1st Act.—The drawing the legs and trunk through the

pelvis and vulva. An incidental part of this act is the care of the umbilical cord.

2nd Act.—The liberation of the arms.

3rd Act.—The extraction of the head.

Before commencing the operation, there are certain preparatory measures useful or necessary to adopt.

The question of *inducing anæsthesia* arises. It would partake too much of the nature of a digression to discuss at length the indications for chloroform or ether as an aid in turning. I will do no more than glance at the principal points.

Anæsthetics are resorted to in the hope of accomplishing two objects :—The first is to save the patient pain; the second is to render the operation easier to the operator. The attainment of both objects is sometimes possible; sometimes not. It is not difficult to render the patient insensible; but you will not always at the same time make the operation more easy. It will commonly be necessary to push anæsthesia to the surgical extent. If you stop short of this degree, the introduction of the hand will often set up reflex action, and you will be met by spasmodic contraction of the vaginal and uterine muscles, and perhaps by hysterical restlessness of the patient. You will have lost the aid of the patient's self-control. You must then carry the anæsthetic further, to subdue all voluntary and involuntary movements, and to lessen the reflex irritability of the uterus. Then, but not always, you will secure passiveness, moral and physical, on the part of the patient; the uterine muscles will relax; they will no longer resent the intrusion of the hand. These advantages are not, indeed, always obtained without drawbacks. A perfectly flaccid uterus indicates considerable general prostration, and predisposes to flooding.

Here we want agents that will produce the opposite effect to oxytocics. It is probable that nitrate of amyle or physostigma will be found more valuable, in such a case as that under discussion, than chloroform, ether, or chloral, which are apt to cause vomiting, and to excite instead of allaying muscular irritability.

To facilitate the passage of the hand is the first great object. The hand is opposed, first, by the muscles of the vulva and vagina, the levator ani and the sphincter. These contract spasmodically when the attempt is made to introduce the hand. Secondly, the cervix uteri may resist in like manner. The difficulty is overcome by anæsthetics, and by free lubrication. Place a large lump of lard in the vagina before attempting to turn.

The *state of the cervix uteri* has to be considered. It is one of the natural consequences of a shoulder-presentation that the cervix uteri is but rarely found dilated sufficiently for turning and delivery until after—perhaps long after—the indication for turning has been clearly present. A shoulder will not dilate the cervix properly. The same may be said of many cases where turning is indicated by danger to the mother, as from convulsions or hæmorrhage. To wait for a well-dilated cervix might be to wait till the child or mother is dead. It follows, therefore, that we must be prepared to undertake the operation at a stage when the cervix uteri is only imperfectly dilated.

What is the degree of dilatation necessary? If the question be simply one of turning, it is enough to have a cervix dilated so as to admit the passage of one or two fingers only.

But since the ulterior object contemplated is delivery—with the birth of a live child, if possible—we must have a cervix dilated or dilatable enough to allow the trunk and head of the foetus to pass without excessive delay. The modes of dilating the cervix artificially have been described in Lecture VII. It is sufficient here to call to mind the two principal modes— viz., by the hydrostatic dilators and by the hand. The water-bag properly adjusted inside the cervix, if labour has begun at term, will commonly produce an adequate opening within an hour. Sometimes the fingers alone will succeed as quickly. In a case where the head presenting could not bear upon the cervix to dilate it because of slight conjugate contraction, I expanded it by the fingers sufficiently to admit the narrow blades of Beatty's forceps within a few minutes. The instrument, however, was not powerful enough to bring the head through. I therefore turned and made the breech and trunk complete the dilatation. The head required considerable traction to bring it through the narrow conjugate; but the child was saved. At the beginning the os barely admitted one finger; yet the patient was delivered within an hour. But we cannot always proceed so quickly. Nor is it commonly possible to effect by artificial means that complete dilatation which is required to permit the head to pass freely.

The average obstetric hand will easily traverse a cervix that is too small to allow the head to pass; so that after all, even in head-last labours, as in head-first labours, the head must generally open up the passages for itself. What we have to do is to take care that the parts shall be so far prepared when the head comes to be engaged in the cervix that the further necessary dilatation may take place quickly, for this is the stage of danger to the child from compression of the funis between the os uteri and the child's neck. The management of this stage will be described further on. It is enough now to point out that a cervix uteri expanded so widely as to admit of three fingers manœuvring without inconvenience is *enough for turning* under the circumstances of the case we have assumed.

The general rule of *emptying the bladder and rectum* applies even more strongly to turning than to forceps or craniotomy operations.

What shall be *the position of the patient?* I have generally performed the operation of turning, under whatever circumstances, the patient lying on her left side. It is of importance, I think, not to raise alarm in the patient or her attendants by adopting any great departure from the usual rules of the lying-in chamber. To place the patient on her back involves very considerable, even formidable preparations. The patient must be brought with her nates to rest on the very edge of the bed; she must be supported at her head; and two assistants must hold the legs. Still, there are cases in which this position may be preferable or unavoidable. There is another position, also in some cases useful—the knee-elbow position. But this generally precludes the use of chloroform. We may obtain all the necessary facilities by keeping the patient on her left side. The nates must be brought near the edge of the bed; the pillows are removed so as to allow the head and shoulders to fall to the same level as the nates. The head is directed towards the middle of the bed, so that the operator's arm may not be twisted during manipulation; the knees are drawn up; and the right leg is held up by an assistant, so as not to obstruct or fatigue the operator's right hand, which has to pass between the thighs to work on the surface of the abdomen.

The *presence or absence of liquor amnii in the uterus* is a matter of accident. If water be still present, so much the better; but you must be prepared to act all the same if it be not there. It is needless to state that the child will revolve more easily if it be floating in water; but it must not only be made to revolve; you have to seize a limb. At some time or other, therefore, the membranes must be ruptured. What is the best time to do this? If you are proceeding to turn in the old way —that is, by passing the whole hand into the uterus before seizing a foot—it is an advantage to follow the plan recommended by Peu, of slipping the hand up between the uterine wall and the membranes until you feel the feet, and then to break through and seize them. During this operation the arm, plugging the os uteri, retains the liquor amnii, and on drawing down the legs, the body revolves with perfect facility.

But if you are proceeding to turn by the bi-polar method, with a cervix perhaps imperfectly opened, the membranes must be pierced at the os. In this case you may perhaps accomplish the first act in version—that is, of removing the head or shoulder from the brim, and of bringing the knees over the os, whilst the membranes are intact. This you can try first, only rupturing the membranes when you are ready to seize the knee. But sometimes an excess of liquor amnii imparts too great mobility to the child; you are unable to fix it sufficiently to keep the pelvic extremity steady upon the os; it will bound away as in *ballottement*, the moment you touch it through the os. In such a case it is better to tap the membranes first, and *allow a part of the liquor amnii to run off*. While doing this you should keep your fingers on the presenting part to ascertain how its position and mobility are being influenced by the escape of the waters and the contraction of the uterus, so as to seize the right moment for proceeding.

Now, if you assent to what I have stated, you will find that you are committed to *the use of your left hand as the more active agent* in the operation. You want the right hand to work outside on the abdomen; therefore, the left hand must be introduced into the vagina. It is a case where ambi-dexterity is eminently required. The left hand in most people is smaller than the right. The patient lying on her left side, the left hand follows the curve of the sacrum far more naturally than the right. It meets the right hand outside, the two working consentaneously with comfort, involving no awkward or fatiguing twisting of the body. Moreover, in the great majority of cases the anterior surface of the fœtus, and consequently its legs, are directed towards the right sacro-iliac joint—that is, inclining backwards and to the right, so that the left hand passed up in the hollow of the sacrum will reach the legs with the utmost convenience.

I strenuously advise every young man who is preparing for Obstetric Surgery to put his left hand into training, so as to cultivate its powers to the utmost. There are a thousand ways of doing this, and I hope it will not be considered idle to mention some. In all athletic exercises or games requiring manual skill, use the left arm as well as the right. It is an excellent practice to dissect with the left hand. Shave the right side of the face with the razor in the left hand; use your toothbrush with the left hand; and, if you now and then come to grief through left-handedness, think how much less is this evil than injuring a woman or breaking down in an operation.

Well, all things being ready, we will proceed to the operation. In the case we have assumed, the head is over the os uteri; the os uteri is open enough to admit the play of two or three fingers; the liquor amnii is still present, or has been only recently and partially discharged. The preparations necessary have been made. One thing more I have to insist upon: it is to avoid all parade or fuss in your conduct. Make your preparations as quietly and unostentatiously as possible. Do all that is essential, and no more. Tell the patient and her attendants that you find it necessary to help the labour. But let your help be so given as to involve the least possible changes from the usual proceedings in ordinary labour.

When the patient is in position, and in anæsthesia, slip off your coat, turn up the shirt-sleeves above the elbows, anoint with oil, lard, or pomade the back of the left hand and all round the wrist; insinuate a piece of lard into the vulva.

The Introduction of the Hand.—Bring the fingers together in the form of a cone; pass in the apex of this cone, gently pressing backwards upon the perinæum, and pointing to the hollow of the sacrum. If you find any difficulty—as you probably will if the case be a first labour—you must watch for the most fitting

opportunity. Wait till a pain comes on. There is good reason
for deferring to the popular idea of "taking a pain." The pain
caused by expulsive action will partly mask that caused by the
manœuvre ; and expulsive action tends to produce sphincteric
relaxation, so that the passage of the hand will be actually
facilitated. A source of difficulty is the tendency of the labia

Fig. 66.

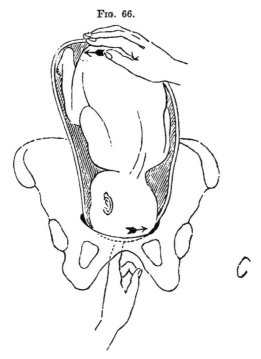

REPRESENTS THE FIRST STAGE OF BI-POLAR PODALIC VERSION.

The right hand on the fundus uteri pushes the breech to the right and backwards, bending
the trunk on itself. The left-hand fingers on the vertex push the head to the left ilium, away
from the brim. The arrows show the direction of the movements given to the two poles of the
child.

and hair to turn inwards before the fingers. This is counteracted
by drawing the labia open by the thumb of the right hand, by an
action similar to that you would use to lift up the closed upper
eyelid. The passage of the vulva is often the most difficult part
of the operation. It is commonly necessary to pass the entire
hand into the vagina : and great gentleness and patience are
required. I have, indeed, turned and extracted a mature child
without passing in more than two fingers, without even turning

back or soiling the cuff of my coat; but the circumstances must
be favourable to admit of this.

We have now got as far as the orifice of the uterus, and it is
an immense improvement in obstetric art that we are able to
complete the operation without passing the whole hand through
this part. The *first act* begins by passing the tips of the first
two fingers through the os to the presenting part, which we

FIG. 67.

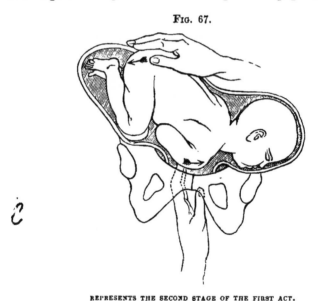

REPRESENTS THE SECOND STAGE OF THE FIRST ACT.

The right hand, still at the fundus uteri, depresses the breech, so as to bring the knees over
the brim, whilst the left hand pushes the shoulder across the brim towards the left iliac fossa.

assume to be the head. We ascertain to which side of the pelvis
the occiput is directed, for it is to that side that we must send
the head. At the same time, an assistant holding up the woman's
right leg at the knee, so as to give you freedom of action, you
apply your right hand spread out over the fundus uteri where
the breech is. And now begins the simultaneous action upon
the two poles of the foetal ovoid; the fingers of the hand inside
pressing the head-globe across the pelvic brim towards the left
ilium, the hand outside pressing the breech across to the right
side and downwards towards the right ilium. The movements
by which this is effected are a combination of continuous
pressure and gentle impulses or taps with the finger-tips on the

head; and a series of half sliding, half pushing impulses with the palm of the hand outside. Commonly, you may feel the firm breech through the abdominal walls under the palm, and this supplies a point to press against. A minute sometimes, seldom much more, will be enough to turn the child over to an oblique or nearly transverse position; the head quitting the os uteri, and the shoulder or chest taking its place.

Now, it is important to keep the breech well pressed down, so as to have it steady whilst you attempt to seize a knee (*see* Fig. 67). This is the time to puncture the membranes, if

FIG. 68.

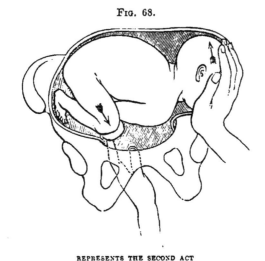

C

REPRESENTS THE SECOND ACT

The trunk being well flexed upon itself, the knees are brought over the brim; the forefinger of the left hand hooks the ham of the further knee, and draws it down, at the same time that the right hand, shifted from the fundus and breech, is applied, palm to the head-globe, in the ilium, and pushes it upwards. The arrows show the opposite directions imparted to the two poles by the two hands.

not already broken. The fingers in the os uteri are pressed through the membranes during the tension caused by a pain, and you enter upon the

Second Act, the seizure of a knee. Which knee will you take? In the particular case we have to deal with, it is not of much importance which you seize, but the further one is, on the whole, to be preferred. You will observe in the diagrams, Figs. 67, 68, that the legs, doubled up on the abdomen, bring the knees near the chest, so that, as soon as the head and

shoulder are pushed on one side, the knees come near the os
uteri.

The knee being seized, the further progress of the case is
under your command. By simply drawing down upon the part
seized, you may often complete version. But it will greatly
facilitate the operation to continue to apply force to the two
poles. You will observe in Fig. 68 that the hands have
changed places in relation to the two poles of the fœtal ovoid.
Although the left hand has never shifted from its post in the

FIG. 69.

REPRESENTS THE THIRD ACT IN PROGRESS.

The right hand continues to push up the head out of the iliac fossa; the left hand has seized
the further leg, and draws it down in the axis of the brim. Version is now nearly complete.
The arrows show the opposite directions given to the two poles of the child by the hands.

vagina, the ovoid has shifted; and the forefinger, drawing
down the left knee, virtually acts upon the pelvic end of the
ovoid. The right hand, therefore, is at liberty to quit this end;
it is transposed to the head-end of the ovoid, which has been
carried over to the left iliac fossa. The palm is applied under
the head, and pushes it upwards in response to, and in aid of,
the downward traction exerted on the child's leg. This outside

manœuvre singularly facilitates the completion of version. It may be usefully brought into play in almost every operation of podalic turning. If it is neglected, as I shall show on a special occasion, you will sometimes fail in effecting complete version ; for the head will not always quit the iliac fossa by simply pulling upon the legs.

The third Act. Continuing to draw upon the leg, as soon as

FIG. 70.

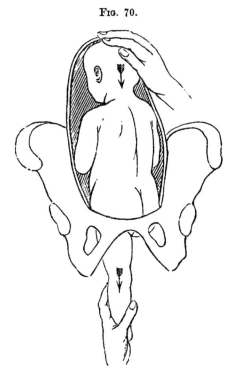

REPRESENTS THE COMPLETION OF THE THIRD ACT.

The right hand still supports the head, now brought round to the fundus uteri. The left hand draws down on the left leg in the direction of the pelvic axis. Version is complete. Rotation of the child on its long axis has taken place, the back coming forward as the breech enters the pelvis. The long axis of the child is now coincident with the axis of the pelvis. If *extraction* is necessary, the forces of both hands, traction and pushing, are exerted in the same line as shown by the arrows.

the breech nears the brim a movement of rotation of the child on its long axis begins, the design of which is to bring the back to the front of the mother's pelvis. This rotation depends upon a natural law of adaptation of the two parts. You are not to trouble yourselves in "giving the turns," as some authors imagine they can. I cordially agree with Wigand when he says, "Nature knows better than we do how to impart the proper turns."

What you have to do is simply this—*to supply onward movement.* If the uterus be doing its own work, propelling the child breech first, we know we may rely upon Nature so to dispose the child in relation to the pelvis as to enable it to pass with the greatest facility. So it is when we supply the moving force from below. If this force is wanted, supply it; but do not attempt to do more. Avoid that fatal folly of encumbering Nature with superfluous help. Keep the body gently moving in the direction of the pelvic axis by drawing upon the leg, and Nature will do the rest. You will feel the leg rotate in your grasp, and the back will gradually come forward.

I have said that, upon the whole, the further knee is the better one to seize; but if you compare Figs. 68 and 69, you will see that, by drawing the nearer or anterior knee, you would directly secure the rotation of the child's back forwards; so that, as I have before said, it is not worth while to lose time in trying to seize the further knee if you find the anterior one the more easy to seize.

This completes version. The breech is substituted for the head. Nature may effect expulsion; but, if she fail, we have it in our power to effect delivery by extraction. We have assumed that extraction is necessary, and will proceed to this operation.

LECTURE XIV.

THE OPERATION OF EXTRACTION AFTER PODALIC VERSION, OR
OTHER BREECH-FIRST LABOURS—THE THREE ACTS IN EX-
TRACTION—THE BIRTH OF THE TRUNK, INCLUDING THE
CARE OF THE UMBILICAL CORD—THE LIBERATION OF THE
ARMS—THE EXTRACTION OF THE HEAD.

THE operation of turning being completed by engaging the pelvic extremity of the child in the brim, we have next to consider the question of delivery. This, as I have already pointed out, is a distinct operation. Nature unaided may accomplish it. It is only in her default that we are called upon to undertake it. It is very desirable that as much of this operation be trusted to Nature as possible. Our duty is to watch the progress of the labour closely, interposing aid when that progress is too slow, or when the interest of the child demands it. As a general rule, the natural forces will carry the child through with more safety than the forces of art. But, even in the most favourable breech-first labours, whether the breech or feet have originally presented, or have been brought to present by art, care on the part of the practitioner is necessary to avert certain dangers incurred by the child in its transit; and in some cases serious difficulty to the transit arises to demand the exercise of active skill.

The description I now propose to give of the operation of podalic extraction will embrace, and apply to, all the cases in which this operation is called for. We will begin with the most simple case—that in which there is no serious complication, in the shape of pelvic contraction, excessive size of the child, or resistance by the soft parts. It is either a case of

inertia, or one in which prompt delivery or the acceleration
of labour is indicated in the interest of the mother or child.

We possess, in our hold upon a leg, a security for the further
progress of delivery, of which we can avail ourselves at pleasure.
In this security consists one of the main arguments in favour
of podalic version. We have divided the operation of extraction
into three acts: drawing down the trunk through the vulva;
liberation of the arms; extraction of the head.

Fig. 71.

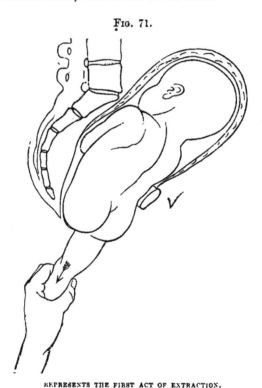

REPRESENTS THE FIRST ACT OF EXTRACTION.

The *first act* is effected by simply drawing down upon the
extended leg in the axis of the brim. Two rules have to be
observed. The first is to draw down simply, avoiding all
attempts to rotate the child upon its long axis. You must not
only not make such attempts: you must even be careful not to
oppose the natural efforts at rotation. This is secured by
holding the limb so loosely in the hand that the limb may
either rotate within your grasp under the rotation imparted to

it by the rotation of the trunk, or that the limb in its rotation will carry your hand round with it. The other rule is to draw well in the direction of the axis of the brim, and especially to avoid all premature attempts to direct the extracting force forwards in the axis of the outlet.

When the breech has come to the outlet, the extracting force is directed a little forwards, so as to enable the hip which is nearest the sacrum to clear the perinæum. This stage should not be hurried. The gradual passage of the breech has been doing good service in securing free dilatation of the vagina and vulva, an essential preparation for the easy passage of the shoulders and head. When the hips have cleared the outlet, you may pass the forefinger of your left hand into the groin, and gently aid extraction by this additional hold; and, at the same time, by pressing the knee forwards across the child's abdomen, you may facilitate the liberation of the leg.

When both legs and breech are outside the vulva, you have acquired a considerable increase of extracting power. You must, however, use it with discretion. You may now draw upon both legs, holding them at the ankles between the fingers and the thumb of one hand.

And if you still want more power, you can grasp the child's body just above the hips with the other hand. It is generally desirable to interpose a thin soft napkin between and round the ankles. It gives a better hold, and lessens the risk of contusion.

Traction may now be greatly aided by pressure upon the fundus uteri, pushing the child down.

Traction must now again be directed in the axis of the brim, in order to bring the shoulders through that aperture. The shoulders will enter in the same oblique diameter, back forwards, as that in which the breech traversed.

As soon as the belly comes to the vulva, your attention will be turned to the umbilical cord. This is apt to be put upon the stretch, by slipping up under the influence of friction as the body is drawn down; and, besides being stretched, it is liable to direct compression. The way to lessen these risks is to seize the cord near the umbilicus and draw down very gently a good loop; this loop should be laid where it is least exposed to pressure, that is, generally, on one side of the promontory of the sacrum; and you must further take care to keep off the pressure of the vulvar sphincter upon it by guarding it with your fingers. From time to time feel the cord, to ascertain if it continues to pulsate. If you find the pulsations getting feeble or intermittent, you have an indication to accelerate extraction.

The observations of May and Wigand upon this point are worthy of attention. Reasoning that the pressure suffered by the cord affects the vein more than the arteries, and hence that the access of blood to the foetus is hindered, whilst the removal of the blood from the foetus is little obstructed, so that a fatal anæmia results, they advise to tie the cord, as soon as the body is born, as far as the navel, and then to complete extraction. The apparent asphyxia so produced is easily remedied by the usual means. Von Ritgen says he has often done this, and affirms that when done there is little need to hurry extraction.

The *second act* comprises *the liberation of the arms.* In the normal position of the foetus the arms are folded upon the breast, and if the trunk and shoulders are expelled through a normal pelvis by the natural efforts, they will commonly be born in this position. But if ever so little traction-force be put upon the trunk, the arms, being freely moveable, encountering friction against the parturient canal as the body descends, are detained, and run up by the sides of the head. Hence often arises a serious delay in the descent of the head, for this, the most bulky and least compressible part of the foetus, increased by the thickness of the arms, forms a wedge which is very apt to stick in the brim. This is one great reason for not putting on extraction-force if it can be avoided. If, however, we find the arms in this unfortunate position, we must be prepared to liberate them promptly, and, at the same time, without injury. It is very easy to dislocate or fracture the arms or clavicles if the proper rules are not observed. What are these rules?

The cases vary in difficulty, and therefore in the means to be adopted. In some cases the arms do not run up in full stretch along the sides of the head. The humeri are directed a little downwards, so that the elbows are within reach. In such cases it is an easy matter to slip a forefinger on the inner side of the humerus, to run it down to the bend of the elbow, and to draw the forearm downwards across the chest and abdomen, and then to bring the arm down by the side of the trunk. But many cases require far more skill.

The cardinal rule to follow is to observe the natural flexions of the limbs, always to bend them in the direction of their natural movements. The arms, therefore, must always be brought forwards across the breast. The way to do it is as follows:—Slip one or two fingers up along the back of the child's thorax, and bend the first joints over the shoulder between the acromion and the neck; then slide the fingers

Fig. 72.

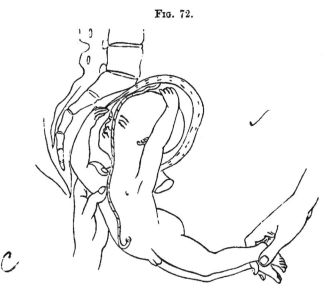

REPRESENTS THE MODE OF LIBERATING THE POSTERIOR OR SACRAL ARM.

forwards, catching the humerus in their course, and carrying this with them across the breast or face. This movement will restore the humerus to its natural flexion in front of the body. Of course, as the humerus comes forwards the forearm follows. Your fingers continuing to glide down will reach the bend of the elbow, and, still continuing the same downward and forward movement across the child's breast and abdomen, the arm is extended and laid by the side of the trunk.

That is what has to be done. But is it indifferent *which arm you shall bring down first?* The most simple rule is to take that first which is the easiest, for when one is released the room gained renders the liberation of the second arm easy enough. Generally there is most room in the sacrum; therefore it is best to take the posterior arm first.

Now I have to describe manœuvres for overcoming the difficulties which not seldom oppose your efforts to release the arms. There are two principal ones. The first is this: You want to bring the posterior or sacral arm within reach of your finger. Carry the child's body well forwards, bending it over the symphysis pubis. (Fig. 72.) The effect of this is a twofold advantage. Space is gained between the child's body and the sacrum for manipulation; and as the child's body

Fig. 73.

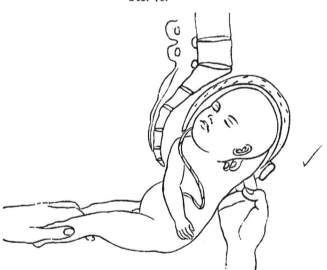

REPRESENTS THE MODE OF LIBERATING THE ANTERIOR OR PUBIC ARM.

revolves round the pubic centre, the further or sacral arm is necessarily drawn lower down, commonly within reach. When the sacral arm is freed, you reverse the manœuvre, and carry the child's trunk backwards over the coccyx as a centre. This brings down the pubic arm. (Fig. 73.)

The second manœuvre may be held in reserve should the first fail. To execute it you must bear in mind the natural flexions of the arms. You grasp the child's trunk in the two hands above the hips, and give the body a movement of rotation on its long axis, so as to bring its back a little to the left. The effect of this is to throw the pubic arm, which is prevented by friction against the canal from following the trunk in its rotation, across the breast. (Fig. 74.) Then, your object being accomplished so far, you call to your aid the first manœuvre,

FIG. 74.

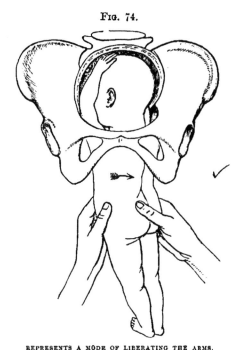

REPRESENTS A MODE OF LIBERATING THE ARMS.

The trunk is rotated an eighth of a circle from right to left, so as to throw the left arm across the face.

and bring this arm completely down. This done, you reverse the action and rotate the trunk in the opposite direction. The sacral arm is thus brought to the front of the chest, and, by carrying the trunk back, your fingers will easily complete the process.

It is desirable, for reasons we shall presently explain, to avoid this rotation if possible; but under certain circumstances of difficulty it is exceedingly valuable. The rotation need not be considerable; an eighth of a circle is commonly enough, and as it is neutralized by reversal, an objection that might otherwise be urged against the manœuvre is removed.

P

A paramount reason why you should be careful in imparting rotation to the trunk, or "giving the turns," is this: the union of the atlas with the occipital condyles is a very close articulation; it permits flexion and extension only. The atlas forms with the axis a rotatory joint, so constructed that if the

FIG. 75.

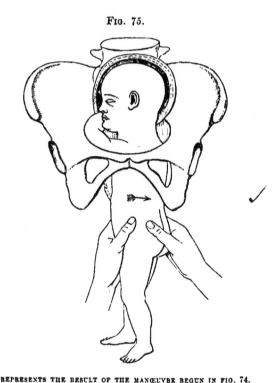

REPRESENTS THE RESULT OF THE MANŒUVRE BEGUN IN FIG. 74.

By rotating the trunk from right to left, the left arm is thrown across the face.

movement of rotation of the head be carried beyond a quarter of a circle, the articulating surfaces part immediately, and the spinal cord is compressed or torn. Thus, if the chin of the fœtus pass the shoulder in turning backwards, instant death results. I have no doubt that many children have been lost through oblivion of this fact.

Sometimes the arm will hitch on the edge of the pelvic brim, or just above the imperfectly expanded os uteri. Never attempt, by direct hooking on the middle of the humerus, to drag it through. You would almost certainly break it. Press it steadily against the child's face, and under its chin, running your finger down as near the elbow as possible, so as to lift this part, as it were, over the obstruction.

The arms liberated, now begins the *third act, the extraction of the head*, often a task of considerable difficulty, and always demanding the strictest observance of the laws which govern the mechanism of labour. This act differs from the two first in that, whilst these are sometimes effected by Nature, the liberation of the head must almost always be conducted by art. When the head is last, and has entered the brim, it is very much removed from the influence of expulsive action. The uterus can with difficulty follow it into the pelvis, and the trunk, unless supported by the hands, would, by its mere *vis inertiæ* and friction against the bed, retard the advance of the head. Moreover, this is the stage of chief danger from compression of the cord. The round head fills the brim and the cervix uteri, so that the cord can hardly escape. It would be folly, therefore, to sit by and trust to Nature in this predicament, at the risk of losing that for which the whole operation of version and extraction has been performed — namely, the child's life. Let us suppose for a moment that the head is in the pelvis, and that you cannot extract it at once. If you can get air into the chest, which, being outside the vulva, is free to expand, there is no need to hurry the extrication of the head. You may sometimes get the tip of a finger in the mouth, and drawing this down, whilst you lift up and hold back the perinæum, you may enable air to enter the chest. In this way I have kept a child breathing for ten minutes before the head was born. Another plan is to pass a catheter or other tube up into the mouth, so as to give, by means of a kind of artificial trachea, communication with the external air; or better still, Richardson's bellows. But I must warn you not to trust to these or similar plans, lest the golden opportunity be irretrievably lost. The real problem is to get the head out of the pelvis.

There are two principal modes of doing this. One is to apply the forceps. This operation I have described. (*See* Lecture V.) It has been advocated by Busch, Meigs, Rigby and others. I have practised it successfully, but think it is

inferior in celerity and convenience to the second mode, by
manual extraction. Remember that the head has to perform

FIG. 76.

REPRESENTS THE EXTRACTION OF THE HEAD.

The dotted line is the curve of Carus, which indicates the direction to be observed in
extraction.

a double rotation in its progress. It must revolve round the
symphysis pubis as a centre; it must rotate in the cavity on

its vertico-spinal axis, so as to bring the face into the hollow of the sacrum. You must then, in extracting, respect these natural movements. You will better follow or guide these movements if you fork the fingers of one hand over the neck behind, and at the same time, holding the legs with the other hand, draw down with careful attention to the curve of Carus. If you carry the body forward too soon, you simply convert the child's head and neck into a hook or crossbar, which, holding on the anterior pelvic wall, will effectually resist all efforts at extraction.

When there is little or no resistance to the escape of the head, it is enough to support the trunk with one hand by holding it at the chest, whilst the other hand on the nucha regulates the exit of the head.

Sometimes it requires considerable force to bring the head through the brim; but whilst force will never compensate for want of skill, it is astonishing how far skill will carry a very moderate force, especially if firm pressure upon the fundus uteri be made to aid traction. To practise this with the greatest advantage, the patient should be on her back. The modes of extricating the head under circumstances of unusual difficulty will be discussed hereafter. But before passing on I must refer to one practice commonly taught, which is, I believe, based on erroneous observation. You are told to pass a finger into the mouth, or to apply two fingers on the upper jaw, to depress the chin, in order to keep the long axis of the child's head in correspondence with the axis of the pelvis. Now this is a piece of truly " meddlesome midwifery," because it is perfectly unnecessary. The chin is not likely to be caught on the edge of the pelvis or elsewhere, unless, by a previous piece of " meddlesome midwifery," you have been busy in "giving the turns." The truth is, Nature has taken care to arrange the convenient adaptation of means to end in head-last labour as in head-first. It is true that the occipito-spinal joint is seated behind the centre. It might, *primâ facie*, appear that the occiput, forming the shorter arm of the head-lever, would tend to roll back upon the nucha. But this is not so in practice. The broad, firm expanse of the occiput, forming a natural inclined plane directed upwards, is surely caught by the walls of the parturient canal as the head descends. The greater friction thus experienced by a larger superficies favourably disposed virtually converts the shorter arm of the lever into the more powerful one; it is more retarded in its course (Fig. 75); and therefore the chin is kept down near the breast, and therefore, again, there is no need for the obstetrist to meddle in the matter.

LECTURE XV.

So long as there is any liquor amnii present in the uterus, and often for some considerable time afterwards, the bi-polar method of turning is applicable. But a period arrives when it becomes necessary to pass a hand fairly into the uterus in order to seize a limb. We will now discuss the mode of turning under the more difficult circumstances of loss of liquor amnii, more or less tonic contraction of the uterus upon the child, and descent of the shoulder into the pelvis.

The contraction of the uterus, naturally concentric or centripetal, tends to shorten the long axis of the child's body. The effect is to flex the head upon the trunk, and to bend the trunk upon itself, reducing the ovoid to a more globular form. This brings the knees nearer to the chest, but does not diminish the difficulty of turning.

I need not pause again to discuss minutely the preparatory measures. It is only necessary here to call to mind that chloroform, ether, or opium is especially serviceable, and that it is important to empty the bladder and rectum.

The first question to determine is, *Which hand will you pass into the uterus?* I have given in Lecture XIII. some of the reasons why the left hand should be preferred. In the majority of cases the child's back is directed forwards; to reach the legs, which lie on the abdomen, your hand must pass along the hollow of the sacrum, and this can hardly be done, the patient lying on her left side, with the right hand, without a most awkward and embarrassing twist of the arm. I need scarcely point out how violent and unnatural a proceeding it would be to pass up the right hand between the child's back and the mother's abdomen, to carry the hand quite round and over the child's body in order to seize the feet which lie towards the mother's spine, and then to drag them down over the child's back. If you attempted this, you would probably get into a difficulty. The child, perhaps, would not turn at all. To avoid this failure, the rule has been laid down to pass your hand along the inside or palmar aspect of the child's arm. This will guide you to the abdomen and the legs. Or the rule has been stated in this way :—Apply your hand to the child's hand, as if you were about to shake hands. If the hand presented to you be the right one, take it with your right, and *vice versâ.*

Rules even more complicated are proposed, especially by Continental authors. Some go to the extent of determining the choice of hand in every case by the position of the child. The fallacy and uselessness of these rules are sufficiently evident from the disagreement among different teachers as to which hand to choose under the same positions. Rules, moreover, which postulate an exact knowledge of the child's position are inapplicable in practice, because this diagnosis is often impossible until a hand has been passed into the uterus; and it is certainly not desirable to pass one hand in first to find out which you ought to use, at the risk of having to begin again and to pass in the other.

The better and simpler rule is this :—*In all dorso-anterior positions, lay the patient on her left side ; pass your left hand into the uterus*—it will pass most easily along the curve of the sacrum and the child's abdomen; *your right hand is passed between the mother's thighs to support the uterus externally.*

In the case of *abdomino-anterior positions, lay the patient on her back, and you may introduce your right hand using the left hand to support the uterus externally.* If the patient is supported in lithotomy position, you can thus manipulate without straining or twisting your arms or body. But it is equally easy to use the left hand internally if the patient is on her back, so that the exception is only indicated to suit those who have more skill and confidence with the right hand.

We will first take a dorso-anterior position. Introduce your left hand into the vagina, along the inside of the child's arm. The passage of the brim, filled with the child's shoulder, is often difficult. Proceed gently, stopping when the pains come on. At the same time support the uterus externally with your right hand. Sometimes you may facilitate the passage of the brim by applying the palm of the right hand in the groin, so as to get below the head and to push it up. This will lift the shoulder a little out of the brim. Or you may adopt a manœuvre attributed to Von Deutsch, but which had been practised by Levret. This consists in seizing the presenting shoulder or side of the chest by the inside hand, lifting it up and forwards, so as to make the body roll over a little on its long axis. This may be aided by pressure in the opposite direction by the outside hand on the fundus uteri, getting help from the bi-polar principle.

Sometimes advantage is to be gained by placing the patient on her elbows and knees. In this position you are favoured by gravity, for the weight of the fœtus and uterus tends to draw the impacted shoulder out of the brim.

The brim being cleared, your hand passes onwards into the cavity of the uterus. This often excites spasmodic contraction, which cramps the hand, and impedes its working. Spread the hand out flat, and let it rest until the contraction is subdued. In your progress you must pass the umbilicus, or a loop of umbilical cord may fall in your way. Take the opportunity of feeling it, to ascertain if it pulsates. You thus acquire knowledge as to the child's life. But you must not despair of delivering a live child because the cord does not pulsate. I have several times had the satisfaction of seeing a live child born where I could feel no pulsation *in utero*. You are now near the arm and hand. They are very apt to perplex. Keep, therefore, well in your mind's eye the differences between knee and elbow, hand and foot, so that you may interpret correctly the sensations transmitted by your fingers from the parts you are touching.

At the umbilicus you are close to the knees. The feet are some way off at the fundus of the uterus applied to the child's breech.

What part of the child would you seize? It is still not uncommon to teach that the feet should be grasped. You will see pictures copied from one text-book to another, representing this very unscientific proceeding. There ought to be some good reason for going past the knees to the feet, which are further off, and more difficult to get at. Now, I know of no reasons but bad ones for taking this additional trouble. You can turn the child much more easily and completely by seizing one knee. Dr. Radford insists upon seizing one foot only, for the following reasons :—The child's life is more frequently preserved where the breech presents than where the feet come down first. A half-breech is also safer than cases where both feet come down. The dilatation of the cervix is better done by the half-breech. The circumference of the breech, as in breech-presentations, is from twelve to thirteen and a half inches, nearly the same as that of the head; the circumference of the half-breech, one leg being down, is eleven to twelve and a half inches, whilst the circumference of the hips, both legs being down, is only ten to eleven and a half inches.

But a knee is even better than a foot. You determine, then, to seize *one* knee; which will you choose? The proper one is that which is furthest. The reasons are admirably expressed by Sir J. Simpson. We have a dorso-anterior position, the right arm and shoulder are downmost; these parts have to be lifted up out of the brim. How can this be best done? Clearly by pulling down the *opposite* knee, which, representing the opposite pole, cannot be moved without directly acting upon the presenting shoulder. If the opposite knee be drawn down, and supposing the child to be alive or so recently dead that the resiliency of its spine is intact, the shoulder must rise, and version will be complete, or nearly so. But if both feet are seized, or only the foot of the same side as the presenting arm, version can hardly be complete, and will, perhaps, fail altogether.

This point is worth illustrating. I have taken Fig. 77 from Scanzoni (*Lehrbuch der Geburtshülfe*, 4th edition, 1867), in order to show you the error in practice which I wish you to avoid. It represents a dorso-anterior position, the right shoulder presenting, and the method recommended by Scanzoni. The operator's left hand is seizing and drawing down the right leg. I have introduced the arrows to indicate the direction of the movements sought to be imparted. You want the shoulder to run up whilst you draw down the leg. Now, drawing on the right leg necessarily tends to bring it towards the shoulder, the line of motion of the leg being more or less perpendicular to that of the shoulder. The body bends upon its side, the leg and shoulder get jammed together, and you have failed to turn.

Contrast this figure with Figs. 78 and 79, which I have designed to show the true method and principle of turning. The arrows, as before, indicate the direction of the movements.

FIG. 77.

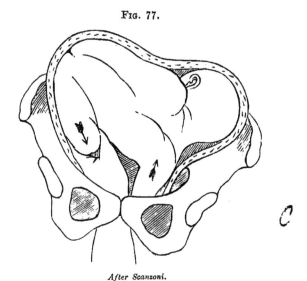

After Scanzoni.

TO SHOW THE ERROR OF ATTEMPTING TO TURN BY SEIZING THE LEG OF THE SAME SIDE AS THAT OF THE PROLAPSED SHOULDER.

By drawing upon the opposite knee to the presenting shoulder, the movements run parallel in directly opposite directions, like the two ends of a rope round a pulley. You cannot draw down the left leg without causing the whole trunk to revolve; and the right shoulder will necessarily rise. To turn effectively, the child must revolve upon its long or spinal axis, as well as upon its transverse axis. Turning, in short, is a compound or oblique movement between rolling over on the side and the somersault.

If you seize both legs, you mar this process. The only cases in which I have found it advantageous to seize both legs are those in which the child has been long dead. Here the spine

FIG. 78.

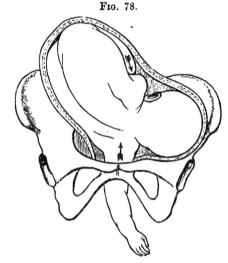

SHOWING THE PRINCIPLE OF TURNING BY BRINGING DOWN THE KNEE OF THE OPPOSITE SIDE TO
THE PRESENTING SHOULDER.

The arrows indicate the reverse movements effected. The object is to carry up the right
shoulder. By bringing down the left knee this is most surely effected.

FIG. 79.

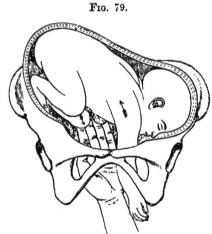

SHOWS TURNING IN PROGRESS.

As the left knee descends, the trunk revolves on its transverse as well as on its long axis,
and the right shoulder rises out of the pelvis.

has lost its elasticity. The body will hardly turn, and there is nothing to be gained for the child in maintaining the half-breech and preserving the cord from pressure.

The seizure of a foot is not seldom a matter of so much difficulty that various instruments have been contrived to attain this object. To draw the foot or feet down, you must grip them firmly— that is, your fingers must be flexed in opposition to the thumb, or two fingers must coil around the ankle. This doubling of your hand takes room. Whereas to seize a knee only requires the first joint of the forefinger to be hooked in the <u>ham</u>. Fig. 80 shows Braun's contrivance for snaring a foot. A loop of tape in the form of a running noose is carried by means of a gutta-percha rod, about a foot long, into the uterus, guided by the hand to the foot. When you have succeeded in getting the noose over the ankle, you pull on the free end, and withdraw the rod.

Fig. 80.

DRAUN'S SLING-CARRIER, TO APPLY A LOOP ROUND THE FOOT, OR TO REPLACE THE UMBILICAL CORD.

Hyernaux, of Brussels, has invented a very ingenious instrument, a *porte-lacs*, or noose-carrier, for the purpose. There are many others, but since they are created in order to meet an arbitrary—I might say a wantonly-imposed—difficulty, arising out of an erroneous practice, they need not be described. It is true that it is often convenient to attach a loop to the foot when brought into the vagina, to prevent it from receding before version is complete. But this can be done by the fingers with a little dexterity. The occasions on which it is necessary to seize a foot which can only be barely touched by the fingers are extremely rare. For these, I think the simple apparatus of Braun, which also serves for the reposition of the umbilical cord, is as efficient as any. The wire-écraseur forms an excellent snare to seize a foot. A loop just large enough for the purpose is made, and guided over the ankle. It can be slightly drawn in to fix the grasp, avoiding, of course, cutting into the limb.

LECTURE XVI.

TURNING IN THE ABDOMINO-ANTERIOR POSITION — INCOMPLETE
VERSION, THE HEAD REMAINING IN ILIAC FOSSA, CAUSES OF
— COMPRESSION OF UTERUS, TREATMENT OF — BI-MANUAL
OR BI-POLAR TURNING WHEN SHOULDER IS IMPACTED IN
BRIM OF PELVIS — IMITATION OR FACILITATION OF DELIVERY
BY PROCESS OF SPONTANEOUS EVOLUTION — EVISCERATION —
DECAPITATION — EXTRACTION OF A DETRUNCATED HEAD FROM
THE UTERUS.

TURNING in abdomino-anterior positions does not differ essentially from turning in dorso-anterior positions. I have already said that the best position for the patient is on her back, and that the right hand may be used. The uterus, as in all cases, is supported externally, whilst you pass your right hand along the inner aspect of the child's arm and behind the symphysis pubis; it proceeds across the child's belly to seize the opposite knee. Drawing this down in the direction of the arrow in Fig. 81, the presenting shoulder rises out of the pelvis.

There is a feature in the history of turning which has not received the attention it deserves. I have found that, notwithstanding diligent adherence to the rules prescribed, turning is not always complete. The head and part of the chest are apt to stick in the iliac fossa, the trunk being strongly flexed. Indeed, I believe that complete version is rather the exception than the rule in cases where the liquor amnii is drained off, and the uterus has moulded itself upon the fœtus so as to impede the gliding round of the fœtus.

The complete version which exists as the ideal in the minds

of most of those who perform the operation is not often realized. Indeed, it can hardly take place unless the bi-polar method by combined external and internal manipulation is carefully pursued. The head may commonly be felt throughout the entire process nearly fixed in the iliac fossa, and sometimes the forearm remains in the upper part of the pelvic cavity. The nates and trunk are delivered as much by bending as by version. The process is something between spontaneous

FIG. 81.

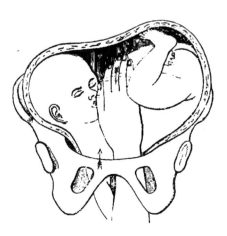

TURNING IN ABDOMINO-ANTERIOR POSITION.

The operator's right hand seizes the upper or left knee. As this comes down, the child's body, rotating upon its transverse and long axes, draws the right or presenting shoulder up out of the pelvis.

version and spontaneous evolution. The two following diagrams (Figs. 82 and 83), taken from memoranda made of a case which occurred to me, will serve to illustrate both this feature of incomplete turning and the importance of the principle of drawing upon the leg opposite to the presenting shoulder.

If the head and shoulders rise enough to permit the breech to enter the brim, delivery will not be seriously obstructed. But it not uncommonly happens in extreme cases of impaction of the shoulder in the upper part of the pelvis, that even when you have succeeded in bringing down a leg into the vagina, version will not proceed : the shoulder sticks obstinately in the brim. In such a case the bi-polar principle must be called into action. It is obvious that if you draw down upon the leg, whilst you push up the shoulder, you would act at a great advantage. But you cannot get both your hands into the pelvis. Sometimes you may release the shoulder by external manipulation, pressing up the head by the palm of your hand insinuated between it and the brim of the pelvis. In cases of real difficulty, however, this will not answer. You

Fig. 82.

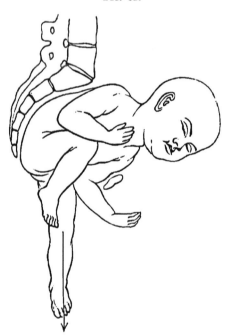

REPRESENTS AN ABDOMINO-ANTERIOR POSITION, LEFT SHOULDER PRESENTING.

Traction was first made upon the left leg, as shown by the arrow. The effect was to bend the trunk and jam the shoulder against the symphysis.

must push up the shoulder by the hand inside. To admit of this, you pass a noose of tape round the ankle in the vagina, and draw upon this. The noosing of the foot is not always easy. To effect it you carry a running noose on the tips of two or three fingers of one hand up to the foot, held down as low as possible in the vagina by the other hand. Then the loop is slipped up *beyond the ankles and heel,* and drawn tight. Often you will have to act with one hand only in the vagina, the hand outside holding on the free end of the tape ready to tighten the noose as soon as it has got hold. Or, whilst holding the foot with one hand, you may carry the noose by help of Braun's instrument. (*See* Fig. 80.) The foot being securely caught, the right hand is passed into the vagina, and the fingers or palm, if necessary, are applied to the shoulder and

Fig. 83.

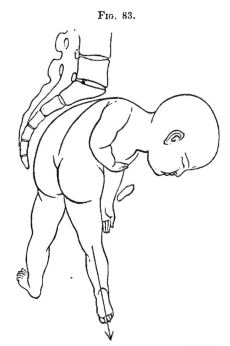

REPRESENTS THE CORRECTION OF THE ERROR COMMITTED IN FIG. 82.

By drawing on the opposite, or right, leg, the trunk was made to revolve on its spinal axis, drawing up the presenting arm from the pelvis, and allowing the breech to descend. Although delivery was now effected, version was not complete, as the head remained in the iliac fossa, and the hand never quitted the pelvic cavity.

chest. Now, you will find it difficult to draw upon the tape and to push upon the shoulder exactly simultaneously. There is so little room that, whenever you push, there is a tendency to carry the leg up as well. The most effective movement is as follows:—Pull and push alternately. Presently you will find the leg will come lower, and the prolapsed arm will rise.

In pushing the chest and shoulder, it is not unimportant
in what direction you push. You cannot push backwards,
or even directly upwards. Your object is to get the trunk
to roll over on its spinal axis. Here, then, is an indication
to carry out the manœuvre of Levret and Von Deutsch. *Push
the shoulder and adjacent part of the chest well forwards*, so as
to make them describe a circle round the promontory as a
centre. The reviewer of the first edition of this work in the

FIG. 84.

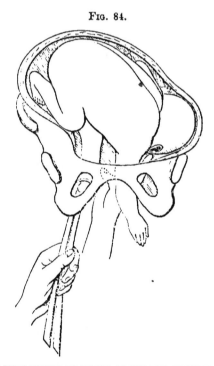

REPRESENTS THE BI-POLAR METHOD OF LIFTING AN IMPACTED SHOULDER FROM THE BRIM OF
THE PELVIS, SO AS TO EFFECT VERSION.

American Journal of Medical Sciences calls attention to the
practice of *pulling upon the opposite arm* to produce rotation
on the spinal axis. I have never tried it, but it is clearly
a resource to be borne in mind.

Various other contrivances have been designed in order to
accomplish this end. Crutches or repellers have been made,
by which to push up the shoulder instead of by the hand.

The objection to these is that you cannot always know what you are doing. But your hand is a sentient instrument, which not only works at your bidding, but constantly sends telegrams to the mind, informing it of what is going on, and of what there is to do.

In the majority of cases of this kind we are justified in attempting to turn, because there is still a prospect of the child being preserved. But there are cases in which matters have proceeded a stage further, in which the shoulder and corresponding side of the chest are driven deeply into the pelvis—in which, consequently, the body is considerably bent upon itself. Now, this can only occur after protracted uterine action, such as is scarcely compatible with the life of the child. Either the child was already dead at an early stage of labour —a condition, especially if the child were also of small size, most favourable to the carrying out of this process of spontaneous evolution—or the child has been destroyed under the long-continued centripetal compression of the uterus.

In the presence of such a case, the first question we have to consider is—Will Nature complete the task she has begun? Will the child be expelled spontaneously? A little observation will soon enable us to determine how far this desirable solution of the difficulty is probable, and when we ought to interpose. If the pelvis be roomy in proportion to the child; if the child be dead, small, and very flaccid; if we find the side of the chest making progress in descent under the influence of strong uterine action possessing an expulsive character, and if the patient's strength be good, we shall be justified in watching passively. But if we find no advance, or but very slow advance, of the side of the chest, the child being large and not very plastic; if the uterus have ceased to act expulsively, and the patient's strength be failing, her pulse rising, we must help. Then comes the second question: In what manner? This must depend upon the circumstances of the case. If a little help *à fronte* to make up for deficient *vis à tergo* promise to be enough, we may imitate the proceeding of Peu, who, in a case in which spontaneous evolution was in progress, passed a cord round the body to pull upon and aid the doubling.

Or we may much facilitate the doubling and expulsion by evisceration. This operation consists in perforating the most bulging part of the chest, and picking out the thoracic and abdominal viscera. When this is done, traction upon the body by the crotchet or the craniotomy forceps, and dragging in the direction of Carus' curve, will commonly effect delivery without difficulty. This operation is represented in Fig. 85.

Fig. 85.

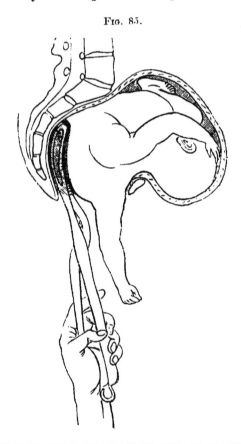

REPRESENTS EXTRACTION AFTER PERFORATION OF THE CHEST, AND DELIVERY IN IMITATION OF SPONTANEOUS EVOLUTION.

Sometimes perforation and evisceration are insufficient in themselves, and another step will be necessary in order to complete delivery. This ultimate step is *decapitation*, an operation of extreme importance, capable of bringing almost instant relief and safety to the mother. It is pointed out by Celsus, and was clearly described by Heister after Von Hoorn.

The recognition, or at least the application, of this proceeding is so inadequate, that I think it useful to state the arguments in favour of it with some fulness.

The late Professor Davis, in his great work,* a work too much neglected by his successors, says :—" It may be considered a good general rule never to turn when the death of the child is known to have taken place." In cases of long impaction he recommends an operation to be performed *upon the child* —namely, "*bisection of the child at the neck.*" Again, he says: "It ought to be an established rule in practice to decapitate in arm-presentations not admitting of the safer performance of turning."

Dr. Ramsbotham† also says :—" It appears to me better practice either to eviscerate or decapitate the fœtus, than to endeavour to deliver by turning, in all cases where the uterus is so strongly contracted round the child's body as to cause apprehension of its being lacerated by the introduction of the hand ; because if such a degree of pressure is exerted on it as to render the operation of turning very difficult, the child must have died, either from the compression on its own chest, or on the funis, or on the placenta itself."

The justness of the rule thus distinctly expressed by Davis and Ramsbotham is attested by the practice of the most eminent Continental practitioners. Decollation has been advocated and practised by l'Asdrubali (1812), by Paletta,‡ by Braun of Vienna, by Dubois, by Lazzati of Milan,§ and by many others.

When in presence of a case such as that described we may select one of two modes of delivery. Passing by the imitation of spontaneous evolution just described, we may (1) bisect or break down the spinal column at the most bulging part. This is called " Spondylotomy." (2.) We may bisect at the neck, that is, decapitate. There is a third plan which it is well to know, although it is less scientific than the two preceding. It is to cut off both arms at the shoulder. We may thus gain space to get at the feet and turn.

1. *Spondylotomy, or bisection of the trunk.*—The principle of

* " Obstetric Medicine."
† " Medical Times and Gazette," December, 1862.
‡ " Del Parto per il Braccio, Bologna, 1808.
§ " Del Parto per la Spalla," 1867.

this operation is to facilitate the complete doubling of the body. It may be likened to breaking a bent stick in the middle, thus destroying the bow or arc, and allowing the two pieces into which the stick has been resolved to come through parallel and close to each other. It is to be preferred when the head is retained above the brim, and access to the neck is rendered difficult by the bulging of the chest and trunk. The spine is to be divided at the most prominent part. This may be done by strong scissors, by the craniotomy-perforator, by a knife, or sometimes by getting a strong cord round the body, and cutting through the whole trunk by a sawing movement. Extraction when the trunk is thus broken is generally not difficult. The base of the obstructing wedge is materially reduced, and still greater facility is gained for compression. But it may still become necessary to extract the two parts of the severed body separately. In this case we should take the lower extremity first, seizing it by the craniotomy-forceps. This extracted, the other end to which the head is attached may be dealt with in like manner.

2. *Decapitation, or bisection at the neck.*—Various instruments have been designed to effect decapitation. Ramsbotham's hook is perhaps best known in this country, and it has served as a model for several modifications made abroad. It is named after the first Ramsbotham. It was described and recommended by Professor Davis. It consists of a curved hook, having a cutting edge on the concave part, supported on a strong straight stem, mounted on a wooden handle. Professor Davis also used another instrument, of his own contrivance—the guarded embryotomy-knife. It consists of two blades working on a joint like the forceps. One blade is armed with a strong knife on the inner aspect. The other blade is simply a guard; it is opposed to the knife, and receives it when the neck is severed.

A plan sometimes resorted to is to carry a strong string round the neck, and then, by a to-and-fro or sawing movement, effected by cross-bars of wood on the ends to serve as handles, to cut through the parts. Dr. C. Bell suggested the use of an instrument like that for plugging the nares in epistaxis, to carry a cord. Dr. Ritchie designed an instrument like an écraseur, with a perpetual screw and chain. Strong scissors have been made for the purpose, which cut through the vertebræ. A good instrument of this kind has been designed by Dr. Mattéi, of Paris. I believe it is a very useful form, as it is sometimes easy to cut through the spine when it is difficult to pass a hook over the child's neck. It resembles the surgical bone-forceps.

It is important to remember that the spine may be divided by piercing the vertebræ with the common perforator, then separating the blades so as to rend or crush asunder the bones. What remains may then be divided by scissors. Failing special instruments, the spine may be divided by strong scissors, or even by a penknife or a Wharncliffe blade.

The favourite instrument in Germany and Italy is Braun's blunt hook or "decollator." Dr. Garthshore performed the operation with an ordinary blunt hook. It is certainly desirable to do away with the cutting edge, which is not without danger to the mother and the operator. Braun's instrument is twelve inches long, including the thickness of the handle; the greatest width of the hooked part is one inch; the greatest thickness of the stem is from four to five lines. Lazzati introduces a gentle curve into the stem near the hook.

The Operation of Decapitation.—It will be best described as consisting of *three stages.* The first stage is the application of the decapitator and the bisection of the neck; the second is the extraction of the trunk; the third the extraction of the head.

The First Stage.—The patient may lie on her left side or on her back. Take Ramsbotham's hook or Braun's decollator. As the instrument should be passed up over the back of the child's neck, it is, in the first place, necessary to ascertain whether the position be dorso-anterior or abdomino-anterior. It is also necessary to determine accurately whether the fœtus is still in great part above the brim lying transversely or obliquely, in which case the head and neck will be in one or other side; or whether, a great part of the chest having descended into the pelvis, the movement of rotation has taken place, in which case the head and neck will be found in front near the symphysis. The observation of the relations of the prolapsed arm and exploration with the hand internally will inform us as to these particulars. The next step is to get an assistant to pull down the prolapsed arm, so as to bring down the shoulder and fix it well. This brings the neck nearer within reach. Should the

assistant in this duty be in the way, you may seize the prolapsed arm by a clove-hitch on cord or tape, upon which the
assistant may pull, keeping quite clear of the operator. The
operator then passes his left hand, or two or three fingers, if
this be enough, into the vagina, over the anterior surface of the

FIG. 86.

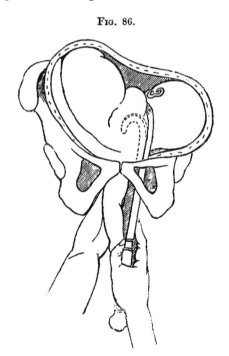

REPRESENTS THE FIRST STAGE OF DELIVERY BY DECAPITATION.

The dotted outline shows how the hook (Ramsbotham's) is introduced, lying flat upon the
back of the child's neck, the beak being then turned over the neck, and meeting the fingers of
the left hand on the anterior aspect.

child's chest, until his fingers reach the fore part of the neck.
With his right hand he then insinuates the hook, lying flat,
as in the dotted outline in Fig. 86, between the wall of the
vagina and pelvis and the child's back, until the beak has
advanced far enough to be turned over the neck. The beak
will be received, guided, and adjusted by the fingers of the left
hand. The instrument being *in situ*, whilst cutting or breaking
through the neck, it is still desirable to keep up traction on the
prolapsed arm. In using Ramsbotham's hook, a sawing move-

ment must be executed, carefully regulating your action by aid of the fingers applied to the beak. If Braun's decollator be used, the movement employed is rotatory, from right to left, and at the same time, of course, tractile. The instrument crushes or breaks through the vertebræ. When the vertebræ are cut through, some shreds of soft parts may remain. These may be divided by scissors, or be left to be torn in the *second stage* of the operation.

The Second Stage: The Extraction of the Trunk.—The wedge

FIG. 87.

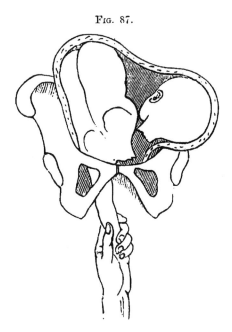

REPRESENTS THE SECOND STAGE OF DELIVERY AFTER DECAPITATION, OR EXTRACTION OF THE TRUNK BY PULLING ON THE PROLAPSED ARM.

The head, no longer linked to the body, is pushed out of the way as the trunk descends.

widening above the brim, that hitherto obstructed delivery, is now bisected, divided into two lesser masses, each of which separately can readily be brought through the pelvis. By continuing to pull upon the prolapsed arm the trunk will easily come through, the head being pushed on one side out of the way by the advancing body. (*See* Fig. 87.) In cases where there was difficulty in extracting the trunk Dr. D. Davis

used a double-guarded crotchet, the two blades of which, fixing
themselves in the trunk, extracted like a forceps.

The Third Stage: The Extraction of the Head.—The problem,
how to get away a detruncated head left behind in the uterus,
is not always easy of solution.   In the case before us, the child

Fig. 88.

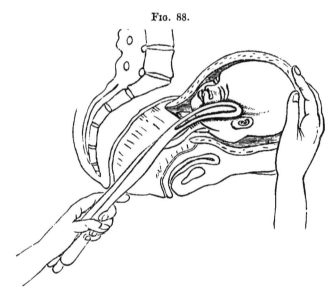

REPRESENTS THE THIRD OR LAST STAGE OF DELIVERY AFTER DECAPITATION.

The head is seized by the craniotomy-forceps, and extraction is made in the direction of
Carus' curve.

having probably been dead many hours, the bones and other
structures have lost all resiliency, the connections of the bones
are broken down by decomposition, and the whole becomes a
plastic mass, easily compressible.   Such a head will sometimes
be expelled spontaneously.   I have taken away a head, under
the circumstances under discussion, by seizing it with my
fingers.   On the other hand, I have on several occasions been
called in to extract a head which resisted ordinary means.
There are four modes of action.   The crotchet, the forceps, the
craniotomy-forceps, or the cephalotribe may be used.   If the
crotchet can be passed into an orbit, or into the cranial cavity,
getting a good hold, this plan may answer.   The objection to
it is the difficulty of getting such a hold, and the risk of the

point slipping and rending the soft parts of the mother. The head being loose, rolls over in the uterus when an attempt is made to seize it. I therefore discard the crotchet.

The *forceps* is better adapted. If the head can be seized, which is not always easy, for it is apt to escape high above the brim, and roll about when touched by the blades, extraction is not difficult. Care must, moreover, be taken to seize the head in such a manner that the spicula resulting from the severance of the vertebræ shall not drag along or injure the mother's soft parts.

I prefer the *craniotomy-forceps*, as being much the most certain and safe. In order to obtain a hold, it is generally necessary first to perforate. The free rolling of the head when pressed by the point of the perforator tends to throw this off at a tangent, missing the cranium, and endangering the soft parts of the mother. To obviate this difficulty, the head must be firmly fixed down upon the pelvic brim by an assistant, who grasps and presses the uterus and head down by both hands spread out upon them. The operator then, feeling for the occiput with two fingers of his left hand, and guided by them, carries up the perforator with his right hand, taking care that the point shall strike the head as nearly perpendicularly as possible. He then, partly by a drilling, screwing, boring motion, partly by pushing, perforates the cranium. The drilling movement avoids the necessity of using much pushing force, and thus lessens the risk of the instrument slipping.* When a sufficient opening is made into the cranium, the craniotomy-forceps is applied, one blade inside, the other outside ; the blades are

---

* It is in perforating under such circumstances that the vice of perforators curved in the blades is most apparent. Such instruments can hardly strike in a true perpendicular, and the point is almost certain to glide off. They are not fitted to economize pushing-force, by drilling or boring, as a good straight perforator is. Indeed, almost all the ordinary perforators sold in the shops are wretched instruments. They will serve in those easy cases where it may be doubted whether craniotomy is necessary or justifiable ; they will fail in cases of serious difficulty. The superiority of a straight powerful perforator like Oldham's will not be disputed by any one who has had to perforate under difficulties. The trephine-perforators are not applicable to the case under discussion, owing to the difficulty of fixing the head. It may be interesting to mention that Celsus distinctly points to the necessity of fixing the head left *in utero*, by firmly compressing the uterus with two hands, so as to push the head down upon the os. Extraction was then made with hooks.

adjusted and locked ; and traction made in the orbit of Carus' curve commonly brings the head away without further difficulty. During extraction, the fingers of the left hand should be kept upon the skull at the point of grasp by the instrument, guarding the soft parts from injury by spicula, and regulating the force and direction of traction.

If there be any likelihood of difficulty in extraction, there is a last and an effectual resource in the *cephalotribe*. When the head is left behind after turning in contracted pelvis, the cephalotribe to crush down the head, having first performed craniotomy, is invaluable.

LECTURE XVII.

TURNING IN CONTRACTED PELVIS AS A SUBSTITUTE FOR CRANIO-
TOMY—HISTORY AND APPRECIATION — ARGUMENTS FOR THE
OPERATION: THE HEAD COMES THROUGH MORE EASILY
BASE FIRST — THE HEAD IS COMPRESSED LATERALLY —
MECHANISM OF THIS PROCESS EXPLAINED — LIMITS JUSTIFY-
ING OPERATION — SIGNS OF DEATH OF CHILD — ULTIMATE
RESORT TO CRANIOTOMY IF EXTRACTION FAILS — THE INDI-
CATIONS FOR TURNING IN CONTRACTED PELVIS — THE OPE-
RATION.

WE now come to a long-contested and still undecided question
in obstetric practice—*Is turning ever justifiable as a means of
delivery in labour obstructed by pelvic deformity?*

The next alternative in the descending scale of operations is
a transition from conservative to what may be distinguished as
sacrificial midwifery, involving the destruction of the child.
It is obviously a matter of exceeding interest to cultivate any
operation that shall hold out a reasonable hope of safety to the
child, without adding unduly to the danger of the mother. So
much may be conceded on both sides. The question, then, may
be set forth as follows:—Do cases of dystocia from pelvic contraction occur in which the child can be delivered alive by
turning, which must otherwise be condemned to the perforator,
without injury or danger to the mother? And, not to blink in
any way the serious character of the inquiry, it is necessary to
append this secondary question to the first—namely:

Assuming that such cases do occur, can they be diagnosed
with sufficient accuracy to enable us to restrict the application
of turning to them? And if we err by turning in unfitting

cases, what is the penalty incurred ?—how can we retrieve our error ?

These questions I will endeavour to illustrate, if not to answer, by the light of the writings of others, and my own experience and reflections.

The choice of an operation in obstetrics will, in many cases that fall within the debatable territory claimed by two or more rival operations, be determined by the relative perfection of these operations, and by the relative skill in them possessed by the individual operator. And in estimating the arguments of different authors, we must bear this law in mind.

The operation of extracting a child through a contracted brim has no doubt often been performed as a matter of assumed necessity, as, for example, when the shoulder has presented ; and contraction of the pelvis is certainly a cause of shoulder-presentation. The observation of such cases, a certain proportion of which terminated successfully for the child, could not fail to suggest the deliberate resort to the operation in cases of similar contraction where the head presented.

Before the forceps was known, and before the instruments for lessening and extracting the head had been brought to any degree of perfection, turning was commonly resorted to in almost all cases of difficult labour. Thus Deventer, who wrote in 1715, as well as La Motte, declaimed against the use of instruments, and recommended turning by the feet in all cases of difficult cranial presentation. The consequence was that the art of turning was cultivated very successfully by some of the followers of Ambroise Paré. It appears to me evident that, in the early part of the last century, turning was better understood and more skilfully performed than it was at the beginning of the present century ; and it is equally evident to me that, by turning, many children were saved under circumstances that are now held to justify their destruction. Of course this gain was not achieved without a drawback. If children were sometimes saved, many mothers were injured or lost by attempts to turn under circumstances which are now encountered successfully by the forceps or by craniotomy.

As instruments were improved, the choice of means was extended. The forceps first contested the ground. The contest, indeed, was for exclusive dominion. The reputed inventor of the forceps, Hugh Chamberlen, did not hesitate to accept the challenge of Mauriceau to attempt to deliver a woman with extreme pelvic contraction by means of his instrument, feeble and imperfect as it was. He failed ignominiously. As science advanced, the contest was better defined. As the obstruction to delivery was due to contraction of the pelvic brim, and the problem was, how to extract a live child arrested above the brim, it is obvious that a short single-curved forceps must fail. It was only when the long double-curved forceps was designed, that the knowledge and the power arose which enabled the obstetric surgeon to bring another means into competition with turning, for the credit of saving children from mutilation.

It is, then, from the time of Smellie and Levret, who perfected and used the long forceps, that the real interest of the inquiry dates. It is not a little remarkable that amongst those who have most distinctly recognized the value of the long forceps have been found the advocates for turning in contracted pelvis. The following words, written by Smellie in 1752, challenge attention now:—"Midwifery is now so much improved that the necessity of destroying the child does not occur so often as formerly; indeed, it never should be done, *except when it is impossible to turn or to deliver with the forceps;* and this is seldom the case but when the pelvis is too narrow, or the head too large to pass, and therefore rests above the brim."

Pugh, of Chelmsford (1754), who advocated the long forceps, says:—"When the pelvis is too small or distorted, the head hydrocephalic or very much ossified, or its presentation wrong, ... provided the head lies at the upper part of the brim, or, though pressed into the pelvis, it can without violence be returned back into the uterus, the very best method is to turn the child and deliver by the feet." He then goes on to lay down the conditions which would induce him to prefer the curved forceps, and states that, as the result of these two modes, "I have never opened one head for upwards of fourteen years." Has not midwifery retrograded since his time?

Perfect (1783), who used the long forceps, delivered a rickety woman whose conjugate diameter measured three inches, the head presenting, and brought forth the first living child out of four, the first three having been extracted after perforation. La

Chapelle (1825) advised and practised the method. She relates that out of fifteen children extracted by forceps (long) on account of contracted pelvis, eight were stillborn, seven alive ; and that out of twenty-five delivered footling, sixteen were born alive, and nine dead.

It is not less remarkable that it is amongst those who reject the long forceps that the strongest opponents of turning in contracted pelvis are to be found. This is the more astonishing when we reflect that this school, rejecting the two saving operations, has nothing to propose but craniotomy for a vast number of children that claim to be brought within the merciful scope of conservative midwifery.

Denman, who used the short forceps exclusively, was, upon the whole, adverse to the operation, although he relates a striking case in illustration of its advantages. He delivered a woman of her eighth child alive at the full period, all her other children having been stillborn. "The success of such attempts," he says, "to preserve the life of a child is very precarious, and the operation of turning a child under the circumstances before stated is rather to be considered among those things of which an experienced man may sometimes avail himself in critical situations, than as submitting to the ordinary rules of practice."

Those who have studied the history of obstetric doctrine cannot fail to see that this dread of encouraging enterprise ·in practice lest disaster should result from unskilfulness, has cramped teaching, obstructed the progress of knowledge, and enforced a slavishly timid, yet barbarous practice, which still persists down to the present time. That the precepts and practice of Smellie and his immediate disciples were infinitely more scientific and successful than those which prevailed in the time of Denman, and in the first half of the present century, cannot be doubted. Thirty years ago or less craniotomy was still frightfully rife in this town and in many parts of the country. Possibly the cautious teaching of Denman and many of his successors was justified greatly by the general imperfection of medical education. They had, as we now have, to teach according to the average capacity and trustworthiness of their pupils. They taught men with the same feeling of reserve with which we should still teach midwives. But surely the day is past for all this. We may safely venture to teach men of a higher standard upon more liberal principles. I am not aware that a similar reticence or restraint has at any time, to a like extent, gagged the teachers of medicine or surgery proper. May we not see in this fact a striking testimony that the practice of obstetrics demands, even more than medicine or surgery, steadiness yet promptitude in judgment, courage under difficulties, and physical skill ?

On the other hand, the forceps has by some been held to be of superior efficacy to turning in contraction of the pelvis ; that is, whilst certain lesser degrees of contraction may be dealt with by turning, the forceps claims the preference in more advanced degrees of contraction. It is needless to say that those who advocate this preference rely upon a very powerful forceps. It is accordingly in Germany, France, and America, especially that the claim for the superiority of the forceps is contended for.

Stein (1773), Osiander the elder (1799), preferred the forceps. Boër was opposed to turning. In France, Baudelocque maintained the same doctrine as Stein and the elder Osiander ; and the recent experiments of Joulin, Chassagny, and Delore with the " appareils à traction," by which a powerful extracting force is added to the forceps, enabling it to bring a head through a greatly contracted passage, seem to strengthen the comparative claim of this instrument.

We will now discuss the question—What is the penalty incurred, or how can we retrieve our error, if we turn and fail to bring the head through the too-contracted brim ? Undoubtedly, the patient will have to go through a second operation. We are driven to perforate after all. We have tried to save the child, and have failed. Is the mother imperilled by this attempt and failure ? This also must be answered by experience. Of course, the mother may suffer if we persevere in dragging the child too long and too forcibly. But we have a right to assume that the attempt is controlled by skill and discretion. The amount of force that can be safely endured is very great—far greater than those who have never seen the operation would readily credit. The violence to which the soft structures are subjected seems to be small in proportion to the traction-force exerted. There appears to be some saving

R

or protective condition. This, I think, is found in the mechanism of the process. I refer to Lecture V. for an illustration and description of the mechanism of labour in contraction of the pelvis from projection of the promontory. This projecting promontory forms the centre of rotation around which the head must revolve in order to enter the pelvic cavity. The side of the head applied to this point scarcely moves at all. The promontory catches the fœtal skull in the fronto-temporal region. If the coarctation be decided, the skull where it is caught bends in. All the onward movement is effected by the opposite or pubic side of the skull sweeping in a circle, which I have called "the curve of the false promontory," until the equator or greatest circumference has passed the plane of the brim, when the whole head slips into the cavity with a jerk. Now, injurious pressure is avoided on the pubic side by the smoothness and flatness of the inner surface of the pelvic brim, and by a gliding movement of the soft parts intervening between the head and the bony canal. Injurious pressure is avoided over the promontory by the yielding or moulding of the head. The temporal and parietal bones will bend in, even break. Children have been born alive after this bending or breaking. Sometimes a large cephalhæmatoma forms at the point of depression. In other cases the child perishes. The observation of these cases shows that the mother will bear with safety an amount of pressure which is sufficient to kill the child.

What follows? This obvious corollary: that the mother will safely bear that lesser degree of pressure which is required to bring through a living child.

The operation, then, is justified in cases of contraction that admit of the passage of a living child. It is further justified in cases of contraction to a certain, though small, degree of contraction beyond this, which admits of the passage of a dead child. We have here, perhaps, carried the experiment to the verge of what is justifiable. Beyond this, there being no possibility of getting a child by this means, live or dead, through the pelvis, it would of course be better not to go. And if all the conditions of the problem could be precisely ascertained beforehand, we should not go beyond this. But, whilst calculating upon an average or standard head, we may encounter a head above the standard in size or hardness, and thus, in our endeavour to save the child, we may find ourselves in a difficulty. The extrication is by perforation. By lessening the head, it is brought within the capacity of the pelvis. This is, indeed, an acknowledgment of defeat; it is beating a retreat. The justification, however, is that we accomplish in the end exactly that which those who reject the operation accomplish, namely, the safety of the mother. We have tried to do more; to save the child as well.

Is there any great difficulty or danger in perforating after turning? I have found none. The child's body is drawn well over to one side by an assistant, so as to facilitate the access of the operator's guiding fingers and the perforator to the head. The best place to perforate is in the occiput; but if that part be not easily struck, the perforator may be run up through the base of the skull. An opening into the cranium being made, the crotchet is passed into it, and the discharge of brain facilitated. Then, resuming traction on the trunk cautiously, the skull will probably collapse enough to pass easily. If not, the craniotomy forceps can be applied; or, better still, the cephalotribe, to crush up the base of the skull. Now, under the postulates of the case, this late recourse to craniotomy must not be considered as a severe or hazardous addition to the risks of the woman. The turning has been performed early in labour—that is, before the liquor amnii has all drained away, whilst the child is still freely movable, and before there is any serious exhaustion of the mother. Under these circumstances the turning, especially if conducted, as it commonly may be, on the bi-polar principle, is not necessarily a long or a severe operation. If we fail in extraction, which is soon ascertained by observing that the head makes no advance, but that its globe expands broadly above the brim of the pelvis, perforation can be performed in good time. In short, the safety of the mother is secured by carrying through both operations whilst her strength is good. If exhaustion had set in we should not have turned at all, but have proceeded to craniotomy in the first instance. To these considerations must be added the result of experience, which is to the effect that the retrieval by the secondary operation of craniotomy is successful.

R 2

What is the chance of saving the child? Dr. Churchill urges that "the life of the child is not secured, and its chance but little increased, even if our estimate of the pelvic diameters be accurate; for, if in turning with an ordinary-sized pelvis rather more than one-third of the children are lost, the mortality will be surely much increased if its diameter be reduced more than one-fourth." I will not stop now to press the preliminary objection I entertain to submit the decision of this or any other question in obstetric practice to à priori arguments drawn from statistics. It would not be difficult to prove that the statistics employed by Dr. Churchill and others are a confused heap of incongruous facts, and that rules to guide practice drawn from them must be stultified by endless fallacies. It is enough to state that the operation is not recommended by any one when the pelvis is contracted more than one-fourth—that is, below three inches—therefore the argument, statistical or other, is beside the question. I am not able to state or to estimate the proportion of children saved or lost under the operation. It is enough to justify the operation if we save a child now and then. I believe, however, that exercising reasonable care in selection of cases, and skill in execution, more than one-half of the children may be saved. And to save even one child out of twenty is something to set against the deliberate sacrifice of all.

Experience here again corrects the foregone conclusion deduced from statistical reasoning. The risk to the child is considerably less than might be fairly anticipated. It is a matter of observation that *in cases of moderate contraction the funis is safer from compression than in cases of normal pelvis.* I have found the cord commonly fall into the side of the pelvis towards which the face looks, and there it is protected in the recess formed by the side of the jutting promontory (*see* Fig. 89); so that if the soft parts are sufficiently dilated not to compress the cord against the child's face, and if the labour can be completed under 5″, or even a little more, the child has a very good chance indeed. This proposition is especially true in the case of premature labour with contracted pelvis. In this case the child may, in the majority of cases, be saved by turning. I have in this way saved many children who still survive to parents who would otherwise be childless. It deserves, I think, to be laid down as a rule in practice, that *where the conjugate diameter measures from 2·75″ to 3″, delivery by turning should be the complement to the induction of labour at seven or eight months*—at least, I have acted on this rule with the happiest results.

Since the design of the proceeding is to save the child, it is obviously useless if the child is dead. How do we know when a child is dead? It is by no means easy to acquire certain knowledge of this fact. Nothing is more common than to read in clinical records " that the pulsations of the fœtal heart being no longer audible with the stethoscope," or, " the pulsation in the cord having ceased," or "meconium having escaped," the death of the child was assumed, and the perforator was used without hesitation. I will not dispute that these are presumptive evidences of death, but I have too often experienced the satisfaction of seeing a child resuscitated after I had ceased to feel the pulsation in the cord, and after the free escape of meconium, to abandon the hope of saving the child without more certain evidence. This is found in great mobility and crackling of the cranial bones; the caput succedaneum falling into loose skin-folds; the coming away of epidermis and hairs. And even these would appear not to be absolute. Thus, D. M. Williams relates (*Brit. Med. Journ.,* 1875) the case of a woman who when eight months pregnant took scarlatina. Her child was born alive at term, the skin desquamating. Dr. Edis relates (*ibid.*) a case of a child weighing eight pounds born alive, all its skin peeling off, either from scarlatina or syphilis.

So long as there is tonicity, rigidity, or firmness of the limbs, life is present; but flaccidity is not a certain sign of death. A sign of threatening imminent death is a twitching or convulsive movement of the leg held in your hand. This indicates an attempt at inspiration, made to supplant the suspended placental circulation. When this is felt, it is a warning to accelerate delivery, and to excite aërial respiration.

The value of turning in moderate degrees of pelvic contraction rests greatly upon the truth of the following proposition:—*The head will come through the pelvis more easily if drawn through base first than if by the crown first.* Baudelocque affirmed this proposition. He said:—" The structure of the head is such that it collapses more easily in its width, and enters more easily when the child comes by the feet, if it be well directed, than when it presents head first." Osiander had maintained the same opinion. Hohl (1845) also pointed out that the bones overlapped more readily at the sutures when the base entered first. Simpson (1847) insisted strongly upon the truth of this proposition, and illustrated the mechanism of head-last labours with much ingenuity. The proposition has, however, been disputed, and that by Dr. M'Clintock.* He says:—"I do not believe that the diameters of the head are more advantageously placed with regard to those of the pelvis, nor can I believe that the head is more compressible when entering the strait with its base than when it does so with its vertex, till this be demonstrated by direct experiment."

It is also contested by Professor E. Martin.† He especially insists that when the vertex presents, moulding may go on safely for hours; but that if the base come first the moulding must be effected within five minutes to save the child.

I venture to submit that I have made such clinical observations as are equivalent to direct experiments. In the first place, let me state a fact which I have often seen. A woman with a slightly contracted pelvis, in labour with a normal child presenting by the head, is delivered, after a tedious time, spontaneously or by the help of forceps; the head has undergone an extreme amount of moulding, so as to be even seriously distorted. The same woman in labour again is delivered breech first; the head exhibits the model globular shape, having slipped through the brim without appreciable obstruction. For examples see my outlines of heads.‡

In the second place, I have on several occasions been called to an obstructed labour, in which the head was resting on a brim contracted in the conjugate diameter. Of course, Nature had failed; the *vis à tergo* was insufficient. I have tried the long double-curved forceps, trying what a moderate compressive power, aided by considerable and sustained traction, would do to bring the head through, and have failed. I have then turned, and the head coming base first has been delivered

* "Obstetrical Transactions," vol iv., 1863.
† "Monatsschr. f. Geburtsk," 1867.
‡ "Obstetrical Transactions," 1866.

easily. Upon this point I cannot be mistaken; and I think this greater facility can be explained. Dr. Simpson has illustrated by diagrams how the head, caught in the conjugate at a point below its bi-parietal diameter, is compressed transversely as traction-force is applied below, causing the mobile parietals to collapse and overlap at the sagittal suture. And surely no one can doubt that the traction-power, and, therefore, the compressing power, acquired by pulling on the legs and trunk, is greater than can be exerted by any ordinary forceps. But

Fig. 89.

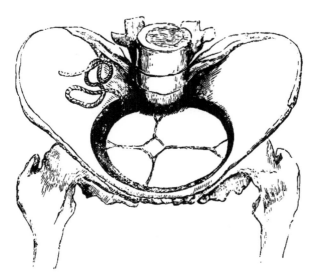

REPRESENTS THE HEAD ENTERING A CONTRACTED BRIM, BASE FIRST.

It is nipped in the small transverse diameter, the greater or bi-parietal diameter and the occiput finding room in the side of the pelvis. The cord lies in the side of the pelvis to which the face is directed, and is protected by the promontory.

there is another circumstance in the clinical history of head-last labours in narrow conjugate which affords a remarkable illustration of this proposition. *The head is rarely, or never, seized in its widest transverse diameter; it is seized by the conjugate at a point anterior to its greatest width—that is, in the bi-temporal diameter;* the bi-parietal and occiput commonly finding ample opportunity for moulding in the freer space left in the side of the pelvis behind the promontory. The head, in fact, fits or moulds into the kidney-shaped brim wherever there is most room. I have given illustrations of this point also in the memoir referred to.* I think, therefore, it may be taken as demonstrated, that the head coming base first passes the contracted brim more easily than coming crown first; and if the head comes through more easily, it may be expected that the child will have a better prospect of being born alive.

Can we define with any precision the conditions as to degree of pelvic contraction that are compatible with the birth of a living child? The question is not easy to answer; nor is it important to be able to answer it very precisely. The great fact upon which the justification of the operation rests is this: many children have been delivered by it alive, with safety to the mother. We know accurately only one element of the problem—namely, the degree of contraction of the pelvis. The other element, the relative size and hardness of the foetal skull, we can but estimate. *We must assume, in many cases, a standard head.* With this assumption the practical question is reduced to this: *What is the extreme limit of pelvic contraction justifying the attempt to deliver by turning?* In other words, this means: What is the narrowest pelvis that admits of the safe passage of a normal head? This is answered chiefly by experience. It is not to be answered by *à priori* reasoning like that urged by Dr. Fleetwood Churchill, who says:†— "The bi-mastoid diameter in the six cases measured (by Dr. Simpson) varied from $2\frac{6}{8}$ to $3\frac{4}{8}$ inches, and a living child can pass through a pelvis of $3\frac{4}{8}$ inches antero-posterior diameter, with or without the forceps. With a pelvis of this size, then, the operation is unnecessary; and if the antero-posterior diameter be less than $2\frac{6}{8}$ inches, the operation would be impracticable. These, then, are the limits of the operation; for us to attempt to drag a child through a smaller space would be unjustifiable."

To this statement of the case serious objections may be taken. The proposition that a living child can pass through a pelvis with an antero-posterior diameter measuring 3·25", with or without the forceps, can only be accepted with considerable

---

* "Obstetrical Transactions," 1866.
† "Theory and Practice of Midwifery," 1866.

qualifications. I claim to speak with the confidence drawn
from large experience, when I say that a head of standard
proportions and firmness will hardly ever pass a conjugate
reduced to 3·25″ without the forceps, and very rarely indeed
with the forceps—that is, alive. I might even extend the
conjugate to 3·50″, and affirm the same thing. The compressive power of the forceps, unless very long sustained, is not
great, rarely great enough to reduce a bi-parietal diameter of
4·00″ to 3·50″ without killing the child. My opinion, then, is
that a standard head, especially if it happen to be a female
head, which is more compressible than a male one, *may be
drawn* through a conjugate of 3″, but not with much prospect
of life; and that the proper range of the operation of turning
is from 3·25″ to 3·75″, at the latter point coming into competition with the forceps. I believe no one advocates resort to
turning when the conjugate measures less than 3″, unless the
child be premature.

A correlative proposition to the foregoing is the following:
*Compression of the head in its transverse diameter is much less
injurious to the child than compression in its long diameter.* The
truth of this is attested or admitted by most authors who have
considered the point. It is insisted upon by Radford, Ramsbotham, and Simpson. It is confirmed by the observation of
the form which the head assumes under moulding in natural
labour, which, as I have shown, is effected by the lengthening
of the fronto-occipital diameter and the shortening of the
transverse diameter. (*Obstetrical Transactions*, 1866.)

Now, it is an almost necessary consequence that when the
head, arrested on a contracted brim, is seized by the forceps,
it is seized by its fronto-occipital diameter, and to the longitudinal compression is added the increased obstruction to the
entry of the head into the narrowed conjugate caused by the
lateral bulging.

The Indications for the Operation.

Assuming a standard head whose base, unyielding, measures
3″, this is obviously the limit beyond which the operation
would be useless; for although the head is caught in the

bi-temporal diameter, a little in front of the greatest transverse or bi-parietal diameter, the base must be exposed in its full width to the narrowed strait. Even if the side of the head be indented by the promontory, no important degree of canting or obliquity of the base can be counted upon. But if the head should fortunately be undersized or unusually plastic, there is a fair prospect of the child being drawn alive through a conjugate diameter measuring 3·00".

Generally, however, from 3·25" to 3·50", or even a little more, is the working range for a child at term. The great majority of those who advocate the operation insist upon this amount of space. It is very important to have a fair oblique or sacrocotyloid diameter on one side; for if the ileo-pectineal margin of the brim incline rapidly backwards, the occiput will not find room.

The operation is also indicated if, the conjugate diameter being 3·50" or more, the forceps have failed.

Velpeau (1835), Chailly (1842), Edward Martin, and others, advise the operation in cases of unequally contracted pelvis where there is more room in one side of the pelvis than in the other—when the thicker or occipital end of the head is not already engaged in this larger side.

I have already shown that the head is always nipped in its small or bi-temporo-frontal diameter, which generally measures about 3", and is more compressible than the bi-parietal diameter. *The mark of pressure or indentation against the jutting promontory is always seen at one end of this short diameter whenever there has been obstruction in delivery.* It follows, then, that for the operation to be successful there ought to be room enough on one side of the pelvis to receive the occiput or big end of the head.

The operation may also be performed as the complement to the premature induction of labour where the conjugate measures from 2·75" to 3·50". Indeed, this I believe to be one of its most valuable applications.

The next condition is, that there be reasonable presumption that the child is alive.

The cervix should be dilated enough to admit the fingers pointed in a cone, and dilatable enough to yield with readiness under the extraction of the trunk. In this, as in most cases where the head cannot press fairly upon the cervix, we are not to expect complete spontaneous dilatation.

The membranes should be intact, or there should be enough liquor amnii present to permit of the ready version of the child.

The *contra-indications of the operation are:*—

1. A conjugate diameter narrowed to less than 3".

2. Firm and close contraction of the uterus round the child, compressing it into a globular shape.

3. Impaction or very firm setting of the head in the brim of the pelvis.

4. Marked exhaustion or prostration of the mother.

5. Death of the child.

It must be remembered that the sudden emptying of the uterus of a woman far gone in prostration, acting as a new shock, is apt to increase the *collapsus post partum.*

The Operation.—The preparatory steps are the same as for the ordinary operation of turning. As the conditions postulated admit of bi-polar action, it is important to avail ourselves of a means that so greatly lessens the force necessary to use, and which further enables the operator to bring a leg and the breech through a cervix that would not permit the passage of his hand. Chloroform will be useful chiefly during extraction.

If exploration by the whole hand in the pelvis satisfy us that the pelvis is symmetrical—that is, that there is equal and sufficient space for the big end of the head in either side —we turn according to the ordinary rules. Finding the head in the first position, or with the occiput to the left ilium, depress the breech towards the right with the right hand externally; push the head across to the left iliac fossa with the fingers of the left hand passed through the os uteri, and seize the further knee. Extraction must be performed at first slowly, so as to allow the half-breech to dilate the cervix. This is especially a case where hurry is misplaced. The extraction should go on slowly whilst the trunk is passing. As soon as the funis is felt, draw down a loop, and direct it towards the posterior wall of the pelvis. So long as it pulsates freely, do not hurry. But if the pulsations flag, lose no time in liberating the arms. The pelvic contraction makes this a little more difficult than under ordinary circumstances. (*See*

Figs. 72, 73.) As soon as the arms are liberated, the real difficulty begins: the extraction of the head. Sometimes the head is delayed by being encircled by the imperfectly dilated os uteri. This is an unfortunate complication, since compression at this point is likely to stop the circulation through the cord. To avoid this risk, it is necessary not to hurry the

FIG. 90.

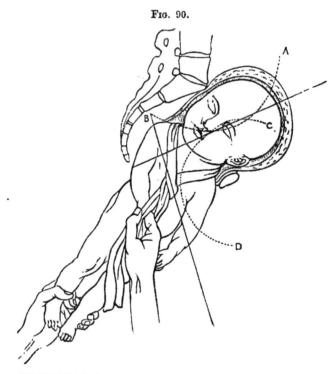

SHOWS THE MODE OF EXTRACTION AFTER TURNING WHEN HEAD IS JAMMED IN THE CONJUGATE DIAMETER.

The right forehead fixed against the jutting promontory is the centre of revolution. The left side of the head, resting on the pubes, sweeps round in the orbit of the false promontory A B. To favour this first movement, traction is made well backwards. As soon as the equator of the head-globe has slipped through the conjugate, the head enters the true orbit C D, revolving round the pubes.

trunk through the cervix. It is above all things necessary to draw at first as much backwards as possible, so as to make the head revolve round the jutting sacral promontory until it clears the strait, when the head can enter its natural orbit, the curve of Carus. Then, traction is changed to the direction of the pelvic outlet. Traction is effected by holding the legs with one hand, and the nape of the neck with the other. Commonly, the force thus obtained is enough; but sometimes more is wanted. This is obtained by crossing a fine napkin or silk handkerchief over the neck, and bringing the ends in front of the chest, and drawing upon them, as in Fig. 90.

Great assistance in extraction may be gained, and traction-force economized, by getting an assistant to press firmly upon the vault of the head through the abdominal walls, thus helping to push the head through the strait. This proceeding was advised by Pugh and Wigand. The possibility of deriving advantage from it should be borne in mind in all cases of head-last labour.

Where the pelvis is unequally contracted, one half being smaller than the other, the object is to throw the big or occipital end of the head into the larger half. Professor E. Martin describes three modes of accomplishing this.

1. *A suitable position of the woman.* Let her lie on that side towards which the forehead is directed. The fundus uteri will gradually sink with the pelvic end of the child to this side; the spine draws the occiput to the opposite side of the pelvis, and the forehead sinks more deeply towards the brim. Martin refers to a case in which he successfully carried this plan into execution, the pelvis measuring three inches only.

2. *The forceps* is a means of releasing the posterior transverse diameter of the head when imprisoned in the pelvic conjugate. This explains the frequent easy extraction when a little traction has been made. Martin admits that we must not be sanguine as to the success of this plan. We must be prepared, he says, to perforate, if there be evidence of exhaustion. My own experience is decidedly adverse to it if the contraction is at all marked.

3. *Turning by the Feet.* How is this to be done? In consequence of the well-known law that, in complete foot presentation, the foot that is drawn down always comes under the pubic arch, if the foetus is not abnormally small, or the pelvis too large, if we draw down the right foot, the child's back, and also its occiput, will come into the right half of the uterus, and *vice versâ*. If, therefore, the right half of

the pelvis is the larger, seize the right knee; if the left side is larger, seize the left knee.

Hohl and Strassmann doubt the possibility of securing this result. If it happens, it does so by accident. I believe, however, the rule and the practice are good and feasible. But the success of the operation is not necessarily imperilled, if even the occiput should fall into the narrower half of the pelvis. I have saved children when this has happened, and Strassmann relates * some striking cases in proof of this proposition.

To determine which side of the pelvis is the more contracted, attention to the following points will help:—

1. If the woman walks straight, and the legs are of equal length, the defect in symmetry will be but slight; but the presumption is that the right side is the larger.

2. If the woman has one hip affected, or one leg shorter than the other, the corresponding side of the pelvis will be the smaller.

3. You may measure and compare the two half circumferences of the pelvis externally from the crest of the sacral spine to the symphysis pubis.

4. The hand in the pelvis may take a very close estimate of the relative space in the two sides.

* "Monatsschr. für Geburtskunde," June, 1868.

LECTURE XVIII.

MECHANICAL COMPLICATIONS OF PREGNANCY AND LABOUR: DYSTO-
CIA FROM DISPLACEMENT OF THE WOMB: RETROVERSION—
HISTORY OF RETROVERSION AND RETROFLEXION OF THE GRAVID
WOMB—ORIGIN OF RETROVERSION AND RETROFLEXION IN THE
EARLY MONTHS, COMPLETE AND INCOMPLETE — INCOMPLETE
RETROFLEXION AT THE END OF GESTATION—DIAGNOSIS FROM
EXTRA-UTERINE GESTATION—FROM PELVIC CELLULITIS—FROM
RETRO-UTERINE HÆMATOCELE—PROLAPSUS AND PROCIDENTIA
UTERI AND HYPERTROPHIC ELONGATION OF THE CERVIX UTERI
—TREATMENT.

To determine the choice of operations, it is desirable here
to study certain complications which may impede labour,
presenting great varieties of difficulty, and demanding a
corresponding range of treatment. The uterus itself may
present abnormalities of development. Tumours may exist in
its walls, or in its neighbourhood, so as to block the pelvic
canal. Both kinds of complications may exert an injurious
influence before the natural term of gestation, and compel the
physician to weigh the question, whether to deal with the
intruding complication or to terminate the gestation. And, if
the term of gestation is reached, the question, how the delivery
can be most safely effected, must be solved—and that without
much time for deliberation. These cases being comparatively
rare, no man can expect to have such an amount of personal
experience as will enable him to form alone safe rules of
practice. The danger, perhaps unsuspected until the crisis
is at hand, culminates quickly. We can only hope to meet
it successfully by careful study of groups of cases, scanning

closely in what the danger consists, and the success of the treatment opposed to it.

By pursuing this course, I hope we shall arrive at certain definite principles which will help us through most particular emergencies. The *first* order of cases I propose to discuss includes certain *abnormal conditions of the uterus.* These I will class in order:—1. *Retroversion and Retroflexion of the gravid womb.* 2. *Hypertrophy of the cervix uteri.* 3. *Prolapsus and Procidentia of the gravid uterus.* 4. *Pregnancy in one horn of a double uterus.*

The *second* order includes *mechanical obstructions to the development of the pregnant uterus,* or to the accomplishment of labour, which arise either in the walls of the uterus or external to it. We may class them in the following order:— 1. *Tumours in the walls of the uterus,* including *polypi* projecting from the walls into the cavity. 2. *Descent of bladder before head.* 3. *Tumours in the vagina,* including *hæmatoceles* of the soft canal. 4. *Abdominal tumours,* including *extra-uterine gestation, ovarian tumours, retro-uterine hæmatocele, hydatid tumours of the liver, cystic disease of the kidneys.* 5. *Pelvic tumours,* including *bony, cartilaginous,* and *sarcomatous tumours* attached to the walls of the pelvis; *ovarian tumours dipping into the true pelvis.*

Retroversion of the Uterus.

The relations of retroversion to pregnancy and labour are very imperfectly related in our text-books. Indeed, many of the most important features are scarcely referred to, or are quite misunderstood. I need do no more than glance at the doctrine of Denman, which attributes retroversion of the gravid womb to distension of the bladder. It is undoubtedly erroneous; an instance of τὸ ὕστερον πρότερον. It is true, the severe symptoms of retention of urine, intolerable pain, straining efforts, and tenesmus appear more or less suddenly; but it does not follow that the displacement of the womb was produced suddenly. So long as the retroverted womb is comparatively small, it does not press upon the rectum and bladder so as to obstruct the function of these organs. It is only when, in the course of development, that the eccentric pressure gradually increasing, at last compresses the structures outside the uterus against the unyielding walls of the pelvis. Then the distress arises, the bladder is jammed against the pubes, and retention ensues. It is an epiphenomenon—a consequence, not a cause of the retroversion.

Retroversion and retroflexion are frequent conditions of the non-gravid womb. They are frequent in single women, still more frequent after child-birth. In many cases these displacements offer no impediment to conception. And having begun under this condition, so pregnancy goes on. I have observed retroversion or retroflexion of the gravid womb under all the following circumstances. *First*, there is the condition of *primitive retroflexion* persisting during pregnancy; *secondly*, it may be a continuation of *retroflexion existing before the actual pregnancy*, the result of a previous labour. W. J. Schmitt (1820) affirmed that retroversion of the gravid, as well as of the non-gravid womb, was a gradual process. He uses the term "reclination." He says it may be overlooked in the earlier stages, and until the severe symptoms arise. This has also been pointed out by Dr. Tyler Smith (*Obstetrical Transactions*, vol. ii.), and adopted by Scanzoni. *Thirdly*, the retroflexion may have been caused by a *tumour in the posterior wall of the uterus*, or by *adhesion dragging the fundus backwards*. *Fourthly*, as I have several times observed, retroversion of the gravid womb succeeds to *prolapsus of the non-gravid womb*. In this last case, the heavy fundus is gradually forced back by the intestines, and being caught under the sacral promontory, the cervix is tilted forwards by a persistent leverage action. In all these four cases the gestation during the early months may be called *pelvic gestation*. The uterus grows within the cavity of the true pelvis, and becomes locked there when it ought to find room by emerging into the abdominal cavity. In several cases (Bailly, Callisen, Boivin) narrowing of the conjugate diameter of the pelvis has been noted as a predisposing cause.

It must not, however, be concluded that retroversion of the gravid womb is always of gradual formation. It may arise, *fifthly, suddenly*. If, for example, a woman, pregnant three or four months, be subjected to great and sudden exertion, as by lifting heavy weights, whilst the bladder is, perhaps, somewhat distended, and thus directing the fundus uteri backwards, the

s

superincumbent pressure of the intestines will bear upon the
anterior surface of the fundus, and the uterus, acting like a
lever under the force applied to one extremity, will roll back,
and may get locked under the promontory. Being there,
tenesmus, uncontrollable efforts at defæcation are produced, and
the pressure of the intestines then acts more directly upon the

FIG. 91.

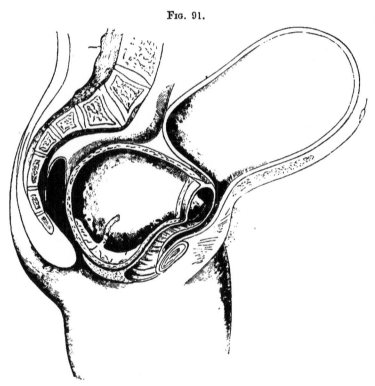

RETROVERSION OF GRAVID WOMB.

The fundus, thrown back, bulges out the perinæum, flattens the vagina, throws forward the
cervix uteri, and drags up and compresses the urethra.

anterior wall of the uterus, tending to maintain and increase
the retroversion. In this fifth case there may have been
originally no flexion; and the accidental displacement will be
the typical retroversion of Hunter and Denman, the only form
at one time known. *Severe repeated vomiting* at this stage of
gestation may also increase or produce retroversion.

There is still a distinction to be made—one of extreme

importance. Retroversion or retroflexion may be *complete or incomplete*. This distinction is hardly yet understood. It has often been observed that retroversion of the gravid womb tends to spontaneous cure. This is generally supposed to arise from the uterus liberating itself suddenly or gradually, as it enlarges, from the pelvic cavity. Now, this is true only in a certain proportion of cases, possibly not in the majority. What really takes place in many cases is this: Down to the end of three or four months of pregnancy, there is pelvic gestation with retroflexion or retroversion. At this stage the effects of eccentric pressure upon the organs surrounding the growing uterus are often felt; they may gradually subside; but examination reveals the existence of the retroflexion. How, then, has relief been obtained? The ovum continuing to grow, pushes out that portion of the uterine wall which looks upwards and forwards to the abdominal cavity. This part is free; and it gradually enlarges to form a secondary sac or pouch accommodating the main bulk of the fœtus down to the normal term of gestation. This secondary pouch, or diverticulum, expands in the abdominal cavity just as the whole uterus does in normal gestation, whilst the posterior portion and fundus are retained in the pelvis throughout. Towards the end of pregnancy the predominance in rate of growth and of bulk of the abdominal diverticulum is so great as to draw the pelvic portion partly out of its lodgment there, producing a partial rectification of the form and relations of the uterus; the os comes nearer to the centre of the pelvis, and there is no obvious obstruction to labour. The phenomena described I have distinctly traced and recorded in several instances; and I regard the process as an ordinary one by which Nature releases herself from the danger of a locked retroflexed uterus. But sometimes the course of events is not so favourable. The development goes on as described in the two pouches, one pelvic, one abdominal; but the pelvic part of the gestation remains so considerable that the os is kept fixed behind and above the symphysis, so that when labour comes on it is found that the pelvic cavity is filled with the uterine pouch, containing, perhaps, the child's head, whilst it is almost impossible to bring the cervix and os down into relation with the axis of the pelvis, so as to afford a channel for the passage of the fœtus. (*See* Fig. 92.)

s 2

Merriman relates two cases which appear to have been of this kind. One of them he attended with Denman. They found the os come down gradually during the pains of labour, as if the uterus were rolling round to its natural position. Merriman believed that retroversion might continue to the end of gestation. He did not seize the fact that retroversion so

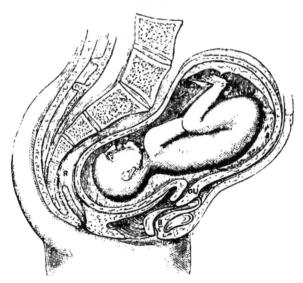

INCOMPLETE RETROFLEXION OF GRAVID UTERUS.

o u, os uteri, bent downwards above symphysis pubis; b, bladder; r, rectum. One part of bi-loped uterus locked in the pelvis contains the head; the other part, developed in the abdomen, contains the body of the fœtus.

continued could only be incomplete. Scanzoni (*Lehrb. d. Geburtsk.*, 1867), says the "partial" retroversion is rather a fault of form than of position. It is always caused by the pressure of the part of the child upon the posterior lower wall of the uterus; and is especially observed when with slight pelvic inclination the fundus of the uterus falls down through the relaxed abdominal walls. The trunk of the child thus sinking forwards, the presenting head pushes the relaxed hinder wall into the sacral hollow in form of a sac. The vaginal-portion (cervix) is pushed forward. This form, he says, is only seen in the last two months. It causes no interruption to pregnancy.

Hecker describes a remarkable case (*Monatssch. f. Geburtsk.*, 1858), observed in the sixth month. A pluripara had suffered repeated attacks of dysuria, at times threatening abortion; at last she took to bed with crampy pains. The fundus was felt about the level of the navel, the uterus contracting; the os uteri was felt, with much difficulty, above the symphysis; the cavity of the sacrum was filled by a smooth, elastic swelling, which compressed the vagina forwards. This was violently pressed down with every pain, so that rupture was feared. After failing to push the tumour out of the pelvis, it at length rose, and at the same time another round elastic swelling—the membranes—came down from behind the symphysis. The child was then delivered. Dr. Oldham relates (*Obstetrical Transactions*, 1860) a still more striking case. The patient, a primipara, was at term. The head was contained in the pelvis, whilst the trunk and breech of the child were contained in the abdominal portion of the uterus; the os pointed downwards, but was high behind the symphysis; the vagina, of course, was flattened against the symphysis. Dr. Oldham contrived to deliver by pulling down the cervix, by pressing the breech down by a hand outside, and pushing up the pelvic mass. A sort of rotation took place, the child being delivered by seizing a foot. Dr. Oldham had seen this woman three years before, when single; she was then suffering from dysuria and retroflexion of the womb.

The following is a condensed account of an interesting case of this kind, which I saw in conjunction with Drs. Hilliard and Brunton. A lady, a week before term of gestation, complained of headache and œdema of the face and legs. The urine was found to contain albumen, blood-discs, and casts. Taken in labour, no os uteri could be found, until we produced anæsthesia. I felt the posterior wall of the vagina closely compressed against the anterior by a rounded, firm mass filling the cavity of the pelvis. Following the vagina I felt the os uteri soft and patulous above the upper edge of the symphysis pubis. Through the os I could feel a hard rounded mass, which I concluded to be a fœtal head covered by the membranes. This head came close to the rounded mass in the pelvis, giving the idea of two heads interlocking. This intra-pelvic mass presented a hard ridge, unlike the head, and Dr. Brunton, on careful auscultation, could only hear one fœtal heart. The abdomen was irregular in shape ; the enlargement was transverse and bi-lobed, suggesting twins. I determined first to try to push up into the abdomen the mass that occupied the pelvis, so as to enable me to seize the head which lay above the pelvis with the forceps. This I succeeded in doing. When the mass was lifted into the abdomen the os uteri came down into the centre of the pelvis. The cervix was then dilated to some extent by bags until I could get the blades of my long forceps over the head. Delivery was completed by traction, extending over an hour and a half. The child small, but mature, was born alive. The placenta was cast. Next day child and mother were doing well. The case proved to be one of incomplete retroversion ; a pelvic pouch contained the breech, an abdominal pouch contained the head. The pressure upon the pelvic and abdominal vessels and upon the bladder was probably the cause of the albuminuria.

The *terminations* of retroversion of the gravid womb are :—
1. *Recovery by reposition*, spontaneous or chirurgical, is, perhaps, the most frequent termination. 2. *Recovery* is also occasionally effected by the safety-valve process of *partial outgrowth of the uterus* upwards into the abdominal cavity. 3. *Abortion* is not uncommon. The immediate diminution in bulk of the uterus, and the cessation of the attraction of blood to the pelvis, bring quick relief. Arthur Farre says * : " The sequelæ, when reposition cannot be effected, are usually premature expulsion of the ovum, or sloughing of the uterine parietes, and slow discharge of the contents by fistulous openings into the vagina, rectum, or bladder." Some of these cases were most probably not cases of retroversion, but of extra-uterine gestation.† 4. Death by *blood-poisoning*, the matters which should be excreted by the kidneys being retained in the system. This I believe to be the most frequent cause of a fatal termination. It was the chief cause in four fatal cases observed by myself. 5. C. Braun relates ‡ a case which ended in death under eclamptic attacks in consequence of *Bright's degeneration of kidneys*, and secondary uræmia. In many cases, conjoined with urinæmia, there is *disease of the bladder*, such as intense con-

---

* " Cyclopædia of Anatomy."
† See a case by Guichard, " Rev. Thérap. du Midi," 1857.
‡ " Clin. der Geburtsk."

gestion, inflammation, even sloughing of the mucous coat.[*]
Dr. Schatz [†] relates and figures a well-marked case, in which
not only the mucous coat but also the muscular coat became
necrosed and entirely separated. Dr. Moldenhauer relates (*Arch.
f. Gynäkologie*, 1874) a similar case. A pluripara, after com-

FIG. 93.

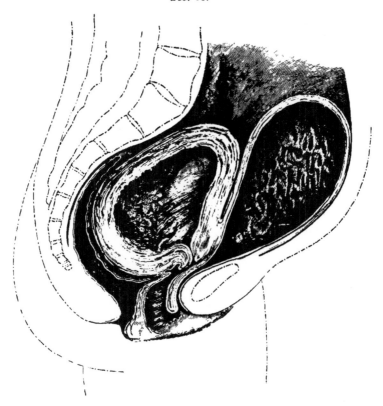

Dr. Chambers' Case.

PARTIAL RETROFLEXION OF UTERUS, CERVIX COMPRESSING NECK OF BLADDER. DRAWN FROM
NATURE (HALF SIZE) FROM SPECIMEN IN ST. THOMAS'S HOSPITAL.

plaining eight days of dysuria, had complete retention. The
bladder was emptied by catheter, and reposition repeatedly
tried. The woman emaciated rapidly, had general prostration,

[*] Haussmann, " Monatssch. f. Geburtsk.," 1868.
[†] " Arch. f. Gynäkol.," 1870.

and was taken into hospital on the 9th October. The cathetor drew off 100 grammes of strongly ammoniacal turbid urine. In spite of many attempts the bladder could not be completely emptied. The last portion drawn was slimy and purulent, and mixed with gangrenous shreds, and so thick that it would barely pass the catheter. An impacted retroflexion of the gravid uterus was made out. Reposition was effected in the knee-elbow position. The woman seemed relieved. Urine flowed involuntarily. On the 15th October she was delivered of a fresh fœtus nine inches long. The uterus contracted to normal size. Vomiting and collapse set in, and death took place on the 17th. *Autopsy :* no mark of recent peritonitis. There were old adhesions binding the fundus of the bladder to the colon and small intestines, forming a bridge which completely obstructed the rise of the uterus out of the pelvis. The contracted uterus was found retroverted. The bladder, distended beyond the size of a child's head, contained a grey-yellow sac, which was the mucous membrane, with a part of the muscular coat, cast off by necrosis. The ureters were not much dilated. The pyramids of the kidneys were very full of blood.

6. Death by *rupture of the bladder*. This is extremely rare. Lynn, of Woodbridge (1771), published a case ending fatally through this cause; Van Doeveren, of Groningen (*Specimen Observationum Academicarum, etc., artem obstetriciam præcipue spectantium*, 1765), gives two similar cases. I have not noted the relation of any recent case. There are three compensating factors which retard or prevent bursting :—1st, dribbling, or overflow; 2nd, stretching of the bladder, to accommodate a great accumulation; 3rd, absorption from the bladder. Moreover, when the retrograde pressure upon the kidneys is great there is diminished secretion of urine, and a diversion is effected to the skin, which throws off abundant perspiration. Death will occur from urinæmia, shock, and exhaustion long before the bladder will burst. The quantity of urine that gathers varies from four to ten or twelve pints. In Dr. Chambers' case the accumulation of urine was enormous. The woman measured thirty-nine inches round the abdomen. The case was at first supposed to be ovarian. Twelve pints of urine charged with blood, and having scarcely any urinous odour, were drawn off. A fœtus of four or five months' gestation was removed. The uterus was

fixed in the pelvis. At the autopsy the bladder was found to contain two pounds by weight of black clot. There was no trace of peritonitis. 7. Death by *gangrene of the uterus or other parts compressed.* This is also extremely rare. Burns says inflammation and gangrene of the vagina and external parts have been produced. 8. *Peritonitis,* which may be fatal, or which may end in adhesions. In none of five fatal cases seen by myself was there peritonitis. Death by urinæmia and the shock of pain usually anticipate the outbreak of peritonitis. Peritonitis is a later event. Mr. Misley relates (*Med. Times and Gaz.,* 1855) a case in which, after reposition, the patient died twenty days later of peritonitis. Adhesions of intestines were found, and the ureters were distended fourfold. In some cases, no doubt, where no post-mortem examination has been made, peritonitis has been inferred from the intense pain suffered during life. But this sign is fallacious, since in cases where peritonitis had been so diagnosed, post-mortem examination refuted the diagnosis. 9. *Shock and exhaustion* enter into every case, but they may be the chief fatal factors. 10. *Rupture of the posterior wall of the vagina,* caused by violence of the expulsive effort, has been noted.

If abortion do not occur during the acute stage it may still follow reposition, some days or weeks later. But in a large proportion of cases, after reposition, gestation goes on to term. After labour the uterus is very liable to fall back, either immediately or in a few days, disposing to hæmorrhage, primary or secondary. Amongst the remote effects, vesico-vaginal fistula, cystitis, chronic catarrh of the bladder, disease of the kidneys, chronic inflammation of the uterus may be observed.

The diagnosis. Seeing that the development of the dangerous state is often gradual, it becomes a question of practical interest to study *the early or premonitory symptoms.* The early *subjective* signs are those of retroversion or retroflexion of the womb in a more or less aggravated degree, preceded by the signs of pregnancy: a sense of dragging, of bearing-down, lumbo-sacral pain, pressure upon the bowel, and, above all, bladder-distress evinced in frequent desire to pass water, with occasional partial retention or dysuria. I think the rule I have submitted elsewhere (*Lancet,* 1875), to examine the state of the pelvic organs whenever there is severe bladder-distress, should be imperative, and especially so when pregnancy is known or suspected to exist. Retroversion, when gestation is but two or three months advanced, is very apt to lead to abortion; and if this danger be escaped from, the more formidable, perhaps fatal, complications may ensue. Thus, I have known a case in which a woman had been attending a surgeon for several weeks on account of partial retention; complete retention supervened suddenly, and she died from impaction of the retroverted gravid womb at the fourth month. Now, these catastrophes can almost certainly be averted by maintaining the uterus in its proper place during the second and third months by a Hodge-pessary. At the end of this time the uterus, having been kept with its fundus pointing forwards, will grow into the abdominal cavity, and henceforward will support itself.

When the *stage of impaction* or *incarceration* has arrived, that is, when the uterus is so large as to press injuriously upon the surrounding structures, the *subjective symptoms* are commonly too urgent to be neglected. The signs are—1. Those of *pressure;* severe pelvic and abdominal pain. 2. *Shock*, the result of pain, interruption of bladder and bowel function, and the local violence due to the displacement of the womb. 3. *Reflex phenomena*, straining or bearing-down, caused by the fulness of the bladder, and the pressure of the fundus uteri upon the rectum. 4. The *secondary or constitutional symptoms, as urinæmia and exhaustion.* In some cases albuminuria has been observed; the legs may swell. 5. Numbness of the legs, suggesting paraplegia, has been described by Bedford.

The *objective signs of retroversion of the gravid womb* ought to lead to a positive conclusion. 1. The abdomen, if the bladder is distended, *is peculiarly prominent;* intensely painful to touch; and percussion will map out *a tense fluctuating tumour occupying the position of the gravid womb* at the sixth or seventh month. Resonance will be made out above it, and in either flank. 2. The *finger passed into the vagina*, instead of going back towards the hollow of the sacrum, *is guided forward close behind the symphysis pubis*, because the posterior wall of the vagina is compressed forward by a firm rounded mass which fills the sacral hollow. Feeling for the *os uteri*, it is found to be difficult or impossible to reach; it is *dragged up* into the displaced roof of the vagina, perhaps high *above the symphysis.*

(*See* Fig. 91.) The finger being in the vagina, we ought, by passing the fingers of the other hand in above the pubes, to meet it and feel it; but the distended and painful bladder prevents this approximation. 3. This leads at once to exploration by *catheter*. And now a characteristic sign is revealed. The base of the *bladder and the urethra attached to the anterior wall of the cervix uteri are pulled up*, and so stretched that the anterior wall of the vagina with the *urethra are smoothed out, the meatus urinarius is dragged up* from its place, and so stretched out that it is sometimes difficult to feel, and how to pass up the catheter may be extremely puzzling. (*See* Fig. 91.) This important sign, which I believe distinguishes *retroversion* of the gravid womb from every other condition, was particularly pointed out by Dr. Roper, at a discussion in the Obstetrical Society, in 1874. In the typical cases there is complete capsizing of the uterus without flexion; the os looks upwards and forwards towards the umbilicus, whilst the fundus points backwards and downwards towards the coccyx or anus. This position of the uterus brings about another symptom scarcely less characteristic, to which my attention was drawn by an old pupil, Mr. Max Simon. This is: 4. *The bulging of the anus and perinæum*, caused by the enlarged uterus, especially manifest if tenesmus and expulsive action are set up. Under this action the bulging of the perinæum and protrusion with eversion of the anus, remind one of the stage of labour when the head is bearing on the perinæum. This symptom has been several times observed. In simple retroversion it is probably seldom altogether wanting. Halbertsma, of Utrecht, relates a case in which the *fundus uteri projected through the anus*, which was widely opened. In Dr. Chambers' case the posterior wall of the vagina was protruded through the vulva. But, occasionally, if certain cases related are not erroneously interpreted, the pressure exerted under the stormy reflex expulsion excited, is so great that actual *bursting of the posterior wall of the vagina may occur*. This accident is so important that the cases must be briefly cited. Grenser relates (*Monatsschr. f. Geburtsk.*, 1857) that a midwife was called to a woman, whom she found forcing down strongly and complaining of sacral pains and obstruction of the bowels. She found a swelling like a lump of flesh projecting from the vulva.

A physician who was sent for found the woman in collapse. Thinking the mass might be a mole, he concluded he must drag it away, and so pulled in various directions. He felt compelled to give in. Another physician arriving, recognized retroversion of the womb. The woman died in four or five hours. At the autopsy, a swelling the size of a child's head projected outside the vulva; this was the uterus, containing an ovum the size of a hen's egg, and the ovaries. At the back part of the vagina was a rent, through which the uterus and ovaries were hanging forth, so that the cervix was still in the pelvic cavity. The case was submitted to Grenser for report. He concluded that the rupture was spontaneous. Dubois relates (*Presse Médicale*, t. i.) a similar case. The gravid retroverted uterus came through a vaginal rent. The woman died a few minutes after reduction. This case is related in detail by Moreau (*Traité prat. des Accouch.*). The doubt may be entertained that a lenient judgment covered ignorant malapraxis. But it seems irrational to deny the probability that the accomplished reporters were right.

When the bladder has been emptied, auscultation in the groins may reveal the uterine souffle. And in some cases the double touch may bring out a sense of ballottement.

Retroflexion of the gravid uterus is not usually attended by such marked symptoms as is retroversion; and thus the *diagnosis* is not so easy. It has been said that in most, if not in all cases, retroversion is the first stage of retroflexion; that is, that as the cervix rises, the dragging of the vagina and bladder upon its anterior wall bends it down. To a certain extent this is true; but I am satisfied that the statement in its general application is based upon theory or very limited observation. In the majority of cases the flexion exists *ab origine*, although it may become aggravated as the body of the uterus enlarges, and as, therefore, the fundus grows downwards. The *subjective early signs* are irritation of the bladder, dyschezia, sense of bearing down, and of weight in the pelvis. Retention is less common than in pure retroversion. The *objective signs* also differ. The posterior wall of the vagina is driven forward by the displaced and enlarged body of the uterus. But since the cervix uteri is not thrown up, the roof of the vagina is not dragged up into an elongated cone to the same degree as in retroversion. For the

same reason the urethra is not so much pulled up, the meatus is
not displaced, and retention of urine is not nearly so frequent.
The os uteri points downwards; it is closely jammed against
the symphysis pubis, and flattened transversely by the pressure.
The finger, pressed into the angle between the cervix and body
of the uterus, may trace the continuity of the organ.   The
bulging of the anus and perinæum will also be .less marked
than in retroversion proper.   (*See* Figs. 92 and 93.)

Up to this point the objective signs are nearly the same as
those of *retro-uterine hæmatocele*, or of *extra-uterine gestation in*

FIG. 94.

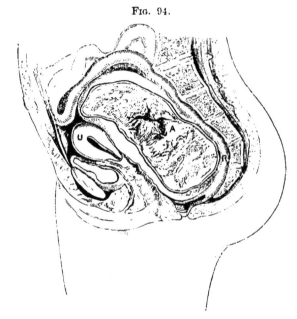

ENCYSTED RETRO-UTERINE HÆMATOCELE, FROM A SPECIMEN IN ST. THOMAS'S HOSPITAL.
A, the hæmatocele, pushing U, the uterus, forward upon the bladder. The os uteri points
downwards, and is easily reached.

Douglas' pouch.   The first condition is that for which I have
most often known it to be mistaken.   The distinction is made
by remembering that in retro-uterine hæmatocele the uterus is
pushed bodily forward by the mass behind, preserving its
normal axis, so that when the bladder is emptied, the finger
of the left hand applied behind the os uteri, and the fingers
of the right hand pressed down above the pubes, the body of
the uterus is felt between.   The position of the uterus is clearly

demonstrated by the sound, which will easily pass up to the fundus for the normal length or more; and thus held supported on the sound, the uterus may be plainly felt close behind the abdominal wall rising above the symphysis pubis. In this case, then, it follows that the mass protruding the posterior wall of the vagina is not the displaced uterus. The *os uteri*, often flattened as in retroflexion, is generally much *lower down.* Again hæmatocele is not often limited to the pelvic cavity; a portion rises into the abdomen. It rarely fills the lower part of the sacral hollow so completely as retroversion or retro-flexion. If of recent formation, it is softer as felt from the rectum. *It fixes the uterus in one mass with itself.* (See Fig. 94).

A small ovarian tumour locked under the sacral promontory may push the uterus forward like a hæmatocele. Here, again, the sound defines the position of the uterus. The tumour is made out by its firm elastic feel; by its globular shape; by its being generally mobile and capable of being isolated from the uterus.

Pelvic cellulitis or peritonitis has some features in common. But the *os uteri is usually lower down* and *nearer the middle of the pelvis;* the peri- or retro-uterine masses are harder, not so uniformly rounded; they embrace the cervix not behind only, but generally on one or both sides as well, fixing it; they rarely fill the retro-uterine pouch so evenly or centrally as the dis-placed uterus or a hæmatocele. The disease may usually be connected by history with labour, abortion, interrupted men-struation, or surgical or other local injury.

A *retro-uterine abscess* may displace the uterus, causing retention of urine, and produce many of the subjective and objective signs of retroversion or retroflexion. The sound will reveal the position of the uterus. The more obvious fluctu-ation will distinguish it from hæmatocele; and the aspirator-trocar, by drawing off pus, and greatly reducing the tumour, or causing it to disappear, will remove all doubt.

An *extra-uterine gestation-cyst* will sometimes get in part into Douglas' pouch, and push the uterus forwards. The aspirator-trocar will be of service here. In a case in which I was at first misled, the woman was calculated to be seven months pregnant when she began to suffer from incontinence of urine, then pain, and a red erysipelatous suffusion spread over the abdomen. The vagina was compressed from behind forward by a round elastic tense tumour partly filling the cavity of the sacrum, and coming from above. The os uteri was close behind the symphysis pubis pointing downwards; it admitted the finger freely, but I could not reach the os internum or feel any part of a child. The sound went readily towards the umbilicus several inches. There were symptoms of peritonitis and prostration. I punctured the elastic swelling in the vagina with a fine trocar, and drew off eight ounces of turbid fluid mixed with blood. This relieved the tension of the tumour and of the abdomen; and the erysipelatous blush vanished almost immediately. I then felt parts of a fœtus through the flaccid walls of the cyst. She died two days after this. *Autopsy.*—Extensive recent adhesions glued the abdominal wall to the intestines. Breaking through the adhesions we came to an irregular cavity from which flowed dirty turbid fluid with some blood. In front of this cavity was the uterus, singularly elongated in its cervical portion, so that the fundus reached nearly to the umbilicus, as ascertained by the sound during life. The body contained a pultaceous membranous investment (decidua ?). The Fallopian tubes and ovaries were undistinguishable in the confused mass of adhesions. Behind the uterus was a cavity, bounded below by Douglas' pouch, behind by the mesentery and vertebral column, above and in front by the liver, intestines, omentum, and abdominal wall. It contained a fœtus of seven months' development, cuticle peeling, and decomposition so far advanced as to suggest that death had taken place a fortnight. The placenta grew to the iliac fossa. The case figured (Fig. 95) is a fair parallel to this case. I have related it in some detail because the retention of urine and the pelvic objective signs closely simulated those of retroversion of the gravid womb, and these coinciding with a history of pregnancy rendered the diagnosis still more embarrassing. It is well known that extra-uterine gestation-cysts commonly encroach in part into Douglas' pouch. The important differences here were the determination of the position of the uterus by the sound, the os being low down, the freer elasticity of the retro-uterine tumour, and the escape of fluid on tapping, and the consequent ability to feel fœtal bones.

One general fact of great service in forming a diagnosis is this: Almost all bodies which get into Douglas' pouch *come from above*, and so push the uterus not only forwards, but at the same time downwards, thus bringing the os uteri within easy reach, and pointing downwards. On the other hand, retroversions of the uterus lifts the os upwards and tends to throw it forward.

In connection with this question I refer to a most valuable series of cases of extra-uterine gestation by M. Depaul, in the

FIG. 95.

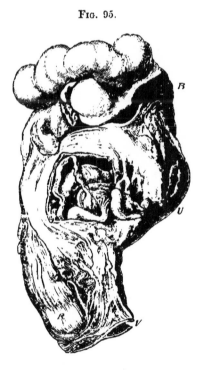

CASE OF RETRO-UTERINE GESTATION, FROM A SPECIMEN IN ST. THOMAS'S HOSPITAL.

B, bowels; U, uterus; V, vagina; R, rectum, forming boundary of gestation-sac, which is opened to expose the fœtus and placenta

Archives de Tocologie, 1874. In the same journal is a history by Dr. Bailly of a case of partial or bi-lobed retroverted gravid uterus, which was taken by the most able observers for extrauterine gestation. The sound passed three inches close behind the pubes, and could be felt through the abdominal walls. It had passed into the upper or abdominal pouch of the uterus. The uterine souffle was heard. In the case figured (Fig. 93), which I saw with Dr. Chambers, also I was at first deceived by the sound passing three inches.

In the cases of *partial retroversion*, with outgrowth of the uterus into the abdominal cavity, there is not so often actual retention of urine. The os uteri will be carried above the symphysis pubis, and it may point downwards or forwards. When the bladder is emptied, the part of the uterus which grows up into the abdomen will be made out by palpation and percussion. The symptoms rarely become urgent until the natural term of gestation. Then the trouble consists mainly in the difficulty of labour. This, no doubt, is the reason why these cases, which, I believe, are not uncommon, have been so long overlooked.

The question of *treatment* must first be considered in relation to the intercurrence of urgent symptoms in early pregnancy. It will first of all be directed to the relief of the imperatively urgent symptoms, retention of urine and pain. The first object and, to some extent, the second, will have been accomplished by *passing the catheter* for the purpose of diagnosis. 1. *Passing the catheter* is not always an easy matter; and, in anticipation of difficulty, it will generally be wise to induce anæsthesia. The first difficulty consists in finding the meatus, which, in retroversion proper (*see* Fig. 91), is drawn up and flattened out behind the symphysis. The point of the catheter—a flexible male one, No. 9 or 10, should be selected—instead of being directed a little backwards under the pubic arch, must be directed close up behind the symphysis. The second difficulty is sometimes found in emptying the bladder, even when the catheter has entered it. It should, in the first instance, be passed in as far as it will go; and then, when urine ceases to flow, withdraw it by slow degrees, when more urine will often flow, as if the catheter tapped fresh pouches of bladder. A very singular phenomenon, sometimes observed, one which I cannot satisfactorily account for, is that, when you have apparently drawn off all the urine, a copious gush will spurt out when the uterus is reduced. This shows, at least, that the obstruction was mechanical, not from paralysis. In some cases, when retention has been severe, the flow of urine is at times stopped by shreds of cast necrosed mucous membrane, or plugs of mucus and blood. 2. *The reduction of the uterus.* The bladder being emptied, and the state of

T

anæsthesia being maintained, careful trial should be made to restore the uterus. Two kinds of difficulty may oppose restoration. First, there may be adhesions binding down the fundus. These are not common. But should they be present, attempts to push up the uterus might be fatal. Blundell relates a remarkable case* of a young lady who had an ovarian cyst, which burst from a fall. Peritonitis followed; on recovery she married; and becoming pregnant, died with an irreducible retroversion of the uterus. Inflammatory adhesions had fixed the uterus in the pelvis. Dr. Moldenhauer's case, related at p. 263, is another example. Dr. Meigs relates a case in which the retroverted uterus was bound down by adhesions associated with a Fallopian gestation. Secondly, reduction may be impeded by the congestion, tumefaction of the parts. The uterus is locked in the pelvis by its bulk. In such cases, and indeed in all cases where it is very difficult to reduce the uterus, the proper course, it appears to me, is to induce abortion. By merely lessening the bulk of the womb, certain relief is obtained, the pressure is taken off, and the stimulus of developmental attraction being arrested, contraction and involution of the uterus set in. Reduction then will generally be easy; and if not effected, the diminished uterus ceases to be a source of danger. The induction of abortion, however, is not always easy. The os uteri may be not only high up above the symphysis, but it may also be directed forwards and upwards so that it is hard to reach; and it becomes difficult, if not impossible, to pass an instrument through it into the uterus. This is especially the case in pure retroversion. In the case of retroflexion the os uteri points downwards, and is more easy to reach. In either case an attempt may be made to pass a curved sound or an elastic bougie into the uterine cavity, to rupture the membranes, or at least to disturb the connections of the ovum. This done, the patient should rest in the prone position, the bladder being emptied every six hours by catheter. If it be found impossible to pass an instrument through the os uteri, if reduction be also impossible, and the symptoms be urgent, it is justifiable to puncture the uterus by the vagina or rectum. The risk of piercing the placenta, which is likely to be attached to the posterior wall of the uterus, should not deter, but must be

* "Obstetric Surgery."

encountered as a lesser evil. The best instrument is Dieulafoy's aspirator-trocar, which is perfectly safe. The forefinger of the left hand applied inside the vagina or rectum determines the most bulging part of the uterus, and the trocar guided by it, is pushed into the uterus, taking special care to enter the uterine wall perpendicularly. The rectum is to be preferred, because puncture there is more certain to tap the body of the uterus, and to keep clear of the cervix. Under the influence of air-exhaustion, liquor amnii should run out of the canula. Some diminution of bulk is immediately gained, and abortion will follow in a few hours. This suggestion, made in the last edition of this work, has been successfully carried out by Anthony Bell. The fœtus was expelled in thirty-six hours. The woman made a good recovery.

Where it has been found impossible to pass a catheter, and reduction being also impossible, the question has arisen between supra-pubic puncture of the bladder and puncture of the uterus by rectum. A case of this kind occurred to Dr. Head at the London Hospital.* He punctured the uterus; it then became possible to empty the bladder; a shrivelled fœtus of five months was passed, and after a very severe illness, the woman recovered. The decision to puncture the uterus first was judicious, since it might still have been necessary to do it after having punctured the bladder. On the other hand, Dr. Münchmeyer relates † a case in which he punctured the bladder; the uterus was then reduced, and recovery followed. And Dr. Schatz relates ‡ a case in which he tapped both uterus and bladder with success. The aspirator-trocar, moreover, greatly lessens the objection to puncturing bladder or uterus. By its use the urine can be drawn off with complete safety. The puncture should be made about 3″ above the symphysis, so as to allow for the descent of the bladder as it becomes empty.

Reduction may be attempted, either leaving the ovum intact, or after abortion. In a fair proportion of cases, if seen early, simply keeping the bladder and rectum empty, and rest in the prone or semi-prone posture, are enough to obtain restoration of the uterus; and unless very urgent distress continue, time should be afforded for this spontaneous cure. When it has

* See " London Hosp. Rep.," 1867.   † " Mon. f. Geburtsk.," 1860.
‡ " Archiv. f. Gynäkol.," 1870.

been decided to try reduction, the plan to be pursued is as
follows :—Empty the bladder and rectum ; induce anæsthesia to
the surgical degree ; place the patient in the prone or semi-prone
position with the nates elevated ; pass two or three fingers, or
even the whole hand, into the rectum ; apply the tips of them to
one side of, and under the uterine globe; push the globe upwards
and *to one side of the promontory* of the sacrum, so as to clear
this projection.   This oblique direction of the reducing force
is a point of great importance.   It was insisted upon by Dr.
Skinner (*Brit. Med. Journal*, 1860).   You adapt the bi-polar

FIG. 96.

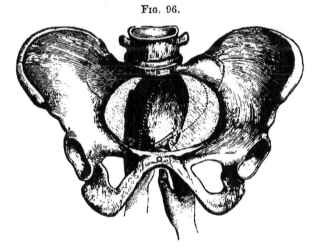

REDUCTION OF A RETROVERTED GRAVID UTERUS BY DI-POLAR DIAGONAL TAXIS.

P, Promontory of sacrum locking F, fundus of uterus ; o, os uteri behind symphysis. The
arrows show the direction of the force applied by the fingers to bring the uterus into an oblique
diameter represented by dotted line in which it is clear of the promontory, and can then rise
out of the pelvis.

principle to the reduction.   First, by finger in the vagina, pull
the cervix uteri backwards towards the right acetabulum, so as
to bring the uterus into the left oblique diameter of the pelvis ;
then, the fingers in the rectum push the fundus towards the left
sacro-iliac synchondrosis. (*See* Fig. 96.) If the patient is not
under chloroform, pressure should be exerted during expiration
only.   I have reduced an impacted retroverted gravid uterus with
comparative facility by attention to this rule, which had resisted
many previous efforts.   If it be found awkward to work with

one hand in the vagina, and the other in the rectum, the cervix uteri may be drawn down by help of a hook.

Various methods have been suggested and tried to facilitate reduction. Halpin applied a sheep's bladder, dilated with air or water, with the view of so lifting up the uterus. Desgranges took up the same idea. Dr. Playfair relates a case in which fluid-pressure by my bags sustained for twenty-four hours in the vagina replaced the uterus. I tried it in one case without success. In fact, a *point d'appui* is wanted below, which the perinæum is ill-adapted to give; the distensibility of the floor of the pelvis allows the bladder to expand downwards. And in most cases it would be difficult to get a bladder into the vagina. But since it is impossible to govern with accuracy the direction of the force, the bladder expanding, presses the fundus uteri directly upwards under the promontory, thus in reality calling for more force than would be necessary if the hand—a sentient and intelligent instrument—were used in the manner already described. It would be better to place the bladder in the rectum as I did. Thus it gets more command over the uterus, and the lateral course of the rectum will direct the pressure to one side of the promontory. Again, position has been much insisted upon. Blundell recommended the knee-elbow position, that is, upon all-fours. The knee-elbow position was tried by Hunter; Moreau says this was the method commonly tried; but that it was bad because the woman could not bear it. It has been supposed to help by gravitation. But this is illusory. The semi-prone position gives all the requisite facility. Godefroy (*Gazette des Hôpitaux*, 1859) placed the patient on the edge of the bed, so that her head and chest were hanging down on the floor, the legs and pelvis only being on the bed. In this position there is no resistance from the abdominal walls. Then, two or more fingers in the rectum act at greater advantage. Négrier, of Angers, introduced the whole hand into the vagina, turned it in supination, and pressed forcibly by the flat surface made by the backs of the first phalanges, and by the thumb brought level with them. This he calls " réduction à poing fermé." He succeeded by it in four cases. It is useful whilst pushing the uterus from below to exert counter-pressure by the other hand applied above the symphysis pubis. Professor Halbertsma relates a case in which a woman died; the uterus was found retroflexed and retroverted. There was no peritonitis. On pressing up the fundus, the uterus immediately rose like a feather into its place. This post-mortem demonstration is especially valuable. E. Rigby and Dewees advised bleeding to facilitate reduction. Where there is great congestion this may be useful. But the commonly existing exhaustion and the impetus given to absorption by bleeding, would be strong contra-indications.

The *management of the case after reduction* is important. Richter, Baudelocque, and Simpson, recommended to apply a pessary to prevent the uterus falling back again. The best pessary for this is a large Hodge's lever.

Looking to the history of retroversion and retroflexion, regarding it as a condition existing before pregnancy, and continuing with advancing pregnancy, the *prophylactic treatment* should have prominent attention. I have on many occasions used Hodge's pessary to support the uterus when retroverted during the first three months of pregnancy, as a precaution against impaction. This precaution should never be omitted in the case of pregnancy occurring in a woman who has once suffered from retroversion or retroflexion, especially if she has had abortions.

Other points are, absolute rest, encouraged by opium; keeping the bladder empty—retention may persist from paralysis even after pressure is removed; and medicines, such as mineral acids, are useful. The bladder requires the most careful attention. The liability of the mucous coat to undergo necrosis must be borne in mind. Shreds of mucous membrane or of fibrinous cast may become detached and occlude the urethra.

The treatment of incomplete retroflexion, that is, when the gestation is partly abdominal, must be mainly expectative. When labour has arrived, should the pelvic portion of the uterus not rise into the abdomen, we must endeavour to push it up by fingers in the rectum, whilst the patient is on all-fours.

If reduction cannot be effected, imitate the practice of Dr. Oldham, described in the case I have quoted. Draw down the cervix uteri, push down the abdominal mass, push up the pelvic mass, so as to effect a rotation of the uterus forwards.

Retroversion, or rather retroflexion, may occasion difficulty even *after delivery* at term. The uterus imperfectly contracted, and therefore enlarged and flaccid, may bend back and become locked in the pelvis. Retention of urine may result. This accident, Merriman says, has principally occurred on the second day after delivery; but he gives a case in which retention of urine happened from this cause on the ninth day. E. Martin says post-partum retroflexion is due to the placenta having been attached to the anterior wall of the uterus, the effect of which is to keep this wall longer than the posterior, and thus to throw the fundus back; and conversely that anteflexion is due to the placenta having been attached to the posterior wall.

But reduction should not be looked upon as the object to be accomplished at all hazards. One cannot think without a shudder upon the description given by Amussat of his supplementing his own strength by that of one or two assistants who helped him in pushing the uterus into its place. When the uterus has shrunk after tapping and abortion, it is no longer the chief source of mischief. Indeed, I have known death to ensue after an easy reduction of the uterus. The woman died from the shock already sustained, the continuing uræmia, and the injury done to the urinary apparatus.

The general rules, then, may be thus summed up :—1. Empty the bladder: 2. Make a gentle attempt at reposition: 3. If this fail, be governed by the urgency of the case; if there be great distress, induce abortion or puncture the uterus by the aspirator-trocar, and wait: 4. When abortion has taken place, make further cautious attempts to reduce: 5. If there be still difficulty, wait again: 6. When the uterus is reduced, support it in position by a suitable Hodge-pessary.

Prolapsus or Procidentia of the Gravid Uterus.

This, like retroversion, may be primitive, or may occur suddenly during the course of pregnancy. Conception may occur in a prolapsed or even in a procident uterus; and continuing to grow, it may retain its vicious position. The simple prolapsus becomes a pelvic gestation, the uterus growing in the pelvis proper. In this case, unless it become retroverted, it is pretty certain to grow up into the abdominal cavity in the fourth or fifth month; and then it is generally too large to sink bodily through the brim again. But, in the case of procident uterus, it is quite possible for the enlargement to go on external to the vulva. The pregnant uterus then, covered by the inverted vagina, forms an enormous mass between the thighs. It has been said also that the entire uterus containing the child has been driven through the vulva during labour. I have no doubt that in some of the instances recorded as of this nature, there was a defect of observation; and that the real condition was simply protrusion of the lower segment of the uterus covering the head, or the hypertrophied cervix, to be described presently.

A curious case which comes under this head is cited by Moreau,* from Chopart. A young woman had procidentia uteri, the result of violence before marriage. This was never reduced; but after twenty years the cervix becoming gradually opened, conception took place. Labour at term went on for twenty-four hours without progress, when the child was dead, and the woman seemed expiring. The surgeon, Marrigue, divided the cervix by incision, and thus was able to extract the child. The mother recovered. It is not clearly stated that the entire uterus and child were external to the vulva at the time of labour. There is a figure in Siebold's *Journal für Geburtshülfe* (1826) of a large mass outside the vulva with a foot projecting; but it is not certain that the whole uterus was outside. A more probable case is that of Portal, treated in conjunction with the first Moreau. A primipara had long suffered from prolapsus, but the uterus had gone up during pregnancy; and only came outside under violent straining in aid of slight labour-pains. The orifice was artificially dilated, and a living child extracted.

Harvey relates a case of conception taking place in a procident uterus. A dead premature child was expelled.

The *Treatment.*—Labour with simple prolapse is generally tedious, from the fundus wanting the full support and impact of the abdominal muscles. If the cervix is slow in dilating, and the expulsive power is deficient, the cervix should be dilated artificially by the water-bags, and the forceps put on the head, care being taken to support the perinæum and vulva well, lest the lower segment of the uterus be drawn through. In the case of complete procidentia the difficulty is aggravated: the uterus is away from all its auxiliary forces. Still, its

* "Traité pratique des Accouchements," 1838.

inherent power may possibly expel its contents. First of all you should try to carry the uterus up into the pelvis. If this fail, delivery must be ended outside the pelvis. Whether the uterus act by itself, or you find it necessary to extract the child by forceps or turning, it is desirable to support the lower segment carefully by means of a square cloth, having a hole cut out of its middle large enough to afford exit to the child. This opening is applied over the os uteri and the four corners are held up around the uterus, so as to counteract the downward traction exerted upon the child. Scanzoni says, the long-continued bruising of the uterine walls against the pelvis may cause metritis or sloughing.

I have not seen a complete procidentia of the gravid uterus at term; but I have several times been called in to see it. This is what I actually saw: an enormous fleshy mass protruded beyond the vulva, of livid colour, and presenting an opening—the os uteri—in its centre. On feeling the abdomen, however, I have found at least a portion of the uterus containing part of the fœtus there. It was clear then that, although the os uteri and cervix might be completely outside the vulva, the whole uterus was not. In one case in which the midwife said the labour-pains were forcing the whole body out, I found the os uteri outside; and, on examining, my finger passed three inches up along the cervical canal till it was arrested by the os internum uteri, above which was the child's head. This condition is represented in Fig. 97.

It is simply a case of *hypertrophic elongation of the cervix uteri*. In a subsequent labour she was seen by Dr. Roper, then my colleague in the Royal Maternity Charity. He found the same condition. There can be no doubt that the elongated cervix was a persistent condition. (See *Obstetrical Transactions*, vol. vii.) In another case I found the head had passed along the hypertrophied cervix and was protruding the os externum outside the vulva. In another, the hypertrophied cervix was enormously distended with extravasated blood—cervical thrombus—a condition to which the hypertrophied cervix is especially exposed. One cause of dystocia in these cases is the hardened state of the os and cervix uteri.

Dr. Roper relates (*Obstetrical Transactions*, vol. xv.) another case which illustrates the influence of labour upon the hypertrophied cervix. The subject was a primipara, aged twenty-two. In labour the cervix protruded through the vulva about three inches, forming a mass in circumference equal to a man's wrist. After reducing

FIG. 97.

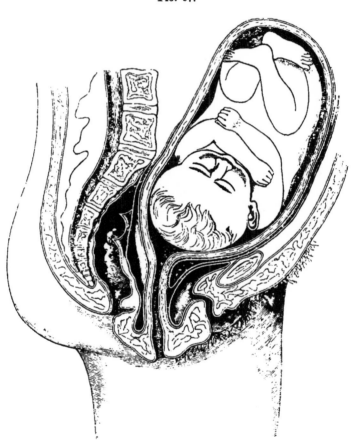

LABOUR OBSTRUCTED BY HYPERTROPHIC ELONGATION OF THE CERVIX.

The head rests upon the os internum; the os externum is outside the vulva. The figure also shows the down-dragging of the bladder and distortion of urethra.

the cervix within the vagina, the head could be felt. The cervix had a firm, gristly feel. I saw her in consultation, and we agreed that free incisions in the os externum were necessary. These were practised, so that the os externum was freely opened up to meet the natural expansion of the os internum. She was then delivered, after an anxious labour of fifty-two hours; both mother and child doing well. Dr. Roper insists upon the importance, in similar cases, of waiting until expansion takes place above, and then meeting it by incisions from below. Two

FIG. 98.

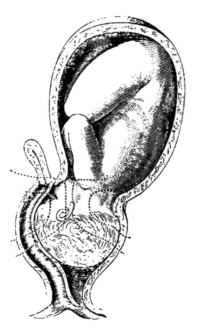

AFTER OLDHAM.

Representing pregnancy in one uterus, a second uterus unimpregnated by the side.

months after delivery, the cervix hung down in the vagina like a shrivelled piece of skin. It was amputated.

It may be necessary to dilate the cervix by means of the water-bags, or even by incision; and to apply the forceps, or to turn.

A considerable number of cases are now recorded where labour has been *obstructed by the hymen.* Where this is found the remedy is incision.

Pregnancy in one compartment of a two-horned uterus is a rare and, therefore, perplexing complication. I was called to a case of severe puerperal convulsions with albuminuria, in which it was considered desirable to expedite delivery. There had been strong pains but no progress. Mr. Garlick, who was in attendance, could touch the presenting head; but when I tried I found a dense fleshy septum between my finger and the head, although it was quite clear that my finger passed into the os uteri. At last, following my friend's guidance, I got my finger also on the head; and here I ascertained that there were two ora, each leading to distinct uterine cavities, one of which contained the child whilst the other was empty. It was necessary to deliver by craniotomy. Dr. Oldham describes a similar case (*Guy's Hosp. Reports*). Fig. 98 is taken from his illustration. Lefort cites a case from F. Tiedmann,* who says there is in the Heidelberg Museum a double uterus with double vagina, taken from a woman who died after labour. At her labour two distinguished physicians attended; one declared the woman was not pregnant, the other that the head was in the os uteri. One had examined the left vagina, the other the right. The same thing happened to two midwives at the Maternité in 1824. My difficulty, therefore, is not without precedent.

Birnbaum relates (*Monatsschr. für Geburtsk.*, 1863) a case of twin-pregnancy, a child in each side of a two-horned uterus. A very distinct saddle-like depression was observed near the umbilicus, the upper boundary between the two horns of the uterus. The summits of these rose to an unequal height. There was one placenta. He relates another case of single gestation in one horn of a two-horned uterus. There is a valuable collection of cases illustrating this subject in Kussmaul, and the subject is referred to in my work on *Diseases of Women*.

In some cases the septum is continued all the way down the vagina. There is then a double vagina as well as a double uterus. In these cases one vagina only generally is used in copulation. If labour were found obstructed by such a septum, there should be no hesitation in dividing it longitudinally with a hernia knife.

---

* "Journ. Complém. du Diction.," t. iv.

MECHANICAL COMPLICATIONS OF LABOUR CONTINUED—DYSTOCIA
FROM TUMOURS: FIBROID OR MYOMA—POLYPUS—OVARIAN—
EXTRA-UTERINE GESTATION-CYST — CONDYLOMATA — RETRO-
UTERINE HÆMATOCELE.

Pregnancy and Labour complicated with Tumours.

THERE are few complications of pregnancy and labour productive
of more perplexity, difficulty, and danger than tumours. They
present great variety of form, size, consistence, position, origin,
and relations, entailing corresponding interference with the due
course of pregnancy and labour.

First, we will discuss the influence of *tumours in the walls of
the uterus.* These will include *polypi* projecting from the walls
into the cavity.

Happily, in a great number of instances, fibroid tumour or
myoma imbedded in the walls of the uterus operates as a bar
to conception; but when pregnancy does supervene, the result
is often disastrous. The tumour interferes with the equable
development of the uterus; and, therefore, frequently determines
hæmorrhage and abortion. And, perhaps, this is the most
fortunate event; for delivery in the latter months brings additional danger. The hæmorrhage of abortion can be stayed by
injection or swabbing with perchloride or persulphate of iron. It
is not often that the tumour is so injured that inflammation
ensues; but looking to future probabilities, we should deprecate
incurring the risk of another pregnancy. I believe every
experienced obstetrician would advise a single woman known
to have fibroid tumours in the uterus to avoid marriage; with
the doubtful exception of tumours seated in the fundus.

Compression of the bladder and rectum may be caused by the combined growth and eccentric pressure of the tumour and uterus, causing symptoms resembling those arising from retroversion of the gravid womb. These may be developed early in pregnancy, and compel prompt intervention. In some cases, indeed, a tumour has caused retroversion.

After five months there is less risk of intra-pelvic compression, because the uterus and fœtus will have grown out of the pelvis. There is some danger of hæmorrhage, but generally danger is henceforth in suspense until the advent of labour.

A myoma growing in the uterine wall induces effects upon the uterus analogous to those of pregnancy. It acts as a cause of development, it attracts blood; it may be said that there is tumour-gestation. If to this be added true embryonic gestation, there is commensurate increase of uterine vascularity and growth. The tumour, or the cellular capsule around it, may undergo inflammation, or a kind of degenerative softening. Pain and even suppuration may result. At least, I think I have observed these events. Dr. Lee relates * a case proving fatal at five months from inflammation and suppuration of the tumour, with peritonitis.

The dangers attending labour are of the most serious kind. *First*, there may be obstruction to labour requiring dangerous operations to overcome it. *Secondly*, the uterus may be rent from the unequal strain of the contracting muscular tissues, or from dragging on the tumour by the advancing child. *Thirdly*, if the tumour be in the anterior wall of the uterus, so as to be compressed between the advancing head and the symphysis pubis, rupture of the bladder or vesico-vaginal fistula may occur. I have known two cases of this accident, one of which was fatal. *Fourthly*, the solid masses in the walls prevent the equable contraction of the uterine muscles during and after labour, and thus give rise to dangerous hæmorrhage, scarcely controllable by the ordinary means. *Fifthly*, the placenta may be attached to the tumour, and lead to inversion of the uterus. Inversion has even occurred where the placenta was not known to have so adhered. *Sixthly*, the bruising to which the tumour and the uterine wall are subjected, by being jammed between the head and the pelvis,

* " Clinical Midwifery," 1842.

is commonly followed by a low form of inflammation in the tumour, tending to necrosis, and giving rise to peritonitis and septicæmic puerperal fever.   There seems little doubt that myoma, being constituted of analogous tissues to those of the uterus in which it arises, grows with the uterus during pregnancy; and hence, at the time when labour comes, the tumour is unusually vascular and more prone to suffer from violence. It also undergoes a process of involution after labour, like the uterine tissue proper.

Myomas, it is known, occasionally get driven outwards, so as to hang loosely in the abdominal cavity by a slender pedicle to the external surface of the uterus.   In this position they are shut off from the influence of the changes going on in the uterus, and may be out of the way of mechanical mischief. I have attended a lady through several labours, who had a tumour which I could grasp in my hand through the abdominal walls; she never suffered in any way in consequence.   These tumours have been known to get detached altogether, so as to roll free in the abdomen.   In this case, they would gravitate into the retro-uterine sac, where they might obstruct labour. The danger and the prognosis are also influenced by the region of the uterus in which the tumours are seated.   Those occupying the lower region, and which are, therefore, most exposed to injury during labour, are the most dangerous; whilst those seated in the fundus above the child are comparatively free from mechanical injury.   They are not, however, altogether free from the danger of inflammation, and are not unlikely to cause hæmorrhage.

In not a few instances a myoma has been spontaneously expelled some days or weeks after labour.   I exhibited a tumour to the Obstetrical Society, expelled in this way some weeks after labour.   It seems that during labour a process of loosening of the attachments, favouring enucleation, takes place.   Danyau and M. Duncan each relate a case in which partial enucleation thus effected was easily completed surgically.   The great decrease in size of the uterus after delivery, and the persistent contractions, tend to cast off the tumour.

Sometimes the tumour disintegrates, or undergoes atrophy, and disappears, wholly or in part, after labour, or at least to such an extent as to escape detection. Dr. Pagan relates a striking case in which a tumour was mistaken for a second child. The woman recovering, the tumour rapidly diminished and disappeared below the pubes. She had observed the same phenomenon in preceding pregnancies. Dr. Leonard Sedgwick relates (*St. Thomas's Hospital Reports*, 1870) two cases in which uterine tumours entirely disappeared after delivery. Montgomery cites similar cases. Hence we see pregnancy may cure uterine myoma. A most important feature occasionally attending this complication is the extreme difficulty in establishing a correct diagnosis. The pelvis may be so blocked, the vagina and cervix uteri so distorted, that access for the examining finger below may be impossible, and at the same time abdominal exploration may be equally unsatisfactory. The mystery may, however, be cleared up by making repeated examinations at sufficient intervals. The comparative observations may reveal changes of form, size, and relation, and possibly auscultatory signs due to the progress of gestation.

Closely allied to the muscular tumour in the uterine wall is the *muscular polypus*, which, attached by a stalk, projects into the uterine cavity, or through the os uteri into the vagina. Such a polypus may impede labour by getting out of the uterus before the head and filling the vagina. A case of this kind occurred in St. Thomas's Hospital Maternity, in 1868. A solid polypus, as large as a full-sized cocoanut, blocked the vagina. It adhered to the os and whole circumference of the cervix. My colleague, Dr. Gervis, being sent for, found that some laceration of the surface of the polypus had occurred, and with every pain it became extremely tense and elastic. The child was delivered by craniotomy. A second child was delivered by turning. The tumour, after labour, protruded through the vulva. It was removed by the écraseur five days afterwards. The woman died on the thirteenth day, symptoms of peritonitis having set in. Peritonitis was found on postmortem examination. The tumour, which is preserved in St. Thomas's Museum, and figured in the *Obstetrical Transactions* (1869), was a myoma enveloped in a capsule of true uterine tissue. Low necrotic inflammation was progressing in its substance. In St. Bartholomew's Museum is a specimen (32.22) of a large polypus removed by excision ; it was five inches long; its base of attachment nearly 1½ inches in diameter. The subject was twenty years old. The polypus was first discovered after labour, and was excised ten days afterwards. She recovered speedily.

Again, the polypus may be above the child's head *in utero*, and not appear until the child is born, as in a remarkable case recorded by Dr. Crisp. The placenta being retained, Dr. Crisp introduced his hand and removed it. He thought he felt another child, but it was found to be a large polypus, causing violent expulsive pains and greatly exhausting the patient. The energetic action of the womb forced the polypus so low down in the vagina as to interfere with the passage of the catheter. The patient died collapsed, worn out with the constant uterine action; there was no hæmorrhage. This symptom: violent expulsive action continuing after the birth of the child, seems one of the dangers attending leaving the polypus attached. It was enough, we see, in Dr. Crisp's case, to cause death. Ingleby and Gooch relate fatal cases. It was very severe in other cases, namely, in one related by Mr. Freeman,* and in one by Dr. Priestley.† It is important to remember another point in connection with labour complicated with tumour. It is that the tumour has been mistaken, as in Dr. Pagan's case, for the head of a child. Such a mistake, leading as it might to attempts to deliver by hand, by forceps, or by perforation, might be fatal. In this way the uterus might even be torn away. And yet how easy the error! It is so natural to conclude that any firm rounded mass, like in size to the fœtal head, is a fœtal head; and so rare, and therefore unthought of, is a tumour, that even experienced men may be deceived. I know of no way of ascertaining the exact conditions, but that of passing the whole hand above the tumour, and tracing deliberately its attachments. Then the question, an anxious one, not free from doubt, arises: how to deal with it? Whether left alone or removed by operation, death may follow. Experience is not decisive enough to justify the laying down an absolute law. My opinion would be generally in favour of immediate removal by the wire écraseur, the galvanic cautery, or enucleation; but in any given case, the surgeon must be governed by circumstances and the means at his disposal.

* "Obstetrical Transactions," vol. v.   † Ibid., vol. i.

U

In a case recently under my care the subject complained of foul discharges, and symptoms of chronic pyæmia three months after labour. I found a large polypoid myoma in a state of decomposition growing from the fundus of the uterus. I removed it by the wire écraseur. The uterus then underwent normal involution, and she made a good recovery.

Since it may be difficult or impossible to make out the nature and attachments of some of the things that may present in the passages by the mere sensation they impart to the touch, it is a good rule in every case of doubt to put the patient in anæsthesia, and to explore with the whole hand. In this way, if we succeed in tracing the walls of the vagina and uterus, getting all round the presenting mass as far as possible, and above it, we can hardly fail to determine whether it belongs to a fœtus or be attached to the uterus.

The sources of danger, then, in polypus, are similar to those of tumour in the uterine wall. The mass gets bruised and undergoes necrotic inflammation leading to metro-peritonitis and septicæmia; imperfect contraction may lead to hæmorrhage; uterine tenesmus or tetanus may cause collapse from exhaustion; and inversion may be produced, as in a case of Dr. Beatty.

The relation of *cancer* of the uterus and vagina to pregnancy will receive incidental illustration in the Lectures on "Rupture of the Uterus" and "Cæsarian Section."

We will further discuss the treatment of these cases when we have gone through the history of complication with other kinds of tumours.

The next class includes tumours external to the structure of the uterus. First, *ovarian tumours*. These are movable or immovable; they may be solid or fluid; be seated below the uterus in the pelvis, or to one side, behind it, or above it; and the issue of the case will depend very much upon which of these conditions is present. The diagnosis is sometimes very difficult, only to be clearly made out by repeated comparative observations.

The influence of ovarian tumours on the course of pregnancy is the first subject of study. It is generally the case that the ovarian tumour exists and has attained a certain size when the pregnancy begins; and as ovarian tumours mostly rise out of the pelvis, the gravid uterus enlarging, finds the tumour above it, probably a little to one side. The tumour being movable, is lifted up by the growing uterus, and pushed aside. If of moderate size, the *pregnancy may go on to term*, and delivery be accomplished without difficulty. Such a case is not uncommon. I have seen several, and some in which a woman with ovarian tumour has been delivered two or three times. But we shall not be always so fortunate. Another issue is *abortion or premature labour*. Sooner or later the tumour presses against the growing uterus, or the double growth becomes too much for the system to bear, and the uterus is driven to expel its contents. When this happens the danger is usually over for a time. But if the danger is not averted by the spontaneous or artificial reduction of the uterus, it is but too probable that the tumour must give way. There is a limit to the distensibility of the abdomen; there is a limit to the pressure which the abdominal viscera, the vessels, and the thoracic organs can endure. In ordinary gestation, provision is made for the accommodation of the uterus. But the system can ill tolerate the simultaneous growth of two masses in the abdomen. To be effected safely, distension and accommodation must proceed at a given moderate rate. When an ovarian tumour is added to pregnancy, the distension is rapid and beyond measure. The tumour grows as well as the uterus. It not only oppresses by its bulk, but it drains the system of part of the supplies wanted for healthy purposes. We may, therefore, find towards the end of gestation such a degree of suffering and exhaustion, that the *patient will sink either before or soon after delivery*.

Or a remarkable circumstance may happen. The uterus growing upwards underneath the tumour, not only lifts it up, but *rolls it round on its axis, elongating and twisting the pedicle.* The consequence of this is strangulation of the vessels which feed the tumour. They burst and pour out blood into the cysts of the tumour, or externally into the peritoneum. Death speedily follows under the shock of the injury, from hæmorrhage or peritonitis. I have recorded a case of this kind in *St. Thomas's Hospital Reports*, 1870. Dr. St. John Edwards, of Malta, records another (*Lancet*, 1861). Or the axial twisting may occur under the expulsive efforts of labour. Rokitansky relates more cases in illustration of this accident.

Another event, more common, is *simple rupture of the cyst*. This may happen at any period of pregnancy, but is most frequent after the sixth month. The accident is not always fatal. The effused fluid may be absorbed, and the pregnancy may even go on. But more frequently the shock is followed by fatal collapse; or, if the patient live to the point of reaction, fatal peritonitis ensues. In cases of this nature which I have seen, death took place from shock, and although intense pain was complained of, there was no evidence of peritonitis on dissection. Whether it be by bursting as above mentioned, or by some other process, there is evidence to show that ovarian tumours may vanish under the influence of pregnancy. Montgomery and Hamilton give examples.

If the patient escape the perilous period of pregnancy, she has to run the gauntlet of even greater perils *during and after delivery*. The tumour may burst or become strangulated under the straining of labour. It may be otherwise injured, so that a fatal result follows in a short time. All this may occur even when the tumour has offered no material obstruction to the passage of the child. But where the tumour is lodged wholly or partly in the pelvis, unless it be movable, it can hardly escape injury, and by encroaching on the pelvic space, it obstructs the birth of the child. Rupture is more likely to happen; the dragging or stretching of the structures to which it is attached may set up inflammation, and the bruising its own structure undergoes may also prove fatal.

I have referred to the familiar fact that many women have gone through several labours without accident under complication with ovarian tumour. Let me insist upon another fact, necessary to moderate the confidence in future immunity which undue consideration of these fortunate cases may inspire: not a few women, after thus escaping once or oftener, have in the end fallen victims to one or other of the catastrophes recited.

Mr. Berry, of Birmingham, relates (*Obstetrical Transactions*, vol. viii.) an extraordinary case, showing what narrow escapes sometimes occur. Labour had been obstructed by an ovarian tumour, and the child had been extracted with considerable force by forceps.   Next day, after coughing, the patient felt something come down.   Mr. Berry being called in, found it to be an ovarian tumour, the pedicle of which was traced through a rent in the upper part of the vagina.   Mr. Berry was of opinion that the rent had been caused by the forceps.   But it is quite possible that in such a case the rupture might occur spontaneously.   A ligature was put on, and the tumour was cut off. The woman recovered.   The preparation is in St. Bartholomew's Museum.

Dermoid tumours, containing hair and fat, may get into the retro-uterine pouch and obstruct labour.   They may feel so hard as to suggest solidity.   But on puncture fat may run out, as in a case of Ramsbotham's (*Path. Trans.,* vol. iv.), and in others by Ingleby.   Hence Ramsbotham says all tumours obstructing labour should be punctured.   Denman cites a case where labour was obstructed by a dermoid cyst between the vagina and rectum.   The head was perforated; but the woman died from the injury to the tumour and surrounding parts.   It would probably have been better to follow Ramsbotham's rule.

Amongst the most remarkable and dangerous forms of tumour that may complicate pregnancy is the *tumour formed by an extra-uterine gestation.*   It must be very difficult to diagnose this with certainty during the uterine pregnancy. The extra-uterine cyst is in great danger of bursting at this time.   If it endure to the term of gestation, it may burst under the violence of labour, or become the centre of inflammation that may prove fatal.   A case of recovery is related in Perfect.   It was communicated by Mr. John Bard, of New York, to Dr. John Fothergill.   A woman, æt. twenty-eight, had had one child and thought she was pregnant again.   At the end of nine months she had some pains which went off, and the swelling grew less.   A hard indolent tumour remained in the right side.   Menstruation returned; she conceived again; at the end of nine months she was delivered of a healthy child.   The tumour still felt as before.   Five days afterwards violent fever, purging, pain in the tumour, profuse fetid sweats set in.   These subsided into hectic and diarrhœa. After nine weeks, fluctuation was manifest in the tumour; it was opened; a vast quantity of fetid matter, and a fœtus of common size, were extracted through the incision.   They imagined the placenta had dissolved into pus. The woman recovered and suckled.

Two cases are recorded by Dr. Greenhalgh.* One of these is also reported at length by Mr. Cooke.† A tumour was found in the pelvis obstructing labour; it was pushed up with some force out of the pelvis, and the child was delivered by turning, dead. The woman died two days after the labour. A full-grown fœtus, contained in its membranes unruptured, was found in the peritoneal cavity. Beneath the tumour the uterus was seen partially contracted, and *unruptured*. The placenta of the extra-uterine fœtus was found attached to the fimbriated extremity of the right Fallopian tube. Dr. Greenhalgh's second case is briefly as follows :—A twin extra-uterine fœtation obstructed labour; the forceps was applied to the uterine child. The woman recovered, with subsequent discharge of fœtal bones. Montgomery has collected some cases of women bearing uterine children several times successfully, an extra-uterine gestation persisting throughout.

Obstruction by *hydatid tumour of the liver* and by *cystic disease of the kidneys* occasionally occurs. It would be better to puncture these before labour than to incur the risk of their bursting into the abdomen.

Sometimes the *distended bladder* is carried down before the head, and presents a tense fluctuating tumour at the vulva. The diagnosis and the treatment offer no difficulty if we have observed the good obstetric rule of passing the catheter in every case of tedious labour. A distended bladder complicating pregnancy has been mistaken for dropsy. Lowder mentions a case where tapping was performed. The trocar traversed the bladder, penetrated the uterus, and pierced the child's head. Death ensued.

The obstruction offered by *tumours in the vagina and vulva* may be great; but these cases are generally less serious, because they are more within reach of operative measures for removal. If they stand in the way, they should be removed before the child comes down. Fibroid tumours may spring from the vaginal walls. *Condylomatous* or *cancerous growths* may form large tumours at the vulva. If these are left until labour

* " Bartholomew's Reports," 1865.
† " Obstetrical Transactions," vol. v.

supervenes they must undergo severe laceration, and may cause rending of the vagina. A young woman was in the London Hospital several months advanced in pregnancy, having an enormous growth of syphilitic condylomata around the vulva. It was removed by the knife, by my advice, by Mr. Curling. The hæmorrhage was controlled without much difficulty, and the pregnancy went on undisturbed. We should now remove such excrescences by the galvano-cautery wire or knife.

Another form of tumour which I have known to co-exist with pregnancy is *retro-uterine hæmatocele*. The mass was expelled by the rectum. Where such a tumour is suspected, it may be punctured by the rectum. Blood-tumours may also form in the os uteri, at any point of the vagina, and in the vulva. If they obstruct labour, it is best to open them before the head comes to pass, otherwise increased laceration and bleeding may result. The diagnosis of retro-uterine hæmatocele and other retro-uterine tumours is discussed in the preceding lecture on "Retroversion." (*See* Figs. 94 and 95.)

The diagnosis of ovarian tumours. In some cases the presence of ovarian disease will be known before the pregnancy has begun. But it is remarkable that in a considerable number, no tumour is suspected until symptoms of distress come on at an advanced period of gestation. Then we are led to examine. The symptoms are mainly those which result from mechanical pressure. Dyspnœa, accelerated pulse, hectic, accompany excessive abdominal tension. This may be due to excess of liquor amnii, to twin-pregnancy, to complication with ascites. With care the two tumours, ovarian and uterine, can generally be distinguished.

By referring to Fig. 99 we see the general outlines of the two masses are preserved. Even through the abdominal walls, the groove or furrow between the two may be felt: it gives the idea of a bi-lobed tumour; the distension of the abdomen is greater transversely, in the flanks, than in simple pregnancy; and the fœtal heart is heard very much on one side, and generally lower than it should be. And I have observed that the spot of greatest intensity of this sound shifts its place as the gestation advances and the uterus is pushed more and more aside. It may or may not happen that the os uteri is displaced, and that a portion of the tumour may be felt in the brim of the pelvis.

A small ovarian tumour, or an early Fallopian gestation-cyst, may be diagnosed by the feeling a tense, elastic swelling in the roof of the pelvis, stretching the posterior wall of the vagina, and carrying the uterine neck forwards, and the fundus to one side. The diagnosis is made clearer, and a good thera-

Fig. 99.

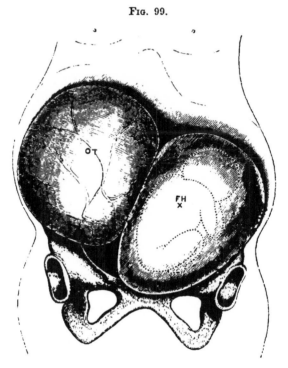

PREGNANCY COMPLICATED WITH OVARIAN TUMOUR.

o t, the tumour lifted out of the pelvis by the uterus, u, which is pushed over to the side and over the brim of the pelvis; f h, spot where fœtal heart may be heard.

peutical indication is fulfilled, by puncturing the swelling with an aspirator-trocar.

The uterus deformed by myoma has been mistaken for a uterus bicornis. Nor is the diagnosis easy, even after the embryo and secundines have been expelled.

The placenta adhering to a polypus may simulate inversion of the uterus. We must be careful therefore to put in practice all the tests for inversion. These I have described under " Inversion."

The *treatment* or *management* of pregnancy and labour complicated with ovarian tumours will be indicated by the compendious history I have traced of the events which have been observed under this complication. The rule of conduct should be based upon the general law of giving primary consideration to the safety of the mother, regarding the fate of the child as of secondary importance. Indeed, a rigid analysis will show that the best hope of rescuing the child will, in many cases, depend upon our success in saving the mother.

The fact which stands most prominently out is, that a main source of danger is the injury the tumour undergoes, especially during labour. During pregnancy even, this danger is great in the case of ovarian tumours. The risk of bursting, of strangulation of the tumour, and of consequent shock, hæmorrhage, and peritonitis, is so serious, and the catastrophe comes with so little warning, that the question whether it is ever prudent to let pregnancy and ovarian tumour proceed together without interference compels attention. To do nothing, because pregnancy has often been completed without mishap, is simply trusting to chance; it is a surrender of judgment but too likely to entail unavailing regret. We have no means of telling whether any particular tumour will burst or become strangulated under pregnancy or labour. The fluid tumours are more likely to burst; but the solid ones are liable to twisting and strangulation and perforation. Both may give rise to unexpected obstruction to labour, and undergo fatal injury during the process. The double burthen itself, apart from mechanical injury, may prove too much for the system to bear. On the other hand, judicious interference does not involve any serious danger.

The observation of a considerable number of cases of severe constitutional or organic disease complicating pregnancy and of ovarian tumours progressing simultaneously, has led me to the general conclusion, that it is better to bring the pregnancy to an end in the first instance, and then to deal with the disease or tumour in its simple state according to its particular indications. This conclusion is the result of three different orders of observation, all converging to the same point. 1st. In a large proportion of cases, Nature solves the problem, takes the case into her own hands, and finds relief by the spontaneous induction of premature labour. 2ndly. In another series of cases, where labour has been induced artificially, immediate relief and safety have been attained. 3rdly. In another series, where labour has not occurred spontaneously, or been induced by art, formidable catastrophes, and even death, have happened. The point to which all reasoning converges, then, appears to be to reduce the case to its simplest expression by eliminating one or other of the complicating elements. Which shall be eliminated? We may tap the cyst and so reduce the bulk of the tumour that it will be brought within tolerance; and even, should premature labour not follow the tapping, and it is very likely to follow, what remains of the tumour may not impede delivery at term. But there are considerations which weigh strongly in favour of eliminating the pregnancy. 1st. The tumour may be in great part solid. 2ndly. Even if obviously in great part fluid, we cannot tell until after tapping what amount of solid residuum there may be to endanger labour. 3rdly. Tapping the cyst may in itself be dangerous, the cyst may fill again; premature labour may set in under unfavourable circumstances. We should, therefore, do better to bring on labour in the first instance, and deal with the uncomplicated disease under the best conditions. This should be the rule. The exception should be to tap or to extirpate the tumour. The propriety of selecting extirpation has been ably discussed by Mr. Goddard (*Obst. Trans.*, vol. xiii.), and illustrated by the history of a case in which Mr. Wells performed ovariotomy on a woman about two months pregnant. She recovered, and was delivered at term. Here, then, as in so many other questions, an absolute solution cannot be stated. We must study any given case in all its aspects, and select that proceeding which is indicated by the special circumstances. If the tumour be in great part solid, we must choose between extirpation of the tumour, or the indication of labour. If the tumour be in great part fluid and of large size, we may tap or induce labour. If the tumour be small and showing no tendency to rapid growth, we may temporize, watch, and perhaps run the risk of letting things take their course undisturbed. That pregnancy is no bar to successful ovariotomy, was proved by the case published by M. Burd (*Med.-Chir. Trans.*, vol. xxx.). The subject aborted three days after the operation, and recovered. She was about four months gone.

Another question must be considered. Take the case where we are surprised by the bursting or strangulation of the cyst, and death by shock, hæmorrhage, or inflammation threatens. If we do nothing, or trust simply to ordinary means, the woman dies. One such case appears to have been saved by decisive action. Mr. Spencer Wells performed gastrotomy, removed the burst tumour, and cleared out the effused fluids (*Obst. Trans.* vol. xi.). The woman made a good recovery, pregnancy being uninterrupted. I am disposed to think that where severe and sudden peritonitis occurs during pregnancy, it is frequently due to rupture of an ovarian cyst. Under these circumstances recovery is scarcely probable. But, as Mr. Wells's case shows, the probability is increased by removing the offending cause. We should, therefore, not hesitate to follow his example.

The aspirator-trocar is likely to prove of great service in cases of small fluid ovarian cysts, and of early extra-uterine gestation-cysts. These cysts may be tapped by the vagina, and the contents drawn off with ease and safety. Professor Thomas has lately operated successfully upon an extra-uterine gestation by opening the cyst through the vagina by the galvanic cautery and removing a fœtus of four months' growth (*New York Med. Journ.* 1875).

Next we have to discuss how to act when labour is obstructed by tumour. Up to a certain point the principles of acting are the same whether the tumour be ovarian or fibroid. We have to bear in mind that our duty is, *first*, to deliver the woman safely; *secondly*, to secure the safety of the child, if possible.

The first question to determine is, *Can the obstructing tumour be pushed out of the way?* Many cases of ovarian tumour, and some of fibroid, are movable, and admit of being lifted above the pelvic brim, so that the child can find room to pass. This may be done by the hand. The operation will generally be facilitated by placing the patient on all-fours. Now and then the tumour rises out of the pelvis in the course of labour, under the unaided action of the descending uterus and child. Beatty and Depaul relate cases in point. In one case, Dr. G. Kidd carried a tumour out of the way by the pressure of one of my bags distended below it in the rectum (*Dublin Quart. Journ. of Med.* 1870). As soon as the pelvic brim is clear, endeavour to seize the child's head with the forceps. Three excellent reasons persuade to this practice : *first*, the head being made to occupy the pelvis, the tumour cannot fall in again ; *secondly*, you save the forces of the mother, by lessening the need of uterine action—you lessen the risk of rupture of the tumour and of the uterus ; *thirdly*, you improve the chances of securing a live child.

This, then, is the simplest course : put aside the tumour if you can.

The second question, one which arises in the event of the first being decided in the negative, is, *Can we diminish the bulk of the tumour so as to give room for the child to pass?* We might greatly simplify this difficult question by deciding at once that we ought not to puncture solid tumours. Little or nothing can be gained in the way of lessening bulk by this procedure, whilst the injury is very likely to be followed by low necrotic inflammation and death. Dr. Charles West has discussed and illustrated this point.* In a fatal case he attributed the death to an attempt to puncture the tumour before trying to carry it above the pelvic brim. There was no general peritonitis, but the wound in the tumour "was gaping widely, the tissue around it was of a black colour, and this discoloration extended thence inwards towards the centre of the tumour." But this rule requires a qualification. The best test of solidity is sometimes puncture. The aspirator-trocar is not open to the objection urged by West.

I would submit this proposition as the rule in dealing with solid tumours : if they cannot be pushed up and aside, puncture by aspirator, but do not incise them, unless you see your way clearly to *removing them altogether.* To enucleate a fibroid tumour from the uterus, during labour, is a hazardous undertaking. But circumstances may occasionally be favourable. Thus, Dr. Hicks relates a case† in which, finding the head arrested by a firm tumour so filling the vagina that delivery by forceps, turning, or embryotomy, seemed doubtful, he made a small incision in the lower part of the tumour, which, under distension, permitted its enucleation and removal. There was

---

* " Diseases of Women." Edition, 1856.
† " Obstetrical Transactions," 1870.

no bleeding. The woman did well. The decision is clearer when the tumour is of polypoid form. If a polypus protrude into the vagina before the child, I strongly advise that it be removed before the child is allowed to pass over it. You may do this by the écraseur directly, or you may apply a whip-cord ligature round the pedicle first, and cut off the tumour below. A better plan still is removal by the galvanic cautery-wire.

In the case of fluid tumours, as ovarian cysts, the propriety of puncturing is established. How shall we proceed ? The

FIG. 100.

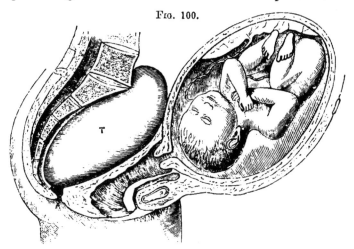

LABOUR OBSTRUCTED BY A FLUID OVARIAN TUMOUR, T, FIXED IN THE HOLLOW OF THE SACRUM, AND PUSHING THE UTERUS FORWARD AND UPWARD.

first thing to do is to rupture the fœtal membranes, and let the liquor amnii drain off. This at once reduces bulk and takes off some amount of tension. Then puncture the cyst. The best place, *if the cyst protrude into the pelvis and fluctuation* be felt there, will generally be the most prominent part of the tumour behind the os uteri in the roof of the vagina, or by the rectum. But if fluctuation be not felt in the vagina, it will be better to puncture in the most prominent part of the cyst in the abdominal walls, carefully determining the position of the uterus by the fœtal heart and palpation. Use a moderate-sized trocar or the aspirator first. The cyst, if punctured by the abdomen, will probably collapse more completely; and you are more likely to avoid the solid basis which is so often found at the lower part of ovarian tumours.

I am disposed to think that even where the ovarian cyst does not materially obstruct labour, it should be tapped. It is a remarkable circumstance that in not a few cases, the cyst has ruptured—or, at least, the evidence of rupture has become manifest—days after labour, causing death. Dr. Clay relates an instance of this.* In such a case as that sketched in Fig. 100, puncture of the tumour by vagina or rectum would at once permit the uterus and child to descend.

Dr. Playfair has examined the question how to deal with ovarian tumours by comparing the histories of nearly fifty instances (*Obst. Trans.*, vol. vii.) of labour complicated with ovarian tumour. He found that of the cases delivered by craniotomy more than half had terminated fatally. It might, he urges, be fairly assumed that had the Cæsarian section been performed in these cases at an early period the mortality of mothers would not have been greater, whilst some of the children would have been saved. This reasoning is difficult to resist; but it clearly does not justify a uniform mode of proceeding. The facts, however, prove the extreme danger of dragging a child past an ovarian tumour; and that, in some cases at least, the Cæsarian section is the safer proceeding. When gastrotomy has been performed with the view to Cæsarian section, it seems a reasonable thing to proceed to the extirpation of the tumour, where practicable, at the same time. Indeed, looking at the fair prospect of a radical cure of a threatening disease, as well as at the solution of the difficulty of effecting delivery, I think the propriety of conjoint Cæsarian section and ovariotomy should be carefully weighed when discussing the treatment of these cases.

The third question arises in the event of the tumour being immovable, and incapable of reduction of bulk. The danger now rises high. There is obstruction to labour; there is the almost unavoidable risk of injuring the tumour. Malignant ovarian cysts may be firmly attached to the pelvic walls, sarcomatous or bony tumours may spring from the pelvic walls, and fibroid tumours of the uterus may be impacted in the pelvis. Our decision as to the course to be adopted must be governed by a careful survey of all the conditions of the particular case. It is assumed that we cannot act

* "Obstetrical Transactions," vol. i.

upon the tumour. The alternative is that *we must act upon the child*. The mode of action will depend upon the degree to which the pelvis is contracted, and upon the estimate we may form of the nature of the tumour and its liability to be injured by the passage of the child. If there remain three inches, or perhaps even less, of space in the conjugate diameter, and the tumour be of a yielding substance, we may possibly deliver by the forceps, but turning is generally to be preferred; if the space be very small, say under two inches, and the possibility of the tumour being seriously bruised is great, we must be prepared, whether we have applied the forceps or have turned, to perforate the child's head to lessen its bulk and solidity. A perforated head will flatten and mould itself in its passage, especially if the cephalotribe be applied; thus plastic and yielding, pressure against the tumour is greatly lessened. In the case of a solid or comparatively firm tumour, leaving barely an inch or so of pelvic space, it may be difficult to reach the head to perforate, or, if perforation be accomplished, it may be impossible to pursue the further steps of crushing the head by cephalotribe, of picking off pieces of the cranial vault, and of seizing the face by the craniotomy forceps. It will, however, be seen in the Lecture devoted to Craniotomy, that this operation admits of being successfully performed in a far narrower space than is commonly deemed possible.

This is especially a case where my new operation of reducing the head by making sections of it by the wire-écraseur promises to be useful. We should exhaust all means of acting upon the child before deciding upon the Cæsarian section. The case should be well studied in all its bearings. If we are of opinion that no amount of mutilation of the child that can be effected will ensure delivery with a reasonable prospect of saving the mother, then we should spare the child and deliver it by the Cæsarian section. Great as is the peril of this operation to the mother, there comes a point at which it holds out the best chance. And to give the best chance we should endeavour to perform the operation as the first step, that is, by election, without having previously damaged the prospect of success by fruitless efforts to deliver by other means. If the obstructing tumour be extra-uterine, bony, or semi-solid, and encroaching to an extreme degree upon the pelvic space, the argument for Cæsarian section will be strengthened. If the tumour be uterine fibroid, the Cæsarian section offers but a slender hope.

Dr. E. J. Lambert, in his valuable *Étude sur les grossesses compliquées de Myomes utérins*, 1870, has collected fifteen cases in which the Cæsarian section was performed. Two women recovered: one operated upon by Mayor, of Geneva, and one by Duclos, whose case is cited by Tarnier.

In the case of *complication with an extra-uterine gestation*, it appears to me that we should first of all determine whether the extra-uterine tumour can be pushed out of the way so as to secure it against violence. This is scarcely likely, for these cysts almost always contract adhesions with the viscera in the lower abdomen. We are then driven to elect between mutilation of the child and gastrotomy. If by means of the former proceeding we can ensure delivery of the uterine child without doing violence to the extra-uterine cyst, we may adopt it. But I should be more disposed, than in any other complication, to resort to gastrotomy. In this case it would be proper to remove the extra-uterine child in the first instance, avoiding opening the uterus unless the passage of the uterine child, *per vias naturales*, should still prove impracticable.

What is best to do in the case of a myoma *in utero* after labour? It has probably been injured; and inflammatory process of a low type, leading to metro-peritonitis and blood-infection, is likely to set in; and the uterus, still in a state of active muscular development and reflex irritability, resents the presence of the tumour as a foreign body. The expulsive pains set up are so severe and exhausting as to be in themselves a source of danger. This I have seen strikingly marked in a case in which I was consulted by Mr. Corner. Cases in which spontaneous or operative enucleation has been effected soon after labour, have done well. The general indication, then, to get rid of the tumour early, seems clear. The mode of proceeding must be determined by the characters of the case in hand. The cervix should be well dilated by introducing a faggot of laminaria tents, so as to give free space for examination and manipulation. Then a hernia, or other convenient knife, carried into the uterus, may be used to divide the capsule of the tumour freely. If the tumour project much into the uterine cavity, it may then be possible to shell it out, partly by the fingers, and partly by a Museux's vulsellum dragging upon it.

If the tumour do not project much, and immediate enucleation be too difficult, further proceedings may be postponed. The uterus continuing to contract may drive the tumour further into the cavity, and in a day or two its removal may be easier. This process may be aided by frequent doses of ergot, or by the subcutaneous injection of ergotine. Hæmorrhage, if it occur, may be controlled by pledgets of lint steeped in perchloride or persulphate of iron; and fœtor may be obviated by lint plugs soaked in carbolic acid oil.

We may sum up the argument in the following propositions, rising from the simplest cases to those of extremest difficulty.

A.—In the case of *tumours complicating pregnancy* :—

    1. Induce premature labour.

    2. If the tumour be fluid ovarian, and distress great, tap it.

    3. If ovarian tumour burst or become strangulated during pregnancy, remove it by gastrotomy.

B.—In the case of *tumours obstructing labour, that is, presenting before the child* :—

    1. Push the tumour aside, if possible.

    2. If the tumour be fluid lessen its bulk by puncture.

    3. If solid, puncture by aspirator-trocar, and if still not diminished in bulk remove it altogether by enucleation or by wire.

    4. If the tumour cannot be advantageously acted upon, reduce the bulk of the child. Turn, perforate, crush the head by cephalotribe, break up the cranial vault, remove it by sections by wire-écraseur.

    5. If neither tumour nor child can be advantageously acted upon, have recourse to the Cæsarian section.

C.—When the *tumours present after the birth of child* :—

    1. If polypoid remove early after labour by wire-écraseur or galvanic cautery wire.

    2. If sessile or projecting in a marked degree from the inner surface of the uterus, more especially if seated in the cervix, or lower zone of the uterus, so that they have been bruised by the passage of the child, remove if possible by enucleation.

    3. If they cannot be so removed, try to promote expulsion by quinine, by subcutaneous injection of ergotine, and watch to counteract septicæmia.

X

MECHANICAL COMPLICATIONS OF PREGNANCY AND LABOUR (*continued*): DEFORMITIES OF THE SKELETON—THE RACHITIC PELVIS—THE KYPHOTIC PELVIS—THE OSTEOMALACIC PELVIS —THE SPONDYLOLISTHETIC PELVIS—DISEASE OF THE PELVIC ARTICULATIONS—NAEGELE'S PELVIS—FRACTURES OF THE PELVIC BONES—THE "THORNY" PELVIS—DIAGNOSIS OF PELVIC DEFORMITY—PECULIARITIES OF LABOUR IN DEFORMED PELVIS.

The Principal Abnormal Conditions of the Skeleton

which offer impediments to labour, are the following :—

1. Those arising from *rickets*. These affect the pelvis most frequently : but often the spinal column is also implicated in such a manner as to influence labour.

2. Those arising from *osteomalacia*. These commonly affect both pelvis and spine.

3. Those arising from *disease or injury of the vertebræ or the vertebral joints*. These, when leading to dislocation of the lumbar vertebræ, produce the deformity called *spondylolisthesis*.

4. Those arising from *scrofula, syphilis, or some other less understood diseases* which lead to *osteophytic* growths, producing nodular or pointed excrescences from the pelvic bones.

Stein, the younger, enunciated the important law, that like causative diseases always produce like forms of pelvis. Hence there is a rickety pelvis, an osteomalacic pelvis, and so on.

It would carry me far beyond the scope of this essentially practical work, were I to attempt a full account of all the varieties and degrees of skeleton-deformities in their obstetric

relations. I limit myself to giving a concise description and
figures of the leading typical forms, which exercise the most
marked influence upon child-bearing, and which determine us
in the choice of operative proceedings.

Fig. 101 may serve as a healthy standard pelvis. Like
Figures 102, 103, 104, 105, 106, it is reduced to one-third of
the natural size for the sake of more precise comparison.

1. *Rickets.* This disease, beginning as it does before maturity, influences not only the shape, but also the development

FIG. 101.

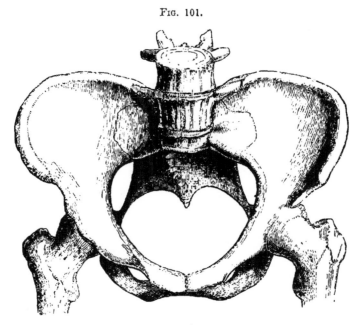

STANDARD NORMAL PELVIS: AD NAT., ONE-THIRD NATURAL SIZE.

of the skeleton. A rickety skeleton is commonly deformed,
and below the average size. This double abnormality inten-
sifies the evils which pertain to each separately.

A.—The most common deformity produced in the pelvis tells
most seriously upon the brim. This is compressed antero-
posteriorly; the promontory of the sacrum is protruded forwards, and the symphysis pubis is flattened in; the conjugate
diameter is shortened, absolutely, and also relatively, to the
other diameters. From cordate, the outline of the brim has

x 2

become transversely oval or _reniform_. Lordosis, or arching forwards of the lumbar vertebræ, often accompanies the pelvic deformity. This spinal curvature is at times so great that the lumbar vertebræ overhang the pelvic brim, impeding the entry of the uterus and fœtus, virtually closing the pelvis. When this occurs, the protrusion forwards of the uterus causes extreme prominence or overhanging of the abdomen. There is a remarkable example of this deformity in St. George's Museum. The woman died after craniotomy. Sometimes the spine is curved laterally—scoliosis. When this occurs there is usually oblique distortion of the brim as well as flattening. A perpendicular drawn from the symphysis pubis will strike on one

Fig. 102.

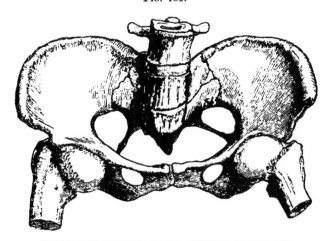

EXTREME RACHITIC PELVIS: AD NAT., ONE-THIRD NATURAL SIZE.

side of the promontory. The general effect upon the pelvis below the brim, that is, in the cavity and outlet, is to expand it. Not that there is absolute increase of room, but the relative dimensions compared with those of the brim are larger. There is often narrowing of the pubic arch, approximation of the tubera ischii, and sometimes incurving of the lower part of the sacrum and coccyx. Still, as a rule, in rickety pelves, the bones at the outlet diverge; there is room for the introduction of the obstetric hand, and therefore for operations upon the child.

B.—Rickets occurring in earliest infancy, before the component parts of the innominate bones have become fused, may lead to the triangular or trefoil deformity similar to that produced by osteomalacia in the adult. Indeed, Hohl contends that rickets and osteomalacia are the same disease. There is a tendency, as development proceeds, to gain the normal form. Hence we rarely find in the adult rickety pelvis a well-marked triangular distortion.

Fig. 102 is an extreme example of rachitic pelvis. The specimen comes from a dwarf upon whom I performed the Cæsarian section, after fruitless attempts at embryotomy, and irremediable injury had been inflicted. The pelvis is in St. Thomas's Museum.

The specimen shows the tendency of this deformity to divide the brim into two parts, one on either side of the much-projecting promontory, so that the available space is reduced to one of the loops of the figure ∞.

Deformities and contractions from rickets vary greatly in degree. The "simple flat rachitic pelvis" of Litzmann, the most common form, is often said to have its transverse diameter absolutely lengthened. In some specimens the basis of the sacrum actually exceeds the normal width. If I may trust the general conclusion drawn from large personal observation in the quarter of London where deformities are most rife, I should say that actual excess of transverse diameter is extremely rare. The rickety pelvis is small generally; the narrowing bearing principally upon the conjugate, but affecting in less degree the other diameters also. Dr. Hicks has formed a similar opinion; and he has kindly had measurements made for me of ten rickety pelves in Guy's Museum. Two only of these measure 5·00" in transverse diameter, that is, attain the ordinary dimension, whilst four measure 4·75", three 4·50", and one 4·25" only..

It is quite probable that deformities vary in kind and degree in different countries, as they certainly do in frequency. In England the poorer classes are better fed, clothed, and housed, than in most countries on the Continent; and our sanitary condition is generally superior. In some districts on the Rhine, and around Milan, osteomalacia is a frequent result of the miserable circumstances under which the labouring classes exist; whilst in England the disease is so rare that many men

in large practice have never seen a case. On the Continent
rachitis also seems far more frequent than with us. It may,
therefore, well be that what we look upon here as rare excep-
tions may be so common abroad as to justify an important
place in a systematic classification.

Narrowing of the pubic arch is not uncommon in London.
By throwing the child's head backwards, it is apt to cause
laceration of the perinæum. I have often found it hinder the
descent of the head and call for forceps.

FIG. 103.

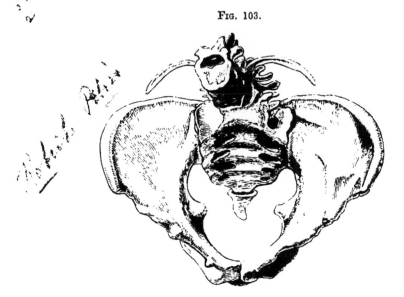

KYPHOTIC TRANSVERSELY-CONTRACTED PELVIS, AFTER HUGENBERGER; ONE-THIRD
NATURAL SIZE.

Spinal kyphosis induces a peculiar deformity of the pelvis,
called by Breisky,[*] and Hugenberger,[†] the "Kyphotic transversely-contracted pelvis;" by Michaelis, the "transversely-
contracted pelvis," "das querverengte Becken." The first
example was described by Robert, of Coblentz, and hence it
is often called "Robert's pelvis." Litzmann also describes
a case in his work.[‡] The illustration I have borrowed from

[*] "Med. Jahrb. Wien.," 1865.
[†] "St. Petersb. Med., Ztschr.," 1868.
[‡] "Die Formen des Becken." Berlin, 1861.

Hugenberger. Robert's case and others appear to be due to defective development of the alæ of the sacrum. This form of pelvis resembles the pelvis of the lower mammalia, of the infant, and of the Bushmen and Malayans of Java. The dimensions of the inlet are reversed, the antero-posterior diameters being lengthened, and the transverse shortened. The effect of this reversal of the normal diameters would be to reverse the position of the head. It would enter more easily with its long diameter in the conjugate direction.

The *funnel-shaped pelvis*, the peculiarity of which consists in the convergence of the turberosities of the ischia and of the sacrum, whilst the brim and cavity are little affected, is not, according to my observation, of frequent occurrence in London. I have, however, seen examples of it in the practice of the Royal Maternity Charity, amongst the weavers of Bethnal Green, and others who, from childhood, spend a great portion of their lives working in a sitting posture. Imperfect nutrition, no doubt, disposes to it. It is frequently referred to by the Dublin School; so much so that I am disposed to call it the "Irish pelvis." But the most marked case I know is that of a lady who shows no other sign of having suffered in health. Two of her children have been sacrificed by craniotomy; in the third pregnancy labour was induced at seven months, the forceps was applied and brought the head without difficulty to the outlet, when the disproportion became too manifest to admit a hope of its passing in this way; I therefore turned, and thus succeeded in bringing through the child, which survives.

In some cases, the sacro-coccygeal joint is anchylosed. This will produce an effect similar to the funnel-shaped pelvis. In such a case we may have to choose between perforating the head and forcible fracture of the anchylosis, so as to allow the coccyx to revolve backwards.

2. *Osteomalacia.*—This disease, occurring after maturity, does not involve general diminution of size of the bones. It produces remarkable deformity. It bears chiefly upon the spine and pelvis, the long bones being comparatively less affected. The bones losing their mineral constituents, become soft and yield under pressure. The pelvis and lumbar vertebræ forming a compressible centre between the upper part of the body and the legs, give way; all the bones sink inwards,

collapsing together; the spine falls downwards and forwards, squatting, the patient losing stature. In many instances the lower lumbar vertebræ dip into the pelvic cavity (there is a remarkable example of this in St. Bartholomew's Museum). The heads of the femora drive in the acetabula and sides of the pelvis; there is universal centripetal falling-in. In some cases the ilia are bent up, distorted like wet pasteboard, as in a characteristic specimen in St. George's Museum. The pubic bones are flattened together by their posterior surfaces so as to form a beak or rostrum, so distinct that the finger and thumb can seize it through the living structures. The effect upon the brim is to produce the extreme cordate shape, some-

FIG. 104.

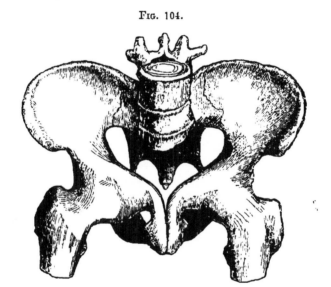

OSTEOMALACIC PELVIS: AD NAT., ONE-THIRD NATURAL SIZE.

times approaching the letter Y, as in Fig. 104. The cavity and outlet share the concentric compression. There is really no pubic arch; the tuberosities are closely approximated; the sacrum is curved forwards; the outlet is almost closed. The obstetric result is that, in marked cases, it is scarcely possible to introduce two fingers; there is little room for manipulation or for working instruments; we are often compelled to get at the child from above, through the abdomen. It must not, however, be forgotten that the plastic condition of the bones may admit of the pelvis being opened out again under eccentric pressure ; and that if labour be brought on before the seventh month by embryotomy, we may secure a fœtus so lessened in bulk and so ductile, that it will traverse even extreme degrees of distortion.

The illustration, Fig. 104, is taken from a characteristic specimen in St. Thomas's Museum.

3. *Disease of the Vertebræ.*—Under various diseases, as scrofula, or injury inducing caries or softening, the articula-

Fig. 105.

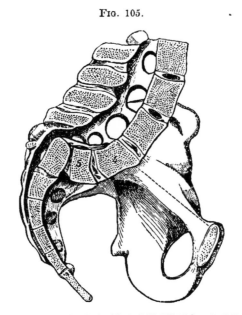

SPONDYLOLISTHETIC PELVIS, AFTER KILIAN ; ONE-THIRD NATURAL SIZE.

4. Fourth lumbar vertebra ; 5. Fifth lumbar vertebra. The virtual conjugate is indicated by the dotted line.

tions gave way, and, the lumbar vertebræ slipping forward, a dislocation results. Hence the name " Spondylolisthesis." This gliding down throws two or more lumbar vertebræ into the cavity of the pelvis, producing a new and false promontory in front of, and lower than the true one ; it not only contracts the conjugate diameter, but by partly filling the pelvis, prevents the uterus and child from entering. Some of the cases appear to have been congenital; but in others certainly the condition arose after maturity. It was first described by Kilian. I have discussed the subject in some detail in the *Obstetrical Transactions,* vol. vi. The general obstetric effect is similar to that produced by a large osteosarcoma in the pelvis. Most of the cases required the Cæsarian section.

The illustration, Fig. 105, is taken from Kilian. It is called the "Prague Pelvis." The original is in the Prague Museum.* In another specimen described by Kilian, the virtual conjugate starts from the *second* lumbar vertebra, the third, fourth, and fifth having all sunk into the pelvis.

4. *Disease of the Pelvic Articulations.*

Inflammation, softening, caries, especially of the sacro-iliac synchondrosis, induce oblique distortion of the pelvis. This is the deformity described by Naegele under the names "Schräg-verengte Becken" and "pelvis obliquè-ovata." The disease affecting one joint, the nutrition of the corresponding side is modified, the relative action of the muscles attached to the two sides, and the pressure bearing upon the two sides, are disturbed. Frequently, also, there is defective development of one side of the sacrum. Hence the symphysis pubis is thrown across towards the sound side; and the affected side is diminished. This condition may arise from congenital causes; but it also undoubtedly arises in childhood and maturity.

Less marked, but still distinct degrees of pelvic obliquity, or asymmetry of the two sides of the pelvis, are produced by lateral curvatures of the spine, and by shortening of one leg, especially if these occur before maturity.

In its obstetric relations we find it preventing the due adaptation of the head, and impeding its descent into the pelvis. The forceps will suffice in the minor degrees: turning, taking care to bring the occiput into the larger or sound side of the pelvis, may answer in cases of somewhat greater severity; and embryotomy with cephalotripsy ought to effect delivery in the most extreme cases known. Cæsarian section can very rarely be necessary.

* "De Spondylolisthesi gravissimæ pelvangustiæ, causa nuper detecta." Bonn, 1853.

The illustration, Fig. 106, is taken from Naegele's classical work.

FIG. 106.

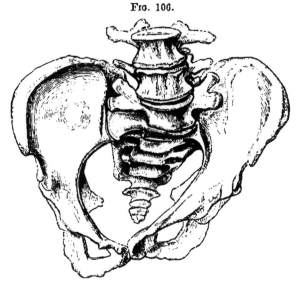

OBLIQUELY-DISTORTED PELVIS, AFTER NAEGELE; ONE-THIRD NATURAL SIZE.

It is usual to describe a *pelvis æquabiliter justo major*, and a *pelvis æquabiliter justo minor*. The first requires no description.

FIG. 107.

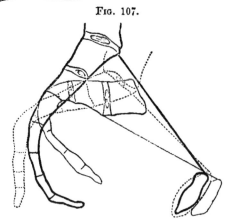

REPRESENTS SECTIONAL VIEWS OF THE NORMAL PELVIS (THE STRONG LINE) CONTRASTED WITH THE OSTEOMALACIC PELVIS (FINE LINE) AND THE RICKETY PELVIS (DOTTED LINE).

It is simply a large pelvis. The existence of the second, as a product of disease, may be doubted. Disease will almost invariably induce deformity, although the degree may be slight. The term might usefully be discarded. A perfectly proportioned pelvis, uniformly reduced in all its dimensions, must be regarded as an *infantile or immature pelvis*, or as a general result of small stature.

Fractures of the pelvic bones from injury may leave various forms of distortion. Cases of the kind are extremely rare, and methodical description of them is scarcely possible.

Certain diseases affecting the osseous system cause sharp ridges, or needle-like projections. If these occur on the linea ileo-pectinea, they will obviously endanger the uterus, cutting or stabbing its walls. They often occur in the rickety pelvis. A form of pelvis is described by Kilian under the name "Acanthopelys," "das Stachelbecken;" *Anglicè*, the "thorny pelvis."

In connection with pelvic disease we may most conveniently refer to osseous, fibrous, and sarcomatous outgrowths from the pelvic walls. These are of various sizes and origin. Perhaps they most frequently spring from the sacrum. Growing, they gradually encroach upon the pelvic space; and in some cases, as in one recorded by Dr. Skekleton, of Dublin, whose drawing is copied into nearly every text-book, so nearly filling the cavity that delivery by the natural passages became impossible. This was the motive of a Cæsarian section at St. Bartholomew's which I witnessed and assented to, having myself vainly endeavoured to deliver by turning, after Dr. Greenhalgh had failed. We both succeeded in touching a foot. The operation was performed by Mr. Skey. The woman died, and the pelvis was found nearly blocked by a sarcomatous tumour. These tumours sometimes grow quickly, so that as in Skekleton's case, the labours become increasingly difficult. In others, they remain stationary, or grow very slowly, as in a case to which I was called by Dr. Rogers, once my Resident Accoucheur at the London Hospital, for the express purpose of performing the Cæsarian section. I however succeeded in delivering by my craniotomy forceps. The subject recovered, and was seen by me occasionally for several years after.

Mr. Berry relates an instructive case (*Obst. Trans.*, vol. vii.) of obstruction from medullary cancer springing from the bones of the sacrum. As it felt fluctuating it was punctured, when a large quantity of florid blood flowed. After great exertions, the child was delivered by craniotomy and turning; but the woman died in an hour. The uterus and vagina were found lacerated.

It is clear from the comparison of the various kinds of deformity that no definite universal rules as to choice of mode of delivery can be drawn from the simple element of dimensions. Shape of pelvis and the nature of the obstruction must also be considered. Still, in the rickety, which embraces by far the larger number of cases of deformity leading to obstructed labour, the space given in the antero-posterior diameter of the brim will generally determine our proceedings. A scheme of the relations of operations to pelvic contractions, in conformity with the improved methods of operating described in this book, is given at p. 82.

The *diagnosis* of the kind and degree of pelvic deformity is easy in proportion to its severity. The rickety pelvis may be suspected from the low stature and ungraceful gait of the subject. It is often attended by marks of imperfect development, and by spinal deviation. In marked degrees there is a peculiar form and expression of the features. In advanced pregnancy the abdomen overhangs the pubes. The sacrum is remarkably flat externally. On internal examination nothing unusual may be noticed at the outlet; but, on pressing the finger backwards, it strikes the sacrum or its promontory, whilst the knuckle is, perhaps, applied to the arch of the pubes. In a healthy pelvis this cannot be done. An experienced hand will from this examination form a very close estimate of the length of the conjugate, and the mode of delivery that must be adopted. Much ingenuity has been expended in the construction of instruments that shall give the pelvic measurements with absolute precision. Nothing surpasses in this respect Van Huevel's pelvimeter. Its application is often of great scientific value. But I imagine few obstetric surgeons of experience rely upon any instrument but the hand. This gives information that can be obtained in no other way. When there is obstruction, or retarded labour, or other cause to suspect pelvic contraction, the most practical course is to induce anæsthesia and to pass the hand into the vagina. In this way you can explore the whole pelvis, take note of the various dimensions, and ascertain the relations of the fœtus.

The oblique pelvis may be made out by the same method, and verified by external measurement from the spinous process of the upper sacral vertebræ to the symphysis pubis on either side.

The osteomalacic pelvis may be suspected by the extreme closure of the outlet making it difficult for the fingers to pass, and verified by the rostrated condition of the ossa pubis.

It was at one time thought that there was a fixed relation between the external and the internal conjugates, and that by measuring the external we might read off the internal diameter. No one would venture to act solely upon the estimate thus deduced. Taking the standard external conjugate to be 7″, it does not follow that, where this is found, we can depend upon an internal conjugate of 4″. Taking two pelves both measuring externally 6″, we may find that one has an internal conjugate of 3″, the other one of 3·50″. There is, in short, no substitute for internal exploration by the hand, and all other modes may be dispensed with. But where the question is as to the induction of labour, we shall hardly be wrong if, on finding an external conjugate of 6″, or even of 6·50″, we advise the induction of labour at seven or eight months.

Mechanism of Labour with Deformed Pelvis.

Pelvic deformity may generally be recognized by the characters it imposes on the course of labour. As a natural pelvis governs the process of labour according to certain definite laws, so do the various forms of abnormal pelvis, each in its own manner, control the process.

The presenting position and the progress of the child are liable to characteristic modifications. In the case of the moderately contracted, flat, or rickety pelvis, the conjugate diameter being narrowed, and the promontory projecting, the head can hardly enter the brim in an oblique diameter ; it must almost necessarily present with its long diameter in relation with the long or transverse diameter of the pelvic brim. The anterior side of the head will in the early stage of labour overlap, more or less according to the degree of conjugate contraction, the symphysis pubis. The broader expanse of the occiput not so easily entering the pelvis as the narrower sinciput, is delayed a little on the edge of the brim; thus the forehead will be at first driven lowest in the pelvic cavity. The driving force increasing up to a certain point with the resistance, the head is gradually moulded by being flattened in its transverse or bi-parietal diameter, inducing great overlapping of the frontal and parietal bones, so as to adapt itself to the flattened brim of the pelvis. The projecting promontory virtually alters the direction of the axis of the brim, depressing it so as to bring it nearer to the horizon; that is, it is made to form with the horizontal datum-line (see Figs. 30, 31) an angle less than 30°. The lower end of this imaginary line would fall, not upon the coccyx, but higher up at some point in the concavity of the sacrum, whilst the upper end would be brought below the umbilicus. There is increased inclination of the pelvis. The consequence of this is that the head must travel more directly backwards under the projecting promontory than when the pelvis is normal. It must in fact travel round the promontory, doubling it, in order to get into the cavity. The anterior side of the head at first occupies nearly all the brim of the pelvis, the sagittal suture running across the brim being much nearer to the promontory than to the pubes. The posterior side of the head is comparatively fixed against the promontory, the vertex pointing backwards towards the sacral hollow. The head describes the first curve, that to which I have given the name of the "false promontory" (see Fig. 31). Having got so far, the occiput usually descends and comes forwards, and enters the normal curve of Carus (see Fig. 32). I may refer back to Lecture VI. for an exposition of the mechanism of labour under this condition.

If the conjugate contraction be more marked, that is, to about 3·50″, the first stage of labour will be more protracted still; the presenting part will remain for a longer time above the brim; the os uteri, wanting the dilating force of the liquor amnii and head, will be more slow in opening; perhaps the cord will come down, and this is especially liable to be the case if the membranes break early, since the head cannot block the brim. At length violent pains may force the head into the strait. The conditions already described will be observed in a more exaggerated degree. It is in these cases and in the preceding

class that some additional *vis à fronte* by the aid of the forceps
is often valuable by aiding and economizing the struggling *vis
à tergo*.

In the still more contracted conjugate, where it is reduced to
3″ or below, the head can hardly enter at all; it rests upon the brim
touching at two or three points only, therefore perfectly moveable except when fixed by the driving power; and there it will
remain unless its bulk be reduced. In these cases, a marked indentation, even fracture, of the frontal or parietal bone, caused
by the long and violent pressure against the projecting promontory, has been noticed. This injury is not necessarily fatal.
It is useful to bear it in mind, because in the event of the
forceps having been used, this might be assigned as the cause.
I have several times known actions for damages threatened on
this ground, even although the child and the mother have done
well.

In the osteomalacic pelvis the head will rest upon the brim
in a similar manner.

The following circumstances should provoke the suspicion
that there is pelvic deformity:—A protracted first stage of
labour; slow dilatation of the cervix; premature rupture of the
membranes; prolapse of the cord; an unduly transverse position of the head, the forehead being as low or lower than the
occiput; an abnormal presentation; failure of the presenting
part to enter the pelvic cavity, although the cervix may be
dilated; approach to a pendulous belly, the fundus uteri being
apparently lower, pointing more forwards than usual.

As after-effects of severe pressure concentrated upon limited
areas or points of the soft parts, fistula, or rupture of the uterus
or vagina may arise; and from the local injury, combined with
the systemic exhaustion and absorption of effete matter, there is
greater risk of puerperal fever.

LECTURE XXI.

RUPTURE OF THE UTERUS, VAGINA, AND PERINÆUM—DEFINITION OF LACERATION, RUPTURE, AND PERFORATION—PREMONITORY SIGNS AND PROPHYLAXIS — THE CONDITIONS UNDER WHICH RUPTURE OCCURS : IN NON-PREGNANT UTERUS ; DURING PREGNANCY ; DURING LABOUR—LACERATIONS FROM OBSTRUCTION TO LABOUR—LACERATION OF THE VAGINA—TRAUMATIC LACERATIONS — THE SYMPTOMS AND COURSE OF CASES OF LACERATION—THE TREATMENT—DIFFICULTIES FROM DIFFICULTIES IN DIAGNOSIS — ILLUSTRATIONS OF SOURCES OF ERROR—DESCENT OF INTESTINE.

RUPTURE of the uterus has this affinity to the Cæsarian section, in that it is sometimes produced by conditions similar to those which determine us to perform the Cæsarian section. Indeed, a great motive for this operation is to avoid rupture. And when rupture has occurred, it is often necessary to open the abdomen in order to remove the fœtus. There are, in fact, cases of dystocia where Nature, unable to effect delivery *per vias naturales*, seems, by rending open the uterus and extruding the child into the abdominal cavity, to endeavour to accomplish that which the surgeon accomplishes by cutting open the uterus after laying open the abdomen. It rests with the surgeon in these cases to meet Nature half way, by performing abdominal section, to get at the child cast out into the abdominal cavity.

There are few subjects in obstetric practice more interesting, or possessing a wider range of relations, than rupture of the uterus. A full knowledge of the conditions under which the accident may arise, of the symptoms and terminations, is of the highest importance in medical and in medico-legal relations.

Y

The accident rarely or never happens without some imputation or suspicion of malapraxis falling upon the medical practitioner. It is of the last importance to know what to do, and what not to do, not alone in the interest of the patient, but also, reflecting on the fearful penalties under which we practise, in our own. It is of the very nature of the accident that it commonly happens suddenly, without warning, and therefore precludes the medical attendant from using means to obviate it. I have been more frequently consulted in criminal charges connected with rupture of the uterus than with any other obstetric casualty. In almost every instance the conclusion that the accident arose from unavoidable causes proved to be the best founded.

The following general propositions may be affirmed :—

1. The non-pregnant uterus may burst.

2. The uterus may burst at any period of gestation independently of labour proper.

3. Any part of the parturient canal may be lacerated during labour.

4. By far the greater number of cases occur during labour at term.

5. The uterus will not burst unless it be in a certain degree of tension, from containing something in its cavity.

6. The uterus may burst in child-bearing women of all ages; in women pregnant for the first time, or in women who have borne one or more children; the greater risk being in primiparæ and in women who have borne many children.

It is desirable to attach definite meanings to certain terms. By so doing, we shall at once effect a natural classification of cases that will much simplify our inquiry.

1. *Rent*, or *laceration*, occurs when a breach begins at the edge of the os uteri, or perinæum, and extends, under the action of labour; or when the structures are torn by the hand or instruments.

2. *Rupture*, or *bursting*, occurs in the body of the uterus, or at the junction of the vagina and uterus, or in the middle of the perinæum. Rupture of the uterus may occur at any period of pregnancy or labour. This is generally the result of spontaneous uterine action.

3. *Perforation*, or *boring-through*, occurs when the tissues give way from change of structure, as from disease, or long-continued

compression of one part, or continued attrition. In these cases the wearing of tissue is the first stage; but the same forces may act as in "rent" or "rupture," and cause extension of the lesion. Rent and rupture are generally produced suddenly, although they may subsequently be extended gradually. Perforation is a slow, gradual process. It may occur from disease in the non-pregnant uterus. Ruptures, then, may be *instantaneous* or *progressive.* They may also be *spontaneous* or *traumatic.*

Before discussing particularly the causes of rupture, we may here conveniently dispose of the important questions: 1. What are the *premonitory symptoms?* 2. What is the *prophylactic treatment?* Since, in a large proportion of cases, the injury occurs without warning, there is no opportunity of preventing it. And, although, starting from the general fact that the uterus may give way if there be impediment to the progress of labour, we may, in some cases, hope by removing recognised impediments, to obviate rupture, there will still remain other cases in which it will not be possible to discover impediments, or even, if discovered, to remove them. As Denman well observes, "Some of the causes are unavoidable, for it is not within the sphere of human abilities to give to some part the principle by which it has the disposition or power to perform any function." The tissues, for example, may be diseased or degenerated.

The explanation commonly given of rupture of the uterus is that it is produced by obstruction to labour. The histories of the great majority of reported cases prove that obstruction to labour was the immediate antecedent. But this explanation can scarcely apply to those cases where the uterus suddenly bursts during pregnancy when there was no labour. The immediate cause is, I think, more comprehensively stated in the following proposition:—*The uterus ruptures because there is a loss of balance and of due relation between the expelling power of the body of the uterus and the resisting power of the parturient canal, the resisting power being in excess.*

Trask, discussing the 417 cases he had collected, says: "*Inordinate voluntary exertion* deserves to be enumerated among the causes of rupture. We believe no case of rupture has yet (1856) been published in which chloroform was used, which may be due to the fact that voluntary effort is greatly suspended

Y 2

under its influence." Tyler Smith says: "In ordinary labour, some amount of voluntary or instinctive action of the muscular system, and particularly of the expiratory muscles, is quite natural during the stages of propulsion and expulsion. In acute or severe labour, these voluntary exertions are productive of great mischief, as lacerations of the uterus, and perinæum and exhaustion.

In many cases no decisive symptoms precede. This uncertainty is a cogent reason for watching closely in every case for the signs of obstructed labour, and for removing any cause of obstruction that may be detected. Obstruction, allowed to persist, leads either to exhaustion or to rupture. In a case of occlusion of the cervix uteri, the pulse rose to 140 ; crampy, painful contractions of the uterus set in ; rupture seemed imminent. Incision of the cervix, allowing expulsion of the fœtus, almost immediately brought relief. This rule must especially be observed in weakly, exhausted subjects. The case is well stated by Dr. Ewing Whittle. When we are attending a patient from whose antecedents and constitution, and from the slow and unsatisfactory progress of whose labour we have reason to fear that we have a feeble, flabby uterus to deal with, our first care should be to soothe the patient, and keep her quiet until the os is fully relaxed. Opiates will here be of service. This sedative treatment should, however, not be trusted too long. The cervix should be dilated by the hydrostatic bags. In these subjects avoid ergot. There is serious danger in goading an exhausted system to put out strength it cannot spare. When the os is dilated, rupture the membranes, and, if labour does not proceed satisfactorily, deliver with the long forceps. Since Roberton has shown that when rupture occurs from faulty pelvis, it generally does so within twelve hours, the indication is clear to watch closely, and to act betimes where the signs of obstruction are present.

Commonly, when the resisting power is in excess, the uterus becomes exhausted, or from other causes ceases to struggle against the resistance. In this way rupture is averted. We may then look upon obstructed or retarded labour, or the signs of it, as summed up in Lecture I., as the *premonitory signs* of rupture. We shall rarely get anything more precise. But the following summary by Crantz is valuable : "When a woman is threatened with rupture of the womb in difficult labour, the abdomen is very high and tense, the vagina is retracted, and the os uteri very high; the pains are strong, leave but short intervals, and are without effect."

We might, then, in cases where we see the uterus struggling impotently against an obstacle, avert rupture by one of two ways. First, we may endeavour to restore the proper preponderance of the uterine force, by lessening the resistance; or, secondly, we may avert or postpone danger by subduing the excessive action of the uterus.

We will dispose of the latter alternative first. Can we persuade the uterus to be passive for a time ? If we can command temporary quiescence, we may gain time and opportunity for the diminution or removal of the resistance. We possess in ergot a great, a dangerous, power of augmenting the force of the uterus. Indeed, this agent has too frequently, by goading the uterus beyond measure, caused it to burst itself. We want an agent endowed with the opposite effect, that will control and suppress uterine action. An agent that could be depended upon to do this would extend, and—if considered in association with the powers we now possess of accelerating labour, namely, the means of provoking labour, the uterine dilator, the forceps, turning, ergot—complete our command over the course of pregnancy and labour. I consulted Dr. Richardson on this point. He tells me the desired power exists in the nitrite of amyle. Three minims of this added to one drachm of ether taken by inhalation is the form he recommends. It does not produce unconsciousness, but it is an anæsthetic as well as a sedative of muscular action. It is the antidote or opposite force to ergot. In it we have the desiderated "*epechontocic agent.*"

We have long been accustomed to give opium with the view of allaying muscular action and postponing labour. Its use in this way is often valuable in gaining time. Chloroform has of late years been frequently resorted to. But it has the disadvantageous property, derived from the chlorine in its composition, of exciting vomiting and muscular spasm. It is only when pushed to extreme anæsthesia that muscular resolution is obtained. I think, therefore, that the nitrite of amyle should have a fair trial.

A most interesting narrative of a cure of tetanus by the

Calabar bean, by Dr. Ringer (see *Practitioner*, *Nov.*, 1874), suggests the probability that this agent might control undue energy of the uterus. The patient was a physician. He took ⅔-grain of the extract every fifteen minutes. The remedy is hazardous. Much smaller doses, say the $\frac{1}{16}$-grain every twenty minutes, might first be tried. Chloral has been extolled for its "amyosthenic" property by Martineau (*Gaz. des Hôp.*, 1873). A woman seven months pregnant took quinine. Uterine action quickly followed. Laudanum had no effect in stilling it. Chloral in fifteen-grain doses allayed all action, and averted abortion. Chloral would be safer than physostigma.

In stormy, tetanoid, or spasmodic action of the uterus, where rupture or exhaustion is threatened, the *first* indication is clear to subdue this dangerous activity; and the *second* is to remove the obstacle against which the uterus is struggling.

The Conditions under which Rupture occurs.

1. *Rupture may take place in the non-gravid uterus.* Duparcque cites cases (*Maladies de la Matrice*, tome 2, 1839). But disease of its tissues, as thinning, softening, abscess, appears to be a necessary condition. Other factors are: closing of the os and accumulation of fluid in the body. Perforation by cancer—a distinct event from bursting—is not infrequent.

We need not dwell in this place upon the cases of direct injury to the uterus by wounds or blows. Such may, of course, occur at any time. It is well to remember that blows inflicted upon the pregnant womb may wound, even kill the child, without causing rupture of the womb. Duparcque gives an example.

It has been stated that sudden violent efforts of the child have caused rupture. The cases cited in support of this proposition are not convincing. It is more probable that the uterus ruptured itself, or that some violence external to the uterus was concerned.

2. The history of *rupture or bursting of the uterus during pregnancy* more immediately concerns us. This event is rare, but well authenticated. Indeed, it is not more surprising than spontaneous rupture of the heart. It is known to have occurred as early as the third month. H. Cooper (*Brit. Med. Journ.*, 1850) saw a pluripara of thirty, who was taken with collapse after dancing, and died next day. The uterus was found torn

at the left side of the fundus, a three months' fœtus projecting through the rent. The tissue at the part was thin, pulpy, cheesy. There was tubercular degeneration.

It has occurred in the fourth month. Dr. McKinlay relates a case (*Glasgow Med. Journ.*, 1861). Without any exertion or injury a woman died after being taken ill the previous night. The uterus was ruptured across the fundus. Its tissue was apparently healthy.

Dr. Harrison (*Amer. Journ. of Med. Sc.*, vol. viii.) relates the case of a pluripara, who in the fifth month, after a long walk, felt a sudden and severe pain " as if something had given way within her." She died in a few hours. Blood and the fœtus in its membranes were found in the peritoneal cavity. There was a transverse rent from one Fallopian tube to the other. There was no thinning or any appearance of disease. Collins relates a fatal case at five months.

Mr. Scott, of Bromley, relates (*Med. Repository*, vol. viii.) the case of a woman in the sixth month who was awakened from sleep by a sudden pain about the umbilicus. Rupture was found at the fundus, through which the fœtus, enveloped in its membranes, had escaped into the abdominal cavity.

Mr. Mitchell relates (*Obst. Trans.*, 1870) the case of a woman in the seventh month, who died after sudden abdominal pain caused by terror from lightning. The uterus had burst apparently under contraction upon the projecting knee of the fœtus.

Other cases will be found in Trask's " Memoirs " (*Amer. Journ. of Med. Sc.*, 1848 and 1856).

In the eight and ninth months spontaneous ruptures are less rare. From the cases above referred to, and others, the following conclusions flow: first, spontaneous bursting of the uterus may occur from weakening of the uterine tissue from thinning, from disease, as tubercular, fibrous, or fatty degeneration. Second, under emotional excitement to uterine contraction where the tissue appears sound. The influence of emotion in determining uterine contraction and hæmorrhage is well known. The cases of Mr. Scott and Mr. Mitchell just cited are examples of rupture from this cause. Francis White (*Dublin Journ. of Med. Sc.*) gives another, that of a woman near the end of gestation. She fainted under terror, was delivered a week after, and died almost immediately. A large quantity of blood was found in the abdomen. Two rents were found in the anterior part of the womb involving the peritoneal coat and some muscular fibres. Thirdly, after violent exertion and fatigue, or injury, as a blow or a fall, or from severe vomiting. Duparcque cites a case of rupture attributed to vomiting. The subject was between two and three months pregnant. The uterus at the seat of rupture, near a tubal angle, was thinner than elsewhere. Another case is cited by Duparcque to show the influence of effort. A woman was carrying a weight on her head, when symptoms of internal injury set in; but she rallied for a time, had another attack, and died. The uterine cavity contained a fœtus of three or four months. There was a fissure in the fundus uteri near the right tube. Fourthly, there is another order of cases in which the pregnancy deviates from the normal kind. For example, Canestrini relates a case in which there was a double uterus. One of the uteri, after some pains in the fourth month, burst. The ovum was found entire in the abdomen. Goupil cites a case, from Payan, of a woman who died under symptoms of shock, causing suspicion of abortion having been procured. Above the proper cavity of the uterus was another cavity formed in the wall. This interstitial cavity had become thinned by the growth of the ovum, and burst. Goupil cites a similar case from Duverney, and several others are recorded. Indeed, interstitial gestations mostly end in fatal rupture before the fourth month. The rupture is nearly always close to the opening of a Fallopian tube.

By these last cases we touch upon another order which may, and no doubt often have, given rise to error. The symptoms cannot be distinguished from those of rupture of a tubal-gestation cyst, and, without dissection, there may be no means of proving what it was that ruptured. The inference that the case is one of extra-uterine gestation is strengthened whenever fœtal structures make their escape externally by abscess and fistulæ.

Spontaneous rupture of the uterus early in pregnancy is so rare that suspicion of foul play is easily excited. No doubt the vagina and uterus have been frequently wounded by instruments used to procure abortion. The characters of the wounds may differ from those due to spontaneous rupture.

Spontaneous bursting most frequently occurs in the body of the womb. Criminal wounds are generally inflicted upon the os, cervix, and lower part of the uterus. They will show evidence of cutting, puncture, or bruising, according to the nature of the instrument used. A careful microscopical examination of the tissues, especially at the seat of injury, should be made.

3. *Rupture or bursting of the uterus at term or on the advent of labour.*—In this order of cases the reversal of the normal relation between active uterine force and passive resistance is more obvious. In a large number of cases there is decided mechanical resistance to the expulsion of the fœtus. But it is remarkable that bursting has frequently happened long before obstruction to labour could be encountered. These cases are similar in their mode of production to those which occur early in pregnancy. There is one striking point of resemblance: namely, the frequency with which the entire ovum is cast out of the uterus into the abdominal cavity. In this respect they differ from the lacerations which occur during obstructed labour, in which the child, or at any rate the placenta, is more commonly retained in the uterus. The explanation of these early ruptures appears to me to be that the uterus is excited to contract suddenly. The resistance is hydrostatic. The contents of the uterus are incompressible ✗ The conditions are analogous to those of the famous Florentine experiment to test the compressibility of water : the water exuded through the metal globe. There is no provision for diminution of the bulk of the contents by the opening of the os uteri and discharge of the liquor amnii. Under these circumstances a moderate contracting force may result in bursting. There can be no doubt that in many cases this catastrophe is averted by the facility with which the membranes burst. Thus abortion may be looked upon as an alternative of rupture. On the other hand, the uterus will often stretch, and in this way bursting is also averted. But stretching is a process that requires time. A *sudden* contraction, such as is induced by emotion, gives no opportunity for this, and so the tissue gives way. Belonging to this order are cases of over-distension of the uterus from excess of liquor amnii, and from the presence of twins or triplets. Under any one of these conditions, although the uterus grows in some measure in

proportion to the increasing bulk of its contents, the rate of accommodation is liable to be outstripped by the distending force. The uterus becomes *thinned out*, stretched, therefore weakened. If the thinning happen to be more marked at one part, rupture at that part is very likely to happen if a sudden contraction occur. And especially is this likely if the tissues have undergone morbid change.

This is the more likely to happen because distension of the uterine fibre is very apt to cause vomiting; and vomiting has been noted as a factor in the production of rupture. That the expulsive action of the auxiliary muscles may conduce to rupture, the following considerations show.

Tyler Smith says: "In cases where the uterus is feebly developed or weakened by disease and exhausted action, the contractions of the abdominal muscles must contribute to the rupture of the organ, by urging the head or presenting part of the child through the os uteri." In the lecture on "Retroversion" (*see* p. 267) cases are cited showing that the vagina may be ruptured by the pressure upon it of the retroverted uterus, gravid or after labour. And so an ovarian tumour, as in Mr. Berry's case already referred to, and in Dr. Dunn's (*see* p. 368), may be so driven into Douglas' pouch as to tear through the posterior wall of the vagina.

Reasoning from the observation of some cases in which the uterine tissue was altered by disease, and partly from the analogy of rupture of the heart, the opinion has lately found favour that *disease of tissue* is a necessary condition to rupture of the uterus. It is urged that a healthy uterus will not rupture. It is scarcely necessary to point out that the supposed analogy of the heart is fallacious. The uterus tears itself by pulling upon a fixed point, as where the head jams the lower segment of the uterus in the pelvic ring. There is no similar condition in rupture of the heart. There is, no doubt, a class of cases in which the uterine tissue gives way from disease like the heart. But to affirm the universal presence of disease as a factor is to sacrifice experience on the shrine of theory. The opinion that alteration of tissue led to rupture was first clearly insisted upon by Murphy. He described a *softening* of the tissue as the result of inflammation during pregnancy, as an indication of which there is frequently pain in a particular spot. Another condition which has been many times observed is *extreme thinness* of the uterine wall. This thinness may be general, or limited to a part. If this part be the lower segment, rupture is very likely to occur. But apparent thinness may occur at the fundus, more especially at the presumed origin of a Fallopian tube. In some of these cases, there can hardly be a doubt that there was gestation in one horn of an imperfectly developed uterus. This seems to me to be the explanation of an "Obscure Uterine Disease" in which fatal rupture took place, published by E. C. Ling (*Lancet*, 1872). Other cases are cited and figured in my work on the "Diseases of Women." Thinness may be the result of distension during pregnancy, as from multiple gestation, excess of liquor amnii, or it may be produced during the labour. In this latter case, the thinning will take place at the point of resistance from the muscles constantly dragging upon this part. Thinning may be the result of disease, as in a case related by E. Whittle (*Liverpool Med. and Surg. Reports*). The child had passed through a rent near the junction of the body and neck anteriorly. For two or three inches on either side of the rent the uterus was thin and soft. The edges of the rent were not more than 0·25″ in thickness. The woman was affected with *secondary syphilis*, and Whittle believed the degeneracy and atrophy were due to this condition. Collins relates a marked case of this kind.

Fatty degeneration is the change most frequently described. In confirmation, reference is made to the normal process of involution of the uterus, in which, at the end of pregnancy, some molecular granular change occurs in the muscular cells. If it be held that the ordinary kind and degree of granular change is sufficient to impair the strength of the tissue to the extent of causing it to rupture, it must be enough to point out the extreme improbability of Nature's so bungling as to weaken the uterine tissues at the moment when the greatest vigour and resisting power are required. If it be urged that this physiological condition may pass into a pathological excess, an argument is used more difficult to refute; the more so because in some cases direct observation seems to prove that excessive fatty degeneration did exist. But to prove that a particular factor existed in a limited number of cases is very different from establishing it as a general or universal law.

And there is abundant evidence to prove that in a very considerable number of cases no such excess did exist. In the first place, many ruptures have occurred during pregnancy, at a period when even the physiological change is rare; in some of these the healthy condition of the tissue was established by competent observers. Secondly, a large proportion of the cases occurring in labour at term were in primiparæ, in whom the presumption is strong against morbid change of tissue. Thirdly, in many cases of rupture at term in pluriparæ the tissue was found healthy. In four cases examined by myself, this was the case. In one of these my conclusion was verified by Dr. Bristowe and Dr. Montgomery. Fourthly, in not a few cases recovery with perfect healing of the uterine wound has occurred. Experiments made upon small pieces of uterus are useless in their application to the problem of the power required to burst the living uterus. In the living uterus various parts may vary greatly in thickness and strength; and, like a chain, the uterus is only as strong as its weakest part.

On the other hand, the frequency of rupture in pluriparæ, in women about forty years of age, who have led a hard life, whose system is generally enfeebled, makes it reasonable to infer that the uterus partakes of the general feebleness or degradation of tissue. In two of the cases I refer to, the muscular fibres of the heart exhibited marked granular change. I found a similar condition in the heart of a woman who died of accidental hæmorrhage; she was also a pluripara, aged about forty, much worn by poverty and labour. In fact, accidental hæmorrhage and rupture are apt to occur in the same class of subjects. In another case, which was the subject of a criminal charge, it was stated by the medical witnesses for the prosecution, that the heart and other organs showed fatty change. In another case which came under my knowledge, a pluripara had albuminuria at the end of pregnancy; the kidneys were found in an advanced stage of Bright's disease.

Klob, however, distinctly says * that he has, in several cases of spontaneous rupture, observed fatty degeneration of the muscular wall at the place of rent.

Dr. E. L. Ormerod,† who minutely examined a case, says:

* "Pathol. Anat. d. Weibl. Sexualogane."
† "St. Bartholomew's Hospital Reports," 1868.

" As far as I have seen, a ruptured uterus does not give way from rupture of fibres which have undergone fatty degeneration; indeed, I am not aware that this form of disease occurs at all in the pregnant uterus. The torn edges display strong fibrous tissue, not degenerated fibres." The subject was in her fourteenth pregnancy. The rent extended from cervix to fundus. He found *follicular disease* about the cervix, and *fibrous degeneration* of the adjacent muscular structure.

C. Braun describes a *peculiar hyperplasia of the uterus,* in which the body and cervix lose all relation to each other.

Cancerous degeneration has led to rupture. Dubreuhl relates (*Lyon Médical,* 1871) a case where the uterus burst along its whole anterior surface from encephaloid degeneration of its walls. In the museums of St. George's and Guy's Hospitals may be seen specimens of fatal laceration from a similar condition. Rigidity of the cervix from *cicatricial tissue* may also cause rupture.

In one case related by H. Cooper,* a pluripara in the third month, aged thirty, died in collapse the day after dancing. The uterus was found torn at the left side of the fundus, a three months' fœtus projecting through the rent. The wall of the fundus was thinned, the tissue was pulpy, soft, cheesy. It is probably an instance of *tubercular degeneration.*

Alteration of tissue produced during labour is a factor, the frequent existence of which cannot be doubted. Where the lower segment of the uterus is long compressed between the pelvis and the child's head, the circulation gets stopped, intense congestion of the parts above and below the compressed ring ensues, the part compressed becomes friable, softened, and may yield. If delivery be effected before this, we see the effect at a later period in sloughing, forming vesico-vaginal fistulæ, and perhaps even gangrene. The points at which this bruising or rubbing-through of the uterus is liable to occur, are chiefly those opposite the promontory of the sacrum, and the symphysis pubis in rickety pelvis. (*See* Figs. 108, 109, 110.) The pectineal ridge may be sharp enough to cut through; sometimes the pubic portion of this ridge folded back as in osteomalacia, gives points of resistance against which the head may be jammed. Unnatural sharpness of the edge of the pelvic brim, or spinous projections are even

---

* " British Medical Journal," 1850.

more dangerous than simple narrowing. (*See* Lecture XX.) Unhappily the two conditions are sometimes combined, as in a case seen by myself. The specimen is preserved in St. Thomas's Museum. Duparcque cites a case in which "the interior part of the neck of the womb separated from one side to the other, and the child passed immediately into the abdomen. The pelvis was a little narrow; the point of the os sacrum passed through the posterior part of the womb. The inner and prominent edge of the pubis and ilia resembled an ivory paper-knife." In deformed pelves there is often found at the joints an *excrescence or exuberant growth of bone projecting inwards*, so that when the fœtus presses the uterus against the pelvic ring, laceration easily occurs. I have found this condition frequently. It is a reason why, in all examinations of contracted pelves, we should sweep the finger carefully all round the brim, when taking note of the kind and degree of the deformity. This bony excrescence is very common at the pubic symphysis behind, and is especially dangerous; but it is also found at the line of the sacro-iliac joints. The direction of the lesion in these cases of rubbing-through, or perforation, is usually transverse, corresponding to the direction of the pelvic brim.

A fibrous tumour in the wall of the uterus has caused rupture. There is a specimen illustrating this in the Middlesex Hospital Museum. In a case I examined with reference to a charge of malapraxis, a tumour in the anterior wall had got compressed between the head and the pelvis, causing perforation of the uterus and bladder. Probably no care could have averted this catastrophe. The accused, an illegitimate practitioner, was discharged.

A stone in the bladder has caused a similar injury. Guillemeau relates a case. Vesico-vaginal fistula resulted.

Rupture has several times occurred in women who had been the subjects of Cæsarian section. In some of these the laceration took place in the seat of the *cicatrix* of the wound.

The *uterus may be torn after delivery* of the child. One lip, most frequently the anterior one, may hang down in a large loose flap, and even protrude at the vulva. This has been mistaken for the placenta, supposed to be adherent; and being pulled upon has been actually torn away. Duparcque relates a case where this error occurred.

Laceration from obstruction to labour.—It is not necessary to do more than refer to the familiar causes of obstructed labour; that is, to narrowing and distortion of the pelvis, tumours blocking the pelvis, rigidity or other diseased conditions of the cervix and vagina, obliquity of the uterus, excessive size of the child, hydrocephalus, to monsters, locking of twins, or malpositions, producing a wedge the base of which is too large to enter the brim, or to traverse the pelvic canal.

We may assume, in the first place, that the *uterus, like the*

FIG. 108.

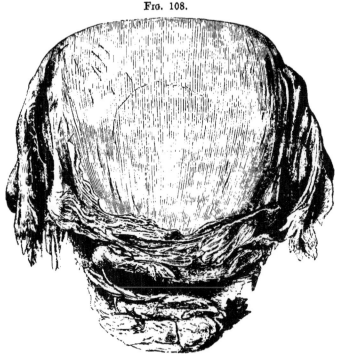

RUPTURE OF UTERUS.

From a preparation GG² in St. Thomas's Hospital; half-size. "The rupture" (or rather rubbing-through) " is in a transverse direction, from four to five inches in extent, and situated at the lower part of the anterior wall. The peritoneal surface is coated with recent lymph. From a woman who had a slight deformity of the pelvis. She had previously borne four children at the full period. The uterus is at the eighth month of gestation. Death in five days. Dr. Waller's case."—*Museum Catalogue.*

heart, ruptures itself. The mode in which rupture is brought about during pregnancy or independently of labour has been pointed out. When laceration occurs during obstructed labour, the conditions differ. The liquor amnii has almost always been discharged. The uterus has contracted upon the child. There is no longer the equable hydrostatic pressure. The necessary condition now is, that some part of the uterus be fixed, whilst the rest of the organ is pulling from that fixed point. It is this point which generally gives way. Duparcque says, " Uterine contractions alone are the most frequent causes of transverse ruptures of the uterine neck." A muscle in active contraction will not tear its own contracting fibres. It tears at its attachments, or at the point it is pulling upon, just as the tendo Achillis gives way rather than the muscles which pull upon it. If a muscle is not strong enough to accomplish the object for which it contracts, it becomes fatigued and relaxes. It is only when a sudden increased strain or injury is inflicted at the moment of contraction that the muscular fibres are liable to be torn. The uterus is no exception to this law. We see how it acts in ordinary labour. It contracts in the direction of its long axis, tending to shorten itself, pulling the os towards the fundus. So acting, the os uteri is partly pulled open, partly dilated by the hydrostatic pressure of the liquor amnii, or of the protruding part of the fœtus. The effect of the hydrostatic pressure in dilating the os is marked in ordinary healthy labour. If it fail through too early discharge of the liquor amnii, and there be obstruction, so that the fœtus is slow in engaging in the cervix, and in descending into the pelvis, dilatation is slow, and is ultimately effected mainly by the continuous pulling-up of the cervix by the action of the longitudinal uterine muscles. In either case, so long as the cervix, the part pulled upon or stretched, yields, there is no fear of laceration. But, if the cervix will not dilate, and the uterus continue to contract, as it will do under excito-motory stimulus, the fœtus being driven down violently upon its lower segment, the uterus will lacerate here. In such a case usually the laceration begins at the edge of the os uteri, and extends upwards longitudinally. Or, if the fœtus be dead, or presentation be abnormal, so that it cannot traverse the os, a limb, as a knee or elbow, forming an angular projection at some part of the body of the uterus, may render this particular part the fixed point upon which the uterine muscles pull; and, this point gradually softening and weakening, gives way. In this case the rent may be longitudinal or transverse, and at any part of the uterine wall. But the direction of the rent is usually determined by the drag of the muscular fibres. Hence, if the rent occur in the sides of the uterus, the rent is longitudinal; if at the fundus, or on the front or back, it is transverse. I have, however, known a case in which the uterus

Fig. 109.

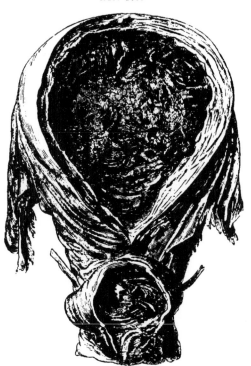

RUPTURE OF UTERUS.

From a preparation in St. Thomas's Museum, one-third size, GG²'. Anterior aspect. There is a boring-through of the anterior wall of the cervix into the bladder. The result of long-protracted compression against the symphysis pubis. Dr. Barnes's case.

was rent along its whole length in the middle of the anterior wall.

A similar explanation holds in the greater number of cases of rupture from contraction of the pelvis. The proneness to rupture is not in proportion to the degree of pelvic contraction. If the contraction at the brim be so great that the fœtus and

z

inferior segment of the uterus cannot enter, the risk of laceration
is less when there is just that degree of contraction which
permits of the descent of the head into the pelvic cavity, but
impedes its progress. This point has been well made out by
Radford (*Obst. Trans.*, 1867) and others. In nine cases out of
eighteen reported by Radford there was slight contraction of the

Fig. 110.

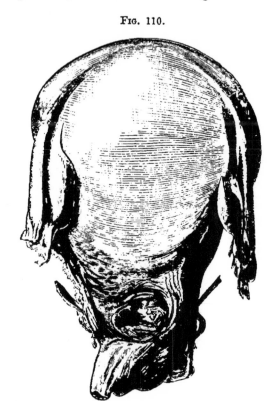

RUPTURE OF UTERUS.

Posterior aspect of same uterus as that represented in Fig. 109. The posterior wall of the
cervix is bored through by long compression against the jutting promontory of the sacrum.

brim. In these cases, when the head is forced down, the uterine
neck and vagina become fixed between the head and pelvic walls.
The uterine muscles continuing to contract, and not being able
to propel the fœtus, pull upon the tissues forming the fixed ring,
which either directly tear, or being first contused and softened,

yield. Hence it is that in these cases the rent is usually transverse or circular. (See Fig. 108.) In some instances, even where there has been no pelvic contraction, but the os has been rigid, the cervix has been torn through all round, or has gradually sloughed off, and been cast as a ring. Not to mention other cases usually cited, I will refer to one related by Dr. Herbert Barker (*Obst. Trans.*, vol. ii.) under the title, "Annular Laceration of the Cervix," and to another referred to me by Dr. Hague. In these cases the peritoneum is not involved; and the danger is not great.

But where the lower segment of the uterus is nipped in the ring of the pelvic brim, the laceration commonly involves all the tissues, at least in some part of the circumference. We then have a direct opening into the peritoneum through which the child may pass out of the uterus.

In some cases of great pelvic contraction, where the head has been long compressed upon the brim, the soft tissues are worn down by rubbing and necrosis, and give way transversely at the two points of greatest pressure, viz., at the anterior wall and bladder opposite the symphysis pubis, and at the posterior wall opposite the jutting promontory of the sacrum. In these cases the lesion is always transverse. This is well illustrated in Figs. 109, 110. The subject was a very deformed dwarf, to deliver whom I was called to one of the workhouses. She had been long in labour, and great efforts had been made to deliver after craniotomy.

In another class of cases, strictly analogous to the preceding, the head of the fœtus has traversed the os uteri and entered the pelvis, when, owing to impaction, a zone of the vagina is nipped. The uterus still striving to drive the child on, pulls upon this fixed vaginal ring, and a transverse, or circular laceration occurs. Of this there are many examples. Duparcque thus describes the process: "The child's head arrested on, or more or less engaged in the pelvic brim, the womb continuing to contract pulls itself, so to speak, away from the child. The borders of the uterine orifice drawn towards the fundus of the organ, therefore rise and abandon gradually and sometimes completely the fixed head. It results that the vagina is subjected to an active traction proportioned to the energy of the uterine contractions, and that, opposing only a passive resistance gradually becoming weaker by stretching, it ends by giving way by bursting." Once begun, these ruptures may, by the continuing or the renewal of the uterine contractions, separate the uterus from the vagina almost completely. Velpeau says he has met with this twice (*Tocologie*, 2 ed. t. ii., p. 194). But the forceps or hand had been used in both cases. "It will be easily understood how the lifting-up of the uterus, by ill-directed efforts, as when the woman throws up her stomach by suddenly throwing the trunk back, by vomiting, or convulsions, may determine a vaginal rupture which was only imminent. Abdominal compression so directed as to lift up the womb when the dilatation of the orifice is complete, may in the same manner provoke rupture of the vagina. The same thing may happen from pushing back the child, as this act adds to the state of tension the canal is already suffering; and this is not one of the least common causes of ruptures of this kind."

The unskilful introduction of the forceps, the blades being pushed behind the os, may cause rupture.

In transverse vaginal ruptures the head rarely escapes into the abdominal cavity; but the body may easily do so, since the uterus, after the rupture, is drawn up away from the child.

When the whole child goes into the abdomen, it may be because the head has been pushed back by attempts to manipulate.

I was consulted in a case in which the circular laceration was complete, so that the entire uterus containing the placenta came away from the body. Malapraxis was charged, but the facts seemed to refute the accusation. The uterus was not inverted; it seemed difficult for the attendant so to grasp the uterus as to tear it away. The case is consistent with experience of the mechanism of spontaneous injuries. Ingleby says,[*] "The rent is usually more or less oblique, sometimes including nearly the entire circumference, and thus virtually dissevering its connection with the uterus." Dr. Collins relates a case where the uterus was almost torn from the vagina. Moulin and Elkington relate similar instances. I have known another case of the kind. Dr. Roberton relates a case in which the cervix was separated from the vagina except by a shred. In the case to which reference is first made, it was objected that the uterus could not have

[*] "Obstet. Med."

rent its ligaments; that these must have been torn by manual violence. But there is no doubt that these ligaments, the round and broad, have been torn during inversion (Casper.) In women who have borne many children, all whose tissues are degraded, the ligaments at the time of labour are not capable of much X resistance. When the vagina has given way, the whole strain will fall on them. A little accidental force, then, as introducing the hand to remove the placenta, the application of the forceps, vomiting, might easily complete the detachment. Such is a summary of the arguments in favour of spontaneous circular laceration of the vagina, which led to the acquittal of the prisoner. Shortly afterwards a singular confirmation of their truth occurred in a strikingly similar case recorded by the late Thomas Paget, of Leicester (*Brit. Med. Journ.*, 1861). A pluripara had a not severe labour lasting three hours. The placenta was removed with but little traction in twenty minutes. To this, however, was attached a large fibrous mass, which was found to be the uterus, with Fallopian tubes and ovaries. The vaginal portion hung loose, and was jagged by laceration. She died in forty-five minutes. It was the decided opinion of Mr. Paget that the accident occurred spontaneously. He says : " For the exertion of the detrusive power of the uterus there is no *point d'appui*, no purchase, except the slender moorings, the round ligaments and sheet-like attachments of the vagina. Who shall wonder that they occasionally give way ? " Boër (*Medic. Obstetr.*), in chapter " De fluxu quodam sanguinis, etc.," describes a most extensive separation of the vagina from its attachments in consequence of an immense effusion of blood into the cellular substance. The uterus was not inverted.

Dr. Miller, in *A Memoir on Inversion of the Uterus*, relates a case in which the whole uterus, placenta included in it, was torn away by a midwife inserting her fingers inside the cervix. There was very little hæmorrhage, and the woman recovered.

These cases of separation by laceration must be distinguished from the entire separation completed after labour by sphacelus as in the instance recorded by Dr. More Madden (*Brit. Med. Journ.*, 1874). A primipara, aged 33, was admitted into the Rotunda Hospital; when admitted she had been eight hours in labour, the os was about the size of a shilling, and the head low down in the pelvis, pressing on the cervix uteri. The pains in the second stage were weak. A stimulating enema and a dose of ergot having been tried, she was delivered under chloroform by forceps. No force was required, a child weighing 6 lbs. was born alive in two minutes. She became suddenly collapsed on the fourth day and died. Autopsy revealed advanced peritonitis. The uterus was intensely inflamed in the cervical portion, which was actually in a state of sphacelus. The ulceration had extended completely through the cervix, so as to have entirely separated the uterus from the vagina; the line of separation being as sharp and clear as if effected by the knife. The vagina was, particularly near the vulva, also in a state of intense inflammation, and had sloughed considerably on its anterior wall. Dr. Jamieson (*Edin. Med. Journ.*, 1872) relates a case in illustration. The woman's spine was short, so that the last ribs squatted below the iliac crests. There was great anteversion of the uterus, which, therefore, acted at great disadvantage. Continuing to contract violently, it tore itself away from the vaginal attachments.

In St. Bartholomew's Museum are two specimens illustrative of this point. "No. 32·46. Uterus and vagina. During parturition the vagina was torn through half its circumference close to the part connected with the uterus." "No. 32·47. Uterus, the neck of which was torn through two-thirds of its circumference during parturition. Child hydrocephalic."

The argument that the uterus ruptured itself, as described, is enforced by Roberton, who draws the conclusion, that, in the majority of those instances of rupture of the uterus, caused by faultiness in the brim, the accident occurs within twelve hours after labour has commenced. This implies unimpaired vigour of contractile force.

But we must not forget that, although complete separation may occur spontaneously, a transverse laceration of the vagina near the uterus having begun, very moderate force may cause it to extend. This is adverted to by Denman, who says: "I was called to a case of a very extraordinary kind, in which that part where the vagina and uterus are united was ruptured; the child remaining in the cavity of the uterus, the os uteri being little dilated. Here my advice was not to attempt to deliver, because so much force would be required for dilating, that it was feared the uterus would be completely torn from the vagina before the hand could be passed into the uterus, at least before the child could be extracted, and then the case would have been more horrible." The woman would, in all probability, have had a better chance if left in the hands of a bolder man, who would have done what Denman dared not to do. But he would have done it at the risk of his own reputation, and of a conviction for manslaughter, with imprisonment as a felon !

In another class of cases there may be no fixing of the lower segment of the uterus, but the cervix rends, beginning at the edge of the os. In almost every first labour this takes place to some extent, and the trace is found in the cicatrices of the os, which tell that the subjects have borne a child. If the os be unusually indisposed to dilate, if the head be large, if the liquor amnii have escaped early, if the os have been carried down into the pelvis capping the head, and if the pains be severe—especially if stimulated by ergot—an initiatory rent of this ordinary kind may easily extend upwards into the uterus, and downwards into the vagina. In these cases, the direction of the rent is longitudinal, and usually in one side. The particular side is determined by the inclination of the uterus. Dubois says the habitual direction of the uterine neck to the left, and the left occipital position of the fœtus, account for the greater frequency of lacerations in the left side.

The most frequent seat of rupture is at the junction of cervix and vagina. It is almost the exclusive region of perforation from pressure, and from external violence. But it is also the most common for spontaneous rent; and in many cases where rupture has been observed in the body of the uterus, it probably *extended* from the cervix. Collins and Clarke never saw rupture at the fundus; and there is good reason to doubt whether some cases reported as such were not really examples of the giving way of a sac of a mural gestation, or of one formed in one horn of a one-sided uterus.

Apart from any detected abnormality of pelvis or of obstructed labour, the uterus has given way under a *suddenly excited increase of contractile energy.* Dr. Tyler Smith particularly insists that the irritation caused by manual examination of the os uteri is sometimes enough to cause rupture. Several cases lend confirmation to this view. In one I have recorded (*St. Thomas's Hospital Reports*, 1870), "when examined,-

the uterus was felt contracting; the os uteri was not reached, but the head was just felt. Whilst under examination the pains stopped ; she said she felt something give way. She walked across the room after this, but collapse followed in five minutes." We may, then, admit that rupture may be caused through the excito-motory system. But the *centric* stimulus to inordinate uterine action is more frequent. In a very large proportion of cases, ergot had been given. It is to be remarked that in this order of cases, the action of the uterus is *excessive, and suddenly evoked,* and probably disorderly in character; in this respect resembling those which occur before labour, in the circumstance that there is no adequate safety-valve process of cervical dilatation to meet the sudden contraction of the body of the uterus upon its incompressible contents.

Rupture has on many occasions taken place, or been extended, during *straining at stool.* This was observed in one of my cases.

Lacerations of the vaginal portion do not always extend through the peritoneum. At the lower segment, the connection of the peritoneum is looser. *Rent beginning at the mucous surface and involving the fibro-muscular tissue* may terminate by stretching the connective tissue, so that the peritoneum is dissected off from the utero-vaginal surface. Another condition favouring this escape of the peritoneum is the great distensibility of this membrane. The cavity of the abdomen is thus preserved. During labour some stretching of the connective tissue around the cervix always takes place, often with some amount of blood-effusion, always with serous effusion. When there is considerable laceration of the cervix, hæmorrhage takes place into this loose tissue, forming peri-uterine hæmatocele, which, when intra-abdominal inspection is made, may be seen forming a bluish tumour composed of coagulated and fluid blood, in a sac formed of the peritoneal investment of the uterus and the broad ligaments. The peritoneum is, in fact, undermined by the effused blood. This may burrow even as far as the lumbar vertebræ upwards, and in the pelvic cellular tissue as far as the labia vulvæ downwards. It cannot be rare, since Collins observes that in nine out of his thirty-four cases, the peritoneal covering did not give way. "Yet," he adds, "death ensued equally speedily, showing that the free admission of air into the abdominal cavity, is not necessarily followed by any increase of danger." The part where the blood is effused forms a bulging prominence on the outer surface of the uterus, and the peritoneum may crack. Simpson has seen it produced by injecting water into the uterus to induce labour. This variety is sometimes called "*partial rupture.*" When moderate in extent, the injury and shock are much less severe than in complete ruptures. Hæmorrhage externally commonly attends. The collapse and ominous change of countenance, the sensation as if something had snapped, may be wanting. Hecker has drawn attention to a sign* depending on the extra-peritoneal thrombus. The pulse, he says, always falls, even at the beginning of the rupture; it is quick and small. Then a hæmatocele forms in the cellular tissue between the bladder and the uterus. The swelling thus formed is *smooth, elastic, and quickly-growing.* They differ from rents in the body in that they do not readily close; they remain gaping. The deficiency of muscular fibres in the lower segment accounts for this. The rapidity and extent of the blood-effusion is accounted for by the great vascularity of the parts wounded. Collins narrates cases that illustrates this point. In one he says, "The peritoneum was not injured, but was raised up and distended with blood underneath, resembling a bladder." Recovery is not infrequent.

Another form of incomplete rupture is that in which *the peritoneal coat is alone or chiefly torn.* Over the body of the uterus the connection of the serous membrane with the muscular wall is so intimate that it is scarcely possible for rent of the peritoneum to take place without involving the muscular wall to a slight extent. When it occurs, some blood is often effused into the peritoneal cavity, where it may excite inflammation. Jacquemier describes splits and scars in the neighbourhood of the Fallopian tubes and round ligaments, the result of mechanical distension. These, say Dubois and Pajot, do not appear to give rise to perceptible symptoms. They may be likened to the lesions of the skin from distension of the abdominal walls. But in the more severe cases of peritoneal laceration death may occur from shock, as in a case narrated by Clarke (*Trans. of the Soc. for Improvement of Med. and Surg.*

* "Monatssch. f. Geburtsk.," 1868.

Knowledge), where only an ounce of blood was found in the abdomen; and in another by Collins. A more frequent cause of death seems to be hæmorrhage, as in cases by Ramsbotham and Francis White.

There is a preparation in Middlesex Hospital Museum presented by Dr. Hall Davis. The subject fell heavily on her back, being near the end of gestation. The under surface of the liver and the opposite surface of the uterus were lacerated. Copious effusion of blood was found in the abdomen. It did not appear that any direct violence was inflicted on the womb.

Rupture may ensue from *obstruction caused by the child*— namely, from excessive size and want of elasticity—hence the explanation of the fact, that rupture is more frequent in labour with boys than with girls. Thus Collins says of thirty-four cases seen by him, twenty-three were with boys; and McKeever, out of twenty, says fifteen were boys. From hydrocephalus: this was the cause in one of my cases; from malposition, the child lying so as to form a wedge, the base of which cannot enter or traverse the pelvis; from death of the fœtus, as in two cases observed by myself. I am disposed to think that this last cause, hitherto unsuspected or overlooked, is not infrequent. Where the fœtus has been dead some few hours, all its elasticity is lost; force bearing upon its breech is not propagated through its spine so as to drive on the head as in a live child, but the trunk collapses, doubles up, the head bends upon the chest, the whole fœtus tends to mould into a ball. It thus offers resistance as effectual as does an ordinary shoulder-presentation.

Lacerations of the vagina have been carefully described by McClintock (*Dubl. Quart. Journ. of Med.*, 1868) and Scanzoni. There are *incomplete lacerations*. Tearings of the mucous membrane and of the sub-mucous connective tissue may happen. Bleeding occurs if the mucous membrane is torn. Thrombus forms if the connective tissue is torn, whilst the mucous membrane remains whole.

Circular laceration of the upper zone of the vagina under the action of the uterus has been described. Spontaneous lacerations of the upper part of the vagina are often the result of extension of laceration of the lower segment of the uterus. The peritoneum behind, or the bladder in front, may be involved. (*See* Figs. 109, 110.)

Ingleby says: "The contractions of the uterus will act upon the child, even to the laceration of the vagina, without producing a breach of its own surface." But we must not forget that similar injury may be inflicted by the hand or by instruments in attempts to deliver. Thus, Ingleby says: "In very difficult turning cases, the extensive separation of the vagina from the uterus has occurred whilst the practitioner was engaged in changing the position of the child, the uterus instantly passing into the abdomen out of his reach. I am acquainted with two such instances. From these and several others of a similar kind, which have been reported to me, it appears that a laceration of the superior portion of the vagina, including nearly its whole circumference, is an occurrence by no means unfrequent." And we must also bear in mind that transverse rupture of the posterior wall of the vagina has been caused by the sheer force of the expulsive efforts driving an ovarian tumour upon it (Berry and Dunn's cases); or the retroverted uterus in early pregnancy (Martin's, Dubois', Schnakenberg, Grenser's cases, and Fehling's), in which the non-pregnant prolapsed uterus was thus protruded. In the same way an extra-uterine gestation-cyst has burst its way through the vagina at the retro-uterine pouch, as in the case of Dr. Thormann (*Wien. Med. Wchnschr.*, 1853). Under expulsive efforts a rent was made in the posterior vaginal wall, and an arm of the fœtus protruded.

As to the consequences of such injuries, Ingleby says: "The absolute separation of the whole uterus from the vagina, as the result of laceration or sloughing, would seem quite incompatible with the preservation of life; but an instance of this kind, which terminated favourably, has been recorded by Mr. Cook, of Coventry. The separation took place the second day after delivery, and the specimen, which embraces the uterus in a state of *inversion*, together with its ligaments, has just been deposited in the anatomical museum in this place" (Birmingham).

Rupture of the *middle part of the vagina* is rare. It can scarcely take place through its own contractions, as the contractile power of the vagina distended during labour is not great. It commonly occurs when the head is in the pelvis; and is therefore probably caused by the action of the uterus either dragging the canal upwards from the impacted head, or driving the head through the distended walls. A dead flaccid child would favour this accident, so would a face-presentation, or an occipito-posterior presentation, in all of which there is a tendency to delay in labour, and to rolling back of the head in extension, stretching and bruising inordinately the posterior wall of the vagina.

Vaginal lacerations mostly take a circular form; they remain patulous; and if the posterior wall is torn, the escape of the fœtus into the peritoneal cavity is frequent; prolapse of the intestine is also not uncommon.

Laceration of the inferior third of the vagina almost always takes place in the posterior wall, and merges in *laceration of the perinæum*. This may be central or vulvo-perinæal. In a considerable proportion of cases I believe the rent begins in the centre of the perinæum, then extends backwards and upwards into the recto-vaginal septum, and forwards through the commissure. There is, in fact, rent of the posterior valve (*see* p. 63), the final obstacle to the expulsion of the head. In almost every first labour more or less laceration of the commissure or fourchette takes place. This is usually insignificant. But it is not uncommon to find lacerations an inch in length. As the parts recover from distension these are reduced, and granulation often substantially repairs the injury. Occasionally, in spite of every care, the rent extends through the sphincter. There is a form of pelvis that seems to predispose to this accident. The symphysis pubis is unusually deep, the arch narrowed so that the head is necessarily directed backwards, and in some women the vulva is extremely small and rigid. Then, again, in lingering labour, the parts may lose a portion of their expansibility, and the forceps becoming necessary, under the strain the perinæum will be more likely to give way. No doubt, if unskilfully used, the forceps may increase the risk of rupture; and that in some cases skilfully used, it may lessen this risk. But still, on the assumption that the use of the forceps is indicated to avoid the dangers to mother and child of retarded labour, the lesser danger of injury to the perinæum must be encountered. Rupture of the perinæum, whether it occur under spontaneous or instrumental delivery, is not evidence of want of skill. Under instrumental delivery, which presupposes unusual difficulty, rent is *à fortiori* more probable. This is important to remember, because threats of legal proceedings to extort money have been based upon accidents of this kind. Sometimes the rupture is limited to the central perforation. Duparcque measured the perinæum as distended by the head. The length was 3·50″ to 4·00″, the breadth 6·00″, being more than twice its ordinary dimensions. It is, besides, excessively thinned. If the head be large, firm, the perinæum at all rigid from obstruction to its circulation, or unyielding as in primiparæ, and especially if the coccyx retreat backwards, or the pubic arch be narrow so as to throw the head backwards, opposing extension round the symphysis, the perinæum becomes enormously distended centrally, and is very apt to be perforated. The child has several times been driven through this perinæal opening, the commissure being preserved intact; but more commonly the rent forwards is completed. Occipito-posterior or face-presentations are especially liable to cause perinæal lacerations. The structures may also give way before stormy tetanic uterine contractions. I have sometimes seen the central opening produced in the way described, but the head has ultimately passed through the vulva. In one case a fistulous hole in the perinæum remained for some time. The perinæum is not seldom torn by instruments. I have, when discussing the use of forceps, dwelt upon this subject. The structures also give way after undergoing a stage of mortification from compression, bruising, and stagnation of circulation, the result of protracted labour. They then become as lacerable as wet brown paper, literally rotten.

Rent sometimes begins at the vulvar edge from extreme rigidity. This may be prevented by timely incisions. Laceration may take place at the anterior edge of the vulva (*see* Tyler Smith quoting my authority); and P. Müller, of Würzburg (*Scanzoni's Berträge*, 1870), on ruptures between the clitoris and meatus. In one case fatal hæmorrhage ensued.

Regarding the process by which perinæal laceration commonly occurs, the mode of guarding against it seems clear. *The perinæum requires support.* It is necessary to obviate the initiatory tendency to central perforation by giving an artificial firm flooring to this part by the palm of the hand, which resting behind upon the end of the sacrum prolongs the pelvic floor, and guides the head forward in extension. Abundant lubrication is useful. And occasionally the long double-curved forceps, by carrying the head well forwards, may preserve the perinæum. The short straight forceps is more likely to split it.

When laceration of the perinæum occurs, it is well to stitch it up at once whilst the edges are fresh. This may be done even as late as the next day; but it is better done at the time. Abroad it is usual to employ *serres-fines* for the purpose. Sutures are better. If the opportunity for doing this be lost, endeavour should be made to promote closure by granulation and cicatrisation. This often takes place spontaneously. But the process is much promoted by placing a strip of lint steeped in solution of carbolic acid or chloride of lime between the lips of the wound, taking care to press it well back to the fork of the wound. This should be renewed twice a day. It preserves perfect cleanliness, and stimulates healthy granulation.

Traumatic lacerations.—These may result from injury inflicted through the abdominal walls or through the vagina. Injuries from without are of endless diversity.

The injuries inflicted from within are of more special obstetric interest. These may arise from the use of instruments or the hand. During pregnancy, wounds of the vagina and uterus are mostly the consequence of attempts to procure abortion. Various stilets or pointed instruments have been used for this purpose. If unskilfully used, as is commonly the case, puncture or laceration of the vagina or cervix uteri often occurs in the attempt to pass the os. In other cases, punctures have been found penetrating the cervix and the wall of the uterus, opening into the peritoneal cavity. In these cases effusion into the peritoneum, and the injury, may result in death; in minor injury, perimetritis and pelvic peritonitis of an acute type is very common, and may also be fatal. The symptoms of abortion of course attend.

Injuries inflicted during labour may or may not be the consequence of unskilful obstetric manœuvres. Not seldom is it very difficult to determine, even in the presence of the most severe and extraordinary injuries, whether accident, avoidable or unavoidable, or direct violence was the cause. They may have begun spontaneously, and have been extended by attempts, perfectly legitimate and perhaps not unskilful, to complete delivery of child or placenta or to return prolapsed intestine. It requires the greatest circumspection not to commit one's self to an opinion which subsequent evidence may prove to be erroneous. It must sometimes be impossible to discover in the wound unequivocal proofs as to the mode of its formation.

Beginning with the perinæum, we of course conclude that severe injury in this part, if noticed before the descent of the child, is the result of external violence. If known to have occurred during the delivery of the child, it would be difficult to prove that it was due to criminal malapraxis, however strong the suspicion might be that it was due to unskilful treatment. Injuries of the vagina mostly occur at the upper part. They may arise from unskilful use of the forceps or perforator. The blade of the forceps may be thrust through the roof of the vagina at the angle of reflection from the cervix uteri; it may thus penetrate the peritoneum, and partially detach the uterus from the vagina. Wounds made in this way are almost always transverse. But so are most of the lacerations of this region which occur under the natural forces. Possibly the edges of the wound may show marks of bruising if made by the forceps. The roof of the vagina has been pierced by the perforator, and this accident is really not so utterly inexcusable as may be supposed. A very projecting sacral promontory, occupying as it does exactly the place where the head ought to be, and presenting physical characters very similar to the touch to those of the foetal head, may easily deceive the inexperienced. Strict observance of the rules given in describing the operation of craniotomy will secure against this error. I believe the accident has also happened from the use of bad perforators. Indeed, it must require an amount of skill far beyond the average so to use some of the vile instruments still in vogue, as to avoid their slipping and doing mischief, under some circumstances of difficulty.

Injuries may be inflicted by the hand, first, in attempts to turn when the child is tightly compressed by the contracting uterus or impacted in the pelvis; secondly, in detaching an adherent placenta.

The first danger in turning is encountered in the endeavour

to get the hand past the presenting part of the fœtus. If this be roughly made, the uterus may be partially torn from the vagina. If this danger be overcome, the uterus may be perforated in the body by the projection of the knuckles, or by violent thrusting forward of the fingers. I must refer to the minute instructions given in Lecture XVI. as to the modes of turning under difficulties, as the best means of avoiding lacerating the uterus. I would especially insist upon the importance of the operation of decapitation where the child is dead, as the most certain way of saving the mother from further suffering and danger. Upon this subject Denman says : " If the uterus be strongly contracted, it may be ruptured by attempts to pass the hand for the purpose of turning a child; but in this case a rupture could only happen when the force with which the hand was introduced was combined with the proper action of the uterus; for the strongest person has not the power to force his hand through a healthy and unacting uterus." This, I believe, is true. But we must admit that it is very possible to tear through the vagina, and also to rend the os uteri, whence the lesion might easily extend into the walls of the body of the uterus, especially if the tissue be abnormally weak.

When the placenta is adherent, there is sometimes disease of the uterine tissue, and it may be impossible to detach all the placenta without injuring the uterine wall. The operation should be done with great gentleness. It is better to leave portions adherent, than to persist too strenuously in tearing them off. This subject is dealt with under "Hæmorrhage." Dr. Dunn relates* a case in which rupture occurred with an adherent placenta. There is a preparation in University College Museum showing the same combination. The caution required in removing the placenta is well enforced by Denman : " I have known two cases in which it appeared that the uterus was ruptured by the very effort which expelled the child. If the placenta be afterwards retained, and it should be thought necessary to extract it, on passing the hand for that purpose, this would be more likely, without the greatest circumspection, to pass through the ruptured part into the cavity of the abdomen than into the uterus, the os uteri being more contracted. It

* " Obstetrical Transactions," 1868.

might then be possible to mistake some of the viscera for the detached placenta. This mistake was actually made with very aggravated circumstances in a late unfortunate case ; and the immediate loss of the patient's life, and the irreparable destruction of the attendant's character, were the natural and unavoidable consequences."

The crotchet, again, may slip and tear the uterine wall, especially near the cervix. Jagged pieces of cranial bone detached by the crotchet may inflict similar mischief. The obvious way of avoiding these accidents is to discard the crotchet, and substitute the craniotomy-forceps or cephalotribe. Injuries of this kind are not necessarily fatal.

Besides lacerations from more or less excusable want of skill, the most frightful injuries have been inflicted under the influence of ignorance, terror, or intoxication. The inverted uterus has been seized by both hands and literally torn away from the body. A specimen of this kind is preserved in the Museum at Birmingham (Ingleby). The woman recovered. Several feet, even yards, of intestine have protruded or been pulled through the vagina and cut away. In a case of the last kind in which I was called to give evidence, it was suggested on behalf of the prisoner that he might have mistaken the intestine for umbilical cord. The difficulties associated with this question will be discussed further on.

The attempt to determine, by examination of the subject after death, how a rupture was produced is surrounded with difficulties. To invoke the history of the case, in order to explain the post-mortem appearances, seems like begging the question. And too often the history will be so defective, drawn from imperfect or unskilled observations, essential facts being overlooked or suppressed, that it will rather mislead than help investigation. The great practical question will generally be, to decide whether the injury was caused spontaneously or by violence from without. But, even if we arrive at the conclusion that the injury was caused by the medical attendant, it will not necessarily follow that it was done through unskilfulness.

There are, however, certain points that may help to form a conclusion :—First, it ought not to be difficult to affirm or to negative the healthiness of the tissues of the uterus. For this purpose a careful inspection of the uterine tissues at different

A A

parts, remote from as well as at, the seat of rupture should be made. They should be tested as to their power of resisting traction, and especially as to their intimate structure, by the microscope. And to throw additional light upon the question of degeneration, it is desirable that a similar inspection should be made of the tissues of the heart, kidneys, and liver. It should be remembered that pregnancy itself occasionally causes organic changes in these organs. Hence it may well be that a similar change may involve the uterus.

Such an examination would reveal, if present, cicatricial, fibrous, cancerous, fatty degeneration or tumours; softening, ecchymosis, breaking-down, necrosis from long continued pressure and friction; the relative thickness of the walls at the seat of injury and elsewhere; the marks of internal or external pressure, friction or dragging. And these latter indications should be carefully studied in connection with the pelvis in its dimensions and other characters, and in connection with the characters of the child and placenta. The pelvic cavity and lower abdomen should also be searched for tumours or other abnormities.

In some cases inspection of the seat of injury alone will throw decisive light upon the cause of injury. Thus, where boring-through has occurred from obstruction to labour, the marks of injured tissue may be much more obvious and extensive, and the breach of continuity may be considerably greater externally than internally (See Fig. 108). So again where there are breaches of surface at different parts of the circumference of the lower zone of the uterus or upper zone of the vagina, the inference that the cause was obstruction at the pelvic brim will generally be correct. As an example of the first proposition, I may refer to a preparation in University College Museum (No. 2872) in which it seems obvious that the lesion was the result of grinding against the pelvic brim; and, as an example of the second proposition, to the case (Figs. 109, 110) in which the rectum and bladder were both perforated by long jamming and grinding against the narrow pelvis.

On the other hand, injury inflicted by the hand or by instruments may be expected to show more extensive marks on the inner aspect of the uterus or vagina; and these marks will consist of bruising, ecchymosis, punctures, rents, or incisions; and if marks of the kind are seen in healthy tissue the presumption that violence from without is the cause will be increased.

I hardly think it would be justifiable to give an absolute opinion as to the manner in which a lesion of the uterus was produced, where a close examination of this kind has been neglected.

The *symptoms and course* will naturally vary with the cause, extent, and seat of rupture. The *common signs* are those of "abdominal shock," indicating sudden severe intra-abdominal injury.

Rupture of the uterus during early pregnancy can hardly be distinguished from rupture of a tubal gestation-cyst. The subjective symptoms will be almost identical. We may arrive at a diagnosis after the accident by exploring the uterus with the sound, especially after dilating the cervix with laminaria tents. We may thus find the uterus intact, and if its size be not much increased, we have additional evidence that the pregnancy was not uterine. There may be external hæmorrhage in both cases, but there will probably be more in uterine rupture.

The *symptoms of spontaneous rupture or laceration early in labour*, as commonly described, and as they occur in a great number of instances, are—sudden acute pain with a sense of rending in the belly, sometimes attended with an audible snap, it is said; quick collapse, marked by pallor, fainting, extinct pulse; vomiting; some hæmorrhage externally, and the signs of anæmia from greater loss internally; cessation of uterine contraction; if the child be cast wholly or in part out of the womb, the abdomen sinks in, there is retreat of the presenting part of the child from the os uteri; occasionally prolapse of intestine in the vagina or beyond the vulva; great pain, especially on palpation, over the abdomen, where irregular, hard projections, which may be identified as parts of the fœtus, when the fœtus has been extruded from the womb. If the effusion of blood be great there is increased and distressing tension of the abdominal walls. Cramp-like or spasmodic pains follow. The flushed face becomes suddenly deadly pale; the eyes lose their brilliancy; the whole surface is covered with a clammy sweat; tremblings of the limbs or repeated faintings announce a profuse internal hæmorrhage. Presently, when reaction comes, the patient complains of feeling a warm

A A 2

fluid pouring out in the neighbourhood of the groins and loins. She sometimes feels the movements of the child when it has escaped into the abdomen; but usually the child quickly dies.

It has often been observed that the symptoms are not so strongly marked. Sometimes very little pain is complained of at the time when it was presumed the laceration took place; the collapse creeps on gradually, the patient being capable of walking about for some time. Denman says: "I have seen one case in which there was reason to believe that the woman walked a considerable distance, and lived several days after the uterus was ruptured before her labour could be properly said to commence."

But sooner or later, almost always within two or three hours, collapse becomes pronounced, and pain is severe. The gradual development of the symptoms is explained in some cases by the gradual progress of the injury. The rent does not, at one stroke, attain its maximum. There is first a moderate rent, possibly not through the peritoneum at first, without much effusion of blood into the abdominal cavity; the rent extends, and blood, and perhaps the fœtus, are extruded. We must not then expect uniformity in the symptoms. They vary considerably, according to the nature and seat of the injury. Thus in vaginal laceration there may be no retreat of the head or other presenting part of the child, although the body may escape into the abdominal cavity.

The symptoms of *partial rupture*, or where the peritoneum has remained intact, have already been described.

The *diagnosis* of laceration early in labour is generally distinct enough if the fœtus have been extruded into the abdominal cavity. Parts of the child are felt by external palpation. This palpation gives great pain. The contour of the abdomen is distorted; it presents irregular prominences. The uterus may be felt contracted and shrunk down towards the symphysis pubis. But the most certain sign is obtained by passing the hand into the uterus, where possibly intestine may be felt coming down into its cavity, or even into the vagina, and the hole may be felt. In one case where the rent was anterior, I could feel my finger, protruded through the rent, by the hand on the abdomen.

Where the fœtus is still retained *in utero*, the diagnosis is less obvious. But here also there is generally some recession of the presenting part.

The *symptoms of boring-through* under long pressure and friction can hardly be distinguished from the extreme collapse and irritative fever which follow upon long-protracted labour. The perforation of the uterine tissue is only the climax, the last stage, of long-preceding injury. The symptoms are mostly continuous. The patient dies of prolonged shock, exhaustion, and deteriorated blood.

Hæmorrhage varies greatly in different cases. In some fatal cases scarcely any blood has been lost. This may be partly explained by the rupture having traversed parts distant from the cervix and sides of the uterus, and from the placental attachment; from the placenta itself not being detached; and from the uterus having quickly contracted. There is usually little or no hæmorrhage in the cases of gradual perforation or separation resulting from long compression.

The symptoms also vary according to the seat and degree of the injury. Laceration of the cervix uteri not extending to the uterus or vagina may give rise to no marked symptoms. It is, moreover, a cause of secondary hæmorrhage.

The symptoms of *laceration of the vagina* are usually less severe than where the uterus suffers. Premonitory symptoms are rare; the shock is moderate; vomiting occasionally occurs; escape of the fœtus and placenta into the peritoneal cavity is more frequent (McClintock) than in uterine rupture. Prolapse of the intestine is not uncommon. But simple laceration without expulsion of the child into the abdomen, or prolapse of intestine may be fatal from shock, as in a case recently (1875) communicated to me by Professor John Clay, of Birmingham.

The *prognosis*, it is needless to say, is in a high degree unfavourable. The successive risks which the subject of rupture of the uterus has to run are—1. *Shock*. This may kill in a few minutes. 2. *Hæmorrhage*. This acts more slowly. But shock and hæmorrhage are often combined and act quickly; and secondary hæmorrhage may occur. 3. *Metritis, gangrene, perimetritis, peritonitis, diffuse suppuration*, leading to 4. *Thrombosis* and 5. *Blood-infection*, prove fatal at variable intervals, extending to days or even weeks. 6. In the perforative injuries, there.

may be *sloughing or gangrene*, especially when the bladder is also involved. Rokitansky mentions that the uterine artery has been opened by sloughs in the cervix, giving rise to fatal hæmorrhage. 7. The agglutinations from inflammation may obstruct the intestine and cause *fatal ileus*. 8. *Psoas abscess* was the cause of death in one or more of Collins's cases.

It is worthy of remark that in several fatal cases, masses of blood decomposing were found in the abdominal cavity.

The most formidable cases, however, do not exclude hope. Numerous instances are recorded of recovery even after the child, extruded into the abdomen, has been extracted by turning through the uterus. Some cases have been reported in which recovery followed, although the child was left in the abdomen. It may fairly be doubted whether some of these were not cases of extra-uterine gestation. On the other hand, it is possible that the child may have become encapsulated by inflammatory deposits, and subsequently discharged on disintegration. Recovery where the child is not cast into the abdominal cavity is more frequent. A moderate quantity of blood in the peritoneum may form a hæmatocele, and the uterine wound may heal. Perimetritis in such a case, becomes a conservative process; and the uterus may contract adhesions with the abdominal walls. The uterus contracting retreats into the pelvis; the wound either closes by a scar, or it has remained unclosed in part, but the opening shut off from the peritoneal cavity by adhesions to the abdominal wall.

It is a gross error to suppose that the uterus and the auxiliary muscles necessarily cease to contract after the shock of rupture. Yet I have heard this unhesitatingly affirmed by medical men in a court of law, their theory being that, since intestine was found protruding, it must have been dragged out by the medical attendant. First, as to the power of the uterus. Leclerc, cited by Duparcque, tells a case of a woman who had had six children. The body of the child escaped into the peritoneal cavity, whilst the head descended into the pelvis. The edges of the rent were firmly contracted about the neck. She died undelivered in twenty-four hours. The fact is that in a considerable proportion of cases the uterus contracts forcibly, just as it does after the analogous injury inflicted by the Cæsarian section. In not a few cases the child is expelled either by the natural passages, or through the rent into the abdominal cavity. Even after the expulsion or removal of the child, contraction has been ascertained by the hand *in utero* as well as by external manipulation. This I have felt myself. In fatal cases the uterus has often been seen, on post-mortem examination, so contracted that a rent through which the child had passed into the peritoneal cavity, was reduced to the length of two or three inches. And if the uterus did not sometimes contract, how are we to account for the cases of recovery? It is because the uterus contracts that prolapse of intestine is not more frequent. The uterus, like the heart, has a certain independent power. It has been known to contract after being almost entirely separated from its attachments, and even after death.

Then as to the diaphragm and abdominal muscles. That expulsive efforts generally cease under the collapse of shock may be true. But there is nothing so dangerous as that fatal tendency of small experience to rush to the absolute "Never" or "Always." If expulsive effort always ceased, how can prolapse of intestine be accounted for in any case? Yet there are many examples of protrusion where manual interference was out of the question. I can vouch for one from personal observation. A frequent consequence of shock is vomiting. What does the act of vomiting imply? Does it not involve expulsive action of the abdominal muscles. The strain of vomiting has even caused rupture of the uterus.

An observation by Prof. R. H. Thomas (Trask, *Amer. Journ. of Med. Sc.*, 1856) is precise. He found a woman in collapse, and felt the child's limbs through the abdominal walls. He introduced his left hand into the vagina, and discovering a large rent on the left and upper part of its connection with the uterus, passed his hand into the peritoneal cavity, seized the feet and brought them into the vagina. "Some bearing-down pains (abdominal muscles) now came to my assistance, and a large dead child was delivered. The uterus was found to be contracted." The contracted uterus was also seen on post-mortem examination.

It may appear superfluous to point out to medical men of ordinary education that it is not the uterus, but the diaphragm and abdominal muscles, that force out intestine. But for want of knowledge of this simple physiological fact, excusable in a lawyer, and for want of due regard to skilled evidence, not excusable in a judge, Lord Coleridge, in the case of the Queen *v.* Peacock (*Warwick Assizes*, 1875), found fault with me for citing Dr. Fehling's case (referred to, p. 365,) as a proof that a large mass of intestine could be spontaneously protruded through a rent in the vagina. Lord Coleridge was unable to understand that the uterus being quite passive in this matter, it could not signify *quoàd* this point whether the woman were pregnant or not.

The Treatment.—The general principle, one not however universally assented to, is thus stated by Duparcque :—

"Whatever be the cause, seat, and direction of the rupture produced in the womb during labour, one first principle and fundamental indication presents itself : this is to deliver. This is necessary to obviate the passage of the child into the abdomen, if this has not already taken place. It is indispensable in every case, in order to save it, and at the same time to rescue the mother from the formidable dangers which threaten her. Time presses : a moment's delay may accumulate obstacles, may render them insurmountable, or allow accidents inevitably fatal to both mother and child to arise. It is especially in these circumstances that the physician must be equal to his mission. Success depends upon his accuracy of perception, upon rapid appreciation, upon action quick, bold, even rough, but always calm and reasoned out."

The *prophylactic* treatment has been already referred to. The *remedial* measures will be dictated by the nature of the case. Where there is spontaneous rupture during pregnancy, and at the onset of labour before the os uteri is expanded, first of all dilate this part, and then explore the cavity of the uterus. This may be done by incisions, if necessary, but certainly by the hydrostatic bags. This proceeding will enable you to extract the fœtus *per vias naturales,* if it be not extruded into the abdominal cavity ; and will, at any rate, afford outlet for blood and liquor amnii. If the head is within reach apply the long forceps, if you are called early and there is hope of the child being alive ; but if the accident has happened more than thirty minutes, if prostration be great, in all probability the child is dead. If by perforating it seem likely to extract the child with less distress to the mother, this method should be preferred. If the head do not present, or be beyond reach of forceps, turn.

Take the first case : that where *the child, or at least the presenting part, remains in the uterus or vagina.* Assuming, and I assent, with only that qualified reserve which a prophetic sense of unforeseen complications dictates, to the dictum of Duparcque, that quick delivery is indicated, we have but to decide upon the best mode of delivering. Our choice will here be limited to the forceps, craniotomy, and cephalotripsy, and turning. If the head present, if the cervix be expanded, and there be no marked obstruction from pelvic deformity, or from the child, the forceps will naturally be selected. Anæsthesia should be induced, I think, notwithstanding the influence of shock. The introduction of the blades will demand extreme care. The points must be carefully kept close to the child's head. An assistant should grasp the uterus on either side between his two hands during the extraction. When the child is delivered, even greater care, if possible, is required in the removal of the placenta. The cord must be tracked up to the placenta; and the greatest possible circumspection will be necessary in order not to mistake this body for anything else, and to avoid drawing down intestine along with it. Although the presenting part of the child may not have escaped into the abdominal cavity, yet the placenta may have done so, either soon after the occurrence of the rent, or during the uterine contraction attending the extraction or expulsion of the child. The cord then may guide you into the abdominal cavity, where the placenta will be surrounded by floating intestines. You will, holding the free end of the cord by one hand, and fixing the fingers of the hand which is in the woman's body round the insertion of the cord in the placenta, draw it gently down and extract it, if detached. If it does not come easily, carefully, gently feel all round the margin of the placenta, and try to ascertain the cause of retention. If the placenta be in the abdominal cavity, the difficulty may be due to contraction of the opening of the rent, more especially if the rent be in the uterine wall. In this case the utmost care will be required to avoid breaking the cord off at the root, and thus leaving the placenta loose amongst the intestines. What shall be done if this accident happens ? If the rent be through the uterine wall, it will be advisable to seek for the placenta, by passing the hand through to the opening, if this can be done *without using force*. But not seldom the uterus having contracted, the opening will be so much reduced, that the hand would only pass by considerable pressure; and this pressure might enlarge the opening, not by stretching, but by increasing the rent, and might even tear the uterus from the vagina. If then this difficulty be encountered, we have to choose between trusting to Nature and the operation of gastrotomy. The first course, that is doing nothing, has high authority; but the balance of recent opinion is in favour of gastrotomy. The placenta is an offending foreign substance likely to set up irritation and inflammation in any case, and since air may get access to the rent, to putrefy. It therefore should be removed. And the argument for gastrotomy is strengthened by the fact, that blood in considerable quantity has been also poured into the abdominal cavity, and can be removed at the same time. If the rent be a large one in the posterior vaginal wall, there is, of course, less liability to closure by contraction, and there will consequently be little difficulty in extracting the placenta, although entanglement with intestines may equally embarrass.

If the pelvis be narrowed, or the child of excessive size, it will be better to perforate the head, to crush it down by cephalotribe, and then to extract. In perforating, the rule laid down in Lecture XXIII. to have the uterus well supported upon the pelvic brim by an assistant is specially to be observed, lest pushing force applied to the head below increase the laceration. If the child present transversely, decapitation or bisection at the trunk should be preferred to turning, as involving less strain upon the injured maternal parts.

Take the second case, that *where the whole child with or without the placenta has been cast into the abdominal cavity.* Sometimes the entire ovum unruptured has been cast out of the uterus, as in one of my cases. If the rent be in the vagina, it may not be difficult to remove the child by the hand passed through the rent; and success has several times followed the removal when the rent was in the uterine wall. Ingleby relates a successful case. A case related by Danyau *(Mém. de la Soc. de Chir.*, 1851), and another by Dr. Bell, of Bradford, were instances of extraction through a vaginal rent. Generally, the attempt at removal, to offer any prospect of success must be done at once ; but Duparcque relates a case in which the fœtus was extracted putrid through a transverse rent at the union of the cervix with the body of the uterus eight days after the rupture, the woman recovering. If, however, the rent be in the body of the uterus the chances of success by operating by this route are much diminished. You may succeed in passing the hand through and in seizing the child ; you may even have brought the child back into the uterus as far as the head only to find it strangled in the opening, and resisting all further effort at extraction, unless at the expense of so much force as may destroy the little chance left of life. Unless, then, the hand pass easily, unless the child's feet—for you must seize both feet—be easily grasped, and unless you feel pretty confident that the child can be brought back without difficulty, it is better not to make an attempt, which if not successful, will be sure to make matters worse. Precise contra-indications are, large size of the child, hydrocephalus, cancerous or other disease of the uterine neck, contraction of the uterus, narrowing of the pelvis.

In the cases where the fœtus and placenta cannot be brought out by the natural passages, we have a resource in *gastrotomy*. The operation is described in the Lecture on the Cæsarian Section. It is very simple. In two cases in which I performed it, although death followed, very sensible immediate relief appeared to be gained ; the shock seemed diminished ; the pulse recovered, and I was satisfied that life was prolonged. In both cases the operation was delayed some hours. Had the opportunity been offered of operating at an earlier period, the prospect of recovery would have been better. In other cases a fair proportion of success has been attained. Where death follows it must be difficult or impossible to assign to the original injury, and to the operation their respective shares in the result. But, looking to the history of ovariotomy, to the exploratory incisions made without completing ovariotomy, we are justified in regarding the operation in itself as of comparatively small moment. It is the original injury that kills. If we fail to save the subject from her extreme peril, we must find comfort in the reflection that we have done the best that art and humanity suggest. Not to cite cases more or less known,

1 may refer to one published by Dr. Crighton (*Edinb. Med. Journ.*, 1864).

Gastrotomy offers the great advantage over all other proceedings, of enabling us to clear the abdominal cavity of placenta and of blood, and the best chance of saving the child.

Shall any attempt be made to close the uterine wound? I am not aware that it has ever been done in any case of gastrotomy for rupture of the womb. The fact is certain that women have recovered after gastrotomy. Therefore the operative closure of the uterine wound is not in all cases necessary. But it may be true, notwithstanding, that some cases that proved fatal would have had a better chance of recovery had the uterine wound been stitched up. In the successful cases, no doubt the wound was closed more or less completely by the muscular contraction of the womb. It may be that in some of the unsuccessful cases, the womb did not contract, so that the opening remained gaping. Might not then closure by suture compensate for want of natural contraction? The affirmative is not a necessary sequence. Contraction is proof of vital power; flaccidity, of prostration from injury that struck deeply. Mere closure of the wound cannot give vital power. The advantages to be gained from closure are: security against protrusion of intestine, and the prevention of the escape of blood or noxious discharges from the womb into the abdominal cavity. When the uterus is flaccid it would be better to sew up the opening. The operation is described in the Lecture on the Cæsarian Section.

Take the case where *intestine protrudes through the rent*. The following propositions may be stated:—Intestine may be protruded by the natural expelling forces, or it may be dragged out accidentally, or through recklessness, or by getting entangled in the limbs of the child. We may next inquire how much, and what portions of the intestine can be protruded spontaneously? The question is not easy to answer. The histories of cases given are wanting in precision on this point. But looking to what we see in operations for ovariotomy; to the analogy of hernial protrusions where there is no rupture in the sense of rent; and to the few cases of rupture of the uterus or vagina where the length of intestine protruded has been more or less precisely adverted to, we may conclude with confidence that more than six feet may be thus driven out; and we are not entitled to deny that very much more may be driven out. The force required to drive out intestine is really very small. The intestines are retained *in situ* by being packed in a closed bag; the mesentery is not wanted to suspend them, and Nature does nothing in vain; the mesentery is a delicate membrane, the chief use of which is to carry vessels and nerves to the intestine. Any one who observes what takes place when the intestines are handled in a post-mortem examination, must have been struck with the facility with which the body may be disembowelled. I have instituted an experiment to determine the force required to draw out intestine. The mesentery being detached at one point of the small intestine a two-pound weight was attached to the coil; it quickly ran down to the ground carrying intestine with it, the mesentery offering but slight resistance.* Now, the expulsive force of the abdominal muscles much exceeds two pounds. It is incontestable that it must equal the weight of the child, for the abdominal muscles alone may expel the child, and the uterus with it. This commonly weighs seven pounds, and may weigh twelve or more. Moreover, this weight is propelled against considerable friction. An approximate idea of the power of the abdominal muscles, even in a delicate woman, may be obtained when we attempt to grasp the uterus through them to cast the placenta. Not seldom, contraction so powerful is excited that the hands of a strong man using his utmost strength, and putting his weight into the effort, are thrown up from the uterus. In a case of ovariotomy I performed (June, 1875) a solid tumour weighing twenty-eight pounds was thrown out of the abdomen by the mere force of the diaphragm under vomiting. Without insisting upon the experiments and calculations of Dr. Haughton, I am satisfied that the expulsive force is not overstated at fifty pounds. A case related by Dr. Fehling (*Arch. f. Gynäk.*, 1874) is a distinct illustration of the simple action of the expiratory muscles in expelling intestine, the more valuable as it occurred in a non-pregnant woman. The subject, a pluripara aged sixty-three, had suffered from a reducible prolapsus vaginæ for thirty years, latterly about as large as a fœtal head. Carrying a bucket of water up four

* Dr. Braxton Hicks and Dr. Goodhart have since made experiments with similar results at Guy's Hospital.

steep steps, the womb came down. She tried to replace it, using some force. She felt something give way, and intestine protruded, forming a mass as large as a man's head. The intestine was followed into the abdominal cavity through a large rent in the posterior vaginal wall. An attempt was made to replace the bowel, but it was unsuccessful; F. punctured the inflated bowel, but with no good result. She died eleven hours after the accident from shock. There could be no doubt, says Fehling, as to the diagnosis of rupture of the peritoneum. The parts removed from the body showed that the posterior wall of the vagina formed a large tumour, into which the rectum protruded (*rectocele*). In this cavity probably a large part of the small intestines had also lain. The obstacle to complete reposition was attributed to the diminished space in the abdomen from long-standing prolapse, and to the string-like stretched mesentery of the protruded intestines which drew them back into the vagina. What was the amount of intestine expelled in this case? The account is not precise; but Dr. Fehling tells me it was probably over twelve feet; and a mass "as big as a man's head" could not be much less. No doubt there was predisposition to protrusion from the old hernial state; but that the main bulk of intestine had never been down before the injury is proved by the facts that the hernia was "as big as a child's head," and the protruded mass "as big as a man's head."

In Cooper's *Surgical Dictionary* (Art. Wounds of Abdomen) a case is related where the whole of the intestines, except the duodenum and the arch of the stomach, had escaped through a wound four inches long on the side of the upper abdomen. The intestines had not been dragged out.

What may happen if protruding intestine be not returned? 1st. It may be gradually reduced by spontaneous processes. The peristaltic action probably, or the passage of air along it may draw the prolapsed coil, if this be of moderate length, back into its place. Examples of this are not rare. I observed one myself. 2nd. It may get strangled in the uterine wound, inflammation and gangrene supervening. Deneux quotes a case from Percy. But recovery has been known. In the case of Dr. M'Keever four feet of intestine sloughed away, and recovery ensued. An artificial anus may form, as in a case seen by Roux. (*See* Duparcque.) It appeared that the surgeon had nipped a loop of intestine along with one lip of the os uteri between his forceps and the child's head, and this sloughing, a communication with the vagina was formed.

As it is a matter of great practical interest to know the possible sources of error in diagnosis and consequent error in practice, I will enumerate those which by experience or reading have been brought under my notice.

1. *The Placenta.* A child may have been delivered, and a placenta has followed. Another placenta may be felt in vagina or uterus. This would naturally lead to the conclusion that there was another child; and as none might be found in the uterus, it would suggest the inference that there was rupture, and escape of a child into the abdominal cavity. I have been called to such a case. The second placenta was a *placenta succenturiata.* There was no rupture and no second child.

2. A rupture may have taken place. The child may have been delivered naturally or artificially, the cord tied; and in searching for placenta, tracking the cord, the hand may be guided through the rent into the abdominal cavity where the placenta may have been cast by the uterus. Dr. John M. Cottle communicated to me a case to which he was called after death. He "found the body of the child, excepting the head and the neck—the head being fixed low down in the pelvis—and *also the placenta were amongst the intestines*, having passed through a large rent at the back of the vagina. Had the child been delivered, the placenta might either have been driven out along with the intestines, or the attendant tracing the cord would have got amongst intestines, and very moderate traction upon the cord would have drawn intestine along with the placenta."

3. A more or less *solid blood-mass* may present or be felt *in utero* after the birth of the child. Such a mass may easily be mistaken for placenta, although no cord be attached to it. And such a mass surrounded by membranes or layers of fibrin may even to the eye resemble the placenta. And such a mass may come, as in a case I published in St. Thomas's Hospital Reports, from the abdominal cavity.

4. Substances other than a child or placenta may be pro-

truded from the uterus. Thus a *fibroid tumour or polypus* may
be thrust out after the child has been born. It may be taken
for the placenta or a firm clot. It is a most puzzling complication. And it is not unlikely to be attended by rupture.

5. An *abnormality of the fœtus*, which might be seriously
embarrassing, is figured (Fig. 111) from a specimen in St.
Thomas's Hospital. It might easily be taken for placenta.
It is a tumour of the sacro-coccygeal gland.

Fig. 111.

6. In the *Virginia Medical Monthly*, for December, 1874, a
remarkable case is related by Dr. Dunn, in essential features
the counterpart of Mr. Berry's. He was called to a woman
suffering from " shock " in labour. On " placing his hand on the
abdomen, the fundus of the womb felt as if it had separated
from the right and right anterior half. The perinæum was
enormously distended, so as to make the impression that the
child's head was in the vulva." He felt *a fluctuating tumour*

in Douglas' pouch. The fœtal head, emerged from the os uteri, was pressing firmly upon the upper surface of the tumour. This pressure had caused it to tear an opening of two inches in length through the upper portion of the posterior vaginal wall. The elevation of the womb due to the resistance to the exit of the head, and the elongated distension of the perinæum caused by the pressure of the tumour upon it, gave the vagina a length of eleven inches as measured along its posterior wall. Dr. D. enlarged the rent in the vaginal wall, and bore firmly upon the perinæum during a "pain," which forced the tumour through the enlarged rent into the vaginal cavity; the pedicle of the tumour was ruptured by the same pain. It was now easy to extract the tumour lying loose in the vagina by forceps. The contraction of the abdominal muscles during the extraction of the tumour through the vulva, forced the fœtus into the abdominal cavity through the opening in the posterior wall of the vagina. He immediately passed his hand through the rent, turned, and delivered the child. The expelled tumour was found to be an ovary which had undergone multilocular cystic degeneration. It weighed thirty-three ounces. The fundus of the uterus resumed its normal position soon after delivery. The hæmorrhage did not exceed three pints. The woman was treated by repeated doses of morphia, and after protracted convalescence recovered. In a narration otherwise precise and consistent, no mention is made of the placenta.

7. *Omentum* may be mistaken for placenta or the membranes. Dr. Hicks relates a case (*Lancet*, 1875). He " was called to see a woman in whom there was considerable prolapse of the omentum from the vulva. He was told a curious modification of the placenta existed, but the medical man fearing all was not right, asked him to assist. Turning was accomplished when this mass came down to the vulva. During examination and manipulation it came outside three or four inches, where it was seen." Had, he observes, this omentum not been looked at, but treated as a portion of the membranes (a not difficult mistake to make), dragged down, and very possibly endeavoured to be peeled off as the membranes sometimes require, the case would not have been without many points of resemblance to that tried at Warwick. (Queen *v.* Peacock, 1875.)

8. *Intestine* may come down into the vagina or outside.

B B

But this is not necessarily proof of rupture. The intestines may belong to a child *in utero*, malformed, having no abdominal wall, as in a case (*Brit. Med. Journ.*, 1875) reported by Dr. Sheehy, in one by Dr. Meadows (*Obst. Trans.*, vol. vii.), and as in a specimen in St. George's Hospital Museum; or the abdomen having burst from decomposition, or from over-distension from ascites, the intestines may have fallen through. Fœtal intestines will be smaller than those of the mother; but the difference may well escape detection by one who in all probability has never felt one or the other before.

9. Maternal intestine then may be mistaken for fœtal intestine. And for what else? It has been mistaken for *umbilical cord*. This, it is said, ought not to occur. It is easy to be wise after the event. We are all under the dominion of habit. We believe the sun will rise to-morrow because it has done so every day hitherto. The obstetrist who has never felt anything else but placenta or cord in the vagina, is instinctively led to conclude that anything he feels there which bears to the touch any resemblance to them is placenta or cord. A mass of omentum consolidated by inflammatory deposit, intestine adhering to it, would hardly be distinguished from placenta. Then, as to the resemblances between intestine and cord. First, what are the characters of umbilical cord after division from the born child? It is a single string; there is the cut end outside; it has no mesentery; it is tolerably firm, not inflated by air; traced up it leads to its insertion in the placenta. What are the characters of intestine? It is a hollow tube, forming loops or coils; to its inner border is attached a thin membrane, the mesentery; it has an elastic feel less solid than the cord; traced up to its source it does not lead to the placenta, but through a rent into the abdominal cavity. These differences are marked enough to enable any skilled person under ordinary circumstances, in the full possession of his faculties, and retaining his delicacy of touch, aided by sight, to distinguish one from the other. But the circumstances are generally, *ex necessitate rei*, extraordinary; and the characters of cord and intestine are not always so peculiar as above stated. The cord, even after being divided, may form loops in the vagina, the cord being carried back into the vagina in searching for the placenta or lost in clots of blood. If the cord be long

—it has been recorded as being five feet long, and I have heard a surgeon declare in a court of law that he had known one fifteen feet long—there is no difficulty in the formation of loops and coils ; and a cord of a second, or even of a third child may come down in a coil, there being no free end. Then as to the feel: some cords are very thick, pulpy, as in a specimen in St. George's Museum, so resembling intestine as to deceive even the eye as well as the touch, especially resembling intestine which has been a little stretched, possibly separated from its mesentery, as it is very likely to be when protruded through a rent ; intestine so stretched loses its distinctive character of a hollow elastic tube ; it becomes a thickish cord ; so that, as Tyler Smith showed me, and demonstrated to a judge, it would require no small coolness and skill to distinguish them. Moderate tension upon the intestine will displace the air in it, reduce it to the semblance of a cord, and may detach it from its mesentery. Practised men have been deceived, at least for a time. Thus Duparcque : " examining I found in the vagina a body which at first I took for the umbilical cord." I have experienced the same difficulty, and in the same case two other practitioners were in like manner puzzled. In a very interesting case narrated by Mr. O. Lowsly, of Reading, the cord gave way in attempt to extract the placenta. " Being obliged to pass his hand into the uterus, there finding something like a portion of placenta, he brought it down, and to his great surprise discovered it was a loop of intestine." It was immediately returned. The woman recovered, and dying fourteen months afterwards of phthisis, a post-mortem examination showed a star-shaped cicatrix in the posterior wall of the body of the uterus. The way in which doubt has been cleared up was by gently drawing down a loop of the presenting body, and inspecting it by the eye.

Dr. Hicks calls attention (*Lancet*, 1875) to the singularly elongated state in which the vagina is often found at the time of labour. "In removing the placenta," he says, "one has had to pass the hand as far as the ribs before the os uteri could be reached ; that is, the uterus receded so readily, that till it was caught and fixed by the external hand, which is not always easy, the internal hand has reached so high. But in most cases the hand will reach to the umbilicus very readily. This is just about the centre of the mesentery, and therefore if a rent exist behind at its upper end there is an opportunity afforded for a prolapse of magnitude. If, therefore, in searching for the insertion of the funis, a coil or two of intestine have protruded through, the entanglement of them in the fingers, and the movements of the latter to avoid them and find the funis, would be sufficient to bring through a large amount, especially if aided by the straining of the woman and much pressure outside."

What are *the motives, or rather the excuses, for cutting off intestine?* I do not think that any adequate surgical reason has been shown for cutting-off protruding intestine in a recent case of rupture of the uterus or vagina. Where a coil has been partially separated by gangrene it may with propriety be removed. But it is a remarkable fact that in several, probably in many cases, the protruding mass has been cut away by scissors. Cases of the kind are referred to in all countries. Instinct and reason alike are shocked when such a case occurs. Indignation is not unnaturally excited. The actor is presumed to have been drunk, ignorant, or reckless. But now and then a case has occurred where the actor has been neither drunk nor ignorant, and where recklessness was not, at least, an attribute of his known character. On what principle then are we to explain such conduct? Two hypotheses may be invoked. 1st. He has been appalled by the sudden appearance of a mass of intestine; his distress has been increased by inability to return it; under the mental shock his judgment and his power of accurate manipulation are paralyzed; in ill-directed efforts more intestine comes down, and very slight force will cause this. Terror succeeds—first alarm for the danger of his patient, then dread of the consequences to himself. Under this temporary mental disorder he may do the most desperate things. He gives the woman up for lost; and, vaguely hoping to escape censure, he may seek to conceal what has, perhaps from no fault of his, taken place. Thus the intestine has been cut away, thrown behind the fire or buried or thrown into a privy.

2ndly. More or less disturbed by an occurrence altogether beyond common experience, he may mistake the protruding intestine for something else; and acting on the broad general rule, that whatever is expelled from the vulva during labour is to be removed, he may remove it in the readiest way. Thus intestine has been cut, thus the uterus itself inverted or detached, has been dragged away.

I do not urge these reflections, founded though they be upon experience, to excuse drunkenness or gross ignorance, or to advance a plea for professional irresponsibility. But it deserves consideration whether it is reasonable to expect from every practitioner, no matter how young, how necessarily limited his knowledge and experience, unaided by counsel, harassed by the bewildering emotions of those around him, under emergencies sudden and terrible, and beyond ordinary experience, the calmness of judgment and the skill in action which rarely come but with maturity in years and familiarity with difficulties. If the object of punishment be not vengeance for a particular crime, but the prevention of crime for the common protection, it may well be doubted whether criminal prosecutions for surgical casualties or blunders really conduce to the public interest. A young man entering on practice will hardly be encouraged under difficulties that may compromise his reputation and his liberty, by the feeling that he is working with a halter round his neck, to keep a cool head and a steady hand. Will not this feeling rather terrify him into the attempt to conceal a mishap, than encourage him to an effort to repair it?

If intestine protrude through a rupture in the uterus, attempts should be made to return it. This is sometimes extremely difficult, even impossible. As fast as you push up one part another comes down. Expulsive reflex action may be set up, so that the very effort you make to replace intestine, excites to the expulsion of more. It would be advisable, I think, to administer large doses of opium, chloral, chloroform, or ether, to lessen reflex irritability, if this is found to oppose reduction. The attempt must not, however, be persisted in too long, lest we increase the prostration and perhaps the injury. If we succeed in reducing, we may next endeavour to excite uterine contraction by keeping the hand in contact with the uterine wall, and in this way to close the opening and prevent recurrent protrusion.

When the rent is in the posterior wall of the vagina, the difficulty in returning the intestine and in preventing it from coming down again is much greater. The opening remains gaping, and all expulsive action bears directly upon it. It has occurred to some that the opening might be stitched up. Of course this presupposes that the intestine has been first reduced. I know of no instance of this proceeding being executed. I have heard of its having been done, but the cases are not published. Even if one were fully prepared for the operation with all appliances and adequate assistance, the carrying it out would be so difficult, tedious, and dangerous to the patient, that it will generally be wiser not to make the attempt.

An easier proceeding, one which might fairly be justified, would be to open the abdomen by an incision large enough to admit of manipulation; to draw back the intestines, taking the opportunity of removing blood clots, and then to stitch up the wound by operating from above. But the courage of the operator might not be appreciated; and there are not many men possessing nerve and skill enough to execute it without the countenance and assistance of a professional brother.

To lessen hæmorrhage there are two principal resources :—1st, to induce contraction of the uterus; 2nd, the topical application of styptics. The first object may be promoted by grasping the uterus; by the application of ice to the uterus. But the injury involves shock; and so long as shock endures it may be impossible to obtain contraction. We are then driven to styptics. A solution of persulphate, chloroxyde, or perchloride of iron may be applied by a swab to the wound. A most interesting case is related by Dr. More Madden (*Brit. Med. Journ.*, 1874). There was a rent so large that his "hand passed into the abdominal cavity. There was little external hæmorrhage, but a considerable effusion of blood was going on into the peritoneal cavity from the open uterine vessels. The patient was now pale and collapsed, and it was evident that any further loss of blood must be fatal; therefore, with Dr. White's sanction, I saturated a sponge in the strong solution of perchloride of iron, and applied it freely to every part of the rent. The effect was instantaneous. The woman complained of intense pain, but the hæmorrhage ceased completely, and the lacerated gaping wound in the vagina and uterus contracted sensibly under the stimulus. The woman was still cold, pulseless, and apparently moribund. Brandy and liquor opii were thrown up into the rectum. I left her, firmly expecting to hear of her death." She was kept under the influence of opium, and recovered.

During collapse, and indeed, for long after, perfect rest is all-important. Give occasional stimulants, as brandy, ether, and ammonia. When re-action takes place, or even before, give opium freely by mouth, rectum, or by subcutaneous injection. If peritonitis arise, leeches to the abdomen may be useful.

LECTURE XXII.

Inversion of the uterus is an accidental complication of labour which may most usefully be studied in connection with rupture. In the suddenness of its occurrence; in its instant danger; in the call for prompt recognition and remedy; and in its medico-legal consequences, it has many points of analogy.

Definition.—Inversion may be simply defined as a turning inside-out of the womb. The fundus falls through the cervix, coming out at the os, so as to form a new cavity, the lining of which is the external or peritoneal coat of the uterus.

Inversion may be complete or partial. Crosse describes three degrees. (*See* Fig. 112.)

1, depression; 2, introversion; 3, perversion or complete inversion.

Inversion is *acute* or *chronic*. In my article on this accident in Samuel Lane's edition of *Samuel Cooper's Surgical Dictionary*, I defined acute inversion as ending with the completion of the involution of the uterus. When this process is complete, the case is chronic. The distinction is based upon the important physiological fact that whilst involution is going on, the muscular fibres are still possessed of some active property, the organ is larger, and the cervix less rigid. During this stage the parts are more yielding, and reduction is comparatively easy. The history of the chronic form is traced with care in my work on

FIG. 112.

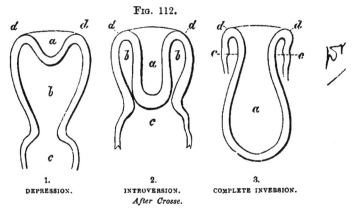

1.
DEPRESSION.

2.
INTROVERSION.
After Crosse.

3.
COMPLETE INVERSION.

a, Inverted outer surface of uterus; *b b,* cavity of uterus disappearing in 1 and 2, gone in 3; *c,* vagina.

the *Clinical History of the Diseases of Women.* Our present concern is limited to the recent or acute form of the accident.

FIG. 113.

DR. CHAMBERS' CASE OF INVERSION OF UTERUS: ONE-THIRD OF NATURAL SIZE.

The Fallopian tubes and round ligaments are seen hanging out of the infundibulum formed by the inversion. The rough surface at the lower part is the placental site. A secundipara, æt. twenty-seven. Half an hour after child's birth hæmorrhage set in, the placenta being not yet expelled. The midwife exerted traction upon the cord; the patient had severe abdominal contraction; uttered a violent shriek of pain; and a round ovoid body protruded from the vulva. The patient was seen moribund. There was no opportunity for replacement.

Inversion again may be *spontaneous* or *produced by external violence.*

Frequency of the accident. It is so rare that many men of large experience have never seen it. In former years the accident was not uncommon; and this may still be said to be the case in those countries where obstetric practice is largely in the hands of women. Like other of the great catastrophes of midwifery, it has become rare in proportion as the art has improved. Thus, it is all but unexampled in the records of the Dublin Lying-In Hospital.* In the practice of the London Royal Maternity Charity, from 1857 to 1874 inclusive, during which time 61,906 women were delivered, only one case of inversion occurred; whilst Ruysch tells us that it was not uncommon in Holland in his day, when midwives were generally employed. But although as a general conclusion it may be true that the frequent occurrence of inversion is indicative of bad practice, it would be wrong to apply this conclusion absolutely to every particular case. Now and then instances occur and will occur under skilful hands, and when approved principles of conduct have been sedulously observed. These general facts lead us to study the modes in which inversion spontaneous, or induced by external force, is brought about.

The following circumstances have been noticed in the histories of cases of inversion associated with labour :—1. It has occurred almost always, if not invariably, immediately after the completion of the second stage of labour, that is, after the expulsion of the child, and during the delivery of the placenta. 2. In a large proportion of cases, the accident followed pulling upon the cord, or other forcible attempts to deliver the placenta. 3. In other cases, the cord is noted as having been unusually short, or twisted round the child's body so that undue traction would be exerted on the placenta during the expulsion of the child. These cases, although resembling in mechanism those in which inversion is caused by dragging on the cord, differ in being spontaneous and presumably unavoidable. 4. In some cases there was adhesion of the placenta, and efforts to detach it have been followed by inversion. 5. In some cases where there is no mention of the placenta having been meddled with, the delivery was very rapid. 6. In some cases the labour was

* Dr. Speedy relates a case which occurred outside the hospital during Dr. Denham's mastership. The woman died although the uterus was quickly returned.

natural but slow, characterized by inertia or want of power. 7. It has occurred in primiparæ as well as in women who have borne many children. 8. In one case it was caused by compression of the uterus by the hand to cast the placenta. Delivery in the upright position has in several cases been an attendant condition. It has happened after delivery by forceps and ergot.

Dr. Woodson relates a case in which inversion followed abortion at four months (*Amer. Journ. of Med. Sc.*, 1860). Dr. John A. Brady (*New York Med. Times*, 1856) relates a case following abortion at five months. Both these recovered after reposition.

It has occurred in non-pregnant women always, so far as I can ascertain, from the dragging or expulsion of a polypus or tumour, except in one case related by E. Clemenson (*Hospital Tidende*, 1865), who attributed the inversion to an altered texture of the organ resulting from fatty regression after labour. But it is quite possible that this case was not exceptional. The history is defective; and the inversion may have taken place at the usual time of labour and have been overlooked.

The fact of inversion has not seldom escaped detection at the time of labour. Hence some have supposed that the uterus might undergo spontaneous inversion hours, or days even after labour. Ané, Baudelocque, and Dubois cite cases. The evidence upon which this rests is mainly the negative fact that no inversion was observed at the time of labour; and then upon the positive fact that an inverted uterus was found at some subsequent date. This kind of evidence is rarely satisfactory. The doubt will naturally arise that the inversion had been produced at the time of labour, and been overlooked.

The essential conditions for the production of inversion are, on the one hand, relaxation of some part or the whole of the walls of the uterus; and on the other, considerable enlargement of its cavity. When the uterus has contracted, its walls are so thick, and the cavity is so reduced—the anterior wall being flattened close in contact with the posterior wall—that inversion cannot take place. The extreme flaccidity of the uterus has been distinctly described by many observers. Thus Smellie relates a case told him by Lucas of a woman whose uterus, after inversion, was immediately reinverted. It was "like a piece of tripe." The uterus has even been inverted after post-mortem delivery, under circumstances which preclude the idea that active contraction of the organ was an efficient factor. In *Guy's Hospital Reports* (1864) Dr. A. S. Taylor relates a case, the details of which were supplied by Mr. Bedford, of Sydney. A woman æt. thirty-seven died in labour undelivered with her seventh child. She had had violent pains, but it was clearly ascertained that the head, although low down, had not been extracted. Inspection of the exhumed body was made a week after death. The abdomen was much distended from decomposition. A male child was found between the woman's thighs. The uterus was inverted, and with the placenta attached to it was lying also between the thighs. There was also a rupture of the uterus a little above the cervix, transverse, and about six inches long. The conclusion arrived at, which the history seems to justify, was that the rupture occurred during life, and was the cause of death; and that the expulsion of the child and inversion of the uterus were caused by the pressure of gases in the abdomen. The uterus was found flaccid.

Hæmorrhage, again, is known to be a disposing cause. Whoever has had his hand in the cavity of a uterus powerless through loss of blood, who has felt its flaccid wall yielding to every pressure, internal or external, like "tripe" or wet brown paper, will understand how easy it would be for such a uterus to be inverted. Indeed, I have often felt partial inversion taking place whilst endeavouring to detach adherent placenta by the fingers. Lazzati precisely says that the uterus was inert at the time of inversion. Whilst in this state it is not difficult to understand that dragging upon the placenta by the cord may easily pull down the fundus and turn the uterus inside out; or that pressure exerted upon the fundus from without, as by the hand forcing it down with the design of provoking contraction, or by the expulsive action of the abdominal muscles, voluntary, reflex, or in the act of vomiting, may drive the fundus down, and even through the os uteri; and, as we have seen, delivery of the child and inversion of the uterus may be effected after the woman's death by the pressure of the gases accumulating in the abdomen.

We may then have spontaneous and artificial inversion in an entirely *passive state of the uterus*.

It is, however, certain that the uterus may be inverted by a *spontaneous active self-inversion*, the uterus being of course *active*. The first distinct enunciation of this process is given by John Hunter. In the museum of the College of Surgeons (No. 2654, *see* Pathological Catalogue) is a specimen of an inverted uterus with a fibroid polypus detached, which had caused the inversion. The accident occurred independently of pregnancy. The polypus had grown from the fundus. A ligature had been applied near the attachment. The tumour had sloughed off just before the patient died. Hunter described the case under the title of " Intussusception," to which he likens inversion. " The uterus," he says, " is liable to inversion from two causes : one is immediately after labour, when it is so large as to admit of its containing itself, and which is commonly from an imprudent mode of disengaging and bringing away the placenta when that substance has been attached to the fundus of the uterus. . . . The second is somewhat similar, namely, the expulsion of an adventitious body although of another kind, and at a very different period in the state of this viscus. It begins to take place when this viscus is small, but becoming gradually large enough (by the very disease that produces it) to admit of an inversion ; so that in the first cause, the uterus is first large, so as to admit of an inversion, and by its contraction to its natural state it, as it were, fixes it. This is done immediately, because its cause is immediate, for this enlarged state of the parts is of short duration ; but the second is gradual, because it is to produce itself by the very action of the uterus in expelling an un-natural body (such as a polypus). The polypus as it grows will gradually fill the cavity of the uterus, and the uterus will be constantly endeavouring to remove it. The action of the uterus will be downwards, and as the body of the uterus acts on this substance it will be gradually squeezed down towards the os tincæ, and the fundus will of course be gradually drawn into its own cavity, and as the polypus is squeezed down so will the fundus follow. When the whole of the polypus has got into the vagina, if it has no length of neck, then will the fundus uteri be as low down as the os tincæ, the upper half of the uterus just filling the lower half ; but I conceive it does not stop here. I conceive the contained or inverted becomes an adventitious or extraneous body to the containing, and it continues its action to get rid of the inverted part, similar to an introsusception of an intestine." It is remarkable that in the subject of this case an introsusception of the small intestine co-existed.

A similar theory was set forth by Crosse. He says, the most powerful predisposing condition to the commencement of inversion (*depressio*), and without which the greater degrees cannot transpire, is partial inertia. He also points out that one of the most constant conditions is attachment of the placenta to the fundus uteri. Then, again, the action of the uterus in increasing an inversion, when once this has commenced, has been admitted and pointed out by many authorities, but by none more pointedly than by Denman, who observes that " if a disposition to an inversion be first given by the force in pull-ing the funis, it may be completed by the action of the uterus." Crosse then states the modern doctrine in distinct terms : " I cannot conceive that the organ itself has any power to *com-mence the displacement*, and to cause simple *depressio*. . . . . But when a commencement has been made, and the case goes on to *introversio*, bringing the fundus within the grasp and influence of the uninverted body of the uterus, this organ will by the natural powers called into action by its sensibility, regard the inverted part as an extraneous mass, and proceed to act upon it instinctively by successive and suitable efforts of its muscular coat, to propel it downward ; whilst the os and cervix will by consent, and, as transpires in the regular process of delivery, become dilated, and thus a part of the uterus will act on the rest, and carry on the displacement even to *perversio extrema*. . . . . The *nisus depressorius* of the abdomen awakened will assist the expulsion."

Tyler Smith has seized the same idea, and expounded it with unmatched felicity and terseness of language.

An important factor in the process has been insisted upon by Rokitansky. Inversion begins at the placental site. This part is liable, says the great pathologist, to paralysis, and being thicker than the other parts of the uterine walls, forms a pro-jection into the cavity. That is the first step ; the first postulate of Hunter. Then if the placenta adhere, and be dragged upon by the cord from below, or if the diaphragm and abdominal walls act, as in a bearing-down effort, the part already disposed to fall inwards is forced further down into the cavity. The external cup-like depression formed by paralysis of the placental site may be felt by examination through the abdominal walls ; and especially is this the case if you drag upon the cord, the placenta adhering. When things have gone thus far further pressure or dragging brings the fundus down upon the cervix. If this part be contracted it may prevent the fundus from coming through ; or, the pressure continuing, the os may yield, and allow it to slip through (as it does in the converse case of artificial reposition) ; or the advancing fundus may find the cervix relaxed, and offering no opposition. Indeed the cervix is very liable to temporary paralysis after labour, and more especially is this the case when, as is not uncommon, it is lacerated. Accordingly, it has been observed that some cases have occurred gradually ; others suddenly.

It may be supposed that the uterine ligaments would offer sufficient resistance to obviate spontaneous inversion ; and the inference from this assumption would be that inversion must be the consequence of direct violence. But in reality the anatomical conditions are easily overcome. During gestation the ligaments become elongated, so as to be easily drawn into the hollow or cup formed by the inverting body of the uterus. The moorings of the fundus are too slack to keep this part up ; but the connections of the cervix to its ligaments, and to the bladder and vagina, hinder the inversion of this part for a time. Casper (*For. Med. N. Syd. Soc. Transl.*, vol. iii.) says laceration of the pelvic ligaments may attend spontaneous inversion of the uterus. Hence one explanation of the greater frequency of the two first stages of inversion than of the third.

The *symptoms and diagnosis* of recent inversion. The symptoms are chiefly those of *shock, indicating sudden severe injury.* They vary with the degree and progress of the inversion. Thus the first degree, or simple *depression,* may be unattended by pain, and be indicated solely by hæmorrhage and a corresponding depression of the vital powers. The hæmorrhage comes from the relaxed introcedent part. As the descent proceeds, and becomes *introversion,* urgent symptoms arise, according to the degree of compression exercised by the uninverted portion upon the inverted portion. A sense of fulness, weight, as of something to be expelled, is felt. The woman has thought another child was passing. Expulsive efforts, both uterine and abdominal, sometimes very violent, follow. Hæmorrhage is not constant. It seems that when the inverted portion is firmly compressed the hæmorrhage is arrested, and that bleeding is a mark of inertia. When the inversion is complete pain and collapse are aggravated. Clammy sweats, cold extremities, vomiting, alarming distress, restlessness, extinction of pulse occur.

During the expulsion the woman has often exclaimed that " her body has come out," or that the attendant "is tearing away her inside ;" and this even although no one may have been touching her. Perhaps convulsions set in; but generally consciousness is retained. The second sign is probably *hæmorrhage*. This is often profuse and continuous.

In the recent state *retention of urine*, owing to the distortion and compression of the neck of the bladder and urethra, is not uncommon. This has been relieved when the uterus was restored. But congestion of the mucous membrane of the bladder, and even retrograde kidney trouble may ensue, as in retroversion.

Examination made, the appearance of an unusual mass is observed outside the vulva. This may be as big as a child's head or bigger. In Middlesex Hospital Museum is a specimen as large as the adult head hanging down from the vulva. There was also rupture ; both said to be produced by the midwife pulling on the cord. The depending mass is fleshy, more or less firm, rounded or pyriform, dark red, bleeding, probably partly obscured by blood-clots. The cord, if the placenta had not previously come away, and if the cord had not been torn off, will be hanging from the mass, and being traced up to its root in the placenta, may give the impression that it is placenta and nothing more. Under the influence of this impression, the natural impulse is to seize the mass and try to remove it, and to this the attendant will be the more impelled by the profuse flooding, which seems to admit of no delay. But deliberate examination will show that the hand cannot be passed up beyond the presenting mass or presumed placenta into a cavity like the vagina and uterus ; but is arrested all round the root of the mass by a groove low down in the pelvis, or even outside the vulva ; feeling above the pubes for the uterus it is not found where it ought to be, but the hand sinks in carrying the flaccid abdominal walls before it quite back to the spine, and even into the pelvic cavity. The uterus then is not where it ought to be, and there is a body of corresponding bulk and shape in a place where no such body is usually found. The inference is strong, if not conclusive, that this body in a strange place is the uterus inverted.

If the *placenta still remains attached*, the mass will be correspondingly larger. It ought to be known whether the placenta has come away or not. If it has not come away, it will naturally be sought for. The cord will afford a ready clue, if it has not been torn off. In all cases of difficulty the doubtful body should be inspected by the eye if possible, as well as by the touch. The ramifications of the umbilical vessels over the surface of the mass, and the torn membranes will reveal the placenta. A placenta in this position is presumably detached, and removable without further force. If it be found that it cannot be so removed, the reason of the resistance must be carefully sought. The membranes should be drawn forward so as to bare the edge of the placenta; it will then be felt, and seen to be attached to a spherical body, which is almost certainly the uterus inverted. I say almost certainly, because cases are known where the placenta has grown to the child's head, or to a polypoid tumour projecting from the uterine cavity. The second case is one of real difficulty. The more so since the polypus may have caused inversion of the uterus. There will then be three masses in very unusual connection: placenta, tumour, inverted uterus. Whilst we are making all this out, the patient may be dying of shock and hæmorrhage. Still, howsoever urgent the need of help, there is more danger in precipitate unreasoned efforts than in the procrastination necessary to take an accurate survey of the situation, and to deliberate on the proper course of action. If you take time you will probably do what is right.

The recently-inverted uterus has a remarkable property which distinguishes it from every other tumour. Contractility is a vital attribute that no polypus possesses. If the tumour, then, hardens and relaxes alternately, we know it is the uterus.

The recently-inverted uterus is also usually very much larger than a polypus. And the polypus traced up to its root

c c

or origin will lead to a cavity—the cavity of the uterus, which cannot be felt in the wholly inverted uterus. The case of incomplete inversion, that is, in the stages of depression or introsusception, may be very difficult to diagnose from polypus. There is a tumour above or engaging in the cervix uteri. The finger or sound will pass into the uterine cavity around the base or neck of the presenting tumour, very much as in the case of a polypus springing from the fundus. The diagnosis will depend upon our being able to feel the inverted cup of the fundus through the abdominal wall, whilst a finger of the other hand is in the uterus. If the womb be inverted, we find on passing a finger or a bougie upwards to the root of the tumour, that it is arrested by a groove or cul-de-sac all round, about an inch or half an inch within the os uteri. The depth of this groove will depend upon the degree of inversion. The greater this is, the shallower will be the cul-de-sac. It is even possible there may be no such cul-de-sac, the os uteri being entirely obliterated; in this case the outline or surface of the tumour will be quite continuous with the vagina all round. The distinctive character between inversion and polypus is that in the first the tumour is all below the groove or cul-de-sac; whereas, in the second, there is still a mass, the uterus itself, above and beyond the tumour, the polypus. These facts are determined by bi-manual palpation, and by the sound. In the case of inversion, a finger applied to the root of the tumour may feel the point of a sound passed into the bladder. The two meet above the inverted uterus.

Should the placenta have come away the case, although so far simplified, is by no means free from danger of mistake. There is a large rounded, fleshy, roughish mass, bleeding, depending perhaps between the thighs. It may be mistaken for a placenta, for some part of a child, normal or abnormal, for a polypus or other tumour, or even for a mass of coagulated blood. If you have been following the uterus down during the extrusion of the placenta, or afterwards to ensure contraction, and suddenly feel the uterus retreat into the pelvis, a firm mass appearing simultaneously in the vagina or externally, you have, with the symptoms of shock, strong presumptive evidence that the uterus has become inverted. You may make this sure by pressing the tumour back a little up the vagina with one hand, whilst you press in the flaccid abdominal wall with your other hand behind the symphysis; you may thus feel the cup-shaped depression or funnel formed by the introcedent fundus uteri.

If there be any possibility of getting the aid and counsel of a professional brother, for your own sake as well as for that of your patient, get it. But whilst waiting your patient may be dying—time is life; you must be prepared to act decisively on your own judgment. And here it may be well to say a few words upon the use of consultations in difficult cases. I will state as a leading proposition that in most difficult cases *two*, if not more, skilled practitioners are wanted to execute, to carry out the treatment. The cases are many in which one man, no matter what his skill, can hardly do all that is necessary; and the cases are more numerous still in which what has to be done will be done much better with efficient help. Nor does the benefit stop here. For two men to act in concert there must necessarily be interchange of thought; both must verify the circumstances; and this implies deliberation, the only security for judicious action. These two conditions, conjoint deliberation and skilled manual help, contribute largely to the apparent superior efficiency of the consulting physician. This I have often felt. And often have I felt that an undue share of the credit of a successful operation has been awarded to me, which I could not have performed alone.

These considerations apply most emphatically to emergencies comparable to that under discussion. Where the best energies of two men, one of them bringing a fresh mind and a steady hand, may be tasked to the uttermost, it is not a matter for surprise that any one man may prove unequal to the occasion.

Terminations.—1. The patient almost immediately, or very soon after the accident, dies from pure shock. The shock attending simple depression has proved fatal. 2. More frequently death occurs rapidly from shock and hæmorrhage combined, that is, within twelve hours. 3. The inverted uterus may be strangled by its own cervix, or by the vulva. This induces, first, continuous or secondary shock; and in some cases, if the patient survive long enough, and it need not be long, gangrene ensues. This may, or may not, be fatal. The uterus has sloughed off: Saxtorph (in *Actis Soc. Med., Hav.*); Deboirier

(*Richter's Chir. Bibl.*) ; Radford (*Dublin Journ of Med.*, 1835).
E. Clemenson relates (*see* my work on *Diseases of Women*, p. 719)
a case of gangrene following spontaneous inversion in a woman
æt. fifty, independently of pregnancy or polypus. In other
cases the strangulation caused by the cervix has ended fatally
before there was time for sloughing (Velpeau).

Death has ensued from strangulation of intestine in the uterus.
Gérard de Beauvais relates a case (*Arch. Acad. de Méd.* 1843).

In some cases, and these not uncommon, the shock and
hæmorrhage are not very severe, or at any rate the woman survives the immediate effects, rallies, and is thought to be all
right. Tolerance, more or less complete, may ensue. But generally metrorrhagia continues, interrupted perhaps at intervals, but
always liable to recur at the menstrual epochs. The losses thus
induced may prove fatal at no distant date.

Where the case is not fatal, and the uterus is not reduced, the
symptoms of chronic inversion succeed. First, the tumour by
its bulk causes distress of the bladder and rectum. Then it is
probably forced externally. I must refer to my work on
Diseases of Women for the further history of this condition.

Meigs and others affirm that spontaneous reduction has occurred. But Crosse says, " There is not a shadow of evidence of
total inversion in the strict sense replacing itself spontaneously."
It is not difficult to imagine that partial inversion, that is ,where
the fundus has not come through the os, may be restored spontaneously.

To make good the fact of spontaneous reposition it is necessary to start with clear evidence that inversion actually existed,
and then to verify the return of the uterus to its normal condition. In some of the cases as they stand recorded, there is the
possibility of fallacy attaching to one or the other of these
postulates.

A fibroid tumour may have simulated the uterus and have
either retreated into the uterine cavity, or have been cast off.
But the possibility of spontaneous reposition seems to be established by the following case of Spiegelberg (*Arch. für Gynäk.*
vol v., part 1). In a woman æt. forty, in her twelfth labour, inversion followed attempts of midwife to hasten delivery by pulling
on the body and umbilical cord of the child. Attempts at replacement failed. At the end of six weeks a rather copious

hæmorrhage occurred; the uterus was found to be still inverted and could not be replaced. Two and a half weeks afterwards the patient came into hospital. The inverted portion was closely embraced by the os uteri; the part which was not inverted measured scarcely $\frac{2}{3}$-inch. As the woman had a severe attack of diarrhœa soon after admission, a fortnight elapsed before a second examination was made with a view to reduction. Now, however, the uterus was found to be completely reduced. According to the explanation offered by Schatz, with which Spiegelberg agrees, the uterus was raised by the continued lying in bed, the round and broad ligaments accommodating themselves to the new position: and during the diarrhœal evacuations the vagina, with the vaginal-portion of the uterus, was pressed down, whilst the ligaments being unable to follow this movement raised the fundus.

Dr. Denham relates (*Dublin Quart. Journ. of Med. Sc.*, 1866) a case of partial spontaneous reduction. A primipara, æt. twenty-three, suffered inversion, from pulling on the cord and pressure on the fundus of the uterus. She was admitted to hospital eleven days afterwards. Under steady pressure the uterus gradually diminished, and was felt to yield, so that the fundus alone remained unreduced; "no amount of force compatible with the safety of the organ could enable us to complete the operation." "On the third morning we were agreeably surprised to find that the fundus had spontaneously returned." The woman made a good recovery.

Treatment.—When we find inversion, the indication is clear to reduce it, and that as soon as possible. The presence of the uterus grasped in its own neck excites contraction, the part gets strangled, swollen, and it becomes more and more difficult to return it. In most cases, probably, the mass might be pressed back with comparative ease, if the attempt were made immediately after the occurrence of the accident, whilst the cervix was still paralyzed, and the uterus itself flaccid. But this opportunity is rarely found. The first question that arises is as to detaching the placenta first or not. To detach the placenta is to lose a little time, to risk flooding and irritating the uterus to contract; if we leave it, there is the greater bulk to pass back through the os uteri. If you have the good fortune to recognize the accident at the moment, you may be able to take advantage of the flaccidity of the cervix, and return uterus and placenta at once; but if this favourable moment is lost, it will be better to detach the placenta first. Look for the margin of the placenta; insinuate one or two fingers between it and the globe of the uterus; supporting this organ by the other hand, continue to peel off the placenta by sweeping the fingers along. When it is wholly detached, proceed to reduction. The mode of manipulation must vary according to circumstances. If the uterus is large, flabby, and the cervix dilated, it may be quickly replaced by depressing the fundus with the fingers gathered into a cone, and carrying the hand onwards through the os. Lazzati says it is better to apply the closed fist to the fundus; this acts better, and we avoid the risk—by no means a slight one—of perforating the soft structure of the uterus. In executing this, two things must on no account be omitted: one is to support the uterus by the other hand, pressing firmly down upon it from above the symphysis pubis, lest we lacerate the vagina. This is a real danger, and to avoid it is better than to turn to account the aid undoubtedly acquired by the full stretching of the vagina, which tends to pull down the os and cervix over the inverted body. It is remarkable how extensile the vagina is after labour. It sometimes elongates under pushing so readily that we may easily fail to realize whether it is stretching or rending. The other thing to observe is the course of the pelvic axes, and the form of the pelvic brim. Pressure will first be made a little backwards towards the hollow of the sacrum; then the direction must be towards the brim, and at the same time *to one side, so as to avoid the sacral promontory.* As in attempts to reduce a retroverted gravid uterus, failure has often ensued from not understanding this latter point. It was first, I believe, insisted upon by Dr. Skinner, of Liverpool. I can testify to the value of the rule from personal experience. By attention to it mainly, I was enabled to reduce an uterus in fifteen minutes, which had been inverted for ten days, defying repeated efforts by other practitioners. When reduction has been completed, the hand following the receding fundus will occupy the cavity of the uterus, and the organ will be grasped between the hand inside and the hand supporting outside. The opportunity should be taken to induce contraction by pressure externally and by excitation internally. But I would not withdraw the hand from the cavity, lest re-inversion take place, until I had taken the following further security. Pass up along the palm of the hand an uterine tube connected with a Higginson's injecting syringe; throw up by means of this six or eight ounces of a mixture composed of equal parts of the strong solution of per- chloride of iron and water, so as to bathe the whole inner surface of the uterus; or you may apply the styptic by means of a swab. The effects of this are, instantly to constringe the mouths of the vessels, to stop bleeding, to excite uterine con- traction, and to corrugate the tissues and narrow the os uteri. When this state is induced there is safety. This is all the more important, because at the moment of reduction there is likely to be considerable collapse and uterine inertia. Lazzati injected ice-cold water.

If uterine action be present, especially if the cervix and os are constringing the inverted part, contracting spasmodically, the difficulty is greater, and it is no longer judicious to com- mence by pushing in the fundus. As Dr. McClintock has well shown (*Diseases of Women*, 1863), to do this is to double the inflexion of the uterine walls, and thus to double the thickness of the mass that has to pass through the os. He advo- cates the method practised by Mont- gomery, which consists in regarding the inversion as a hernia, and *in replacing that part first which came down last.* The tumour must be grasped in its circum- ference near the constricting os ; firmly compressing it towards the centre, and at the same time pushing it upwards, for- wards, *and to one side.* The pressure must be steadily kept up, as it is sustained pressure that wears out the resistance of the os. After a time the os is felt to

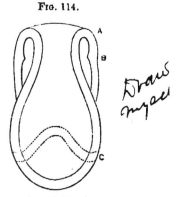

Fig. 114.

After McClintock.

A B, Cervix uteri ; c, in- verted fundus, the dotted line showing effect of doubling in the fundus to reduce it.

relax, the part nearest is pushed through, and then, generally suddenly, the body and fundus spring into position. (Dr. J. G. Wilson, *Glasgow Med. Journ.*) Two things facilitate this operation: anæsthesia and a semi-prone position of the patient. McClintock compares the process to that of treating a paraphymosis. Two movements are required: 1. to squeeze the uterus, 2. to press the uterus slowly up.

If the opportunity of reducing within a few hours be lost the difficulty increases through advancing contraction and involution of the uterus, and especially contraction of the os. But you are not to be discouraged by the teaching, only till recently held, that after a few hours, inversion of the uterus is all but irreducible. Reduction is simply a question of time: keep up pressure long enough, and the constricting os must yield. Re-inversion, however, does not always ensure recovery. Many patients have died soon after re-inversion. In some instances, perhaps, further injury has been inflicted during the operation. But it must not be forgotten that laceration of the vagina or uterus may have been spontaneously produced at the same time as the inversion. And in some cases, it is certain that the reposition has been accomplished without any such injury.

The after-treatment will consist in absolute rest; opiate suppositories, one or two grains; or chloral enemata, containing one drachm of chloral; ice, oxalate of cerium, hydrocyanic acid, bismuth, to restrain vomiting; and light diet, with sparing use of stimulants. As soon as the process of involution has begun, there is reasonable security against return of the inversion.

In subsequent labours increased disposition to inversion may be anticipated. There are several cases in which inversion has happened in successive labours.

LECTURE XXIII.

CRANIOTOMY—THE INDICATIONS FOR THE OPERATION — THE
OPERATION—TWO ORDERS OF CASES—PERFORATION SIMPLE;
AND FOLLOWED BY BREAKING-UP OR CRUSHING THE
CRANIUM, AND EXTRACTION—EXPLORATION—PERFORATION
—EXTRACTION BY CROTCHET, BY TURNING—DELIVERY BY
THE CRANIOTOMY-FORCEPS, USE OF, AS AN EXTRACTOR, AS A
MEANS OF BREAKING UP THE CRANIUM, USE IN EXTREME
CASES OF CONTRACTION—DELIVERY BY THE CEPHALOTRIBE,
POWERS OF THE CEPHALOTRIBE, COMPARISON WITH CRA-
NIOTOMY-FORCEPS—THE OPERATION—DR. D. DAVIS'S OSTEO-
TOMIST—VAN HUEVEL'S FORCEPS-SAW—DELIVERY BY THE
AUTHOR'S NEW METHOD OF EMBRYOTOMY BY THE WIRE-
ÉCRASEUR—INJURIES THAT MAY RESULT FROM CRANIOTOMY.

WE have lingered long on the border-land between Conservative and Sacrificial Midwifery, unwilling to abandon the hope
of saving mother and child; striving to set back, as far as the
dictates of science and the resources of art will enable us, those
limits where the death, certain or probable, of child or mother
must be encountered. We must at length pass the boundary,
we must lay aside the lever, the forceps, and turning, and take
up the perforator, the crotchet, the craniotomy-forceps, the
cephalotribe—instruments the use of which is incompatible
with the preservation of the child's life. A law of humanity
hallowed by every creed, and obeyed by every school, tells us
where the hard alternative is set before us, that our first and
paramount duty is to preserve the mother, even if it involve
the sacrifice of the child. As, therefore, we have striven to
give the highest possible perfection to the forceps and turning

in order to save mother and child, so it now behoves us to exhaust every effort in perfecting the means of removing a dead child, in order to rescue the mother from the Cæsarian section, that operation which Professor Davis justly called, "the last extremity of our art, and the forlorn hope of the patient."

The Indications for Craniotomy.

These are generally—1. *Such contraction of the pelvis or soft parts as will not give passage to a live child, and where the forceps and turning are of no avail.* Contraction of this kind may be due to contraction or distortion of the pelvis, which is most frequent at the brim; to tumours, bony, malignant, or ovarian, encroaching upon the pelvic cavity; to growths, fibroid or malignant, in the walls of the uterus; to cicatricial atresia of the cervix uteri or vagina; to extreme spastic contraction of the uterus upon the child, forbidding forceps or turning. Craniotomy and cephalotripsy are the means of effecting delivery in cases where labour at term is obstructed from disproportion, the pelvic contraction ranging from 3·25" as a maximum to 1·50" as a minimum. If labour occur at seven months, these means may certainly be applied to contraction measuring less than 1·50". F. H. Ramsbotham considered that a full-grown child might be extracted through a pelvis measuring 3-in. in the lateral, and 2-in. or even 1¾-in. in the conjugate diameter. Ingleby, representing a school possessing imperfect intruments, and unexercised skill in these operations, dissented, saying, "it was no easy matter to fix the instruments so as to insure the naso-mental diameter passing first."

We are not hastily to assume, because a woman has been delivered on previous occasions by natural powers or by forceps, that it is therefore unnecessary to resort to craniotomy. It is, indeed, ample reason to pause and to examine anxiously. But it is a matter of experience that some women bear children with a constantly increasing difficulty. This may be from two causes —1. Advancing pelvic contraction; 2. Increasing size of the children. I can affirm the reality of the first cause from repeated observation. I have had to record the histories of many women whose first labours may have been natural, and whose subsequent ones exhibited difficulties increasing in a kind of accelerated ratio, rising from the forceps to turning and craniotomy. The second cause may be independent of, or aggravate, the first. D'Outrepont says he has constantly observed that in fruitful women whose first children were small, subsequent ones become bigger and bigger.

Professor Elliot makes the apposite observation, that "the same degree of deformity admits of varying results in successive pregnancies." For example, the child may at the time of labour have attained a different degree of development; it may present differently; there may be accurate dip of the head in one labour and not in another,

2. *Certain cases where obstruction to delivery is due to the child* —as some cases of face-presentation; some cases of locked twins, in which the lessening one head is necessary to release the other; some cases of double monsters; excessive size of head, as from hydrocephalus; cases where there is obstructed labour, the head presenting, and the child is dead.

3. *Conditions of danger to the woman, rendering it expedient to deliver her as speedily as possible,* and where craniotomy is the quickest way, involving the least violence. Amongst these are *some* cases of convulsions; *some* cases of hæmorrhage; great exhaustion; some cases of rupture of the uterus; and generally where, delivery being urgently indicated, the cervix uteri is not sufficiently dilated to admit of other operations.

An important question is—at what stage of labour shall we begin? As most of the dangers flow from exhaustion, it is obviously proper to begin as soon as the indication for the operation is clear. On the Continent especially, it is still urged by many that we should wait until the child is dead. Now, if it be admitted, and the conditions of the case involve these postulates—1. That the child cannot come through alive; 2. That the operation is undertaken in order to save the mother—waiting till the child is dead is opposed both to reason and to humanity. It seems a refinement of casuistry to distinguish between directly destroying a child, and leaving it exposed to circumstances which must inevitably destroy it; and it is risking the very object of our art to wait for this lingering death of the child until the mother's life is also imperilled. If, then, we see the woman drifting into the dangers of "lingering or obstructed labour," and we have clearly determined by our knowledge of the patient, by exploration, by trial, that the child cannot come through the pelvis by spontaneous action, by forceps, or by turning, it is our duty at once to adopt the best means of securing the safety of the mother. There is no need to wait for the far advance of labour. We should not wait long after the rupture of the membranes. It would, in many cases, be useless to wait for complete dilatation of the cervix uteri. It is one of the necessary results of contracted brim that the cervix uteri dilates slowly and imperfectly. The head-globe resting by two points on the contracted brim cannot bear upon the cervix. It is not, therefore, often desirable to wait for more opening than is enough to admit two or three fingers to guide the perforator. When the head collapses and comes down into the pelvis, it bears upon the cervix, which then yields gradually.

Although it is a good general rule to perform every operation as early as the indication for it is clearly recognized, it is not desirable, in minor degrees of contraction, to arrive at once at the conclusion that perforation is necessary. Some time and opportunity should be given to Nature. The head may be small or plastic, and occasionally even a full-sized head will, under continued action of the uterus, become so moulded as to admit of delivery either spontaneously or by aid of the forceps.

Perforation should be the first step of all operations for lessening the bulk of the head. The necessary condition for full collapse of the bones of the head is that the support given by the brain and the integrity of the cranial vault shall be broken down. Until this is done, you may obtain, with considerable expenditure of time and force, some amount of moulding or alteration of form, but no diminution of bulk. It is astonishing what resistance to compression the unopened head possesses. The most powerful forceps, and even cephalotribes, may be bent in the attempt to crush it in. Whereas, break the arch of the cranial vault, allow the contents to escape, and very moderate compression will cause collapse, more or less complete. Besides, more room is required if we apply the cephalotribe to the whole head. It is remarkable that not a few Continental obstetrists practise cephalotripsy without perforating.

Craniotomy, being used as a general term to include all the proceedings for reducing the bulk of the head, may be divided into three principal orders :—

The *first* includes those cases of minor disproportion in which perforation is enough to allow of such an amount of collapse of the head under the natural forces as will effect delivery.

The *second* order includes those cases of major disproportion in which perforation must be supplemented by breaking-up, removal of parts of the cranium, or by crushing down, and followed or not by extraction.

The *third* includes those cases of extreme disproportion in which the head has to be removed in sections, as by my New Method of Embryotomy.

The *preparations* are generally the same as for other operations.

Position.—The patient lies on her left side, with her knees well drawn up, near the edge of the bed, and with the head supported on a *low* pillow, directed towards the middle of the bed; or if the bed is sufficiently high, she may be placed on her back, the legs being supported by assistants.

Exploration.—The left hand of the operator is introduced, if necessary, into the pelvis, so as to explore thoroughly the shape and dimensions of the brim, and the relations of the head and cervix uteri. Three points especially should be clearly made out: First, the projection of the promontory, which in extreme cases has been mistaken for the head; secondly, the head; thirdly, the os uteri. The finger passed inside the os should be made to sweep all round the circumference of the head, taking note of the outline of the brim.

Perforation.—The point to be selected for perforation is that which presents most centrally. It is easier to strike; it offers a better resistance to the point of the instrument; the opening made allows a free exit to the contents of the skull; and it affords greater facility for the introduction of the crotchet or the blades of the craniotomy-forceps, which have to follow. Two most essential things to be attended to are : That an assistant shall support the uterus and child externally, pressing them firmly down towards the pelvis, so as to fix the head upon the brim, and obviate the retreat or rolling of the head under the impact of the instrument. The other thing is, to take care that the instrument shall strike the head perpendicularly. If it strike at an angle, the point will be apt to fly off at a tangent, or to make the head roll over, at the risk of wounding the mother. I have seen so many wretched

FIG. 115.

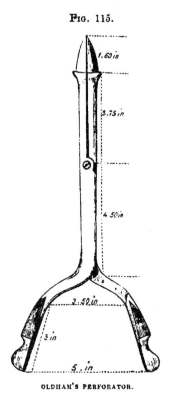

OLDHAM'S PERFORATOR.

perforators, and knowing that they are not only liable to fail in doing what they ought to do, but to do mischief, that I here introduce a figure of the best and most efficient instrument I have ever handled. Those usually sold as Naegele's, Holmes's, Smellie's, etc., are almost all bad. The curved point is especially vicious. They answer, perhaps, well enough in slight cases where it may be doubted whether craniotomy is necessary. They deceive and fail in cases of real difficulty.

Sometimes in cases of great deformity the uterus is so twisted from its normal direction that reposition is necessary before the os can be brought near the centre of the brim to allow of safe perforation.

Two fingers of the left hand then are passed up to the head, keeping the os uteri at their back; the instrument is run up in the groove formed by the fingers; the point having struck the part desired, the perforation is effected by a movement combining boring and pushing. When the skull is pierced, push the blades in up to the shoulders; then open the blades to enlarge the aperture, turn the handles at right angles to the first position, and open the blades again, so as to make a free crucial opening. This breaks the continuity of the arch of the cranium; allows free discharge of brain, and ample entry for the crotchet or craniotomy-forceps.

Now you may wait a little, to afford opportunity for spontaneous compression and collapse; or you may at once pass in the crotchet. This should be carried in as deeply as possible, and moved freely round in all directions, to break up the tentoria and the brain. This proceeding greatly facilitates the evacuation and collapse of the skull.

If the disproportion is not great, and the powers of the patient are good, it commonly happens that uterine action sets in as soon as the bulk of the head is a little diminished, and the compression and propulsion resulting will often suffice to expel the child. Reasonable opportunity should be given for this spontaneous process.

Should no advance be made, the case falls into the second order, and we must proceed to extraction, or to artificial compression of the skull.

Extraction may be accomplished in several ways.

1. By the *crotchet*. This instrument was long preferred by the Dublin school. It was very naturally resorted to in preference to bad craniotomy-forceps; and some practitioners of great experience, who have acquired exceptional skill in the management of it, accomplish delivery by its aid in cases of great disproportion. Until I had contrived a good craniotomy-forceps, I myself trusted to it entirely. But for safety and expeditiousness it is very inferior to the craniotomy-forceps, or the cephalotribe. The way of using it is as follows:—Two

fingers of the left hand guide the end of the crotchet into the
hole in the skull. The ends of the fingers are then passed up
outside the skull, to serve as a guard and support to the sharp
point of the crotchet, which is fixed into the bone inside. The
part to which the crotchet is first applied is not perhaps very
important, since, if there be any great resistance, the part will
be broken away, and the instrument will have to find fresh
hold. This may have to be repeated several times, pieces of the
parietal, occipital, or frontal bones being successively torn out.
Whenever a piece of bone is detached, it is wise to remove it
altogether. This may be generally done with the fingers. By-
and-by—for the process is apt to be tedious—when the cranial
vault is much broken up, if a good hold can be obtained in the
occipital bone or in the foramen magnum, collapse or falling-in
of the skull takes place, and extraction is successful. In very
difficult cases, when the vault is well broken up, it is better to
take hold in the orbital region, fixing the point of the crotchet
either inside the skull, under the sphenoid on one side of the
sella turcica, or in the eyeball. In this way the base is brought
into the brim edgewise or end on.

In the last century it was a recognized plan to perforate and
leave the evacuation of the brain and the compression of the
skull to the action of the uterus. The process was usually slow
and tedious; commonly, some degree of decomposition had to
take place before the bones would collapse sufficiently, and
exhaustion, or even inflammation of the uterus, and fatal prostration, would sometimes ensue. The late Professor Davis
records that, even in his time, this mode of proceeding was still
followed by the disciples of one school in London, and that he
was often called in to witness the most disastrous consequences.
Indeed, when it is considered that craniotomy is not often performed except at an advanced period of labour, and after much
suffering has been endured, it will hardly appear justifiable to
throw upon an enfeebled system a task entailing further exhaustion, and under which it may sink. It is our duty to relieve
Nature, and not to leave her to struggle through unaided. This
plan of waiting upon Providence was no doubt imposed by the
wretched instruments in use. This necessity is now gone.
Operations which formerly lasted twelve hours or more, requiring several sittings, exhausting alike the patient and operator

are now accomplished with safety and comparative ease in an hour or less.

2. *Delivery by turning.*—When the cranial arch is broken the bones will readily collapse, if the skull be drawn through the contracted brim base first. In certain cases turning is a very efficient method of completing delivery. The torn scalp during extraction is drawn over the aperture made by perforation, and sheaths the jagged edges of the bones. But this plan will be generally inferior to the use of the craniotomy-forceps. The child, being dead, does not always lend itself readily to turning. It may be necessary to pass the hand into the uterus, which is moulded upon the child, and through a brim so contracted as to oppose considerable difficulty. Turning, however, must be regarded as a valuable resource in certain cases of exceptional emergency.

3. *Delivery by the Craniotomy-forceps.*—The use of this instrument is twofold: it will seize and extract the head; it will seize and remove portions of the bones of the cranial vault. The first use is adapted to minor degrees of disproportion.

What part is best to seize? If the head is found to collapse well, and the disproportion is not great, it is enough to seize the forehead, which, being generally directed to the right ilium, is the easiest to do. But if there is any great difficulty, it is better to quit the forehead and seize by the occiput. The head will not come down well, face presenting, unless the vault and occiput are in a condition to be crushed in against the base. In this proceeding, compression of the skull is effected by its being drawn through the narrow passage formed by the soft parts supported by the pelvis. The head must therefore be ductile enough to admit of the necessary compression and elongation. If the skull be too unyielding, or the passages too small for this process, a totally different principle must guide us. Portions of the vault must be removed, and then we get the most remarkable advantage.

Dr. Osborn contended that by canting the base of the skull, so as to bring it edgewise into the brim, it was quite possible to deliver a full-sized child through a conjugate diameter measuring an inch and a-half only. His contention was hotly disputed by Dr. Hull. Dr. Burns came to the same conclusion as Dr. Osborn, and showed that by removing the calvarium,

D D

reducing the skull to its base, and bringing it through as in
a face-presentation, nothing was opposed to the conjugate but
the distance from orbital plates to chin, which is rarely much
more than an inch.   (*See* Fig. 116.)   Thus an inch and a-half to
an inch and three-quarters' conjugate diameter, with a transverse
diameter of three inches, is enough ; and degrees of contraction
beyond this, requiring the Cæsarian section, are rare indeed.

Fig. 116.

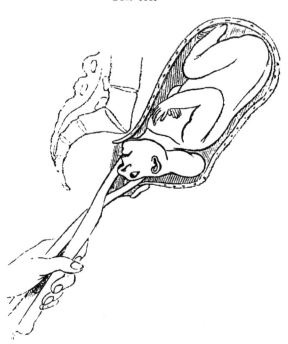

SHOWS THE BONES OF THE CALVARIUM REMOVED, AND THE BASE OF THE SKULL, GRASPED BY
THE CRANIOTOMY-FORCEPS, DRAWN THROUGH THE CONTRACTED BRIM EDGEWISE, FACE FIRST.

This question has been investigated and illustrated anew by Dr.
Braxton Hicks.* He describes very fully the mechanism of
the proceeding. Having removed the calvarium, he grapples
the orbit with a small blunt hook, the hook of which is hard,
and the stem soft, so as to admit of easier adaptation. The
face is then gently drawn down, turning the chin forwards, as

* " Obstetrical Transactions," vol. vi. 1865.

occurs in ordinary face-labour. A fresh hold, in the mouth or under the jaw, is then taken for traction. I, however, prefer the craniotomy-forceps. The proceeding I practise is as follows :—I pass the inner or small blade into the cavity of the skull as usual, then the outer blade in between the portion of bone to be removed and the scalp. Then, a considerable piece of parietal or occipital bone being seized, by a sudden

Fig. 117.

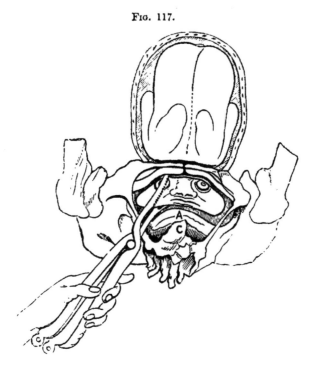

SHOWS THE REMAINS OF THE SKULL DRAWN THROUGH THE CHINK OF THE BRIM, FLATTENED LIKE A CAKE.

The calvarium being removed, the head resembles that of an anencephalous fœtus. A, projecting promontory of the sacrum; c, coccyx.

wrench is broken, and then cautiously torn away under the guidance and protection of the left hand in the vagina. If the distortion is not extreme, it may be enough to break away two or three pieces, say an angle of each parietal and of the occipital. This destroys the arch of the calvarium, so that the remains of the walls easily fall in upon the base, forming a flat

D D 2

cake, when the head comes to be compressed in the chink of
the brim. When enough has been taken away to admit of this
flattening in, the blades of the forceps are made to seize the forehead and face, the screw working at the ends of the handles helping to crush in the frontal bones and to secure an unyielding
hold. (*See* Fig. 117.) The craniotomy-forceps, in fact, here acts
like, and fulfils the chief function of, the cephalotribe, possessing the advantage of taking up less room. Then traction is
made, carefully backwards at first, in the course of the circle
round the false promontory. As the face descends it tends
to turn chin forwards, and this may be promoted by turning
the handles of the instrument. It is not necessary that this
turn should take place, for the case differs entirely from that
of the normal head. There is no occiput to roll back upon the
spine between the shoulders. The head comes through flatwise
like a disc by its edge.

If the pelvic deformity be very decided—say to 2·50″ or 2·00″
or under—it will be wise to take away the greater part of the
frontal, parietal, squamous, and occipital bones before beginning
traction.

By adopting this method, a full-sized head may be delivered
with safety to the mother through a pelvis measuring even
less than 2·00″ in the conjugate, provided there be 3·00″ in the
transverse diameter. I go further, and declare that it is
perfectly unjustifiable to neglect this proceeding, and to cast
the woman's life upon the slender chance afforded by the
Cæsarian section.

The late Professor Davis, relying upon his method of embryotomy, used these words:—"There are few pelves with
superior apertures so small as not to furnish from 1″ to 1½″ in
the conjugate diameter, or at least of antero-posterior diameters
across some part of the brim. In any such cases it would
be the practitioner's duty to avail himself of the use of the
osteotomist, and to undertake delivery by the natural passages.
It will have the effect of reducing almost to zero the necessity
of having recourse to the Cæsarian section."

4. *Delivery by Cephalotripsy.*—What are the relative powers
of cephalotripsy and of craniotomy as just described? I doubt
much whether cephalotripsy can carry the possibility of safe
delivery at all beyond the point attained by craniotomy. The

craniotomy-forceps takes less room than the cephalotribe, and it effects the same object—by a different process it is true—*i.e.,* of bringing the base of the skull through the contracted brim edgewise like a disc. Cephalotripsy nevertheless possesses considerable independent advantages under many circumstances, and may lend much help to other proceedings.

The Powers of the Cephalotribe.—The all-essential point is that it shall be able to compress and crush down the remains of the calvarium upon the base of the skull, so as to bring the flattened skull into relation with the chink of the inlet. A secondary property which it is desirable to possess is that of holding during extraction. The crushing power can be attained in sufficient perfection, and with a gain in the facility of handling, if the instrument be made much less formidable in bulk than are most of the Continental cephalotribes. Three good modifications have been constructed here. Sir James Simpson's is the best known. He insists upon a pelvic curve in the blades, as being less likely to slip than straight blades. Dr. Kidd's, of Dublin, is the best type of a straight-bladed cephalotribe. Dr. Kidd insists strongly upon the advantages of long straight blades on the following three grounds :— First, straight blades admit better of the head being rotated whilst in the grasp ; secondly, they are easier to introduce ; and lastly, they hold more securely. Dr. Braxton Hicks has modified Sir James Simpson's cephalotribe, producing a very handy and efficient instrument. He preserves a moderate pelvic curve, and adapts a very convenient screw to the handles as a crushing power. I believe that to seize a head above the brim, as is necessarily the case where crushing is required, the blades should be curved ; but this curve should be very slight, otherwise the inconvenience in rotating or shifting the relation of the instrument to the pelvis referred to by Dr. Kidd will be felt. I think the blades of Dr. Hicks' instrument are too short by an inch ; and a little too much curved.

When the instrument is applied to the perforated head, it may be made to partly crush the base, imparting great plasticity ; then the base is tilted edgewise, and the skull is flattened down by the pressing the squamous and parietal bones on to the base. To facilitate the breaking-up of the base,

Goyon begins by perforating it. This can only be necessary in
the most extreme cases. Under the proceeding described the
head can be so flattened as to allow the blades to meet, and, as
the instrument then measures only 1·50″, the obstacle is reduced
to that degree. It is generally desirable to repeat the crushing,
which is done by taking a fresh hold in a different direction,
and then compressing again. This increases ductility. Two
crushings will generally be enough.

What are *the limits of application of the cephalotribe?*
The maximum, of course, is not difficult to determine. It
may be usefully employed in almost any case of minor
disproportion. But what is the least amount of space admitting of its use? This must depend somewhat upon the form
and size of the particular model adopted. In a discussion held
at Berlin, the majority of the speakers thought a minimum
of 5·4 mm. = 2·0″ conjugate diameter, was necessary. Lauth*
says the application begins at 8 mm., or about 3 inches, at
which point the forceps and turning are not available, and
ends at 5·0 mm., or a little under 2·0″. But Pajot goes beyond
this, and contends that it ought to be applied where there is
only 1·25″ conjugate diameter. Credé thinks it should be
used if only there is room enough to apply it. I have used
it with perfect success in a case of extreme rickety deformity, at St. Luke's Workhouse, aided by Messrs. Harris,
Rogers, and Sison, in which the conjugate certainly did not
exceed 1·50″. In this case, after the first crushing, I removed
some pieces of the cranial vault which had cropped up, by
means of my craniotomy-forceps. The delivery was completed,
without hurry, in an hour. I have arrived at the settled
conviction, that cephalotripsy is quite practicable with a pelvis
measuring an inch and a-half in the conjugate diameter; and
that the risk to the mother is inconsiderable compared with
that attending the Cæsarian section.

Other conditions are—

The os uteri must be sufficiently dilated; but this can be
readily effected by the caoutchouc water-bags.

The head must be previously perforated. Abroad, sometimes the ordinary forceps is put on to hold the head during

* "De la Céphalotripsie." Par J. F. Ed. Lauth. Strasbourg, 1863.

perforation ; or the blades of the cephalotribe are first adjusted, and then perforation is effected. But this is sometimes not feasible for want of room, and is never necessary.

Position.—The patient may lie on her left side ; or on her back, the breech brought well to the edge of the bed, and the legs flexed, held by an assistant on either side.

Operation.—The rules laid down for the long forceps will generally apply to the application of the blades, and it is equally unnecessary in either case to have an assistant or a "third hand." The lower or posterior blade is passed first, guided by the left hand passed well into the pelvis if possible. This blade is passed along the hollow of the sacrum until the point approaches the brim and touches the head-globe, when the handle is raised, and the point, turning into the left ilium or to the left sacro-iliac synchondrosis, travels over the head. It is passed high up, for the point of the instrument must get beyond the base of the skull. This being *in situ*, the second or anterior blade is introduced also at first in the hollow of the sacrum, crossing the handle of the first blade. When the point approaches the brim, the handle is lowered and carried backwards, and the point rises over the head-globe into the right ilium, or opposite the right cotyloid cavity, when it falls into opposition with the first blade. Being locked, the screw is turned slowly and steadily, the hand in the vagina taking note of the work done. (*See* Fig. 118.) If spicula crop out of the scalp, they should be picked away by the fingers or the craniotomy-forceps. Indeed, the removal of portions of the cranial vault in this way much facilitates the subsequent transit of the head. When the base is crushed in the direction first seized, you may use the instrument as a tractor. If there be any marked resistance, it is better to take off the blades, to re-apply them in the opposite oblique diameter, and repeat the crushing ; then, rotate the head by turning the handles about a quarter of a circle, to bring the flattened head into relation with the transverse diameter of the brim before extraction, so as to bring the head, flattened like a disc, to correspond with the chink of the pelvic brim. It is not, indeed, always necessary to give this rotation. I found, in the extreme case referred to above, that the necessary adaptation took place almost spontaneously. Extraction may be made by the cephalotribe, taking care to allow time for the dilatation of the cervix uteri and vulva.

Pajot (*Arch. Gén. de Méd.*, 1863, *and Preface to French Edition of this Work*) has practised a method analogous to that formerly employed in this country in craniotomy.   He performs

FIG. 118.

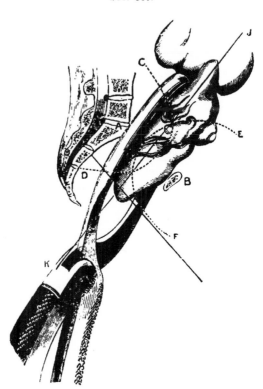

THE CEPHALOTRIBE SEIZING THE PERFORATED HEAD.

The head is seized somewhat in the right oblique diameter of the pelvis. It is partly crushed in; but the base is mainly adapted to pass the narrow brim by being canted. A is the projecting promontory, the centre of c D, the false curve which the head must first take ; B, the symphysis, the centre of the true or Carus' curve, E F, which the head must enter and follow to emerge ; G H, axis of the outlet ; J K, axis of inlet.

(The handle of the instrument is drawn too large.)

what he describes as " céphalotripsie répétée sans tractions "— that is, he first crushes the base by one operation ; he then gently tries to effect a slight rotation of the instrument so as to bring the crushed sides of the head into relation with the contracted diameter. If there is any resistance, he desists, and leaves the case for two or three hours for the uterus to mould the crushed head to the brim. He then repeats the crushing, and again gives two or three hours to Nature. One or two crushings suffice for the trunk. (*See* also "*Osservazioni di Cefalotrissia,*" by Dr. Chiara, Turin, 1867, for a good case in illustration.) Pajot places this method in distinct competition with the Cæsarian section. The cases related by him lend weight to his recommendation; but I cannot help thinking that the operation may and generally ought to be finished at one sitting. The great extent to which the head compressed in the cephalotribe will expand again when the compression is removed is a point to remember. The resiliency retained is considerable. A head flattened in the grasp of the cephalotribe may not measure more than one and a-half inches, yet, on taking off the blades, it will spring out to more than two inches. Why not, then, keep the blades on ?

When the head is extracted, there may be some trouble with the shoulders and trunk. The shoulders will generally be disposed obliquely in the brim—that is, one will be anterior to the other. By keeping up traction on the head backwards, this anterior shoulder will be brought a little down, so that a finger or the blunt hook or crotchet can be fixed in the axilla to pull it through. When this is done the head is dragged down forwards, so as to enable the same manœuvre to be repeated with the posterior arm. If this cannot be readily done, it is a good plan to crush in with the cephalotribe.

In extreme cases it is sometimes convenient to use both cephalotribe and craniotomy-forceps. For example, the cephalotribe having crushed and seized the presenting part, and served to draw it down to a certain extent, the instrument may slip a little, or the fœtus may show signs of giving way above the point gripped. The craniotomy-forceps may then be made to take a fresh hold at a higher point. In a case of extreme contraction, in a dwarf, which I saw last year (1874) with Dr. Corbet, of Surbiton, where the shoulders offered great resistance, I thus renewed the hold several times, and succeeded in extracting without injury to the woman, who made a good recovery.

If turning has been practised after perforating or cephalotripsy

OBSTETRIC OPERATIONS.

the arms fall in upon the crushed head and offer no serious obstruction.

To save the assistants the ghastly sight of the mangled head, wrap a napkin round it as soon as it is born. If traction is necessary in delivering the trunk, it is easier to hold when so treated.

5. *Delivery by the Forceps-saw.*—This instrument, introduced by Van Huevel in 1842, may be said to be the distinctive feature of the Belgian school. It is figured in the Obstetrical Society's Catalogue of Instruments, 1867. Dr. Hyernaux, who had been assistant to Dr. Van Huevel at the Maternité of Brussels, in his "Manuel Pratique de l'Art des Accouchements" (Bruxelles, 1857), rejects in its favour all crotchets and cephalotribes as comparatively dangerous and inefficient. It is therefore used in all cases where embryotomy is indicated. It consists of a powerful long forceps with the pelvic curve, the blades of which are grooved along the inner aspect in order to carry a chain-saw. When the head or other part of the child is seized by the forceps, this chain-saw is worked up from the point whence the blades spring, by means of cross handles attached to the two ends; thus travelling up the grooves, the saw crosses the head and cuts through it. For extraction Van Huevel contrived a pair of forceps toothed on one blade to seize the most convenient part of the child. Notwithstanding the formidable and complex appearance of the forceps-saw, it seems to have made its way into use. Professor Faye, of Christiania, a man of singular judgment and ability, says it is the only instrument fitted to cut through any part of the fœtus; but he appears not to be aware of the power of the wire-écraseur. He has simplified the instrument considerably, and extols it as safe, easy, and effectual. It is also used in Germany; and in Italy, Dr. Billi has modified it and introduced it into practice. We cannot refuse to lend favourable consideration to an instrument so recommended. It is certainly a new power; and its claims to compete with or displace other methods of facilitating delivery by embryotomy should be tested in practice. It appears to me, however, who have not yet used it, that it is more especially adapted for those minor degrees of pelvic contraction which can be dealt with satisfactorily by perforation and the craniotomy-

forceps; and that in extreme cases, where the conjugate diameter is 2″ or less, where the craniotomy-forceps is still available, and in which the cephalotribe can do good service, the
forceps-saw could hardly answer, owing to the size of the blades
and the necessity of getting them to lock accurately in order to
work the chain. It is capable of being most useful in dividing
the neck or other part of the body in cases of impaction of
shoulder-presentation.

6. *Delivery by the Author's New Method of Embryotomy.*—
I have now to describe a new method of embryotomy, designed
by myself, to effect delivery in the most extreme cases of pelvic
contraction. It had long appeared to me that, if the problem,
how to break up and extract such a body as the mature fœtus
through a chink measuring an inch wide and three or four
inches long, were proposed to a skilful engineer, he would find
a solution. The chief difficulty seemed to be in finding an
instrument that occupied little space. I thought I saw in
the wire-écraseur the means of effecting the object in view.
I had found no great difficulty in snaring an intra-uterine
polypus of considerable size with a wire-loop passed through
a cervix uteri whose aperture was much smaller than the
tumour, guided only by one or two fingers. Why should
not the fœtal head be seized in a similar manner and cut in
pieces? I performed several experiments with a very diminu-
tive and delicate rickety pelvis, measuring an inch in the
antero-posterior diameter, and scarcely more in the sacrocotyloid diameter, and I will now repeat the operation before
you.*

As in cephalotripsy, but not so urgently, it is desirable, first
of all, to perforate the head. It further facilitates the operation
to twist off a portion of the parietal bones by the craniotomy-
forceps, so as to destroy the sphericity of the head. The
wire-loop thus buries itself more readily in the skull, a smaller
loop is required, and it cuts through the base more readily.
If the sphericity of the head is not first destroyed, the wire-loop
is apt to glide off the head-globe, seizing only the scalp when the
screw is worked. (*See* Fig. 119.) The crotchet is next passed
into the hole made by the perforator, and held by an assistant

---

* I also demonstrated this operation at the meeting of the Obstetrical
Society of the 2nd June, 1869.

so as to steady the head. A loop of strong steel wire is then formed large enough to encircle the head. The elasticity of the wire permits of the loop being compressed by the fingers so as to make it narrow enough to slip through the cervix

FIG. 119.

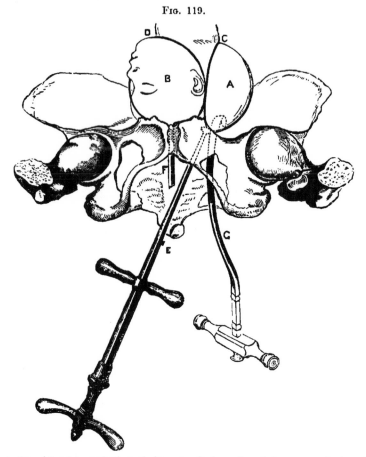

E, stem of écraseur carrying loop of wire over occiput. F, stem of écraseur carrying loop of wire over lateral segment of head. G, crotchet, the point of which is passed into the opening in the cranium made by the perforator, and held by an assistant to steady the head whilst the wire is being applied. A, the occipital segment of the head seized by the wire-loop at c, which buries itself in the head. B, a lateral segment of the head seized by the second application of the wire-loop at D.

uteri and the chink of the pelvic brim. The loop is thus guided over the crotchet to the right side of the uterus, where the face lies. The compression being removed, the loop springs open to form its original ring, which is guided over the anterior part of the head, as in Fig. 119. The screw is then tightened. Instantly, the wire is buried in the scalp; and here is manifested a singular advantage of this operation. The whole force of the necessary manœuvres is expended on the fœtus. In the ordinary modes of performing embryotomy, as by the crotchet especially, and in a lesser degree by the craniotomy-forceps and cephalotribe, the mother's soft parts are subjected to pressure and contusion. The child's head, imperfectly reduced in bulk, is forcibly dragged down upon the narrow pelvis, the intervening soft parts being liable to be bruised, crushed, and even perforated. And this danger, obviously rising in proportion to the extent of the pelvic contraction, together with the bulk of the instruments used, deprive the mother in some cases of extreme contraction, of the benefit of embryotomy, leaving her only the terrible prospect of the Cæsarian section. When the anterior or posterior segment of the head is seized in the wire-loop, a steady working of the screw cuts through the head in a few minutes. The loose segment is then removed by the craniotomy-forceps.

In minor degrees of contraction, the removal of one segment is enough to enable the rest of the head to be extracted by the craniotomy-forceps. But in the class of extreme cases in which this operation is especially useful, it is desirable still further to reduce the head, by taking off another section. This is best done by re-applying the loop over the occipital end of the head as seen in A, Fig. 119. It thus accomplishes what the cephalotribe does not, viz., it breaks up the base of the skull.

The small part of the head still remaining attached to the trunk offers no obstacle. It is useful as a hold for traction. The craniotomy-forceps now seizes this firmly, and you proceed to deliver the trunk. If the child be well developed, this part of the operation will require considerable skill and patience. An assistant draws steadily on the craniotomy-forceps, directing traction to one side, so as to bring a shoulder into the brim. The operator then hooks the crotchet into the axilla, draws it down, and with strong scissors amputates the arm at the shoulder. This proceeding is then repeated on the other arm. Room is thus gained to deal with the thorax. You perforate the thorax. Introduce one blade of a strong pair of scissors into the aperture, and cut through the ribs in two directions. Then, by the crotchet eviscerate the thorax and abdomen, until the trunk is in a condition to collapse completely. This done, moderate traction will complete the delivery.

I have imagined a proceeding by which the arms can be amputated even more easily. A curved tube, shaped like Ramsbotham's hook, may be made to carry a strong wire under the axilla, and the end being brought out, and the tube removed, the wire can be attached to the écraseur, which then cuts through the limb with ease and security. Decapitation may be conveniently performed in the same way.

This operation is particularly adapted to extreme cases of narrowing of the pelvic brim from rickets, in which there is commonly left a moderate amount of space at the outlet for manipulation. Indeed, I believe a case of rickety deformity will rarely be found so great as to compel resort to the Cæsarian section. No doubt the operation I have recommended is more difficult, demands more skill and richness of resource than the Cæsarian section—an operation which cuts the Gordian knot with despotic simplicity, not perhaps unpleasing to the operator, but certainly full of extremest peril to the mother.

The operation is, I freely admit, less practicable in extreme cases of osteomalacic deformity. Here the pelvis is deeper than in rickets; and the deformity bearing in an aggravated degree upon the outlet, leaves insufficient room for manipulation. Where two fingers can barely pass between the tuberosities of the ischia, it will be scarcely possible to guide the écraseur through the pelvis, and to get the loop over the head. But in these cases, as I have already stated, the bones will often open up, under pressure applied within. Professor Lazzati told me he relied upon this dilatability in all cases of osteomalacia, seldom or never resorting to the Cæsarian section, except in the worst cases of rickety distortion.

Certain dangers attend the operation of craniotomy. What are these? Certain injuries may be inflicted upon or result to the mother.

1. The perforator has been known to strike the promontory of the sacrum, or to lacerate the cervix uteri.

2. Spicula of cranial bones resulting from perforation may scratch or tear the soft parts.

3. The crotchet may slip and lacerate the soft parts.

The above, of course, may be avoided with care.

4. But serious evil is likely to result from deferring the operation too long—*i.e.*, until after exhaustion has set in—and under a too protracted operation in an unsuitable case. Long-continued dragging of the head upon a brim which it cannot pass, jamming the soft parts, more especially at the two points of greatest projection, the promontory and the symphysis, ends by stopping the circulation in the parts compressed, bruising them, actually grinding through them. In this way, after severe operations, it has been found that a large hole has been made through the posterior cervix uteri. (*See* p. 337.) Such injury, added to the shock and exhaustion of the system, may be fatal. Under such circumstances I have noticed that the soft parts, *i.e.*, the vulva and perinæum, greatly deprived of vitality, lose elasticity and power of resisting distension. They are deep brown or black from congestion, they break down, or tear like wet brown paper. When the tissues have arrived at this point, I doubt if the patient ever recovers. It is the result of operations which, by forcing the child against the mother's soft parts, arrest the circulation in them. It is a form of acute necrosis. Proper modes of delivery, as removal of cranial vault, cephalotripsy, and the wire-écraseur, acting on the fœtus, if timely used, avoid this serious danger.

5. If the immediate injury above described do not occur, the long-continued pressure may cause mortification of a limited portion of the neck of the uterus. Thus, in the course of a few days, a slough is formed between the vagina and bladder, resulting in vesico-vaginal fistula.

In the possibility of attendant danger, craniotomy differs essentially from the forceps. Whilst under craniotomy, mischief or death may ensue; the forceps, if used rightly and in suitable cases, is an innocuous instrument. Statistics, professing to show that the mortality from the use of the forceps is at the rate of 1 in 20, are flagrant examples of the fallacy of arguing "*post hoc, ergo propter hoc.*" Properly speaking, the mortality from the forceps is *nil*. Women die because the instrument is used too late.

LECTURE XXIV.

THE CÆSARIAN SECTION—THE INDICATIONS FOR—THE MORAL
ASPECT OF THE OPERATION AS BETWEEN MOTHER AND
CHILD—THE CONDITIONS THAT RENDER THE OPERATION
NECESSARY—QUESTION BETWEEN CÆSARIAN SECTION AND
TURNING IN DYING AND DEAD WOMEN TO SAVE THE CHILD—
THE TIME TO SELECT FOR THE OPERATION—PREPARATION FOR
OPERATING—THE OPERATION—THE DANGERS ATTENDING IT,
AND THE PROGNOSIS—SYMPHYSEOTOMY.

The Cæsarian section occupies a doubtful place between Conservative and Sacrificial Midwifery. It is conservative in its
design, in its ambition; it is too often sacrificial in fact. It
is resorted to with a feeling akin to despair for the fate of
the mother, which is scarcely tempered by the hope of rescuing
the child. It is looked upon by the great majority of obstetricians as the last desperate resource, as the most forcible
example of that kind of surgery which John Hunter regarded
as the reproach of surgeons, being a confession that their art
was baffled. On the other hand, it is regarded by some
enthusiastic practitioners, dazzled perhaps by its false brilliancy,
as an operation deserving to be raised into competition with
turning, craniotomy, or cephalotripsy. At different times and
in different countries it has been looked upon with favour,
because promising salvation to the child. The child is weighed
against its mother; and conscience is silenced with the reflection that the possible or probable rescue of the child may
rightly be purchased by subjecting the mother to the most
imminent peril. It is held that this double chance, made up
of odds in favour of the child and against the mother, is to be

preferred to the single chance afforded by an operation which gives up the child for the sake of odds in favour of the mother.

The situation is painful, and may well perplex those who are not steady in their allegiance to those moral laws which ought to rule over all professions, and which certainly recognize no exception here.

None of us will claim exemption from these laws; but some may interpret them differently. And here, again, we must call to mind a lesser law to which reference has before been made—namely, that the choice between two operations will be influenced by the comparative skill in them which the operator happens to possess. Under this influence the favoured operation will be more and more cultivated, and its competitor more and more neglected. Thus, to apply this law to the present discussion: the man confident in his skill in the extraction of a dead child by the natural passages with safety to the mother will be disposed to assign the narrowest possible limits to the Cæsarian section; and, on the other hand, the man who has not this confidence will be disposed to prefer the Cæsarian section, an easy operation.

The operation must be studied under two aspects—*first*, as one imposed by *necessity*, as the *only* means of effecting delivery; *secondly*, as one of *election*, deliberately chosen as the *best* means of effecting delivery.

The distinction is very important to be borne in mind; for, under that fatal fascination which seems to oppress the reasoning faculty in statisticians, conclusions drawn from figures representing the most dissimilar facts are accepted and put forth as the legitimate deductions from experience. With our present materials, I believe it is the most idle and unprofitable waste of intellect and of time to seek to draw rules for practice from statistical calculations. It seems to me impossible in a great number of cases to distinguish or to estimate the relative shares in causing death that arise from causes antecedent to the labour, from causes arising during and in consequence of the labour, and from the operation itself. All cases, therefore, which were not selected—that is, deliberately operated upon, under simple conditions, especially freedom from dangerous disease and from protracted labour—must be put aside or considered apart. How many simple cases of this kind

E E

do we possess? Certainly not enough to deduce a law of mortality to estimate the risk to life from the Cæsarian section. M. Pihan-Dufeillay,* indeed, declared that the operation performed under favourable circumstances, as early as the impossibility of delivery *per vias naturales* is recognized, gives nearly 75 per cent. of recoveries. But what assurance have we that an undue proportion of successful cases are not recorded, unsuccessful ones remaining in the dark? Pajot in the preface with which he has honoured the French edition of my work on the Diseases of Women (1875) affirms that "this operation has cost the lives of *all* the unhappy, ignorant women who have undergone it, in Paris, since the beginning of this century. And," he adds, "it is still practised!"

And if this element of the comparison, the risk to life from Cæsarian section, is so defective, vague, and uncertain, it follows that comparison or ponderation in the statistical scales is impossible, even assuming that the other element, the risk to life under craniotomy, were determined. But do we know more of this second element, the risk to life in cases of exactly similar pelvic deformity under other modes of delivery? Obviously, we cannot recognize fatal cases of craniotomy in extreme deformity, say of conjugate diameter reduced to 2″ or to 1·75″, unless the operation was begun under selected circumstances—that is, before exhaustion had set in—and conducted with due skill, and after the most approved methods. We are fairly called upon to reject all fatal cases in which craniotomy was performed with bad instruments, in which the skull was either not crushed down by the cephalotribe, or the calvarium not removed, so as to leave nothing but the base to bring through the brim, edge on, or the head and trunk not reduced by sections, as by my method. The point to determine is, what is the limit of contraction that admits of this proceeding, or some better proceeding, being carried out with a reasonable presumption of safety? When that is determined, it follows, as a logical necessity, that this proceeding ought to be adopted in cases falling within that limit.

To dispute this proposition—and it has been disputed, the disputants not seeing the dilemma prepared for themselves—

* "Archives Gén. d. Méd.," 1861.

is to dispute the propriety of performing craniotomy in any case. For, why do we perform craniotomy or cephalotripsy in the case of a pelvis contracted to 3″ or 3·50″? Is it not because, possessing means of extracting a dead child through such a pelvis with a reasonable prospect of safety to the mother, we acknowledge it to be our duty to use those means? Now, assume that we also possess means of extracting a dead child through a pelvis narrowed to 2·00″ or 1·50″ with a reasonable prospect of safety to the mother, who will venture to dispute, whether as a matter of logic or of morals, that the mother shall be denied that prospect?

The case, then, as far as relates to pelvic contraction, stands on the old ground. It may be stated broadly as follows :— Embryotomy stands first, and must be adopted in every case where it can be carried out without injuring the mother. The Cæsarian section comes last, and must be resorted to in those cases where embryotomy is either impracticable, or cannot be carried out without injuring the mother. There is, therefore, no election. The law is defined and clear. The Cæsarian section is the last refuge of stern necessity.

Certainly, those who advocate the Cæsarian section in cases of contraction below 2·25″, may deny the possibility of extracting a dead child through the natural passages with a reasonable prospect of safety to the mother. But this denial, of course, applies only to the experience of those who deny. I can but refer to the testimony of Ramsbotham and a host of English teachers, to Pajot and many Continental authorities, as to the positive fact as regards cephalotripsy. (*See* Lecture XXIII.) It is right, however, to state that Scanzoni* declares the limit generally adopted in Germany to be 2·50″. But I repeat, with all the emphasis that conviction based upon experience dictates, that delivery by the natural passages, either by cephalotripsy, by the craniotomy-forceps, or by my new method of embryotomy, if the conjugate diameter measures 1·50″, is perfectly practicable, and with a presumption of safety to the mother much greater than that attending the Cæsarian section. I am even confident that the same may be predicated with a conjugate diameter reduced to 1·25″ or even to 1″.

* "Lehrbuch der Geburtshülfe," 1867.

E E 2

Infant Mortality attending the Cæsarian Section.

The probability of saving the child must obviously vary according to circumstances. Sometimes the operation is determined upon without any consideration for the child, simply as a means of delivering the woman. The child may be known to be dead. The most practical question is—What is the probability of saving the child when the operation is performed under the most favourable selected circumstances ? Statistics, even upon this point, are not conclusive. But it may be assumed that the prospect of the child is less than in the case of ordinary labour. The experience of the Royal Maternity Charity gives rather more than 3 per cent. of still-born children to the total births. Dr. Radford says the risk to the infant of delivery by Cæsarian section is not much greater than that which is contingent on natural labour. Scanzoni* ascertained the fate of 81 children out of 120 operations performed between 1841 and 1853; 53 children, or 60 per cent., were born alive. In all probability several of these died within the first week. The mortality, then, is probably greater than in ordinary labour.

But assuming the probability of saving the child to be ever so high, are we justified by law, or by religion, the basis of law, in taking the woman's life as it were into our own hands, and deliberately subjecting it to the most imminent hazard for the sake of probably saving her child ? It has been urged that, the woman having a prospect of a miserable existence for a few months or weeks only, whilst the child is likely to be saved and live to maturity, the child's life is the more valuable, and ought therefore to be preferred. But this only applies to a limited number of cases.

Or, taking the case of a woman who cannot give birth to a living child by the natural passages, calling upon the obstetrician time after time to deliver her by craniotomy, it is asked, Ought we not at length to refuse craniotomy and insist upon the Cæsarian section ?

Dr. Denman states the question perspicuously thus:—" I

* " Lehrbuch der Geburtshülfe," 1867.

cannot,* however, relinquish the subject without mentioning another statement of this question, which has often employed my mind, especially when the subject has been actually passing before me. Suppose, for instance, a woman married, who was so unfortunately framed that she could not have a living child. The first time of her being in labour, no reasonable person could hesitate to afford relief at the expense of her child; even a second and a third trial might be justifiable, to ascertain the fact of the impossibility. But it might be doubted in morals whether children should be begotten under such circumstances; or whether, after a determination that she cannot bear a living child, a woman be entitled to have a number of children (more than ten have sometimes been sacrificed with this view) destroyed for the purpose of saving her life; or whether, after many trials, she ought not to submit to the Cæsarian operation, as the means of preserving the child at the risk of her own life. *This thing ought to be considered."*

The question is indeed a trying one. We may easily go astray in the labyrinth of casuistry, unless we hold steadily to the clue laid down by the moral law. I think it will not be disputed that, in law, *he who accelerates death is held responsible for having caused death.* We are not justified in regarding a life as less sacred because we believe that, in the ordinary course of disease, it will not last beyond a very short time. We are, therefore, not justified in preferring the Cæsarian section to craniotomy in a case where the latter operation offers a fair prospect of safe delivery, because the woman is suffering from osteomalacia, which is commonly, but not always, a progressive and fatal disease. We have no right to lessen the woman's chance of life because it is already small. We cannot, even in the case of osteomalacia or cancer be certain of our prognosis. I have known women live for many years with osteomalacia, and even recover.

Then take the case put by Denman. The conduct of the woman is assumed to be culpable, and we are assumed to be in the position of accomplices or abettors in her fault if we repeatedly relieve her by craniotomy. But are we entitled to take upon ourselves the office of the Judge? Are we to make ourselves the ministers of Justice? Vengeance, punishment, is

* " Introduction to Midwifery."

not ours. When did Medicine ever withhold her merciful hand from the degraded, the sinful, the criminal? Shall we dare to put a mere vegetative life—that of an unborn child—into the scale against that of a being like ourselves accountable to the Almighty? Can we take upon ourselves the awful weight of deciding that the wretched woman was wrong—criminal, in becoming a mother? She is subject to her husband. If punishment is due, must it fall upon her? and are we to inflict it? I cannot, therefore, hesitate in expressing my conviction that we should be traitors to our trust if we were to perform the Cæsarian section when craniotomy is safer for the woman, because, in our judgment she was culpable in becoming a mother.

The Conditions that render the Cæsarian Section necessary.

The most frequent condition is *deformity with contraction of the pelvis*. The operation is justified whenever the contraction is such as to render it impossible to extract a dead child through the natural passages. This may be stated at 1·50″ conjugate diameter and below, at the higher limit of 1·50″ coming into competition with craniotomy. But I think we may reasonably hope to carry the minimum to 1·00″. Cases may also occur in which a conjugate diameter of 2·00″ may call for Cæsarian section if the pelvis is much distorted, so that the diagonal and transverse diameters offer insufficient compensation for the narrow conjugate. This is more especially the case when, as in osteomalacia, the outlet of the pelvis is also so contracted and rigid as to render the necessary introduction and manipulation of instruments impossible.

The most frequent form of distortion calling for the Cæsarian section is that which arises from osteomalacia. In this disease the sides of the triangle forming the brim of the pelvis are all pressed inwards, and are more or less convex. The result is that the brim is practically divided into two parts, neither of which is available for the passage of the head. Rickets, also, will sometimes produce a pelvis that will leave no alternative. The slipping down of the lumbar vertebræ—spondylolisthesis —into the pelvic cavity, if to any great extent, leaves no other resource.

The next most frequent causes are tumours of various kinds growing into the pelvis, such as bony or malignant tumours springing from the wall of the pelvis; tumours of the ovary descending into the pelvic cavity, and getting fixed there. Dr. Sadler, of Barnsley, has recorded a case* in which the operation became necessary from the pelvis being filled up with an enormous *hydatid cyst* springing from the liver. Other exceptional causes, chiefly remarkable on account of their extreme rarity, have been observed. Atresia of the cervix uteri and vagina may be so extensive and unyielding that the Cæsarian section may be less hazardous than the attempt to open a canal through the cicatricial tissues.

We are sometimes driven to the operation after having exhausted other modes of proceeding—for example, when craniotomy may have failed to deliver.

The Cæsarian section, or simple abdominal section, is indicated in certain cases of rupture of the uterus, when the child cannot be extracted with advantage through the pelvis, and in some cases of extra-uterine gestation.

It is resorted to when the mother has died undelivered, in the hope of rescuing the child. In this way several children have been saved. The success, of course, will depend greatly upon opening the uterus very soon after the mother's death. How soon? It is difficult to assign a precise limit beyond which the child's life may be regarded as lost. Harvey said, "Children have been frequently taken out of the womb alive hours after the death of the mother." I am not aware of instances to justify this statement. But Burns says: "The uterus may live longer than the body; and after the mother has been quite dead the child still continues its functions." The fœtus in lower animals will live some time if the ovum is not opened. It seems not improbable that a modified degree of placental circulation may continue for some little time after the mother's general circulation has stopped; and it is, further, even more probable that the child may survive for some little time in a state of asphyxia capable of restoration on being brought into the air. Certainly, children have been extracted alive ten minutes after the mother's death. Wrisberg cites three cases of infants born enclosed in the membranes; they

* "Medical Times and Gazette," 1864.

lived, one seven minutes, and the two others nine minutes whilst thus enveloped. Dr. Brunton relates a case (*Obst. Trans.* vol. xiii.) of a seven months' child expelled in intact membranes, from which it was taken out alive after an estimated lapse of fifteen minutes. Dr. George Harley relates (*Med. Times*, 1850) a case in which he performed post-mortem section, and withdrew a live child "three or four minutes" after the woman's death. It was saved with difficulty by artificial respiration. Dr. Hoschek (*Arch. f. Gynäkol.*, 1871) relates an interesting case in point. A woman about eight months pregnant, dying of phthisis, begged him to save her child by Cæsarian section. The child was extracted about ten minutes after the poor woman's death. It cried lustily, and survived about four months, dying of small-pox. On the other hand, children have been extracted dead when the operation has been delayed fifteen minutes. This has occurred twice in my own experience. The chance of preserving the child by the Cæsarian section *post mortem* will be much influenced by the circumstances of the mother's death. If she dies by sudden injury, the child may survive a little longer. If she dies from hæmorrhage or rupture of the uterus, the child's death is likely to have preceded that of the mother. If she dies from phthisis or other gradually exhausting disease, the child may survive some minutes.

The probability of the child *in utero* surviving its mother for a short time seems to have been impressed upon the popular mind from the remotest ages. Mythical, pagan, Jewish, and Christian creeds have alike sanctioned attempts to rescue the unborn child from its dead mother. The operation is indeed associated from its origin with the history of medicine. When Coronis had died by the hand of Apollo, and was laid on the funeral pyre, the poet says :—

> "Non tulit in cineres labi sua Phœbus eosdem
> Semina, sed natum flammis uteroque parentis
> Eripuit, geminique tulit Chironis in antrum."

The offspring so rescued was Æsculapius.

The Cæsarian section after death comes into competition with forced delivery *per vias naturales*. Sometimes, if the cervix is dilated, the child may be extracted by turning as quickly as by Cæsarian section, and it has this advantage—that, being less likely to shock the friends, it may be practised when the Cæsarian section would be rejected, or resisted until too late.

This view has recently found favour in the Italian schools. Professor Rizzoli proposed to deliver in this manner as soon as the woman was dead. The late Professor Esterlé[*] related a case occurring in 1858, in which he successfully executed this proceeding in a woman dying of cerebral apoplexy. One argument he adduced in support was the fact that in similar cases the child survives its mother's death so short a time that to save it delivery must be effected whilst the mother is still living. He says that, in his judgment, this is the proper course to adopt—of course requiring that all possible gentleness shall be practised, and that, if necessary, the plug and vaginal douche be used beforehand.

Cases in support of this practice are recorded by Belluzzi,[†] Ferratini, and others. Out of three cases, Belluzzi saved two children. Ferratini [‡] saved the child of a woman dying of phthisis. The subject is also referred to by Dr. Hyernaux, of Brussels, in the second edition of his *Treatise on Obstetrics*.

A condition which has several times led to the Cæsarian section has been *disease, mostly malignant, of the lower segment of the uterus*, preventing its due dilatation, and involving the danger of laceration (*See* p. 333). In this case, the opportunity is commonly presented of inducing labour or abortion. Or in some cases it is possible to remove the greater part of the obstructing diseased mass by the galvanic cautery-wire. The conditions of choice are often perplexing. If labour be induced before seven months, the prospect of a viable child is small, but the uterus may be able to dilate sufficiently without injury. On the other hand, the life of the mother is probably doomed to be of short duration from the progress of her disease. Any injury to the diseased structures,

---

[*] " Annali Universali di Medicina," 1861.

[†] " Nuovi fatti in appoggio dell' estrazione del feto col parto forzato durante l' agonia delle donne incinte, onde salvare più facilmente il feto stesso in sostituzione a tale operazione, o al taglio cesareo post mortem." Bologna, 1867.

[‡] " Un nuovo fatto in appoggio," etc. Genova, 1868.

as even by premature labour, is liable to accelerate her death. Is it not then better, both in her interest and in that of the infant, to let things go on to the natural term of gestation? She may live two or three months longer, and the child will undoubtedly have a better chance.

What is the best time to select for the operation?

1. Sometimes, of course, all choice is denied us, or the range of time offered is extremely limited. If called to a woman in labour at term, and under the conditions assumed to require the operation, it ought to be performed without delay. There is, if possible, greater reason than in conditions requiring other less formidable proceedings, to anticipate the exhaustion and local injury that follow upon protracted labour. It is a misfortune, tending fatally to compromise success, to be obliged to operate when the system is prostrated; when the structures that have to be wounded are so worn and injured that the power of reaction and repair is seriously reduced; and when the blood is deteriorated by the products of nervous and muscular overwork. It is, then, a clear indication to operate early in labour.

2. If the patient come under observation early in pregnancy, we have the double opportunity of considering the propriety of inducing labour, with the object of avoiding the operation, and of selecting the time for its performance should it be unavoidable. Is it an advantage to operate during labour?—that is, does the process of labour conduce to the success of the Cæsarian section? Amongst Continental practitioners, and indeed generally here, the Cæsarian section being regarded as a mode of delivery, it is held to be a primary indication to respect the laws of parturition, and to enlist the natural powers in our aid as much as possible. It has been, therefore, almost universally considered proper, in cases where the ultimate necessity of resorting to the Cæsarian section is recognized, to postpone the operation until the advent of labour. It is presumed that the epoch which Nature fixes for labour is that when the most favourable conditions for the process and for the recovery are present. The whole organization is better prepared; and the uterine muscles having acquired their highest stage of development, and contraction having actually set in, it seems reasonable to anticipate that

the wound made in the uterus will close better, and that the necessary changes attending delivery will be more safely carried out.

Dr. Ludwig Winckel,* whose experience in this operation is the greatest, says the most favourable time is the end of the second stage of labour, when the membranes are ready to burst. He advises not to rupture the membranes. The escape of liquor amnii into the abdomen does no harm, and the extraction of the child is more easy if the membranes are kept entire until the moment of seizing the child.

But admitting spontaneous labour at term to be a favouring condition, may not labour induced artificially before term be an equally favourable condition? The arguments in support of the affirmative deserve attention. Dr. Braxton Hicks brought on labour a fortnight before term as a preparation for the Cæsarian section, influenced by the opinion that by so doing, the uterus, taken at a period prior to the highest degree of degeneration of its muscular fibres, would heal better. I think this is a physiological error. The degree of fatty change observed in the mature uterus is no impediment to reparation. There are, at any rate, too many examples of complete repair after section by the knife, and even after rupture, to admit of a doubt that the mature uterus is in a condition not unfavourable for the operation. On the other hand, it cannot be doubted that the uterus at seven or eight months is also capable of complete repair after injury. We may, then, very properly consider whether, assuming things to be equal, *quoàd* the uterus, there may not be other circumstances that may rightly turn the scale in favour of premature delivery. Such, I think, do exist. For example, if we wait for the advent of natural labour, we may be called upon to operate in the middle of the night, and surrounded by many difficulties, all concurring to lessen the prospect of success. By selecting our own time, we may have daylight, the assistance of colleagues, and every appliance that may be thought useful.

The best time to select, then, would be as near the natural term of gestation as possible, and this may be determined

---

* "Monatsschrift für Geburtskunde," 1863.

approximately by taking some day in the estimated last fort-night of gestation.

Then comes the question, Shall we start labour before operating, or proceed to the operation at once without exciting any preparatory action of the uterus? I think the prepon-derance of reason is in favour of operating upon a uterus already in the act of labour. The *first step* will be to pass up an elastic bougie into the uterus overnight. This will excite some degree of uterine action. Next day, the hour of operating being fixed, say, for 1 p.m., we may in the morn-ing ascertain to what extent labour has proceeded. If the os uteri is not open more than enough to allow a finger to pass, it will be useful to dilate it a little more with the caoutchouc bag No. 2. This will probably induce further contraction of the uterus, and secure one most desirable object —namely, a free outlet for liquor amnii and other discharges by the natural passages.

The labour then being started thus far, we are ready for the operation. The question between general and local anaes-thesia arises. In all abdominal operations, the vomiting so liable to attend or follow chloroform is a serious drawback. The violent straining is apt to open the uterine wound, to stretch the abdominal wound, to destroy the "rest" which is such an important condition of repair, and thus to compro-mise the success of the operation. The ether spray is at least free from this objection; and possibly, further experience will show that it ought to be preferred. Dr. Keith, whose success in ovariotomy is so conspicuous, has entered a decided protest against chloroform in this operation. "Had chloroform," he says, "never been heard of, I doubt if humanity would have suffered from the want of it." He now uses anhydrous sul-phuric ether, made from methylated alcohol, administered through Richardson's apparatus. The reasons for preferring it to chloroform for the analogous operation of Cæsarian section are cogent. This opinion, Dr. Keith tells me, his enlarged experience down to the present time (1875) strongly confirms.

The Instruments and Assistants.

The *instruments* required are:—1. A sharp bistoury. 2. A

bistoury having a blunt end. 3. A director such as is used in ovariotomy. 4. A large probang armed with sponge. 5. Artery-forceps and ligatures in case of bleeding from the abdominal wound. 6. Clean sponges, wrung out with solution of carbolic acid or iodine. 7. Ice. 8. Two powerful apparatuses for inducing local anæsthesia by congelation; or an apparatus for inducing anæsthesia by ether. 9. Needles. 10. Silver or silk-sutures for uterine and abdominal wounds. 11. Lint. 12. Many-tailed bandage and adhesive plaster. 13. Carbolic acid oil.

Assistants.—Skilled assistants should stand one on each side of the patient. Another should be free to hand instruments and assist in sponging, etc. A nurse or two to help will complete *the necessary staff.*

Preparation.—The bowels should be emptied by castor-oil or by enema on the morning of the operation.

The Operation.

Position.—The patient is laid on a table on her back, with the head and shoulders slightly raised. The operator stands on the patient's right or in front.

The *catheter* is introduced to empty the bladder. If the case be one of osteomalacia, explore for the last time carefully to ascertain if the pelvis can be opened up by dilatation by the hand. In this way, several times, the operation has been avoided. It is related that a young surgeon, burning with the ambition to perform the Cæsarian section, invited his *confrères* to assist. Everything was ready, when Osiander requested permission to examine. He opened the pelvis, turned, and extracted a live child; and thus, says he, "I saved the woman from the Cæsarian section, and from death." Dr. Tyler Smith has related a case in which he was able to dilate the pelvis with his hand, and deliver after craniotomy. Olshausen relates a similar case (*Berl. Klin. Wochenschr,* 1869). I have already adverted to the testimony of Professor Lazzati, the late distinguished Director of the Lying-in Hospital of Milan, who informed me that, although osteomalacia is very frequently observed there, he did not often find it necessary to resort to Cæsarian section, even in cases of great deformity. The common practice is to turn, and the bones yield to admit the child, either mutilated or not. When the child has passed, the bones collapse again. He has more frequently been called on to perform the Cæsarian section on account of extreme distortion from rickets.

In cases of great distortion, the uterus is not seldom found considerably displaced. Where there is great prominence of the sacral promontory, and squatting of the chest down upon the flanks, the uterus is necessarily thrown much forward, sometimes so as to overhang the symphysis. With this there is occasionally marked lateral obliquity; and, what is less common, but still likely, a twisting of the uterus on its long axis, bringing one of its sides to look more or less forward.

It is also desirable to take note, by auscultation, of the seat of attachment of the placenta. Assistance may also be obtained by laying the hand flat on the uterus, when, if the walls are thin, a peculiar thrill or vibration marks the seat of the placenta, which is confirmed by feeling the part bulging a little, as if a segment of a smaller globe were seated on a large spheroid. This has been pointed out by Dr. Pfeiffer (*Monatsschr. für Geburtsk*, 1868).

The uterus is then brought into proper relation with the linea alba, so that the two incisions may correspond, unless it be found that the placenta can be avoided by cutting a little on one side of the median line. The abdominal incision is best made in the *linea alba*, extending from below the umbilicus to within about three inches of the symphysis pubis. When nearly through it is desirable to get a finger through a small opening, and, using this as a director, to cut from within outwards, so as to avoid scratching the uterus. The assistants support the abdominal wall on either side, looking out to prevent the escape of intestine. The *uterine incision* is made in the middle line, sparing the fundus and lower segment as much as possible, as these parts are not well adapted to close by contraction, and large vessels are more likely to be divided. Circular fibres predominating near the cervix tend to make the wound gape. A manœuvre described by Winckel, and forced upon him by his being often called upon to operate with insufficient assistance, here deserves attention. An assistant hooks the forefinger of each hand in the upper and lower angle of the uterine wound, and, lifting them up, fixes them in contact with the corresponding angles of the abdominal wound. This shuts out the intestine effectually, and tends to prevent the blood from running into the abdominal cavity.

If the placenta is found directly behind the wound, the hand of the operator is insinuated between the placenta and the uterine wall, detaching it until the edge is felt, when the membranes are pierced, and *the child is seized by the feet*. This point obstetricians will pardon me for prominently emphasizing, because I have seen a surgeon pull at the arm and fail to extract the child. Sometimes the neck is tightly grasped by the uterine wound. If the constriction does not soon yield, it is better, says Scanzoni, to extend the incision than to drag overmuch, lest the womb be torn.

When the child and placenta are removed, attention is required to watch the bleeding; this may take place from the cut sinuses in the uterine walls, and from the inner surface from which the placenta has been separated. Hæmorrhage is best checked by direct compression of the uterus with the hand; if the uterus contract well, the hæmorrhage ceases. Ice may be applied to the wound, and a piece placed in the uterine cavity. If hæmorrhage persist from the placental site, it will be necessary to swab it with perchloride of iron. Dr. Hall Davis and Mr. de Morgan induced contraction of the uterus by galvanism.

Before closing the uterine wound, *thrust the probang through the os uteri and vagina*, to make sure that the natural passages are clear for the discharges from the uterus.

The Closure of the Wound.—When the bleeding has ceased, and any blood that may have found its way into the abdomen is removed, we have to consider the question of applying sutures to the uterine wound. It is a matter of observation that in many fatal cases the edges of this wound have been found flaccid and gaping. But in the majority of, if not in all, these cases the operation had been performed on women exhausted by protracted labour; and, on the other hand, in women operated upon at a selected time, when the powers are unimpaired, the uterus commonly contracts well. But we must remember that just as we find it in post-partum hæmorrhage, the uterus may relax again after having been apparently well contracted. Winckel says he has never lost a case from hæmorrhage, and has not stitched the uterine wound. The history of closure of the uterine wound by suture is traced by Dr. Rodenstein (*Amer. Journ. of Obstetrics*, 1871). He was present at a case in which hæmorrhage was stopped by sutures. He cites Lauverjat as stating that Lebas de Mouilleron applied the " suture sanglante " to close the uterine wound in a successful Cæsarian section in 1769. Kilian condemned it as a most rash proceeding. In the *Gazette Médicale*, 1840, Godefroy reports a successful case in which he used sutures to the uterus. In 1863 Sir James Simpson used three iron sutures to control an "immense hæmorrhage." The patient died. In the same year Spencer Wells reported a case (*Obstet. Soc.*) He used a long piece of silk as an uninterrupted suture, leaving one end hanging out through the vagina; by pulling on this end the suture was removed after several days. In Dr. Grigg's case, performed at Queen Charlotte's Hospital this year (1875), at which I was present, Mr. Wells put in a continuous silken suture, not tying it at either end. It was found after death that the suture had worked its way out at the upper half. This was probably effected under the disturbance of severe vomiting and the relaxation of the uterine wall.

Greater, if not complete, security against effusion into the peritoneum is attained by stitching the uterus, *and* uniting it by the sutures to the abdominal walls. In this way the formation of adhesions is favoured. Dr. Hicks and M. Tarnier have each treated a case on this principle, and effusion was prevented. In the carrying out, it is necessary to bear in mind the contraction of the uterus, which going on for several days causes the organ to retreat towards the pelvis, gradually sliding down away from the abdominal wound. If, then, the uterine wound were united to the abdominal wound exactly on the same level, and tightly, there would soon be dragging upon the uterus, which might interfere with its involution, and excite inflammation. I think, however, this difficulty can be obviated.

Uterine suture should meet the following conditions :—1st, it should stop hæmorrhage from the cut surfaces of the uterus ; 2nd, it should secure fair apposition of the two lips of the uterine wound ; 3rd, it should keep the anterior wall of the uterus in apposition with the abdominal wall, so as to favour adhesion without causing dragging; 4th, it should admit of the easy removal of the sutures when they have done their duty.

I have designed a method of suture which I believe will answer all these indications. Fine silver wire is the best material. The needle, armed, is carried perpendicularly through the uterine wall about half an inch from the edge of the wound towards its upper part, so as to enter above any bleeding sinuses. The wire is then brought back through the wall from within outwards below the sinuses. This leaves a loop on the internal aspect. The effect of this, when the two ends are pulled upon, is to compress the sinuses somewhat after the manner of Simpson's acupressure. Next, the opposite side of the wound is pierced on the same level. This suture is made to pass through the loop of the first suture, before piercing the lower part of the wound to bring it out of the uterus. There will now, then, be two loops of wire which intertwine on the inner aspect of the uterus; and the four ends come out on the outer aspect to be carried presently through the abdominal walls. Before proceeding to this stage, it is necessary first to pass the loop of a silver wire over the crossing of the loops of the uterine sutures, and to carry the ends down through the os uteri and out by the vagina. This can easily be done by means of an eyelet probe like a long bodkin. The object of this is to keep a hold upon the sutures with a view to their subsequent removal.

The union of the uterine wound to the abdominal wound may now be effected. The four ends of the uterine sutures are now carried by needles through the abdominal walls, crossing each other, *i.e.*, the two ends emerging from the right side of the uterine wound are taken to the left side of the abdominal wound. The effect of this is, that when the sutures are drawn upon and secured by twisting outside the abdomen, not only is the uterus drawn up close against the inner abdominal wall, but the uterine wound is closed. To obviate the subsequent dragging from diminution of size of the uterus, it is better to pass the uterine sutures through the abdominal wall at a lower level, that is, nearer the pubes than where they emerge from the uterus.

These utero-abdominal sutures should not be made fast until the proper abdominal sutures are all *in situ*. Then, when

F F

you are prepared to close the abdominal wound, you may adjust
the utero-abdominal sutures. Should any dragging arise on
the second or third day, the utero-abdominal sutures can be
untwisted and slackened. They may be recognized by being
kept of greater length than the proper abdominal sutures. To
remove the utero-abdominal sutures, which may be done on
the seventh or eighth day, get an assistant to draw gently
upon the clue-wire brought down from the uterus by vagina

Fig. 120.

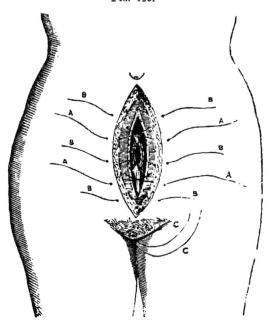

SHOWING THE UTERO-ABDOMINAL SUTURE.

A A, utero-abdominal sutures crossing in the cavity of the uterus. B B, abdominal sutures.
c c, guide-thread passed over crossing of utero-abdominal sutures, and coming out by vagina.

whilst a finger of the left hand follows it up to its connection
with the uterine loops, which can then be divided with
scissors worked by the right hand. The sutures can then
be withdrawn by gentle traction upon the ends which rest
upon the abdominal surface. To ensure accurate linear union
of the skin superficial sutures are necessary. The dressing is
the same as after ovariotomy.

After-treatment.—A full dose of opium should be given immediately, either in form of pill or suppository. Light nourishment and perfect repose are the things to be observed. The dressings should not be removed for five or six days. To obviate foulness, sprinkling with Condy's fluid or weak carbolic acid may be resorted to. The sutures may be removed on the seventh or eighth day. The bowels may be relieved by enema on the fourth or fifth day. It is desirable to wear a well-adjusted abdominal belt for some months after the operation.

The Dangers of the Operation; the Prognosis.

What is *the risk to life attending the Cæsarian section*, nume-rically expressed? I have already made some remarks upon this, the statistical aspect of the question. I doubt if any satisfactory answer can be given. Can any quasi-analogical deduction be drawn from the mortality attending ovariotomy? Take this to be one death to two or three recoveries; may we expect a similar result from the Cæsarian section—I mean, of course, when performed at a selected time? It would be rash to expect an equal success; and there is a consideration which I think, is generally neglected by statisticians. It is this: we are not justified in treating a given patient as an abstract entity, a mere arithmetical unit. Her fate is not to be decided by what are called statistical laws, but which are in reality too often nothing better than the accidental issues of blind gropings. We must study case by case; compare them; analyze clinically. "Non numerandæ, sed perpendendæ sunt observationes." We must weigh carefully all the conditions of the patient who is before us.

The principal risks run are as follows :—

1. If the operation is performed as the last resource after protracted attempts to deliver by other means, the woman is liable to sink from shock and exhaustion within a few hours; or if she survive beyond a few hours, there is the risk of hæmorrhage, of metritis, of peritonitis, and of puerperal fever. It may be said that the prospect of recovery, when the operation is performed under these circumstances, is very small.

2. If the operation is performed at a selected time, the

woman escapes the shock attendant upon the protracted labour, and encounters the shock of the operation with unimpaired strength. Still, the shock is very great, and is not seldom fatal *per se*. This is the first and most pressing danger. Could it be in any way modified or controlled, the Cæsarian section might be undertaken with more confidence. But shock necessarily attends all severe abdominal injury. It affects different persons in different degrees. Nor can we readily predicate of any given person that she will bear shock well or badly. It is an uncertain element, and must probably ever perplex all calculation as to the result of the Cæsarian section in any particular case.

3. The next danger is *hæmorrhage*, and as hæmorrhage is often associated with prostration as cause and as effect, the danger is serious. This may come on within a few hours. It might be expected that hæmorrhage would be liable to come from the inner uterine surface, as after ordinary labour, but the more common source is probably from the sinuses divided in the uterine wound. The quantity lost may be enough to cause a fatal anæmia. But the more common evil is from the irritation caused by the blood collecting in the abdominal cavity. This was the probable cause of death in a case I saw at the *Hôpital de la Faculté*, at Paris, in 1843, performed by Danyau and Malgaigne. There was no contraction of the uterus. It reached to the umbilicus. This may, to a great extent, be obviated by closing the uterine wound by suture, and uniting the uterus closely to the abdominal wall. By averting this danger we avert one cause of

4. *Secondary shock* and *peritonitis*. That secondary shock precedes peritonitis I have no doubt. Intense pain, even tenderness on pressure, rapid small pulse, accelerated and impeded breathing, suggest the diagnosis of peritonitis. This condition I have described as "Abdominal Shock." If at this stage the patient die and be examined, probably no trace of peritonitis, as revealed by redness or effusion, is discovered. Peritonitis may come on the day following the operation. It may be met by fomentations to the abdomen, by opiate suppositories; and the prostration soon ensuing must be combated with wine, brandy, beef-tea, chicken-broth. Salines are often useful, especially at first.

5. If the patient escape the preceding dangers, there is still the risk of septic infection, of *septicæmic puerperal fever*. The source of this is the absorption of septic matter from the cavity of the womb or from the edges of the wound; or it may arise from general blood-dyscrasia resulting from the accumulation in the circulation of effete matters which the excreting organs are unable to dispose of.

6. In addition to the dangers incident to the operation and to the puerperal state, there is the danger inherent to the disease which rendered the operation necessary, liable in some cases, as in cancer, to be aggravated by the operation, which may accelerate the fatal issue.

Winckel says that osteomalacia is much more unfavourable than rickets, in connection with Cæsarian section. Still, osteomalacic patients bear wounds well, and the power of repair is often great.

The uterus often contracts adhesions with the abdominal wall during repair. These adhesions do not appear to entail any serious inconvenience; and should pregnancy again occur, and the Cæsarian section be again necessary, they render the operation less dangerous (Meigs). The peritoneal cavity is shut off; the incision through the abdominal wall leads directly through the adhesions to the uterus. Thus the dangers of hæmorrhage, of effusions into the abdomen, are eliminated, and it is even probable that the shock is less. This most desirable result would be favoured by uniting the uterus to the abdominal wall at the time of the operation. On the other hand, no adhesions may be found, and the wound in the uterus may heal so completely that years afterwards no trace of cicatrix is found (Radford). Again, there may remain a marked cicatrix, with freedom from adhesions, as in Dr. Newman's case. (*See* Fig. 121, p. 438.)

Several cases are now known in which the Cæsarian section has been performed twice, thrice, and even four times on the same woman.

These cases of repeated success would seem to indicate a special tolerance of severe injury in the subjects, and cannot wisely be taken as evidence, absolute or cumulative and statistical, in reduction of the danger of the operation. This is further illustrated in the following history:—Dr. Freericks

(*Nederl. Tijdschr. v. Geneeskunde*, 1858) performed the section on account of contracted pelvis. Mother and child recovered. When again pregnant, premature labour was induced about the eighth month. When labour had begun, collapse set in; the uterus had ruptured. The child was removed by abdominal section. Vomiting caused extrusion of the intestines. To

FIG. 121.

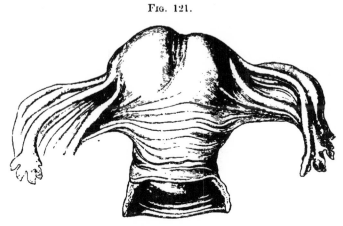

DR. NEWMAN'S CASE OF CÆSARIAN SECTION.   HALF-SIZE.

Deep sulcated cicatrix remained seven years after the operation.   The preparation is in St. Thomas's Hospital Museum.

effect reposition, numerous pricks were made in them to let gas escape, without effect, until an incision was made with a bistoury, and much thin pappy matter escaped. The intestines were then replaced, and the wound was closed. She recovered completely. How many women would be as tolerant?

Symphyscotomy.

Symphyseotomy is the operation by which the pubic joint is divided, so as to permit of the enlargement of the pelvic brim and cavity, by forcibly separating the bones. Its *rationale* rests upon the assumption that the joints soften and yield under the influence of pregnancy. We find the natural analogue of this proceeding in the guinea-pig, whose pubic joint opens widely during parturition. The operation is never mentioned in English works, unless to be condemned. And I am not aware that it has of late years been performed anywhere. The subject is well discussed by Scanzoni and by Lovati (*Del Parto meccanico*). Sigault, who gave his name to the operation, said that an inch was gained by it. Leroy gained two and a half inches. Baudelocque and Desgranges, in their experiments, gained a space between the pubic surfaces of two and a half inches. They say that every half inch of distance between the pubic surfaces gives one line of conjugate diameter; therefore two inches gives four lines; then, if the parietal protuberance is made to project into the space between the pubic bones, you get two lines more; the oblique diameters gain eight lines; the transverse gains half the distance between the pubic bones, that is, one inch. The operation might therefore be available in cases where a gain of six lines will enable a live child to pass, or where a head impacted in the pelvis might be liberated without perforation. In either case extraction by forceps would still be necessary.

Symphyseotomy would stand in competition with the Cæsarian section and with embryotomy, its object being to save the child. Its dangers have probably been exaggerated. Experience is too limited to warrant any precise estimate of its merits. It will probably not again be brought into practice as an alternative for embryotomy or the Cæsarian section. Scanzoni says the only case in which it is justifiable is where the mother dies in labour, the child being partly born, and extrication difficult without enlargement of the pelvis.

LECTURE XXV.

THE INDUCTION OF PREMATURE LABOUR—THE MORAL BEARING
OF THE OPERATION—THE FITNESS OF THE SYSTEM AND OF
THE GENITAL ORGANS FOR PREMATURE LABOUR—THE IN-
SUFFICIENCY OF SIMPLY PROVOCATIVE MEANS—TWO STAGES
OF PREMATURE LABOUR ARTIFICIALLY INDUCED—THE PRO-
VOCATIVE; THE ACCELERATIVE—DISCUSSION OF THE VARIOUS
PROVOCATIVE AGENTS: DANGER OF THE DOUCHE—ACTION
OF THE VARIOUS DILATORS—THE MODE OF PROCEEDING—
PROVOCATION OVERNIGHT—ACCELERATION AND CONCLUSION
OF LABOUR NEXT DAY—DESCRIPTION OF CASES DEMANDING
INDUCTION OF LABOUR—QUESTION OF STARVING THE MOTHER
—MODE OF DETERMINING EPOCH OF GESTATION—PROCEEDING
IN CONTRACTED PELVIS OR OTHER MECHANICAL OBSTRUC-
TIONS — IN CASES OF URGENT DISTRESS OF MOTHER, AS
CONVULSIONS, HEART DISEASE, PHTHISIS—IN CASES WHERE
THE CHILD IS DEAD—MISSED LABOUR—CARE OF CHILD WHEN
BORN ALIVE.

WE now come to an operation which carries us fairly back
within the domains of Conservative Midwifery. The induction
of premature labour is designed to save the mother and child,
or at least the mother, from those perils which one or both
would have to encounter at or before the natural term of gestation. In many cases those perils increase in an accelerated
ratio with the advance of gestation. By anticipating the ordinary epoch of delivery, by selecting a time when these perils
have either not yet arisen or are still comparatively small, we
may make the labour auspicious, indeed, natural in everything
but in the moment of its occurrence. In many other cases we

may obtain an equally auspicious result by commanding the entire course of the labour, overcoming certain difficulties by appropriate proceedings. I shall show that it has been too much the custom to consider that our resources are limited to the first order of cases, to leave too much to accident; and that, by taking the whole conduct of labour into our own hands, we may greatly extend the application of this most beneficent operation, save much suffering, and greatly add to the probability of preserving the lives of mother and child.

We have thus three great conservative operations. Two of these, the forceps and the induction of premature labour, have been contributed by the London School.

Denman tells us that he learned from Dr. C. Kelly, that "about the year 1756 there was a consultation of the most eminent men at that time in London to consider the moral rectitude of, and advantages which might be expected from, this practice." It met with their general approbation; and under this sanction the operation was resorted to with success in many instances.

It is happily no longer necessary to prove the moral rectitude of the practice. Its justification rests upon the same basis as that from which the whole art of Medicine derives its authority. Its design and its general effect are to save life, in many cases the lives of both mother and child; and in the rest, where the child cannot be rescued, to increase at least the chances of safety to the mother. The moral aspect of the question is now reversed; the accuser and defendant have changed places; it rests with those who neglect that which will rescue a woman and her offspring from impending danger, who suffer one or both to drift to destruction, to justify their neglect.

We may therefore proceed at once to discuss what are the advantages to be derived from the practice.

A preliminary question arises, the solution of which is necessary to the right appreciation of what can be gained by inducing labour prematurely, and of the means of accomplishing this purpose. This question, one which has been almost wholly neglected hitherto, is—What is the condition of the uterus and the system generally in reference to its fitness to assume the work of parturition prematurely?

First, as to the fitness of the general system to enter upon labour and the puerperal state. Upon this point little need be said, since we must be content to accept the conditions as they exist at the time of our selection for the operation. We cannot materially modify them. Experience, moreover, has amply shown that the system is fairly competent to assume the duties cast upon it at any time after the end of seven months. To apply the physiological formula of the Genesial Cycle so beautifully described by Tyler Smith, we observe that when gestation is brought to a term, the breasts enter upon their office, milk is secreted for the nourishment of the infant, and the uterus, thrown out of work, undergoes involution. These processes are usually carried on with scarcely less efficiency when labour occurs prematurely.

Almost all the consequences of labour at term may follow premature labour and even abortion : inflammation and abscess of the breast ; peritonitis, including pelvic peritonitis and pelvic cellulitis ; thrombosis, including phlegmasia dolens ; and all the forms of puerperal fever. But experience does not seem to indicate that these complications are in any sensible degree more likely to attend premature labour.

Secondly, as to the fitness of the parturient organs. Here the case is widely different ; for, any lack of efficiency in these must be made up by art. *When labour comes prematurely, the uterus is overtaken in an imperfect state of development.* It is taken by surprise. This implies imperfection in the contractile power of the body of the uterus, and greater resistance in the cervix. And the normal supply of nervous energy stored up in preparation for the work of labour is not yet complete. (See my "Lumleian Lectures," *Lancet,* 1874). Hence there is another source of difficulty in provoking contraction. It is true that the child, the body to be expelled, is smaller, and that in this way the balance between power and resistance is to some extent restored. But this is certainly not always so. It frequently, nay commonly, happens that the uterus is slow to respond to the unexpected call made upon it. It is but reasonable to anticipate that help will often be useful. And help can be given both to facilitate the dilatation of the cervix, and to supplement the contractile energy, if this cannot be brought into play.

Now, the expediency of giving this help, and the means of

doing it, have been almost entirely overlooked. Action has been limited to attempts at provoking the uterus to expel its contents, leaving the rest very much to chance. The consequence too frequently has been that the child has been born at some unforeseen, inopportune time, before aid could be procured, and has perished from one of those accidents, such as preternatural presentation or descent of the cord, which are so likely to occur in premature labour. Thus, supposing it was determined to bring on labour at the eighth month by detaching the membranes, by puncturing them, or by inserting a bougie in the uterus: this done, it has been considered that there was nothing to do but to wait patiently until active labour should set in, when the medical attendant should be sent for. Now, this may come to pass in twelve hours, in twenty-four hours, or in two, three, four days, or even later. There is no certainty about it. When labour comes, the child is expelled with little warning, almost suddenly, and before the medical attendant can be fetched. And it has to run the gauntlet of all those perils which especially surround premature labour unaided.

Does it not follow that it is desirable to keep a control over the whole course of labour, to take care that nothing adverse to mother or child shall happen in our absence, to substitute, in short, skill and foresight for accident? Few, perhaps, will hesitate to answer this question in the affirmative. But another question must follow. Can we so regulate a provoked labour throughout as to limit and define the time expended; and to conduct the delivery so as to give more security to the child and to the mother? This also I am prepared to answer in the affirmative. Repeated experience justifies the declaration made by me in 1862, "That it is just as feasible to make an appointment at any distance from home to carry out at one sitting the induction of labour, as it is to cut for the stone." The operation may be brought entirely within the control of the operator. Instead of being the slave of circumstances, waiting anxiously for the response of Nature to his provocations, he should be master of the position.

Assuming, then, that it is both desirable and possible to control and regulate the entire course of a labour prematurely induced, let me describe the method after which the proceeding should be conducted.

The act of artificial labour may be divided into two stages.

The first stage is provocative and preparatory. This includes some amount of dilatation of the cervix uteri, and implies a certain amount of uterine action, and lubrication of the cervix and vagina.

The second stage is the accelerative or concluding stage. It consists in the expulsion or extraction of the fœtus and placenta.

The ordinary modes of conducting an induced labour almost ignore the last stage, or the means of accelerating delivery.

Dividing the agents at our command for effecting delivery at will into the *provocative* and the *accelerative*, let us first examine *the means we possess of provoking labour.* These are numerous. In a course of lectures designed to be practical rather than historical, it is not desirable to discuss them in detail. I have endeavoured to do this in memoirs "On the Indications and Operations for the Induction of Premature Labour, and for the Acceleration of Labour" (*Obstetrical Trans.*, 1862, and *St. George's Hospital Reports*). It may be stated, as a general fact, that all the means employed act by stimulating the spinal centre to exert itself in causing contraction of the uterus. Some of these agents act directly upon the spinal marrow, being carried thither in the blood. Such are ergot of rye, borax, cinnamon, and other drugs. Quinine will provoke labour. Dr. Lewis A. Sayre (*American Practitioner*, 1871) says he brought on labour by it. Dr. Angelo Monteverdi says Peruvian bark is superior to ergot. Quinine in five-grain doses every half-hour provokes uterine contractions and labour. The diastaltic system is probably more active in Europeans in hot climates. We seldom find quinine efficient in provoking labour in this country; but my friend Dr. Cockburn, who has had long experience in India, says it is very unsafe to give quinine to pregnant women there. It is a popular belief, shared by some medical men, that iron will provoke labour, and charges of attempt to procure abortion have been founded upon the administration of this metal to pregnant women. Sound clinical observation does not support this belief; and certainly iron cannot be depended upon to provoke labour. Some agents evoke the energies of the diastaltic system by stimulating various peripheral nerves. Such are rectal injections, the vaginal douche, the colpeurynter, the carbonic-acid douche, probably the irritation of the breasts by sinapism and the air-pump, the cervical plug, the separation of the membranes, the placing a flexible bougie in the uterus, the intra-uterine injection, the evacuation of the liquor amnii, and galvanism.

The artificial dilatation of the cervix, the evacuation of the liquor amnii, and the intra-uterine injection act in a more complicated manner, and not simply through the diastaltic system. Some of the above agents are altogether uncertain and untrustworthy; some are in a high degree dangerous; and some are both efficient and safe. Ergot, borax, cinnamon, and all other drugs may be dismissed on account of their uselessness or uncertainty. Ergot is not only uncertain, but when it acts it is liable to prove fatal to the child. Rectal injections may be harmless, but cannot be relied upon. Irritation of the breasts often fails, and it is liable to be followed by inflammation and abscess. The *vaginal douche* (Kiwisch's plan), which consists in playing a stream of water against the cervix uteri, is often tedious, and is not free from danger. It requires to be repeated at intervals during one, two, or more days. It is liable to cause congestion of the lower segment of the uterus. Serious shock, metritis, and death have followed. It was advocated by Tyler Smith in this country, until he encountered a fatal result, which he communicated to me.

The *intra-uterine douche*, sometimes described as Kiwisch's plan, was in reality recommended by Schweighäuser, in 1825, and practised by Cohen in 1846.* It is known in Germany as Cohen's method. It was recommended by Schweighäuser as a better means of detaching the membranes than the use of the finger or sound adopted by Hamilton. Cohen thought the injected fluid acted, not by detaching the membranes, but through its being absorbed by the surface of the uterus. Professor Simpson† says that he at first used the vaginal douche of Kiwisch, but "he soon found it a simpler and more direct plan to introduce the end of the syringe through the uterine orifice;" he became convinced that the douche was liable to

* "Neue Zeitschrift für Geburtskunde," Band xxi.
† "Obstetric Memoirs and Contributions," vol. i., 1855.

fail, unless the injected fluid accumulated in, and distended the vagina, so as to expand that canal and enter the os uteri; and that its efficiency was great in proportion to the extent to which it separated the membranes.

The intra-uterine douche, although more certain, is even more dangerous than the vaginal douche. Lazzati relates two fatal cases. Taurin saw, in January, 1860, in Dubois' Clinique, such grave symptoms follow, that death was apprehended. Salmon, of Chartres, related to the Académie de Médecine (July, 1862) a fatal case. Depaul communicated to the Parisian Surgical Society (1860) a case of death occurring suddenly from the uterine douche. A gurgling noise attended the use of the instrument. Air escaped on cutting into the uterus for the purpose of extracting the child by post-mortem Cæsarian section. The uterine tissue was bright red; the blood frothy. Blot had to deplore a similar accident in the Clinique d'Accouchements. Tarnier relates two similar cases. Esterlé relates a case[*] in which serious obstruction to the cardiac circulation, ending in death, occurred. It may be asked, How is it that the injection of a stream of water into the vagina or uterus can prove fatal? The cases cited, and they are by no means all that are known, leave no doubt as to the fact. It seems to me that danger results in three ways. The first is by *shock*. When water is injected into the gravid uterus, it can only find room by stretching the tissues of the uterus. This sudden tension is the cause of shock. It has been supposed that some of the fluid finds its way through the Fallopian tubes into the peritoneum. And the following case, related by Ulrich,[†] suggests another solution:—

"H. W., aged twenty-nine, was, at the end of her second pregnancy, carrying twins. Three vaginal douches were used to accelerate labour, the last one by a midwife. The temperature of the water was 30° R. The 'clysopompe' was used. Eight hours after the injections had been going on, the patient got up in bed, and instantly fell down senseless, and died in a minute at most, with convulsive respiration-movements and distortion of the face. Five minutes afterwards, crepitation was felt on touching the body. Venesection was tried in the

[*] "Annali Universali di Medicina," March, 1858.
[†] "Monatsschrift für Geburtskunde," 1858.

median vein. Only a few drops of blood came. On section, the cerebral sinuses were found full of dark fluid blood; the membranes not very hyperæmic; the brain normal. The heart was lying quite transversely, the left ventricle strongly contracted, the right ventricle quite flaccid; the coronary vessels contained a quantity of air-bubbles. The left heart contained scarcely any blood; the right had a little; it was quite frothy."

This is illustrated by a case told by J. C. Dalton (*Brit. Med. Journ.*, 1860). A gutta-percha tube was used to puncture the membranes and procure abortion. The woman fell back and died. Air was found in the veins and heart. It was believed by Mr. Dalton that air was blown in.

It is, probable, then, that air may get into the uterine sinuses.

Sir J. Simpson relates the following :*—"He had been greatly alarmed by seeing a patient faint under an injection, probably from some of the fluid getting into the circulation. And he had seen two more alarming cases still, where both the patients died. In both, only a few ounces of water were injected; and yet rupture of the uterus took place. The occurrence of the rupture was to be explained by the fact that the uterus, being already fully distended, could not admit a few ounces of fluid without being stretched and fissured to some extent; and during labour these slight fissures might easily be converted into fatal ruptures. In one case the patient died before labour was completed; in the other, in twelve hours after its termination."

Another objection urged by Dr. Simpson is, that in injecting water we have no control over the direction it will take in the uterine cavity, and that the placenta may be detached. Cohen's cases show that this accident may happen.

It is also apt to displace the head, and cause transverse presentation.

Of course no degree of efficiency could justify the use of a method fraught with such terrible danger. But the douche does not possess even the merit of certainty. It has been repeated many times during several days before labour ensued. Lazzati, having tried it in thirty-six cases, found that the number of injections required ranged from one to twelve; the quantity used was about forty pints; the duration of the

* " Edinburgh Medical Journal," 1862.

injections was from ten to fifteen minutes at a time ; the temperature of the water 28° to 30° R. The time expended from the first injection to labour varied from one to fourteen days, the average being four days.

It has also been found that a large proportion of the children were lost.

The douche, therefore, whether vaginal or intra-uterine, ought to be absolutely condemned as a means of inducing labour. I think it necessary to repeat this emphatically, because, notwithstanding the warnings conveyed by many fatal catastrophes, I find that the use of the method is still taught and practised.

Mr. James, formerly surgeon to the City of London Lying-in Hospital, described (*Lancet*, 1861) a plan of intra-uterine injection which he had practised since 1848. He passes an elastic male catheter to the extent of four or five inches through the os, between the uterine wall and the membranes, and then injects about eight ounces of cold water. Of eight children, only two were still-born. Lazarewitch, of Charkoff (*Obst. Trans.*, 1868) has explained, modified, and given more precision to this method. He proves by observations and experiments, that the nearer to the fundus of the uterus the irritation acts, the more sure and speedy is the result, and *vice versâ*. He contends that the frequent failure of the douche was due to the stream not being carried much beyond the os. He found that when the stream was carried up to the fundus, one injection was commonly enough. He therefore introduces a tube as near to the fundus as possible, and then injects several ounces of water. The cases he relates (twelve in number) sufficiently establish his proposition, that this method is more sure than other modes of applying the douche ; but they are too few to prove that it is more safe. I feel very sure that, if it be at all frequently adopted, fatal catastrophes will ensue.

It may, moreover, be doubted whether, in cases managed according to the principle of James and Lazarewitch, the injection of water was not really superfluous. The passage of a catheter five or six inches into the uterus detaches the membranes along its course, and this, it has been seen, is usually quite enough to provoke labour. Why not, then, rest satisfied with that portion of the proceeding which is efficient and safe, and discard that which is superfluous and dangerous?

It is instructive to compare the histories of some cases of intra-uterine injection with those of accidental or intra-uterine hæmorrhage depending upon detachment of the placenta. Sudden severe pain in the abdomen at the seat of effusion, shivering, vomiting, collapse, are all observed in both cases. In the case of hæmorrhage, these are certainly not in proportion to, or due alone to, the loss of blood. They seem to be the direct effect of injury to the uterus from sudden distension of fibre. The uterus will *grow* to keep pace with developmental stimulus of a body contained in it; but *it will not stretch* to accommodate several cubic inches of fluid suddenly thrust into it. Yet this is what it is called upon to do when water is injected. If the water escape as fast as it enters, the shock may be avoided, but then the operation is liable to fail in inducing labour.

The injection of *carbonic acid gas* or even *common air* seems more dangerous still than the injection of fluids. Scanzoni has related two fatal cases from the injection of carbonic acid, and Sir James Simpson relates one where the patient died in a few minutes after the injection of common air.

Another agent is *galvanism.* Herder suggested this as a direct stimulant, to cause the uterus to expel its contents, in 1803. In 1844, Hörninger and Jacoby brought on labour by this agent. Dr. Radford showed the value of galvanism in labour and in controlling hæmorrhage. In 1853, I published* a memoir on this subject. I succeeded, in three cases, in inducing labour by it. But the method is tedious, and sometimes distressing to the patient. I have, therefore, abandoned it.

Another exciting or provoking agent consists in the insertion of some form of *plug or expanding body in the os or cervix uteri.* A great variety of contrivances for this purpose have been proposed and tried. It is unnecessary to describe the greater part of them. Those most in use are the sponge-tent, the laminaria-tent, and the elastic air or water dilator. There is no doubt labour can be induced by these agents. But it appears to me that their use to provoke labour is not based on

"Lancet," and " L'Union Médicale."

a rational view of the physiological or clinical history of the process. I agree with Lazarewitch, that irritants applied to the cervix are slow and uncertain. In most cases some further means, such as rupturing the membranes, will be necessary. The laminaria-tent is, however, extremely useful in expediting the dilatation and evacuation of the uterus in some cases of abortion.

The method known as Professor Hamilton's, which consists in *detaching*, by means of the finger or sound, *the membranes* of the ovum from the lower segment of the uterus, has the recommendation of safety; but it is uncertain in its operation.

The success that commonly attends the plan of *introducing a bougie into the uterus* between the ovum and the uterine wall is perhaps evidence of the truth of Lazarewitch's proposition, that irritation should be applied to the fundus. I find that the bougie should be passed at least six or seven inches through the os uteri in order to insure action. Probably, in many cases where it has failed, the bougie has only penetrated a short way. By passing the bougie gently, letting it worm its own way, as it were, it will naturally run between the membranes and the uterus where there is least resistance, turning round the edge of the placenta.

Some use an elastic catheter supported in its stilet, and withdraw the stilet when the catheter has been passed. The stilet converts the catheter into a rigid instrument, which is objectionable. An elastic bougie answers perfectly.

If a rigid instrument be used, there is great likelihood of rupturing the membranes; and, although this may happen at some distance from the os uteri, premature escape of the liquor amnii may follow. The bougie owes part of its efficacy, no doubt, to the necessary detachment of the membranes from the uterus; but not all, since it is found that labour more surely supervenes if the bougie be left *in situ* for several hours.

I believe this method is now the one most generally adopted. No other method combines safety and certainty in an equal degree.

Puncturing the membranes as a provocative of labour is practised in two ways. The direct puncture at the point opposite the os uteri is probably the oldest method of inducing labour. It is the surest. The immediate effect of draining off the liquor amnii is to cause concentric collapse of the uterine walls, diminishing its cavity in adaptation to the diminished bulk of its contents. This involves some disturbance, probably, in the utero-placental circulation. The parts of the fœtus come into contact with the uterine wall. Hence uterine contraction is promoted both by diastaltic excitation and by the impulse given by the concentric collapse.

In certain cases, the puncture of the membranes is the most convenient, as where the object is to lessen the bulk of the uterus, and ensure labour quickly. But it is open to the following objection :—It is an inversion of the natural order of parturient events. Some uterine action, lubrication, and expansion of the cervix ought to precede the evacuation of liquor amnii. If this order be not observed, the child is apt to be driven down upon the unyielding cervix, and the uterus still contracting concentrically, compresses the child and kills it. And this is all the more likely to happen in premature labour from the greater liability to shoulder-presentation and descent of the funis.

This objection is to some extent obviated by a modification of this method. Hopkins* recommended to pass the sound some distance between the ovum and the uterine walls, and then to tap the amniotic sac at a point remote from the os. By this mode it was sought to provide for the *gradual* escape of the liquor amnii. This operation may be regarded as a compromise between the direct evacuation of the liquor amnii and Hamilton's method of detaching the membranes. It is an important improvement, and is still successfully adopted in this country and in Germany.

Vaginal Dilatation.—In 1842,† Dr. Hüter described a method for exciting labour by placing a calf's bladder, smeared with oil of hyoscyamus, in the vagina, and distending it with warm water. This proceeding he repeated every day until labour set in, which usually happened in from three to seven days. Professor Braun‡ substituted a caoutchouc bladder, to which, from the purpose to which it was devoted, he gave the name

---

* " Accoucheur's Vade Mecum." Fourth edition. London, 1826.
† " Neue Zeitschrift für Geburtskunde," 1843.
‡ " Zeitschrift für Wiener Aerzte," 1851.

of *colpeurynter*. Von Siebold, Von Ritgen, Germann, Birnbaum, and others, adopted this modification. Another form of vaginal dilator is the air-pessary of Gariel. The earlier trials with this instrument seem to have been especially unfortunate, since six mothers died out of fourteen; and Breit saw inflammation of the genitals and death caused by it. I do not think these dangers are inherent in the method, if carefully pursued; but the principle of vaginal dilatation and excitation is certainly untrustworthy.

Direct Cervical Dilatation.—For the last fifty years various contrivances for mechanically dilating the cervix have been tried. The idea of dilating the cervix by sponge-tents was announced by Brünninghausen in 1820. This was again advocated in 1841 by Scholler. It has since been in constant employment at home and abroad. From personal observation I am in a position to affirm that this method is very uncertain as to time. Symptoms like those of pyæmia have ensued from the absorption of the foul discharges caused. This accident may possibly be obviated by the use of tents charged with antiseptic agents.

Osiander, Von Busch, Krause, Jobert, Dr. Graham Weir, Rigby, invented other forms of dilatatoria more or less resembling the urethral dilators which have lately come into use. These numerous contrivances attest the strength and prevalence of the opinion that it was desirable to possess a power of dilating the os and cervix uteri at will. The subject attracted the attention of Dr. Keiller, in Edinburgh, early in 1859, and in March of that year he, Dr. Graham Weir assisting, accelerated a labour which had been provoked by other means, by introducing within the uterus the simple caoutchouc bag, and gently distending it.

The case of Mr. Jardine Murray* is the first published case I am acquainted with in which fluid pressure was used to dilate the uterus to accelerate labour. It was a case of placenta prævia. Mr. Murray first detached the placenta from the cervical zone after my method, then introduced a flattened air-pessary between the wall of the uterus and the presenting surface of the placenta, and inflated by means of a syringe.

* " Medical Times and Gazette," 1859.

Dr. Storer published a case in 1859,* in which he introduced "the uterine dilator" within the cavity of the uterus. He especially insisted that the dilatation "was *from above downwards*." I saw inconveniences in the use of elastic bags, expanding inside the uterus, even more serious than those attending the vaginal dilator or colpeurynter of Braun. The cervix it was that required dilating, and a bag expanding below it in the vagina, or above it in the uterus, could only act upon it indirectly, imperfectly, and uncertainly. Besides, the uterine dilator seemed unsafe; during dilatation it must distend, stretch the uterine walls at the risk of injury and shock, and it was very likely to displace the head from the os uteri.

I had long felt the desirability of bringing the further progress of labour with placenta prævia, after having arrested the hæmorrhage by detaching the placenta from the cervical zone, under more complete control. I had always strongly insisted upon the danger of forcibly dilating the cervix with the hand, and before Mr. Murray's case was published I was engaged in devising an elastic dilator capable of expanding the cervix with safety. The first form I devised was an elastic bag, with a long tube mounted on a permanent flexible metal tube, having apertures at the end inside the bag. The metal tube served as a stem to introduce the bag inside the cervix, to keep it there, and to carry the water for distension. This form was modified and adopted by Tarnier, of Paris, and others, when I had abandoned it. I at length realized my idea by perfecting the fiddle-shaped bags, which are now in general use, and I have never yet seen any bags for cervical dilatation but those of my own contrivance. The constriction in the middle is seized by the cervix, whilst the two ends expanding serve to prevent the instrument from slipping up or down. This instrument imitates very closely the natural action of the bag of membranes. By its aid it is very possible, in many cases, to expand the cervix sufficiently to admit of delivery within an hour, although generally it is desirable to expend more time. I have completed delivery in five hours, in four hours, and even in one hour from the commencement of any proceedings. In many cases of placenta prævia where there was scarcely any cervical dilatation, I have

* "American Journal of Medical Science," July, 1859.

effected full dilatation in half an hour. And there is now accumulated a large body of evidence from numerous practitioners who have had equal success by aid of my instruments.

In the paper in the *Edinburgh Medical Journal* (1862) I proposed that the first step in the induction of labour should be the full dilatation of the cervix uteri, and after that to proceed to further provocation and acceleration. I related cases in which I began with dilating the cervix, afterwards rupturing the membranes, further dilating, and turning. I am now convinced that, although this rapid method is very feasible, and is even proper under some circumstances where prompt delivery is urgently indicated, it is desirable, under ordinary conditions, to prepare the uterus by some preliminary excitation.

The proceeding recommended.—Having discussed the various methods of provoking labour which have been practised, we are now in a position to select the most safe, convenient, and efficient. The plan I have successfully practised for some years is the following :—First, overnight pass an elastic bougie, No. 9 or 10, as far as it will go into the uterus, and coil up the remainder of the instrument in the fundus of the vagina ; it will thus keep *in situ*. Next morning some uterine action will have set in. The uterine neck and vagina will be found soft, and freely lubricated with mucus, and some degree of cervical expansion will have taken place. The bougie should be kept in its place until the child is ready to pass. In the afternoon, at an appointed time, you may proceed, if desirable, to *accelerative* measures.

Before rupturing the membranes, adapt a binder to the abdomen, and let this be tightened, so as to keep the head in close apposition to the cervix. This will often prevent the cord from being washed down by the rush of liquor amnii. Dilate the cervix by the medium or large bag, until the cervix will admit three or four fingers. Then rupture the membranes, and, before all the liquor amnii has escaped, introduce the dilator again, and expand until the uterus is open for the passage of the child. If the presentation is natural, if there is room, and if there are pains, leave the rest to Nature, watching the progress of the labour. If these conditions are not present, and one or other is very likely to be wanting, proceed with accelerative methods,—that is, to the forceps or turning; or, in cases where the passage of a live child is hopeless, to craniotomy. By pursuing this method we may predicate with great accuracy the term of the labour. Twenty-four hours in all—counting from the insertion of the bougie—should see the completion of the labour. The personal attendance of the physician during two hours is generally enough. But the mode of proceeding must vary according to the conditions of the case.* In many cases, it is desirable, being clear that the child is not imperilled by delay, to allow dilatation and expulsion to be effected as far as possible spontaneously. We must recollect that the safety of the child is best secured by obtaining such full dilatation of the soft parts that it may pass through easily and quickly.

What are the conditions that call for the induction of labour?

Gestation may be divided arbitrarily into two parts. During the first part, terminating at 6½ or 7 months, or at the end of 180 or 200 days, it is scarcely probable that a viable foetus will be expelled. To induce labour within this period is really to bring about abortion. It is, therefore, only done under the pressure of conditions that preclude waiting till the child is viable, and out of regard solely to the safety of the mother. Between 200 and 230 days is a stage of very doubtful viability, and the physician will still endeavour to postpone interference until after the latter date, when the operation may be undertaken with more confidence of saving both mother and offspring.

In a large proportion of cases we are able, within certain limits, to select our time. For example, where there is moderate pelvic contraction, admitting of the safe passage of a child a little below the full size, we may be justified in waiting until the end of eight months—say 250 days. The difficulty is to determine the starting point of the pregnancy. There is a very probable range of error of at least 15 days. If we count 15 days too many, we reduce the duration of pregnancy to 235 days—that is, we run the risk of falling within the first part, when the child is of doubtful viability. If, on the other hand, we count 15 days too few, we run the contrary risk of approaching

---

* For a series of cases illustrative of this practice, see "St. George's Hospital Reports," 1868.

too near the natural term of gestation, and of having a child too large to pass the narrow pelvis alive.

The best way, perhaps, of avoiding these two rocks is to reckon the pregnancy from the day after the cessation of the last menstrual period, the most probable time of conception. Count 230 days from that epoch, and add 20 days for a margin of safety. This will leave a full month, or 30 days, to complete the development of the child. The cases are few, if all the resources in the acceleration of labour are turned to account, in which a child of 250 days may not be delivered alive. But if we fall upon a child of 215 days or less, the chances of its surviving are very slight. I regard the error of procrastination as being generally of less moment than the error of anticipation. Of course, if the pelvic contraction is great—say to 2·50″—it will be prudent not to calculate beyond 240 days, but rather to incur the risk of bringing a non-viable child.

In every case the *first question* to weigh is whether you can postpone interfering until the child is viable. The cases of pelvic contraction which do not permit of this postponement are extremely rare. The answer is more difficult in those cases where the expediency of inducing labour comes under consideration on account of local and constitutional disease of the mother. In the first case, by waiting too long, the mother encounters greater danger, and we may have to sacrifice the child. In the second case the risk of sacrificing the mother is even greater.

It will be convenient to enumerate first those conditions which, in the interest of the mother, and disregarding the child, demand the interruption of gestation during the first part.

These are, A. Certain cases of extreme contraction in the bony or soft parts—*e.g.*, distortion and narrowing of the pelvis below 2·00″; the encroachment of considerable tumours, especially if they are unyielding, upon the pelvic canal; some cases of advancing and extensive cystic disease of the ovary; great contraction from cicatrices of the os uteri and vagina, not admitting of free dilatation; retroversion or retroflexion of the uterus not admitting of reduction; some cases of carcinoma of the uterus or vagina; some of tumours of the uterus.

B. Certain cases of urgent disease of the mother, depending upon and complicating pregnancy—*e.g.*, obstinate vomiting, with progressive emaciation, and a pulse persistent for some days above 120 ; some cases of advancing jaundice, with diarrhœa ; some cases of albuminuria, convulsions being present or apprehended ; some cases of insanity or of chorea ; hæmorrhages producing marked anæmia, especially if depending upon commencing abortion or placenta prævia ; some cases of disease of the heart and lungs, attended with extreme dyspnœa ; such are aneurism, great hypertrophy, valvular disease, œdema of the lungs, pleurisy.

If, in the presence of any of the foregoing complications, we have been fortunate enough to carry the patient over the first part of pregnancy, reaching the period when the child is viable, we may still be compelled to induce labour. The indications from disease beginning in the first part, as hæmorrhage, convulsions, cardiac distress, vomiting, jaundice, may grow more urgent, or they may arise during the second part.

My experience leads me to conclude that in cases of urgent disease there is more frequent occasion to regret having delayed the operation too long, than having had recourse to it too soon. When through obstinate vomiting, for example, nutrition has long been arrested, the starved tissues craving for supplies, and falling into disintegration, feed the blood with degraded and noxious materials ; the system feeds upon itself and poisons itself; the poisoned blood irritates the nervous centres, and these centres, wrought to a state of extreme morbid irritability, respond to the slightest peripheral, uterine, or emotional excitation. All nervous energy is thus diverted from its natural destination, and exhausted in destructive morbid action. Irritative fever ensues, the pulse rises to 140 or more. No organ in the body is capable of discharging its functions, for the pabulum of life is cut off at the very source. At this point labour, whether it occur spontaneously, as it often does, or be induced artificially, comes too late. The tissues are altered, the powers are impaired beyond recovery, and death soon follows delivery.

The most generally recognized indication is the presence of such a degree of pelvic contraction as to forbid the birth of a live child at term.

It was long thought that the end of inducing labour prematurely, namely, that of securing a child whose size should be
kept within the capacity of the pelvis, might be attained by
starving the patient, in order to arrest the development of the
child. The question has been even recently discussed in Italy.
Experience shows that little reliance can be placed upon this
method. Leonard Sedgwick relates (*St. Thomas's Hospital
Reports*, 1870) a case of obstinate vomiting, in which nutrition
was reduced to the lowest point compatible with life. Pregnancy ended in labour at term, when a healthy child was born
quite unaffected by the severe regimen of its mother. It
seems that the organism *in utero* will attract to itself all
it can get, even at the sacrifice of the parent. This is often
illustrated in phthisical women, who, almost dying of exhaustion, emaciated to the last degree, bring forth plump
children.

No one, I believe, disputes that, where we have the choice,
induction of labour should be performed where the ultimate
alternative is the Cæsarian section; and this rule should hold
whether the proceeding hold out a hope of saving the child or
not. It should also be resorted to for the sake of avoiding
craniotomy.

In the great majority of cases, we are led to determine upon
the expediency of inducing labour by the history of antecedent
labours. Where craniotomy has been performed on account of
contracted pelvis clearly recognized, there can be little ground
for doubt. But why should one or more children be sacrificed
in order to teach the physician that the pelvis is too small? Is
there no other gauge of the capacity of the pelvis than a child's
head? Of course it will be admitted that a woman pregnant
for the first time is equally entitled to the benefit of the premature induction of labour, if it be known that her pelvis is too
small. The difficulty is to know this. In this country, and
generally in private practice, the opportunity of making an
obstetric estimate of the pelvis before labour is very rarely
afforded. The first labour at term is, therefore, the common
practical test of a woman's aptitude for child-bearing. But on
the Continent, where a very large proportion of women are
delivered in hospitals, where they are received one or two
months before the end of pregnancy, examination of the pelvis

is made on admission, and thus they and their children come within the benefit of this proceeding.

The object to be attained is the reduction of one of the factors of labour into due relation with the other. The pelvis being a fixed quantity, the alternative is to bring the child through it at an early stage of its development. By referring to the tables giving the relations of the different degrees of contraction to the different operations at term and at seven months (*see* p. 82), it will be seen how the scale of operations, arranged in the order of their severity, may be slided down so that, when applied to labour at seven months, spontaneous labour supersedes the forceps, the forceps turning, turning craniotomy, craniotomy the Cæsarian section; so that nothing remains for the Cæsarian section.

The modifications proper to be adopted in different cases are as follows :—

1. In the case of pelvic deformity not admitting the birth of a live child at term.

There are three degrees of contraction to be considered. The *first* or least degree, say, giving a conjugate diameter of 3·50 in. In such a case a child of seven or eight months' development will probably pass without difficulty. Here it may be enough to provoke the labour, and watch its course, as in ordinary labour.

The *second* degree, giving, say, a conjugate of 3·00″. In such a case, unless the child prove very small or timely aid be given, its head may be delayed so long in the brim that it will be lost. Here it will be proper to provoke the labour by inserting the elastic bougie overnight; to accelerate the labour by dilating the cervix, rupturing the membranes, applying the forceps, or turning.

The *third* degree, giving, say, a conjugate below 3·00″, may admit the forceps, but it may be necessary to accelerate the labour by turning, or possibly by craniotomy.

A double advantage is gained by bringing on labour prematurely when the pelvis is greatly contracted. We not only secure a fœtal head that is smaller, but also one that is more compressible. During the last month of gestation, ossification of the cranial bones proceeds rapidly. Taking two heads, the one at eight months and the other at the full term of gestation, of equal size, the head of eight months' gestation will, on account of its less perfect bony development, come through the same contracted pelvis with more ease, or may even come through alive when the equal-sized head at term would have to be perforated. This is especially seen in those cases where turning is resorted to as an accelerative proceeding.

The course to pursue is as follows:—If the uterus act with sufficient power, and the pelvic contraction be not so great as to impede the passage of the child's head, and the cord do not fall through, watch and let Nature do her work. But if the head be delayed, or the cord fall through, we must intervene. There are two alternatives. We may first try the forceps. But if the conjugate is reduced to 3·00″, or below, turning is the true accelerative means. If I may trust my own experience, I should, without hesitation, say the prospect of a child being born alive under the conditions postulated is much better than under any other mode of delivery, and even better than is the prospect under turning in ordinary circumstances at the full period of gestation. The explanation is this:—The smaller and more plastic head is caught at the smaller or bi-temporal diameter between the projecting promontory and the symphysis pubis; the jutting promontory leaves abundant room on either side in the sacro-iliac region of the brim for the cord to lie protected from pressure; and, if care be taken that the cervix uteri be adequately expanded, the head comes through so quickly that the danger of asphyxia is not great. The mode of turning deserves attention. The object being to secure a quick delivery, the soft passages must be well prepared. We might turn by the bi-polar method, without passing more than two fingers through the os uteri. But I have found that, although it is always well to avail ourselves more or less of the bi-polar principle, it is desirable, in this case, to pass the greater part of the hand through the cervix to grasp the further knee. The reason is this:—The cervix that will admit the hand will, in all probability, permit the ready transit of the child. We thus secure adequate dilatation.

When the turning is completed extraction must follow. It should be performed gently, drawing upon the one leg until the breech has passed the outlet; the extraction of the trunk should be slow; and a loop of cord should be drawn down to take off tension. If the arms run up by the sides of the head they must be quickly liberated. The rules given for this operation (*see* pp. 206 *et seq.*) are here of extreme operation. By observing them, if the cervix is fairly dilated, they can be brought down in a minute or two. When the arms are liberated another difficulty arises; the neck of the child is in danger of being nipped in the circle of the cervix. This is the moment for acceleration. The two legs are held at the ankles by the left hand, whilst the right-hand fingers are crutched over the back of the neck. The head is sure to enter the contracted brim in the transverse diameter. It has then to describe the circle round the point of the jutting promontory, which I have defined (*see* Fig. 90, p. 252) as the "curve of the false promontory." Traction must, therefore, be at first carefully exerted in the direction of this curve or orbit—that is, well backwards—so as to bring the head round and *under* the promontory. When it has cleared the strait, and is in the pelvis, the occiput commonly comes forward, and traction is changed to the direction of Carus' curve, to bring the head through the outlet. Unless rigorous attention be paid to the above rule for bringing the head through the brim, so much time may be lost as to imperil the success of the operation.

In cases of extreme deformity, in which it is difficult or impossible to perforate or to seize a leg, if we have induced labour at six months, the fœtus may still pass, if we give time. After making a reasonable attempt to snare a foot, by manipulation and the wire-écraseur, if we leave the uterus to act for twelve or twenty-four hours, the child having perished and become moulded, some part of it, a foot or shoulder, will come within reach. This can be drawn down; the head can be perforated, and then traction will deliver. The placenta should also, if not following readily, be left for two or three hours to be expelled. By adopting this method, I delivered, in St. Thomas's Hospital, a six months' fœtus, in a case of great osteomalacic deformity, in which it was impossible to get two fingers through the brim. By thus calling Nature to our aid, and practising a little "masterly inaction," we avoided the Cæsarian section, and saved the woman. This course applies especially to cases of osteomalacia in which some amount of yielding or unfolding of the pelvic bones may be generally obtained.

In dealing with cases in which the induction of labour is indicated by urgent distress in the mother, we must again be governed by a careful estimate of the circumstances. There is no common rule.

Take first the case of *convulsions*. It has been seen over and over again that the convulsions have ceased soon after the uterus has been emptied. Everything conspires to prove that the convulsions are due to conditions arising out of the pregnancy. What, then, more logical than to terminate the pregnancy as soon as possible? Yet experience suggests caution as to the mode of acting. In not a few cases the completion of labour has failed to put an end to the convulsions. In other cases death has followed labour, whether this have occurred spontaneously or have been induced. Is the unfortunate issue the consequence of procrastination in inducing labour, or of over-haste of want of precaution in the mode of proceeding? I believe it is due sometimes to one cause and sometimes to the other.

The question of inducing labour before the actual outbreak of convulsions—that is, during the conditions that lead up to convulsions—does not often come practically before us. We have, therefore, mainly to do with the question how best to carry out a labour the indication for inducing which is clear. Is it to be done *citissime?* Is it to be done slowly and deliberately? I believe the latter principle is the more judicious. The proceedings should involve the least possible manual or other operative interference. The detachment of the membranes or the insertion of a bougie is too slow in results. It is better to puncture the membranes. This at once lessens the bulk of the uterus, and diminishes the pressure upon the abdominal vessels. If the convulsions remit, we may leave the labour to Nature. If urgent symptoms persist, we may dilate the cervix carefully by the cervical dilators, and accelerate by forceps, by turning, or even by craniotomy, according to the special indications. Before proceeding to any operation, even making an examination or passing the catheter, it will be wise to induce anæsthesia. In uræmic eclampsia the diastaltic susceptibility is so inordinately increased that this precaution is of extreme importance.

A similar rule applies in almost all cases where the induction of labour is indicated by urgent distress of the mother, such as heart disease, or chorea. In the case of dangerous vomiting in the early months, it will be useful, as a preliminary measure, to insert a laminaria-bougie as far as it will readily pass into the uterus. This will answer the double purpose of detaching the ovum and dilating the cervix.

In retroversion of the uterus, irreducible, and with urgent symptoms, the puncture of the membranes is the proper course. We immediately gain relief by permitting the concentric diminution of the volume of the uterus.

There is a series of cases in which the indication is *simply or primarily to save the child*. There are certain conditions which tend to destroy fœtal life before the term of gestation. If we can bring the child into the world before the anticipated period of its death *in utero*, we may hope, by bringing it under fresh influences, to save it. Denman gives the case of a woman who lost her children about the eighth month, a rigor preceding. He suggested the induction of labour. There are various diseases which are known to endanger the child as they advance. Such as hydrocephalus, syphilis of the child, fatty degeneration, hypertrophy, dropsy of the placenta. In cases where there has been no sufficient opportunity of treating the mothers before or during pregnancy, and where there is a history of labours ending in the birth of dead children, the induction of labour is indicated.

There are cases in which the wisest medical and ethical judgment is required. A woman pregnant about six months is dying of phthisis. Would it prolong her life or improve her condition if labour were induced? and should we be justified in sacrificing the child with that object?

A woman pregnant about seven months is dying of phthisis. The child is assumed to be viable; its life hangs upon the fragile thread of its mother's life, which may break before the natural term of gestation is accomplished. Are we justified in inducing labour to rescue the child, disregarding the mother? or is such a course likely to prolong her life, or to accelerate her death?

The decision in such cases is both perplexing and painful.

Observation of the course of things when pregnancy is complicated with phthisis, lends material help in arriving at a solution. It was long thought, and I believe some people still think, that pregnancy is antagonistic to the advance of phthisis. If this were true, the decision would be obvious. Let the pregnancy alone. But experience, I think, is adverse to the opinion that pregnancy exerts any beneficial influence upon phthisis. I am sure I have seen in numerous instances phthisis advance with accelerated speed towards a fatal issue when complicated with pregnancy, the sufferer either dying before the term, or sinking rapidly after labour.

It is an idea founded more, I am afraid, on imagination than on facts, that Nature, in her solicitude to perpetuate species, will struggle with unwonted energy to sustain the life of the expectant parent until the embryo is matured. Faith in this hypothesis would lead us to procrastinate.

Putting aside poetry not supported by facts, there are two considerations that offer material aid in arriving at a decision. First, pregnancy is commonly less trying to a phthisical patient than labour and childbed. The puerperal state especially throws such an increase of work upon the circulation, that the system often breaks down at this period. It is therefore desirable in the interest of the mother to postpone labour as long as possible.

Secondly, the prognosis in phthisis, even in cases apparently the most desperate, is often open to grave fallacy. Who has not seen patients whose days, whose hours almost were counted, survive for months and years? In the interest of mother and child, then, it is not wise to take precipitately the irrevocable step.

Fixing of the uterus by peritonitic adhesions, so that laceration or other mischief is apprehended during the development of the uterus, may necessitate induction of labour. There is a case in illustration which occurred at St. Bartholomew's, reported in the *Lancet* (May, 1871). But it is surprising how these adhesions may disappear under gradual stretching. Their vitality is quickly destroyed, and they undergo rapid atrophy.

Lastly, there are cases in which the induction of labour is indicated *to remove a dead child retained in utero.* Usually expulsive action will set in spontaneously within two or three days of the death of the foetus. It will rarely be postponed beyond three weeks; but cases are known where a much longer time has elapsed. There is a tendency to postpone the act of labour, no matter at what period the embryo may have perished, until the expiration of nine months. And in some rare cases of so-called "missed labour," one remarkable example of which came under my care, labour may be indefinitely postponed. In these cases the uterus will commonly have lost much of its active contractility and other properties peculiar to the pregnant state. Simple provocative means may be totally inefficient. The cervix may have to be dilated by fagots of laminaria-tents, and the child be extracted piecemeal, perhaps at several sittings.

Care of the child.—Where we expect to deliver a live child we must be prepared with means of rescue from asphyxia. (*See* Lecture XI.) And if we have the good fortune to secure a living child, special care will commonly be necessary to rear it. Born before its time, its development is necessarily imperfect. It has a diminished power of resistance to external impressions. Its heat-producing capacity especially is defective. It will depend very much more than the mature infant upon a supply of warmth from without. It should be carefully surrounded with good non-conductors of heat; kept in bed by the side of its mother or nurse, or by the fire; care should be taken that the air it breathes should be warm—not less than 70° Fahr. If it must be kept in a cot, put a hot bottle by its side, and do not let the nurse expose its surface to the cooling influence of evaporation by washing it too soon or in the usual way.

I will conclude this subject by recalling attention to the rule urged by Denman and his contemporaries, namely, that the artificial induction of labour should only be undertaken after deliberate consultation. When we consider the many and weighty medical, legal, and moral questions involved in the arbitrary interruption of pregnancy, we shall see abundant reasons for seeking assistance in avoiding possible clinical error, and for sharing serious professional and social responsibility.

LECTURE XXVI.

UTERINE HÆMORRHAGE—VARIETIES: CLASSIFICATION OF—FROM
ABORTION—CAUSES OF ABORTION: MATERNAL, OVULINE,
AND FŒTAL—COURSE AND SYMPTOMS OF ABORTION: TREAT-
MENT—THE PLACENTAL POLYPUS—THE BLOOD POLYPUS—
THE PLUG—PERCHLORIDE OF IRON—PROPHYLAXIS—HYDA-
TIDIFORM DEGENERATION OF THE OVUM.

THE management of flooding involves so much operative treatment, that a course of lectures on obstetric operations cannot
be complete without a history of this subject. Hæmorrhage
is certainly one of the most frequent and most perilous of all
the accidents that threaten the pregnant woman. Often occurring without warning, in impetuous torrents, quick death from
shock and exhaustion may ensue. If not immediately fatal,
the drained and enfeebled system, ill adapted to resist injurious
influences, may sink in a few days from some form of puerperal
fever, from thrombosis, or other complication. And, if these
secondary dangers be escaped from, there are still to be
encountered the remote effects, sapping the strength, pervert-
ing nutrition, and predisposing the patient to various protracted diseases. I have seen almost complete blindness,
deafness, hemiplegia, and other forms of paralysis persist as
consequences of hæmorrhage.

The action of the physician must obviously, in the first place,
be fixed on the endeavour to save life from the imminent peril.
But his duty is far from ending here. He has to take measures to prevent the return of the flooding; to rally the patient
from the depression already present; and to secure her against

the secondary and remote consequences. He will draw the most valuable indications in practice from a careful clinical study of the various conditions upon which uterine hæmorrhage depends. He must act on the aphorism that every ounce, every drop of blood that he can save, will be useful in averting subsequent evil. I insist upon this, because I have observed that many men, perhaps most, are more afraid lest the energetic use of the most efficient agent in stopping bleeding may be more hazardous than the continuance of the bleeding itself. This dread, I believe, I know, is overstrained ; and I earnestly beg those who feel it, and whose hands are thereby paralyzed, to reflect that the dangers of hæmorrhage are great and certain, that the time for action is brief, that lost opportunity is irretrievable ; whilst the dangers apprehended from the prompt use of the great hæmostatic agent, salts of iron, are at most problematical. Whilst we are hesitating, the woman dies. The emergency justifies some risk. Even those who distrust the remedy should remember the Celsian maxim : " Anceps reme- dium melius quàm nullum." It is eminently applicable here. We should not choose to founder against Scylla because Charybdis lay beyond. We should try and clear Scylla at all hazards, trusting to skill and Providence to clear Charybdis too. If, therefore, means be pointed out of almost infallible efficacy in stopping hæmorrhage, thus avoiding the first rock, is it rational to withhold those means for fear of shock, air in the veins, thrombosis, septicæmia—rocks ahead indeed, but not necessarily in our course ?

For clinical purposes, hæmorrhages may be most conveniently divided into those which occur during pregnancy, and those which occur during and after labour.

The FIRST order of cases, *those occurring before labour*, in- cludes these three principal forms :—

1. The hæmorrhages during pregnancy, including those of abortion.

2. The hæmorrhages from placenta prævia.

3. The hæmorrhages, so-called " accidental," from premature detachment of the placenta, when not prævia.

1. *The Hæmorrhage of Abortion.*

Hæmorrhage occurring during early pregnancy is both a cause and a symptom of abortion.

Under normal conditions, the relations between the ovum and the uterus, both in structure and function, are so harmoniously balanced, that the abundant supply of blood attracted to the uterine vessels by the developmental stimulus of the embryo finds its natural employment. The demand keeps pace with the supply, and the structures being healthy, there is no extravasation. But if this healthy correlation be in any way disturbed, hæmorrhage is likely to occur. Whatever produces a state of hyperæmia predisposes to hæmorrhage. In pregnancy this hyperæmia exists in a high degree. The developmental nisus acts as a *vis à fronte*, attracting blood powerfully to the vessels of the uterus. The hyperæmia is so great that a flaw anywhere will almost certainly lead to an escape of blood. The following defect is the most frequently observed :—*A morbid condition of the mucous membrane*, combined or not, with a morbid condition of the muscular wall of the uterus, and frequently attended by *abrasion or loss of epithelium.* The part thus affected is commonly the vaginal-portion of the cervix uteri. The morbid action here is aggravated by the physiological developmental stimulus; and at the menstrual epochs the additional stimulus of ovulation, attracting still more blood, the vessels, ill-protected by the diseased tissues, break down, and hæmorrhage results. Simple emotion, or physical shock, will often act in like manner, by directing a sudden excess of blood to the uterus. So long as the blood only comes from the os and cervix uteri, the embryo may be safe; but the hæmorrhage itself may be injurious, and a remedy should be applied. In such cases, lightly touching the abraded congested surface with nitrate of silver a few times at intervals of a week will commonly effect a cure. It may be feared that such local treatment will provoke abortion instead of preventing it. Experience, however, corroborates what physiological reasoning suggests, namely, that the cure of a diseased action in the organs of parturition will contribute to the security of the pregnancy. It is not desirable to let disease go on because a woman is pregnant.

An analogous form of hæmorrhage is that which arises from *the decidual cavity* during the first six weeks of pregnancy, *i.e.*, before the decidua reflexa and decidua vera have coalesced. The mucous membrane of the uterine cavity, being unhealthy, pours out blood which escapes externally. This involves far more danger to the embryo, because the extravasation is not likely to be strictly limited to the free surface of the decidua; it will probably spread to parts in connection with the chorion and young placenta. The embryo then perishes; and blood being retained in the substance of the placenta, and some insinuating itself between the decidua and the uterine wall, the uterine fibre, put suddenly on the stretch, is irritated, and spasmodic contractions, tending to expel the contents, are set up. This spasmodic action, itself a new source of local hyperæmia, adds to and keeps up the hæmorrhage.

There are, of course, many causes of abortion; but whatever the primary or predisposing cause, the immediate or efficient cause of abortion is extravasation of blood into the decidua, and between the decidua and uterine wall, leading to partial separation of the ovum. When things have gone so far that the integrity of the ovum is impaired by extravasation of blood into the structure of the decidua and placenta, abortion is commonly inevitable; the indication then is to accelerate the entire removal of the ovum.

I cannot in this place discuss minutely the numerous causes, maternal and ovuline, which lead to abortion. In dealing with threatening or present abortion, whatsoever the cause, the course of action is very much the same. But, as a guide to prophylaxis, a knowledge of these causes is important. They may be classified as follows :—

A. MATERNAL CAUSES OF ABORTION.

I. Poisons circulating in the mother's blood.

a. Communicated = Heterogenetic : malaria, fevers, syphilis, various gases, as CO, CO_2, lead, copper, mercury, and certain vegetable poisons.

β. Products of morbid action = Autogenetic : as bile - matters, albuminuria, carbonic acid from asphyxia, and in the moribund, etc.

II. Diseases degrading the mother's blood. { Anæmia, obstinate vomiting, over-suckling, albuminuria, lithiasis, cholæmia.

III. Diseases disturbing the circulation dynamically : as liver-disease, heart-disease, lung-disease, excess of vascular tension (the vessels of the uterine mucous membrane most naturally give way), engorgement of the portal system.

IV. Causes acting through the nervous system.
{
α. Some nervous diseases, as epilepsy, chorea.
β. Mental shock, physical or emotional excitement.
γ. Diversion or exhaustion of nerve-force: as from obstinate vomiting or convulsion.
δ. Excess of nervous tension.

V. Local disease.
{
α. Uterine disease: as fibroid tumours, inflammation, hypertrophy, imperfect involution, diseased decidua.
β. Mechanical anomalies: as versions, flexions, pressure of tumours external to uterus, adhesions of uterus.

VI. Abortion artificially induced. Blows, squeezing, wounds, direct interference with the ovum. Generally, the causes which produce rupture ; abortion is an escape from rupture.

VII. Adolescent and climacteric abortion.

B. The Fœtal Causes of Abortion.

I. Diseases of the membranes of the ovum, primary or secondary upon diseases of the maternal structures or blood, as—

Fatty degeneration of the chorion or placenta.
Hydatidiform degeneration of ,, ,,
Inflammation, congestion of ,, ,,
Apoplexy of ,, ,,
Fibrinous deposits in ,, ,,
Vicious attachment of the ovum, as in placenta prævia.

II. Diseases of the embryo : Faults of development—
α. Malformation.
β. Inflammation of serous membranes.
γ. Diseases of nervous system.
δ. ,, of kidney, liver, etc.
ε. Mechanical, as from torsion of the cord.

In short, anything causing the death of the embryo. Often the causes are complicated, arising partly from the maternal, partly from the fœtal side ; and often it is difficult to unravel these, or to discover the efficient cause.

The careful study of the etiology of abortion has a use beyond that of guiding prophylaxis. In a given case of threatening abortion, it will often furnish important indications. For instance, in some cases it is desirable to endeavour to avert the abortion, to arrest the process ; but where certain diseases, as fever, severe blood-change, or conditions causing dynamical disturbance of the circulation exist, abortion is a means by which Nature seeks relief from the double burthen.

This double burthen may overwhelm the powers of the system. By eliminating one burthen, the pregnancy, the case is reduced to its most simple expression ; and the patient may successfully struggle through the uncomplicated disease. In such a case, the indication is to second Nature by accelerating the accomplishment of the abortion.

A similar indication arises in those cases in which there is local disease rendering the continuance of pregnancy hazardous or impossible, as retroversion, tumours, etc.

Also in certain cases of embryonic disease, where the previous history, or our knowledge of the state of the uterus or its contents, establishes a strong presumption that the embryo is either dead or is not likely to survive.

Bearing these things in mind, let us trace the *ordinary course and symptoms of abortion.* These will vary somewhat, according to the cause and condition of the ovum and uterus. The two great symptoms are pain and hæmorrhage. There is reason to apprehend abortion if, after a woman has missed one or more monthly periods, she complains, about a menstrual epoch, of unusual heavy aching lumbar pains, followed by spasmodic or colicky pains in the lower abdomen and pelvis ; and still more if a discharge of blood takes place. But these symptoms may pass away under rest, and the pregnancy go on. When this occurs, we may infer that there is no disease of the fœtus or its membranes incompatible with its life ; and that the symptoms are due to the recurrence of the ovarian stimulus which is the physiological cause of menstruation.

The hyperæmia attending the menstrual nisus may be the cause of extravasations of blood into the decidua and between the villi. There, coágulating and hardening, through the removal of the watery part of the blood, it may help to form a solid mass, which may retain connection with the uterus for some time longer, and be eventually expelled as a *fleshy mole.* No trace of an embryo may be discovered ; but a careful microscopical examination will reveal the elementary constituents of the ovum, especially chorion-villi. This minute examination is of great importance. The naked eye may fail to distinguish a compressed ovum infiltrated with blood from a fibroid tumour or polypus, or from a compressed blood-clot. I have known a fibroid expelled spontaneously to be mistaken for an abortion. It is needless to point out how grave might be the social consequences of such an error.

If things are going on to abortion, the hæmorrhage will continue, perhaps in great abundance. Blot observes that where there is albuminuria the hæmorrhage is greater. The plasticity of the blood is probably impaired. Pains of a forcing or expulsive character persist, until amongst the discharges, the embryo, enclosed or not in the amnion and chorion and decidua, is found. In early abortions, occurring at from six weeks to two months, the embryo will often pass enveloped in the chorion. The decidua, or mucous membrane of the uterus, may adhere for some time longer; and so long as it remains, the hæmorrhage and pain, and the danger last.

In abortions occurring at three or four months, the embryo is sometimes expelled alone in the first instance, the ovum being burst under the contractions of the uterus. Then, the membranes, amnion, chorion and decidua, and placenta follow. But it may happen that the ovum is not burst, and then the whole ovum will come away in one mass. This is more likely to be the case when the embryo has perished some time before its expulsion, and when the process of retrogression of the media between uterus and placenta has made some progress. Some diseased ova are the most liable to be thrown off in this entire form. But if the abortion be the result of hydatidiform degeneration of the chorion, it is likely that only a part of the diseased structure will come at a time. Then, hæmorrhage will still continue. You must carefully examine what has passed, to see if there is any appearance of parts being torn off, giving reason to infer that more remains behind. You must also, in every case, examine *per vaginam* to ascertain if there is anything in the uterus.

The first question of a practical nature is, *Can the abortion be averted?* Can gestation continue?

If you think abortion may be averted, the methods of allaying nervous excitement must be called into exercise. I must refer to the lecture on "Rupture" (Lecture XXI., p. 325) for a discussion on the value of "epichontocics" or "amyosthenics." These agents may here find useful application.

If you have found any portion of the embryo or membranes in the matters expelled, you must give up the hope of this.

If you find very active expulsive pains, attended with free hæmorrhage, affecting the patient's strength, and if the os uteri is opening so as to admit the finger freely, you may, even if no portion of the ovum has been expelled, give up the hope of averting abortion.

The sooner the ovum, or the remains of it, are voided, the sooner the patient will be out of danger. This, then, is the first indication. *Empty the uterus.* How? Sometimes you will feel the ovum projecting partly through the os. It feels like a polypus. In this case it is probably detached, and under proper manipulation it will come away. The method of proceeding is as follows :—The patient placed on her back or on her side, with the thighs flexed so as to relax the abdominal walls, you press with the palm of one hand above the symphysis pubis upon the fundus of the uterus, so as to depress the organ well into the pelvic cavity. By this manœuvre you carry the os lower down, making it more accessible to the finger passed internally, and you support the organ by providing counter-pressure. This singularly facilitates the penetration of the cavity by the finger, which must generally pass to the fundus, in order to get hold of the ovum and scoop it out.

Generally, the uterus, in abortion, is low in the pelvis ; and the vagina being relaxed, it is not difficult to reach the os. There is a characteristic dilatation of the fundus vaginæ present whenever the uterus is acting to expel something with hæmorrhage. This is of importance in diagnosis. But not seldom, although the os can be reached, it may be necessary, in order to command the uterus and its contents, to pass the hand into the vagina. This is a very painful proceeding, and calls for anæsthetics. This is more especially the case if the ovum adheres ; then, the finger must be swept well round the cavity of the uterus—even at the fundus. The decidua may thus be broken up, and only pieces will come at a time.

Various ovum-forceps have been devised for the purpose of seizing the ovum and bringing it away. Levret and Hohl, amongst others, invented forceps of this kind, and Stark contrived a form of spoon. I have used several of these instruments; but, after trial of them, I prefer the hand. The hand gives you information of what you are doing as you go on; and it acts by insinuating the finger between the ovum and the uterus, peeling one from the other, not by avulsion, as the forceps must do. Besides, you know, if you use your fingers, exactly what has been accomplished, if any part of the ovum has been left behind, and you avoid the risk of injuring the uterus. But where the vulva is small, and the ovum is projecting through the os externum, forceps guided by the finger may be useful.

I have said that there is no security against return of hæmorrhage whilst the ovum or a portion remains behind. But there is an exception to this rule. I have frequently observed in cases where it was not possible to bring away the whole ovum, that the remains, if much broken up, did no harm, and that the hæmorrhage ceased. The *débris* gradually become disintegrated; and involution goes on unimpeded.

Sometimes a portion of decidua and placenta adheres so intimately that it cannot be removed by the fingers. Projecting into the uterine cavity, it forms the *placental polypus*. If hæmorrhage continue, this may be removed by the wire-écraseur, which will shave it off smoothly from the surface of the uterus. I believe this is the best and safest way of dealing with these masses. There is another condition to be considered here. Sometimes, blood collecting in the cavity of the uterus gets compressed, so that the fluid portion being squeezed out, the fibrin forms a firm body, which may adhere to the wall of the uterus. This is the *blood or fibrin polypus*. Kiwisch thought it might arise from a kind of uterine apoplexy, independent of conception. But Scanzoni is probably right in regarding it as consequent on abortion. The stalk is commonly formed of a shred of decidua adhering to the uterine wall. Just like an ovum or bit of decidua, it is a cause of hæmorrhage by irritating the uterus, inducing spasm or colic. It can commonly be scooped out by the finger, although preliminary dilatation of the cervix by laminaria-tents may be necessary.

When the embryo has been expelled, the uterus will often close upon the membranes, the cervix contracts, imprisoning

them; and the uterus in its imperfectly-developed state having
little expulsive power, there is *retention of the membranes.*
Retention is often complicated with *adhesion.* Then I have
often found fibrinous deposits in the placenta. These two
conditions sometimes require time to overcome. The consulting practitioner here occasionally reaps credit which is
scarcely his due. He is called in, perhaps, on the third day,
or later, when the adhesion of the decidua to the uterus is
breaking down. He passes in his fingers and extracts at once.
But had he tried the day before he might have failed, like the
medical attendant in charge.

1. The first difficulty to overcome is the narrowing of the
cervix. Whilst using means to dilate the cervix, you may use
the ordinary means to control the hæmorrhage. There are medicines, as ergot, cinchona, strychnine, turpentine, hamamelys,
which are to a certain extent useful. The most frequent
resource is cold, either in the form of cold water applied
to the abdomen and vulva, or in the form of ice in the
vagina.

The most efficient internal remedy is turpentine; but, like
all others, it is apt to fail. It is better not to rely upon these
agents, but to proceed at once,

2. *To plug.*—The usual plan is to plug the vagina; but this
is really unscientific and illusory. In a short time, the vagina
contracting, squeezes the plug, compressing it to a calibre that
no longer fits the canal; blood flows freely past it or collects
about it. The proper way to plug was pointed out to me some
years ago by Dr. Henry Bennet. You must plug the cervix
uteri itself. This may be done through a speculum, pushing
small portions of lint or sponge into the cavity by means of
a sound. It is well to tie a bit of string round each bit of
plug, to facilitate removal. This is a true plug; it arrests
hæmorrhage and favours the dilatation of the cervix.

But in the sea-tangle, or laminaria digitata, introduced by
Dr. Sloan, of Ayr, we possess an agent that surpasses all others
for the purpose.

The tangle may be used either perforated as a tube two or
three inches long, or solid. For ordinary purposes, especially
in the non-pregnant state, where the os and cervix are very
small, the tubular form is most convenient. To introduce a

plug of this kind, it may be mounted on a wire attached to a
stem provided with a canula. Thus mounted, the plug is
carried into the uterus like a sound ; and when in, the wire
is withdrawn, and leaves the plug *in situ*. The instrument-
makers sell an instrument of this kind designed by me. But
a flexible catheter, cut down at the end so as to leave about
two inches of the wire stilet bare, upon which the tangle-tube
can be mounted, answers very well. If desirable, three or
four tubes can be introduced together, forming a fagot. By
this means more rapid and complete dilatation of the cervix
can be obtained. In cases of abortion, however, I have found
it more convenient to take a solid, smooth piece of laminaria
about four inches long, the calibre of a No. 8 or No. 9 bougie,
and to give it a slight curvature at one end. This curve very
much facilitates introduction, and the length of the plug
enables one to pass it as easily as a sound. It may be passed
as far as it will go without obstruction, that is, generally
about three inches within the os uteri, leaving an inch projecting into the vagina. Thus it is maintained *in situ*. In a
few hours it will have swollen, so that, if passed overnight,
next morning, on removing it, the finger will be admitted
with ease. During its expansion, the hæmorrhage will have
stopped ; and on its removal you will generally find the ovum
fit to be detached. During the stretching of the cervix by the
laminaria, vomiting occasionally occurs. For this it is well to
be prepared. Whether sponge-tents or laminaria. be used, they
should be impregnated with iodine to obviate foulness. My
caoutchouc water-dilators are sometimes used for this purpose ;
but they are hardly adapted for the early stages of pregnancy.
The laminaria is far more convenient.

3. When the cervix has been well dilated, and the uterus has
been emptied of its contents, there is rarely any more hæmorrhage ; the patient is then generally safe. But if there should
be any return, I strongly urge the application of an iron styptic
to the inner surface of the uterus. This acts as an immediate
styptic, and may be relied upon to check any further bleeding.
The mode I prefer in these cases, where the cervix is widely
open, is to soak a piece of sponge, fixed on a whalebone stem,
in a mixture of one part of the strong liquor ferri perchloridi
of the British Pharmacopœia with three of water, and to pass   X

this into the cavity of the uterus as a swab. A difficulty sometimes experienced is, that the moment the charged swab touches the cervix, the constriction caused arrests its further progress. Where this is the case, it is necessary to pass a small tube like a catheter, and to inject *very slowly* a small quantity, say half an ounce, of the solution through it. I have recently contrived a more convenient instrument. It consists of a large tube of vulcanite, having large holes at the uterine end. At this end it is charged with pieces of sponge soaked in the styptic solution. The styptic is thus carried quite into the uterine cavity, none escaping by the way. When *in situ*, the fluid is squeezed out by ramming down the sponge with a piston-rod. In this way all *injecting force* is obviated.

I must caution you not to rely upon the topical use of per-chloride of iron *before* the uterus is cleared. There is no substi-tute for, no evasion of, this primary duty.

It must be remembered that a patient who has suffered abor-tion is liable to all the puerperal affections that assail women who have been delivered at term. To most of these affections loss of blood strongly predisposes; hence the importance of stopping the hæmorrhage as quickly as we can.

The after-treatment of abortion is :—1. *Immediate,* directed to obviate or mitigate the concomitants or consequences of abortion, that is, anæmia, the risk of inflammation, and other puerperal accidents.

The first kind of treatment is *restorative,* and the great con-dition of restoration is *rest.* The woman who has suffered abortion should be kept for ten days in bed, and made to conform to all the rules that govern the lying-in chamber. The neglect of this precaution is the source of much immediate danger, and of many ulterior affections that sap the health, and lead to protracted misery. Women are apt to think lightly of a miscarriage ; and many medical practitioners who have not seen the more severe cases countenance this error. They can scarcely be persuaded that abortion may cause death. It is a common belief that the hæmorrhage, however profuse, will stop in time, and that the patient is sure to rally; but this is not the experience of those who are largely consulted in difficult cases. I have known not a few deaths from primary hæmorrhage and shock, not a few from septicæmia, some from inflammation, and I have seen many women who have, indeed, escaped with their lives, but only to suffer for years afterwards from anæmia and other disastrous consequences.

2. *Curative and prophylactic,* that is, aimed at the causes which led to the abortion. This is much neglected in practice. In every case there is an efficient cause. The hope of preventing recurrence of abortions depends upon our tracing it out. Do not sit down disarmed and passive under the influence of that ignorant dogma which asserts that women have a habit of aborting, that they are labouring under an "abortive diathesis." This is no more than saying that women abort because they abort, which is not very instructive. The "habit of aborting" is real thus far: quick recurring impregnations find and keep the uterus in a state of chronic sub-involution. But, then, this is the efficient cause of the abortions.

The management of hæmorrhage from hydatidiform degeneration of the ovum requires a little special consideration. As I have already said, hæmorrhage may recur in repeated attacks, alternating with watery discharges, and only portions of the ovum may come away. The discharges, if carefully examined, will show shreds of membranes having the characteristic cystic enlargements. Sometimes these can scarcely be recognized without a magnifying glass. Whenever they are recognized, the indication is clear to proceed to empty the uterus. A strong presumption that the case is one of hydatidiform, or of some other degeneration of the ovum, is justified when such a degree of enlargement of the uterus is verified as does not correspond with the calculated date of the pregnancy. The uterus is commonly smaller than it should be, and it is often softer, *more pulpy* to the feel. There is little or no prospect of a live fœtus when cystic disease has once begun. In the majority of cases, no embryo at all is found. But it is an error to suppose, with Mikschik and Graily Hewitt, that this disease is always secondary upon the death of the embryo. Many instances of living fœtuses with a cystic placenta have been recorded, for example, by Perfect, Martin of Berlin, Villers, and Krieger. There is every reason to believe that, in many cases at least, it is the advancing cystic degeneration of the chorion that destroys the embryo. In many cases the cause of the chorion-change must be sought in a primary disease of the decidua. Virchow and others describe the decidua as hypertrophied. There is hyperplasia with intimate adhesion of the mucous membrane to the uterine wall. In this state, the chorion-villi shoot into the substance of the morbid decidua; and not only that, but may actually penetrate deeply into the wall of the uterus itself. When this happens, there is such continuity of structure that complete detachment of the ovum is impossible. Volkmann relates (*Virchow's Archiv.* 1868) a case of this kind. In consultation with Dr. Hassall, of Richmond, I saw a remarkable example. These extreme cases are, of course, exceptional; but they illustrate the common condition, that there is disease, hypertrophy, and more or less intimate adhesion of the decidua. The lesson in practice is this: although we ought always to endeavour to detach the whole ovum by passing the fingers, or the hand, if necessary, into the uterus, we must not persevere too pertinaciously in this attempt, lest we inflict injury upon the uterus. Take away all that will come readily, and then, if hæmorrhage return, apply the perchloride of iron. Sometimes the diseased ovum will come away in one mass, moulded to the shape of the cavity of the uterus. Still, in these cases, the decidua will show evidence of being hypertrophied. The surface will be covered with thick shreds; and decidua may be traced into the substance of the diseased mass.

In every case of abortion, I think it is important to apply a firm bandage to the abdomen afterwards. This, by depressing the fundus of the uterus and promoting contraction, lessens the danger of sucking in air, and consequent septicæmia. It is sometimes desirable, also, when the discharges are at all offensive, to wash out the vagina with Condy's fluid, weak iodic solution or carbolic acid.

But I would submit it as a rule, that whensoever the discharges are offensive, the presumption is that there is something in the uterus that ought to be removed. To clear up this point, never fail to make a thorough exploration. You may be satisfied that *the uterus is clear* if—1. There is no offensive discharge; 2. If the os uteri is closed; 3. if the sound shows the uterine cavity to be only 2·50″ long, *i.e.* that involution has proceeded to its normal extent.

LECTURE XXVII.

PLACENTA PRÆVIA — HISTORICAL REFERENCES : MAURICEAU, PORTAL, LEVRET, AND RIGBY — MODERN DOCTRINES EXPRESSED BY DENMAN, INGLEBY, CHURCHILL, AND SIMPSON—THE OLD AND THE AUTHOR'S THEORIES OF THE CAUSE OF HÆMORRHAGE IN PLACENTA PRÆVIA—THE PRACTICE OF FORCED DELIVERY, DANGERS OF—THEORY OF PLACENTAL SOURCE OF HÆMORRHAGE, AND METHOD OF TOTALLY DETACHING THE PLACENTA—THE COURSE, SYMPTOMS, AND PROGNOSIS OF PLACENTA PRÆVIA—THE AUTHOR'S THEORY OF PLACENTA PRÆVIA, AND TREATMENT—THE PLUG—PUNCTURE OF THE MEMBRANES—DETACHMENT OF PLACENTA FROM CERVICAL ZONE — DILATATION OF CERVIX BY WATER-PRESSURE — DELIVERY—SERIES OF PHYSIOLOGICAL PROPOSITIONS IN REFERENCE TO PLACENTA PRÆVIA—SERIES OF THERAPEUTICAL PROPOSITIONS—THE SO-CALLED "ACCIDENTAL HÆMORRHAGE."

Placenta Prævia.

GREATLY-extended personal experience, and the testimony of many excellent observers in this country and abroad, confirm the soundness of the theory of placenta prævia which I first enunciated in the *Lancet* of 1847, more fully unfolded in the *Lettsomian Lectures* of 1857, illustrated in various subsequent publications, and in the preceding editions of this work. From this theory, as from preceding theories, distinct rules of practice logically follow. To obtain a clear appreciation of the principles which should govern us in the difficult task of managing a case of placenta prævia, it is therefore desirable to begin with a statement of the physiology of the subject. This I will make as brief as possible, bearing chiefly in mind the practical purpose before us.

Before the time of Levret and Rigby, there could scarcely be said to be a clearly-reasoned-out theory of placenta prævia. The treatment was purely empirical, using the word in the Celsian sense. Although it was known to obstetricians, especially to Giffard and Portal, that the placenta might be implanted over the cervix uteri, the distinction between hæmorrhages depending upon this, and upon other conditions was not accurately defined. In all hæmorrhages of a severe character occurring during pregnancy, the indiscriminate practice inculcated by Guillemeau, Mauriceau, and their successors, of emptying the uterus, was blindly followed. This practice, rightly called the *accouchement forcé*, consists in forcing the hand, if necessary, through the cervix, in seizing and extracting the child, and in removing the secundines as quickly as possible. The guiding theory was very simple : it was a rough deduction from observations showing that the hæmorrhage generally stopped when the delivery was complete, and that unless delivery were effected, the hæmorrhage went on.

Levret and Rigby, observing Nature more closely, grasped with more precision the special bearing of the implantation of the placenta over the cervix as a cause of hæmorrhage. They believed that so long as the labour continued the hæmorrhage would go on, even increasing. The logical conclusion again was, that it was necessary to empty the uterus as quickly as possible. Rigby at one time drew a broad distinction between the management of hæmorrhage depending on placenta prævia, which he called "unavoidable," and hæmorrhage arising on detachment of placenta growing at its normal site, which he called " accidental." In the first, he says, " Manual extraction of the fœtus by the feet is absolutely necessary to save the life of the mother ; in the second species such practice is never required." Later, however, he admitted that " assistance might be required in accidental hæmorrhages."

Almost all subsequent authorities concurred in accepting the doctrines, and in adopting the practice of Levret and Rigby.

This is Denman's opinion : " It is a practice established

I I

by high and multiplied authority, and sanctioned by success, to deliver women by art, in all cases of dangerous hæmorrhage, without confiding in the resources of the constitution. This practice is no longer a matter of partial opinion, on the propriety of which we may think ourselves *at liberty* to debate ; it has for near two centuries met the consent and approbation of every practitioner of judgment and reputation in this and many other countries."

The following quotation from Ingleby may be taken, as expressing the theory almost universally recognized until recently :—" And thus the placenta will undergo a continuous separation corresponding to the successive expansion of the neck, *until nearly the whole of the surface* is dissevered from its uterine connection. From this it is evident that when the placenta is affixed either to the cervix or os uteri, *whether wholly or partially,* the vessels will become exposed on each successive detachment, and the ultimate safety of the patient will depend upon delivery by turning the child, excepting, perhaps, in two peculiar states, in which rupture of membranes is the only treatment offered to us in one case, and the safest, and therefore the most eligible, in the other. *Pain,* efficacious as it is in the accidental form of hæmorrhage, unless adequate to the expulsion of the child, *is neither to be expected nor to be desired,* to any material extent, in the unavoidable form, as it only renders the effusion more abundant; for though a certain degree of relaxation is necessary, it must be remembered that in exact ratio as the cervix uteri is successively developed, and the os internum progressively dilated, will an additional mass of placenta be detached from its connecting medium, and hæmorrhage necessarily be renewed."*

Churchill† speaks no less distinctly: " The flooding is the necessary consequence of the dilatation of the os uteri, by which the connection between the placenta and uterus is separated, and the more the labour advances, the greater the disruption and the more excessive the hæmorrhage. From this very circumstance it follows that the danger is much greater than in the former cases (of accidental hæmorrhage), and also that what in them was the natural mode of relief is

---

* " Uterine Hæmorrhage," 1832.
† " Theory and Practice of Midwifery," 1855.

here an aggravation of the evil, and cannot be employed as a remedy."

What theory can be more hopeless? Pain, that is, contraction, is the thing wanted, but if it comes it brings danger with it! Expansion of the cervix uteri is a necessary condition of labour, but the cervix cannot expand without causing more hæmorrhage! Nature is utterly at fault. She is condemned without appeal. Art must take her place.

Those, therefore, who accepted the theory of Levret and Rigby, were confirmed in the old practice of delivering at once at all hazards. The *accouchement forcé* received the sanction of modern science. But absolute laws seldom endure long without question or mitigation in practice. Mauriceau had long ago discovered that in some cases of partial placental presentation, rupture of the membranes and evacuation of the liquor amnii was enough to arrest the hæmorrhage. Subsequently, Puzos defined with more precision the cases in which rupture of the membranes might be practised. Wigand and D'Outrepont avoided the *accouchement forcé* as much as possible, trusting to the plug and rupture of the membranes. Dr. Robert Lee* says: "The operation of turning, which is required in all cases of complete presentation, is not necessary in the greater number of cases in which the edge of the placenta, passing into the membranes, can be distinctly felt through the os uteri. If the os uteri is not much dilated or dilatable, the best practice is to rupture the membranes, to excite the uterus to contract vigorously, by the binder, ergot, and all other means, and to leave the case to Nature."

Since the time of Levret, says Caseaux, the insertion of the placenta over the neck of the uterus has been considered as an inevitable cause of hæmorrhage during the last three months of pregnancy and during labour. The loss is then, says Gardien, of the very essence of pregnancy, and especially of labour. The current opinions apply to the separate conditions of pregnancy and of labour. The explanation of hæmorrhage during pregnancy is as follows:—Until the fifth month the body only of the uterus undergoes any considerable change; but from that date the neck participates. The shortening it undergoes is attended by enlargement of its base, at

* "Lectures on the Theory and Practice of Midwifery," 1844.

the head of the os internum. The placenta, fixed and immovable on the spot where it is planted, cannot follow the expansion of the upper part of the neck; hence the connections between placenta and uterus are torn, and hæmorrhage necessarily results.

This doctrine, passed on traditionally, and accepted almost without question, is undoubtedly founded on an anatomical and physiological error. Stoltz clearly showed that the cervix proper contributes in no way to the reception of the ovum. I have repeatedly felt the cervical canal entire closed above by the narrow os uteri internum at the end of pregnancy. The freedom of the cervix has also been ascertained by dissection of women dying after labour. And if the explanation referred to were true, hæmorrhage would be far more frequent; for the rapid expansion of the cervix, if physiological, should be a constant condition, and the uterus must infallibly always grow away from the placenta.

Then, as to the hæmorrhage occurring at the time of labour, which is attributed to the active expansion of the cervix, casting off placenta. Is this consistent with clinical observation? If you reflect upon what is observed, you will see that it can be only partly true. It is an indisputable fact that hæmorrhage, frequently at least, and I believe commonly, begins before there is any expansion of the cervix at all. The true explanation is, I submit, the very reverse of that generally accepted. What is the part endowed with the most active growth? Is it not the ovum, the placenta? The growth of the uterus is secondary; it is the result of the stimulus of the ovum. The first detachment of placenta, then, arises from an excess in rate of growth of the placenta over that of the cervical zone, a structure which was not designed for placental attachment, and which is not fitted to keep pace with the placenta. Hence loss of relation; the placenta shoots beyond its site, and hæmorrhage results. Hæmorrhage is most common at the menstrual epochs, and has not necessarily anything to do with labour. At the menstrual nisus there is an increased flux of blood to the uterus and to the placenta. This over-filling of the placenta makes it too big to fit the area of its attachment; it breaks away at the margin of the os, and blood escapes. Under the irritation of this partial detachment, the infiltration of some blood into the substance of the placenta, which increases the bulk of the organ, and the insinuation of some blood, perhaps clotting, between the placenta and the uterine wall, active contraction of the uterus may be excited. Then the retracting cervical zone may detach more placenta. But this is secondary. And what I shall presently set forth will abundantly prove the error of the theory that "the placenta will undergo separation corresponding to the successive expansion of the neck, until nearly the whole of the surface is dissevered from its uterine connection." The severance will never go beyond the cervical zone.

The view just enunciated, that the first loss of relation between the placenta and uterus is due to the excess in rate of growth and periodical hyperæmia of the placenta, is strengthened by the analogous case of Fallopian gestation. Like tubal gestation, gestation in the lower segment of the uterus is an example of *error loci;* both are alike in being instances of ectopic gestation. In both cases the ovum grows to a structure not adapted to harbour it. This want of adaptation consists in unfitness to grow with the advancing growth of the ovum. Hence, in the case of tubal or interstitial gestation, there comes a time when the growth of the ovum is so rapid that the sac, not able to keep pace with it, bursts. This catastrophe, too, commonly happens near a menstrual period, when increased growth is aggravated by increased afflux of blood. I have drawn attention to the fact (*Clin. History of Diseases of Women*), that before the rupture of a Fallopian sac, a discharge of blood by vagina often takes place. This is evidence of the ovum outstripping its habitat, and getting partially detached. It is exactly what happens in placenta prævia.

To return to our history. Still the prevalent practice was to deliver at once; and this was, and is, often done without much regard to the fitness of the parts to undergo this severe operation. So imperious is the dogma of " unavoidable," persistent hæmorrhage, that the difficulty presented by an undilated os uteri is overcome by a special hypothesis, which assumes that in these cases of flooding, the os uteri is, by the flooding, always rendered easily dilatable. Unfortunately, this is not true. Proofs of laceration, of fatal traumatic hæmorrhage from the injured cervix, as the penalty for forcing the hand through the presumed dilatable cervix, abound; but the error is so prevalent, that I think it useful to adduce some evidence upon the subject. Leroux says (Edition, 1810): "Before Puzos' time, the 'accouchement forcé' was generally performed. The operation was difficult, very laborious, and often followed by a fatal issue." The late Professor D. Davis says he had met with many examples of even fatal hæmorrhage unaccompanied by any amount of dilatation of the orifice of the womb. He relates a case where very profuse hæmorrhage had occurred, yet the orifice of the uterus was but slightly dilated, and as rigid as if no hæmorrhage had been sustained. Labour was induced, taking four or five hours to expand the os. Living twins were delivered. On the fifth day after the labour, profuse flooding set in, and caused death. No rupture was found; but the long-continued bearing incident to the introduction of the hand had produced contusion, inflammation, and suppuration of the os uteri, and a portion of its tissue, about the diameter of a sixpence, had sloughed off, and left behind it a deepish ulcer; several branches of arteries were found in the depth of it, and thus was rendered evident the cause of the fatal hæmorrhage. The preparation is in Middlesex Hospital Museum.

Dr. Edward Rigby says:* "Cases have occurred where the os uteri has been artificially dilated, where the child was turned and delivered with perfect safety, and the uterus contracted into a hard ball; a continued dribbling of blood has remained after labour; the patient has gradually become exhausted, and at last died. On examination after death, Professor Naegele has *invariably* found the os uteri more or less torn."

Collins and others relate examples in point. The truth is, that so far from the os uteri in placenta prævia being in a state favourable to dilatation, the conditions are often the reverse. The implantation of the placenta involves an enormous increase of vascularity of the parts, and this, added to the imperfect muscular development attained when labour comes on prematurely, renders the dilatation of the cervix especially difficult and dangerous. Hæmorrhage from laceration is not the only danger. One, a little more remote, but scarcely less formidable, is that of inflammation, of pyæmia. Some of the worst cases of puerperal fever I have seen, have, in my opinion, been the

* "System of Midwifery," 1844.

direct consequences of the bruising of the vascular tissues of the cervix, caused by forcible delivery for placenta prævia. I will show presently how this injury to the uterus can almost always be averted.

What is the source of the blood? Next there came a new theory, based upon a presumed physiological basis. Levret believed that in this form of uterine hæmorrhage the placenta supplied a portion of the blood. Rawlins, of Oxford,* says: "The blood proceeds more from the vessels of the detached portion of the placenta than from the denuded vessels of the uterus." The late Professor Hamilton entertained the same opinion. Kinder Wood, Radford, and Simpson adopted this opinion; but they did not agree in their reasoning from it. Simpson drew from it the practical conclusion, that to stop the hæmorrhage we should completely detach the placenta. Radford, who had previously adopted this practice, was led to it mainly by the clinical observation that the hæmorrhage had ceased on the spontaneous separation and expulsion of the placenta.

I do not think this hypothesis of the placental source of the blood is now entertained by any one of authority. The little evidence adduced in its support is entirely fallacious. In my Lettsomian Lectures I pointed out that where flooding ceased on the spontaneous total detachment of the placenta, it was because the detachment was effected by contraction of the uterus; and this contraction closes the utero-placental vessels just as it does in *post-partum* hæmorrhage. And then again there are cases in which the hæmorrhage did not stop on total detachment of the placenta effected spontaneously or artificially. This was because the uterus did not contract. That any great amount of blood cannot flow quickly and in gushes through the placenta, was proved by the Hunters, who showed that the cavernous structure of the maternal placenta could only permit of a slow uniform current back from the utero-placental arteries to the sinuses. Legroux relates (*Arch. Gén. de Méd.*, 1855) two cases of fatal hæmorrhage from placenta prævia, in which dissection showed that the fœtus had long been dead. In these, he reasoned, the placental circulation must have ceased, and that the source of bleeding must have

* "Dissertation on the Obstetric Forceps," 1793.

been the uterus. The late Dr. Mackenzie made experiments bearing directly on this question. 1. Having opened the uterus of a pregnant bitch, and detached the placenta, he observed that the blood flowed freely from the uterus, and that it was *arterial*. 2. Having partially detached the placenta in a woman, he injected defibrinated blood into the hypogastric arteries. He again observed that the blood flowed exclusively from the uterus, and from the *utero-placental arteries*. 3. He adduces the recorded observations of many practitioners to show that the blood in women flooding from placenta prævia is arterial in colour.

The blood has often been seen flowing direct from the uterine surface in cases of Cæsarian section and of inverted uterus, both in partial and complete separation of placenta. The late Dr. Chowne collected a mass of evidence of this kind. I may further observe that the languid current through the placenta must be quickly stopped by coagulation. The detached part of the placenta is always found hard and impermeable from this cause.

Disregarding all this evidence, Simpson, dwelling exclusively upon the placental hypothesis, thought he was imitating Nature by totally detaching the placenta, so as to prevent blood from getting into it, and therefore out of it.

How is the total detachment of the placenta accomplished? The method prescribed is to pass the hand, if necessary, into the vagina, then to pass two fingers through the cervix uteri, and with them to separate the placenta.

The fatal objection I urged against this proceeding, false in theory, was that it was impracticable. In by far the greater number of cases the placenta extends higher than the equator of the uterus, often reaching the fundus. The fingers are not long enough to reach even half-way towards the further margin of the placenta. The diameter of the placenta is nine or ten inches; the fingers will barely reach three inches. In the greater number of cases, therefore, in which the directions prescribed have been followed, the placenta has not been wholly detached; and the result, when successful, cannot be attributed to an operation which was not performed. This is further proved by the history of some of the cases narrated as examples of this practice. The child was born alive. Is it consistent

with our knowledge of the conditions upon which the child's
life depends, to suppose that the child will survive if the whole
placenta be detached, unless the birth follow the detachment
immediately? And this condition, under the postulates of the
hypothesis, is wanting. This objection, so obvious when simply
stated, was not suspected by Professor Simpson and his disciples until it was formally set forth by me. I am not aware
that the Professor ever published an acknowledgment of the
error of his theory and practice; but I have been informed that
he admitted the accuracy of my views.

The entire detachment of the placenta has been urged, on
the ground that it can be executed at a stage when the dilatation of the cervix is insufficient to admit of turning. But
if it cannot be executed without passing the whole hand
into the uterus, that is, without forcing the hand through
the undilated cervix, in what respect is the operation less severe
than that of delivery by turning? Is it not reasonable to
conclude that, since the forcible entry has been effected, the
seizure and extraction of the child, as well as the detachment
of the placenta, had better, to give the child a chance, be
completed at the same time?

If so, the special character of the proceeding vanishes; it
is even more severe than turning, which does not require the
hand to be passed through the cervix.

But clinical observations, it is contended, prove that
hæmorrhage has stopped on the total detachment of the
placenta. These observations are partly true, partly fallacious.
The true observations are those in which the placenta has
been *spontaneously* cast off and expelled before the birth of
the child. These cases are not numerous. They do not
justify the conclusion drawn, that the *artificial* total detachment of the placenta will be equally followed by arrest of
hæmorrhage. There is a fundamental physiological distinction
between the two cases. When the placenta is cast off spontaneously, it is because the uterus contracts powerfully. This
contraction stops the bleeding. When the placenta is detached
artificially, there may be, and probably is, defective uterine
contraction. The bleeding will be likely to continue. There
is no independent virtue in the mere detachment of the placenta,
as *post-partum* hæmorrhage abundantly proves.

Here, then, we have true observations erroneously inter-preted. The fallacious observations are of the following nature. The physician, seeking to imitate the spontaneous detachment, obeys the precept to pass two fingers into the uterus, and concludes that he has wholly detached the pla-centa. In this he has, in all probability, deceived himself. The fingers can only detach as far as they will reach, that is, for an area of two or three inches from the os uteri. The operator has unconsciously failed in doing what he tried to do; he has unwittingly done very nearly what he ought to do. The hæmorrhage stops; he sees in his success a proof of the truth of the theory that total detachment of the placenta is the security against hæmorrhage. But he has not wholly detached the placenta; he is unconsciously giving proof of the truth of a very different theory: namely, that the hæmorrhage ceases when that part of the placenta which had grown within the lower zone of the uterus has been detached.

The memoir in the *Lancet* (1847) contains the earliest distinct enunciation of the physiological theory that there is a natural spontaneous arrest of the hæmorrhage attained when that part of the placenta which had grown within the lower zone had been detached, provided uterine contraction concurred; and of the correlative law of treatment by the artificial detachment of this part of the placenta, supplemented by means for ensuring uterine contraction. In 1855 (*Arch. Gén. de Méd.*) Legroux, in a memoir of signal merit, pointed out that the hæmorrhage is diastolic, and hemostasis systolic; and hence he concluded that the proper course was "to provoke fearlessly an energetic action of the uterus."

I will now endeavour to give a succinct view of my theory of placenta prævia, and of the principles that should guide us in the treatment of this complication.

The chief points in the physiology of placental attachment are expressed in the diagram (Fig. 122) taken from my Lettso-mian Lectures (1857):—

The inner surface of the uterus may be divided into three zones or regions, by two latitudinal circles. The upper circle may be called the upper polar circle. Above this is the fundus of the uterus. This is the seat of fundal placenta, the most natural position. It is the zone or region of safe attachment.

The lower circle is the lower polar circle. It divides the cervical zone or region from the equatorial zone. The equatorial space comprised between the two circles is the region of lateral placenta. This placenta is not liable to previous detachment. Attachment here may, however, cause obliquity of the uterus, oblique position of the child, lingering labour, and dispose to retention of the placenta and *post-partum* hæmorrhage.

Below the lower circle is the cervical zone, the region of dangerous placental attachment. All placenta fixed here

FIG. 122.

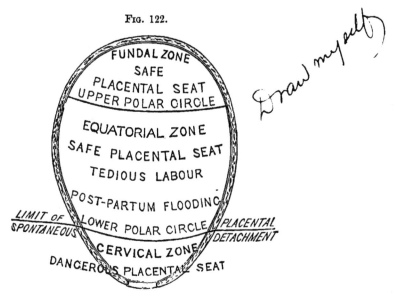

DIAGRAM EXPRESSING THE AUTHOR'S THEORY OF PLACENTA PRÆVIA.

whether it consist in a flap encroaching downwards from the equatorial zone, or whether it be the entire placenta, is liable to previous detachment. The mouth of the womb *must* open to give passage to the child. This opening, which implies retraction or shortening of the cervical zone, is incompatible with the preservation of the adhesion of the placenta within its scope. In every other part of the womb there is an easy relation between the contractile limits of the muscular structure and that of the cohering placenta. Within the cervical region this relation is lost. The diminution in surface of the uterine tissue is in excess.

The lower polar circle is, then, the physiological line of demarcation between prævial and lateral placenta. It is the boundary-line *below* which you have spontaneous placental detachment and hæmorrhage; *above* which spontaneous placental detachment and hæmorrhage cease.

When the dilatation of the cervix has reached the stage at which the head can pass, and all that part of the placenta which had been originally adherent within the cervical zone is detached, and if, as is the constant tendency of Nature to effect, the intermitting active uterine contractions arrest the hæmorrhage, a stage is reached when the labour is freed from all prævial placental complication; the lateral portion of placenta retains its connection, supporting the child's life; the labour is, in all respects, a natural labour.

This is the course which Nature strives to accomplish. I have verified it frequently at the bedside. Many cases are recorded by old and recent authors in which this course was successfully accomplished, although the narrators failed to interpret the phenomena correctly. If observations in point are not more abundant, it is simply because men, acting servilely under the thraldom of the "unavoidable hæmorrhage" theory, fear to let Nature have a chance of vindicating her powers. The instant resort to the *accouchement forcé* interrupts the physiological process; we can see no more.

Mercier seems to have been so struck with the occasional absence of hæmorrhage, that he wrote an essay under the title, "Les accouchements où le placenta se trouve opposé sur le col de la matrice sont-ils constamment accompagnés de l'hémorrhagie?"* He thought the exhalant vessels were in a state of constriction which opposed the passage of blood. Moreau and Simpson thought the previous death of the fœtus, and the consequent diversion of blood from the uterus, explained the absence of hæmorrhage. This to a certain extent is true; but I am in a position to affirm, from repeated observation, two things: first, that hæmorrhage occurs when the child has long been dead (I have cited two from Legroux, p. 487); secondly, that there may be no hæmorrhage although it is born alive.

* "Journal de Médecine," vol. lv.

Caseaux* also says, " The hæmorrhage which has been generally regarded as inevitable under these circumstances may nevertheless not occur, not even during labour ; and the dilatation of the neck may be effected without the escape of a drop of blood."

In an appendix to my work on "Placenta Prævia " (1858), I have related the histories of several cases, some quoted from various authors and some occurring under my own observation, illustrating this point. Indeed, it was the reflection upon a case of this kind, which I saw in 1845, which led to the working out of my theory. The cases in the appendix referred to, prove the fallacy of the explanations before offered as to the cause of the cessation or absence of hæmorrhage. They show that it is not accounted for by the death of the child ; nor by the pressure of the head or other part of the child upon the bleeding surface. They afford strong presumption, at least, that the hæmorrhage ceases when the physiological limit I have described has been reached, tonic or active contraction of the uterus concurring. I am far from affirming that Nature is generally equal to the carrying out of this process. Not seldom she is utterly unequal to the task. It is important to know what are the causes of her break-down ; knowing these we may know how to help her.

There are two conditions commonly present in flooding from placenta prævia. The *first is an immature uterus.* Flooding frequently occurs before the term of gestation is complete. The uterus is therefore taken by surprise, before its tissue is developed, before it has acquired its normal contractile power. Besides this, the tissue is ill-adapted to expand. The *second is the loss of blood* itself, impairing the vital power, causing shock and prostration. The several or joint action of these conditions is often a powerless labour, the absence of contraction. Hence the continuance of hæmorrhage. We feel we cannot depend upon contractile force when all force is ebbing away with the blood. We are compelled to act, to assist Nature in her extremity.

Let us now endeavour to determine the exact position of the lower polar circle or boundary-line between hæmorrhage and safety. This can be done with considerable accuracy.

* " Traité théor. et prat. de l'Art des Accouchements." 6ème éd., 1862.

The lower segment of the womb must open to an extent corresponding to the circumference of the child's head in order to permit its extrusion. By noting, therefore, the amount of necessary recession or shortening of the lower segment of the womb to reach this extent of expansion, we shall obtain the exact measure of the original depth of the cervical zone, the region of prævial placental attachment. Fig. 123 will serve to illustrate this position.

This point may be further demonstrated by the following

FIG. 123.

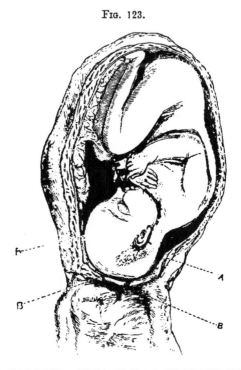

SHOWING A LATERAL PLACENTA DESCENDING TO A A, THE BOUNDARY-LINE BETWEEN THE EQUATORIAL AND CERVICAL ZONES.

The placenta descends to the very fullest expansion of the os, and therefore remains just within the limit of safe attachment. The space between A A and B B is the range of orificial expansion necessary to permit the passage of the head. The uterus is adapted from one of William Hunter's engravings.

simple proceeding:—Take a fœtal skull, and marking the left parietal protuberance for a centre, stretch a vulcanized india-rubber ring over the circle of greatest circumference of the skull, preserving it at equal distance from the centre. This ring will represent exactly the os uteri at the utmost stage of expansion necessary for the passage of the head. To this extent the os uteri *must* expand; beyond this it *need not, and will not* expand. It therefore marks the limit between the orificial and equatorial portions of the uterus.

If we now measure the distance between the presenting parietal protuberance of the fœtal head and any part of the line of greatest circumference, we shall have the utmost extent of the cervical zone. In a full-sized fœtal head this is about three inches. If we now describe a circle within the womb at three inches distance from the undilated os internum, we shall have drawn the lower polar circle. This is nearly exactly what the fingers introduced inside the womb can do.

There is another demonstration of the extent of the orificial zone, derived from examination of the state of the placenta after its delivery. The part which came nearest to the os uteri, and which may have been felt during labour, is exactly indicated by the rent in the membranes being close to it, it being obvious, as was pointed out by Levret, Hugh Carmichael,* and Von Ritgen,† that as the membranes must burst or be ruptured at the os uteri when the child passes, so, by measuring the distance of this rent from the lower edge of the placenta, we shall obtain exactly the elevation of the placenta from the os uteri.

This part of the placenta, which had been detached during labour, is found infiltrated with extravasated blood, making a thick, firm, and black flap, quite distinct in appearance from the rest of the placenta, which, growing above the lower polar circle, had preserved its adhesion in the ordinary way until after the expulsion of the child. The area of this ecchymosed portion corresponds to the area of the orificial zone.

I believe, however, that the boundary line of safety is often practically reached before the expansion of the mouth of the womb has reached the full diameter of the child's head. I have observed that the hæmorrhage has completely stopped when the os uteri had opened to the size of the rim of a wineglass, or even less.

* "Dublin Quarterly Journal," 1839-40.
† "Monatsschr. f. Geburtsk," 1855.

We may now consider

The *Course and Symptoms of Placenta Prævia*. When the placenta grows wholly or in part within the cervical zone, its relation to the uterine wall at this part is always liable to be disturbed. At any time during the course of pregnancy, hæmorrhage from this cause is apt to take place. It is most probable that some cases of presumed ordinary abortion at the third or fourth month are in reality due to this cause. At least, in aborted ova of this period, I have frequently verified the fact that there was placental structure within the lower zone; but, usually, placenta prævia is not recognized as such earlier than the end of the fifth month. At and after this time, the woman may be overtaken without warning by a smart flooding of fresh florid blood. This often occurs when the patient is at rest, in the night, even asleep; sometimes when she is out of doors, or away from home, so little is she prepared for any accident. Sometimes it comes after unusual exertion, on emotion, or coitus. These attacks of hæmorrhage are usually quite independent of labour or of uterine contraction. They occur most frequently at the menstrual epochs; thus it is not uncommon to observe a recurrence of these hæmorrhages at intervals of about a month. It cannot be doubted that the disturbance of relation between placenta and uterus is brought about by the increased afflux of blood brought to the uterus and placenta at those periods. Sometimes the bleeding subsides, and the patient is reprieved for a time. It is even possible that she may go the full term of pregnancy, after having suffered several attacks of hæmorrhage, and be delivered naturally with little loss. But most frequently, premature labour will be excited either at the first or second attack. The seventh and eighth months are frequent critical epochs. The bleeding having begun, some blood is extravasated between the uterine and placental surfaces, the lower part of the placenta gets thickened and hardened with extravasated blood, and thus the uterus is excited to contraction. When once the uterus is set in action, the termination in labour is highly probable.

The Diagnosis.—The os uteri internum is often found scarcely, if at all, dilated more than is usual in pluriparæ—and it is in pluriparæ that placenta prævia most frequently occurs—if examination is made early during the flooding; but the cervix or vaginal-portion is commonly thicker than ordinary. The finger passed into the os will miss the head or other presenting part of the child, especially if the case be one of placenta centralis. In this case, also, ballottement will be difficult or not made out; but, instead, you may feel· the quaggy, spongy placenta, or a blood-clot. The cervix is generally more tender to the touch, and pain is often felt during gestation at the lower segment of the uterus and on the side to which the placenta grows. Levret says, the uterus, instead of being rounded or pointed, is flattened, as if divided into two parts as in twin-pregnancy, but the division is more on one side, causing oblique irregularity of form; and in the first months the patient has been conscious of a swelling, with pain and hardness, in one side. Gendrin says, a pulsation, not synchronous with the mother's pulse, may be felt at the os uteri. The stethoscope will, as Hardy and McClintock* have shown, often determine the seat of the placenta at once. But these points are sometimes difficult to realize, and are therefore of doubtful value.

It is usual to teach that in accidental hæmorrhage the bleeding is arrested during a pain, whilst in placenta prævia the hæmorrhage, although continuing during the intervals, is greatly increased during the pain. Nothing, I believe, can be more illusory than trusting to this distinction. As Legroux pointed out, the hæmorrhagic act is diastolic; the *apparent* hæmorrhage is systolic, the blood previously effused being then expelled; the hæmostatic act is systolic. Certainly, at the onset, there is often no pain in placenta prævia; and, as the case proceeds, active pains will often stop the hæmorrhage.

Varieties of Placenta Prævia.—The precise physiological definition of placenta prævia I submit is: attachment of any portion of placenta within the orificial zone of the uterus. Any portion, or the whole of the orificial zone, may be covered by placenta. The placenta may dip down to the edge of the os uteri internum : this is *partial placenta prævia ;* or it may cover the os uteri internum : this is *complete placenta prævia,* or *placenta centralis.*

The Prognosis.—It is often repeated that partial placenta prævia is much less formidable than the complete form. This

* " Practical Observations."

may be so sometimes. But I am sure no rule of prognosis or treatment can safely be based upon this assumption. Collins says: " I have seen the hæmorrhage as profuse where there was merely a portion of the edge detached, as where the great bulk was separated." My experience entirely confirms this statement. The prognosis applies to three principal questions: 1. What is the immediate danger of the patient? 2. What is the ultimate danger? 3. Will the case go on at once to the completion of delivery, or will the symptoms subside?

1. The immediate danger to life from shock and loss of blood is serious, if the hæmorrhage be profuse, if the cervix remain unexpanded, and if delivery and contraction of the uterus be not secured within a short time. Whenever the loss is rapid and great, and the patient is clearly affected by the loss of blood, the indication is strong to abandon at once the prospect of postponing labour, and to proceed immediately to accelerate delivery.

2. The ultimate dangers, supposing immediate sinking from hæmorrhage is cleared, arise from anæmia. The secondary effects of hæmorrhage are: malnutrition; nervous disorders; the local injury to the cervix uteri during labour; the contusion, laceration, dispose to inflammation, and blood-infection from the necrosis of tissue about the mouths of the uteroplacental vessels; phlegmasia dolens is not uncommon, and is sometimes of very severe, even fatal type, being complicated with more than the usual degree of blood-infection; all the other forms of puerperal fever are more common after placenta prævia; secondary hæmorrhage from laceration of the cervix; and lastly, there is the prospect of imperfect involution of the uterus, and of chronic endocervicitis.

In 1864, I had occasion to review my experience of the terminations in 69 cases of placenta prævia. The deaths were six, i.e, 1 in 11½, a proportion much smaller than that usually given in statistical tables. But upon this point I think it idle to dwell, for general statistical tables, drawn from miscellaneous sources, are utterly untrustworthy in this matter. Of the six fatal cases, one died, three weeks after labour, of pyæmia; one died in a few days of pyæmia following forced delivery, performed by a practitioner who prided himself on his promptitude in the treatment of these cases; two were moribund when first seen; one died of exhaustion (she had had eleven children); one died of puerperal fever, aggravated by ill-usage.

I feel very sure that if we could always see these cases at the earliest stage of hæmorrhage, and if they were treated on the principles I have laid down, the mortality would be brought very much below anything hitherto known. In the cases above referred to, there occurred a series of twenty-nine successful cases uninterrupted by a single death.

3. Will the case go on to delivery? If the hæmorrhage is moderate, if the os does not dilate, if there is little or no sign of uterine action, there is the probability of the utero-placental relations being so little disturbed that the pregnancy may go on. But this question is often practically settled by the physician, who, governed by his estimate of the strength of the patient, the stage of pregnancy, and the urgency, absolute and relative, of the symptoms, may resolve to accelerate the labour. If the pregnancy have advanced beyond the seventh month, it will, as a general rule, I think, be wise to proceed to delivery, for the next hæmorrhage may be fatal; we cannot foretell the time or the extent of its occurrence, and when it occurs, all, perhaps, that we shall have the opportunity of doing will be to regret that we did not act when we had the chance.

The Prognosis as to the Child.—This will depend very much upon the conduct of the case. But under any method of treatment the risk to the child is great, so great, indeed, that Simpson and Churchill have expressed the opinion, that the hope of saving the child ought scarcely to influence the treatment. In this view I cannot concur. It is true that in many cases the child is dead before there is any opportunity for treatment. The child dies from asphyxia, the result of the mother's loss of blood; this blood, which is the means of aëration of the child's blood, comes in too small quantity and too feebly to effect the necessary change. Again, the child is frequently premature, sometimes not even viable; frequently there is transverse presentation; frequently the cord is prolapsed, and then the child has to run the gauntlet of artificial modes of delivery. Exposed to these perils, it is not surprising that the child will often perish. But still the broad fact remains, that a considerable proportion of children are born alive; and nothing can be more certain than that some of the sources of peril to the child may be lessened or altogether avoided, and that not only without adding to the danger of the mother, but even increasing her security. Out of sixty-two cases of placenta prævia which had been noted in reference to this point, but which had not all been treated after my method, or under favourable circumstances, twenty-three children were born alive. I am confident that a better result than this may be obtained, if the proper principles of treatment are applied, beginning at the onset of the flooding.

The total detachment of the placenta before the birth of the child is almost necessarily fatal to it. The precipitate forcible delivery of the child is scarcely less hazardous. The method I recommend is calculated to give the child every possible chance, and at the same time conduces in the highest degree to the safety of the mother.

Treatment of Placenta Prævia.

The treatment must vary according to the nature of the case; and cases of placenta prævia vary greatly. I have already adverted to the distinction drawn by many authors between partial and complete placental presentation, and to the fact that a different rule of treatment applies in the two cases. It is generally taught that in partial presentation there is less danger, and that less active treatment is called for. As a general proposition, this is true. But I have frequently seen the most severe flooding attend partial presentation, so as to call for action as energetic as any flooding from central placenta. I think a more practical division of cases is into—

A. Cases of placenta prævia, whether partial or complete, in which there is active contractile power in the uterus, with spontaneous dilatation.

B. Cases in which the contractile power of the uterus is absent, with or without dilatation.

The first question, we have seen, is to decide whether the pregnancy can be allowed to go on. If the pregnancy is only of five or six months, the os not dilated, all pain absent, and the hæmorrhage very moderate, we may temporize; watching, however, most vigilantly. But if the hæmorrhage be at all profuse, and there be any sign of uterine action, no matter what the stage of pregnancy, act at once ; accelerate the labour. Above all, do not trust to the weak conventional means of keeping the patient in the recumbent posture, in a cool room, with cold cloths to the vulva, to mineral acids, and acetate of lead. This is but trifling in the presence of a great emergency. Commonly, they do no good whatever; always, they lose precious time.

The great hæmostatic agent is contraction of the uterine fibre. To obtain contraction is, therefore, the end to be sought. Although the powers of the system may still be good, the uterus will not always act well, especially in premature labour, whilst it is fully distended. To evoke contractile energy, it is often enough to puncture the membranes. This done, some liquor amnii runs off; the uterus collapsing, is excited to contract, and being diminished in bulk, it acts at advantage. Labour being active, the cervix expands promptly, the placenta gets detached from the orificial zone, the bared uterine vessels get closed by the retracting tissue and by the pressure of the advancing head ; the hæmorrhage ceases spontaneously.

1. *The puncture of the membranes* is the first thing to be done in all cases of flooding sufficient to cause anxiety before labour. *It is the most generally efficacious remedy, and it can always be applied.* The patient lying on her side, a finger is passed up into the os uteri, guiding a stilet, quill-pen, or porcupine's quill to the membranes, whilst the uterus is supported externally.

2. At the same time *apply a firm binder over the uterus.* This further promotes contraction ; and by propelling the child towards the os uteri, it accelerates the expansion of the os and moderates the hæmorrhage.

3. Puncture of the membranes is in many cases enough of itself. But if the hæmorrhage continue, and especially if the patient show signs of exhaustion, the os uteri being undilated, the *plug* may be tried. Leroux, Dubois, Chailly, extol the plug. It has been extensively used on their authority. It often renders great service. The best plan is to plug the os uteri itself with laminaria-tents. This will expand the os, and prepare for further proceedings. The various forms of vaginal-plugs, such as the silk-handkerchief, tow, lint, even Braun's

colpeurynter, are treacherous aids, requiring the most vigilant
watching.   Do not go to sleep in the false security that,
because you have tightly packed the vagina, the bleeding will
cease.   The plug you have introduced with so much pain to
the patient soon becomes compressed, blood flows past it or
accumulates above or around it, and the tide of life ebbs
away unsuspected.   Never leave a patient more than an hour,
✗ trusting to the plug.   Feel her pulse frequently; watch her
face attentively; examine to see if any blood is oozing.
Remove the plug in an hour, and feel if the os uteri is
dilating.   If it be dilating, and the hæmorrhage have stopped,
you may trust Nature a little further, watching her closely.
The labour may now go on spontaneously, probably issuing in
the birth of a living child.

4. You must not, however, be surprised to find that the
hæmorrhage continues, that the os uteri does not expand, and
that there is no active labour.   Expectancy has its limits; the
time has come for more decided measures.   Will you force
your hand in and turn?   Remember what I have said of the
dangers of this proceeding.   There are two means of accomplishing the end in view, without violence, with more certainty
and with more safety to the patient and her child.

Consider, what are the ends to be attained?

We want to check the hæmorrhage at once;

We want security against the renewal of hæmorrhage.

To attain these ends, the uterus must be placed in a condition to contract.   The essential steps towards this consum-
mation are—first, the free dilatation of the cervix; secondly,
the completion of the labour.

✗   The first difficulty is to *effect the dilatation of the cervix*.
Under any process, this must take a little time.   Can anything
be done in the meantime to moderate the bleeding?

Something very effectual may be done.   If the physiology
of placenta prævia, which I have laid down, be correct, it may
be expected that by accelerating the completion of the necessary process of *separating all the placenta which adheres within
the orificial zone*, we shall, at any rate, get over the period of
danger more quickly.   If we can do this, we shall gain other
advantages: we remove an obstacle to the dilatation of the
cervix, for the adherent placenta acts as a mechanical impedi-

ment; and we lessen the risk of laceration of the placenta, an accident very likely to happen in the ordinary course and under turning, and which, by rupturing the fœtal vessels, adds to the peril of the child.

The operation is this: Pass one or two fingers as far as they will go through the os uteri, the hand being passed into the vagina, if necessary; feeling the placenta, insinuate the finger between it and the uterine wall; sweep the finger round in a circle, so as to separate the placenta as far as the finger can reach; if you feel the edge of the placenta where the membranes begin, tear open the membranes freely, especially if these have not been previously ruptured; ascertain, if you can, what is the presentation of the child before withdrawing your hand. Commonly, some amount of retraction of the cervix takes place after this operation, and *often the hæmorrhage ceases.* You have gained time. You have given the patient the precious opportunity of rallying from the shock of previous loss, and of gathering up of strength for further proceedings.

If, the cervix being now liberated, under the pressure of a firm binder, ergot, or stimulants, uterine action return so as to drive down the head, it is pretty certain there will be no more hæmorrhage; you may leave Nature to expand the cervix and to complete the delivery. The labour, freed from the placental complication, has become natural.

5. If, on the other hand, the uterus continue inert, the hæmorrhage may not stop; and you must proceed to the next step, *the artificial dilatation of the cervix.* This is accomplished by the use of my hydrostatic dilators. Insert the middle or largest-sized bag into the os; distend with water gently and gradually, feeling the effect of the eccentric strain upon the edge of the os. When the bag is fully distended, keep it *in situ* for half-an-hour or an hour if necessary. During this time, the hæmorrhage is commonly suspended; probably the intra-uterine portion of the bag presses upon the mouths of the bared vessels; certainly retraction or shortening of the lower segment of the uterus goes on, which is the direct means of closing these vessels; and under the combined effect of pressure from below by the dilator, and from above by the binder, the contents of the uterus are kept in close contact with its inner surface, thus keeping

pressure on the vessels of the cervix, and stimulating the entire organ to contract. Legroux, guided by his observation that hæmorrhage occurs during uterine diastole, puts the patient *in the vertical position* during uterine relaxation, so as to maintain the pressure of the contents of the uterus upon the cervix. He relates a case where this practice perfectly succeeded. But it is obvious that this position disposes to the danger of syncope. The bags accomplish the purpose better and more safely. When you find the cervix freely open, you may withdraw the bag.

Again, you may wait and observe if Nature is able to carry on the work. If contraction persist, if the head present, the labour is henceforth essentially normal, and may be allowed to go on spontaneously. You must nevertheless watch attentively.

If contraction is inefficient, if hæmorrhage goes on, if another part than the head is presenting—a condition very frequent in placenta prævia—we must carry our help further. We must do what Nature cannot do : *we must deliver.* This is done by seizing the child's leg, and extracting. Now, this can almost always be accomplished without passing the hand into the uterus. The bi-polar method of turning here finds one of its most valuable applications. It avoids the danger and difficulty of forcing the hand through an imperfectly-expanded os, through imperfectly-developed and abnormally-vascular structures. Having seized a leg, it must be drawn down gently, so as to bring the half-breech into the cervix. Traction must be so regulated as to bring the trunk through with the least amount of force necessary for the purpose. Whilst delivery is going on, the hæmorrhage is generally arrested. Rapid extraction involves a certain amount of violence and shock. Gentle extraction, giving the cervix time to dilate gradually, avoids this mischief.

As soon as the child is born, re-adjust the pressure upon the uterus; and if there is no hæmorrhage, allow three or four minutes for the system to rally, before attempting to remove the placenta. If hæmorrhage occurs, and the placenta does not come on moderate compression of the uterus and traction, you must pass the hand into the uterus to detach it. The portion growing to the equatorial zone is not always readily cast. Examine the placenta carefully on its uterine surface, to see if it is entire.

In every labour, the cervix, having to suffer great distension and contusion under the passage of the child, and possessing less contractile elements in its structure than the rest of the uterus, is liable to paralysis for a time. This state is more likely to occur in labour with placenta prævia, and it is doubly dangerous because the cervix is so near the placental site. Here is another reason for sparing the cervix to the utmost. Hence it may be desirable to swab the part freely with a solution of perchloride or persulphate of iron.

The chief facts in relation to placenta prævia may be summed up in the following series of physiological and pathological propositions. In the *Obstetrical Transactions* and elsewhere, I have published the histories of cases in illustration and proof; and it may be said that the medical journals of recent years abound with similar cases recorded by many practitioners who have adopted the principles expounded by me.

A.—*Series of Physiological Propositions.*

1. That, in the progress of a labour with placenta prævia, there is a period or stage when the flooding becomes spontaneously arrested.

2. That this hæmostatic process does not depend upon total detachment of the placenta; upon death of the child; upon syncope in the mother; nor upon pressure on the cervix bared of placenta, by the presenting part of the child, or distended membranes.

3. That the one constant condition of this physiological arrest of the flooding is contraction, active or tonic, of the muscular structure of the uterus.

4. That this physiological arrest of flooding is neither permanent nor secure until the whole of that portion of the placenta which had adhered within the lower zone of the uterus is detached—that being the portion which is liable to be separated during the opening of the lower segment of the uterus to the extent necessary to give passage to the child.

5. That, when this stage of detachment has been reached, there is no physiological reason why any further detachment

or flooding should take place until after the expulsion of the child, when, and not till then, the remainder of the placenta, which adheres to the middle and fundal zones of the uterus, is cast off, as in normal labour.

6. That the rent made in the sac of the amnion to give passage to the child, being necessarily at the part opposite to the os uteri, marks the spot at which the placenta was attached to the uterus; thus, in cases of partial placental presentation, the rent is commonly found close to the edge of the placenta. (Levret.)

7. That attachment of the placenta to the lower segment of the uterus at the posterior part is a frequent cause of transverse presentations. (Levret.)

8. That, in the case of partial placental presentation, where an edge of the placenta dips down to near the edge of the os uteri internum, the umbilical cord commonly springs from that edge; and that thus is explained the liability to prolapsus of the funis in cases of placenta prævia. (Levret.)

9. That adhesion of the placenta to the os uteri internum impedes the regular dilatation of the part, and, consequently, whilst such adhesion lasts, the proper course of labour is hindered.

10. That inflammation of the uterine structures, particularly of the cervix, is especially likely to supervene upon delivery attended by placental presentation. That one of the purposes intended by Nature in fixing the seat of the placenta at the fundal and middle zones of the uterus, is the preservation of the parts rendered highly vascular by connection with the placenta, from the distension, pressure, and contusion attending the passage of the child.

B.—Series of Therapeutical Propositions.

1. That the greatest amount of flooding generally takes place at the *commencement* of the labour, when the os is beginning to expand. That the os is always, from its being near the seat of the placental attachment, highly vascular, and is frequently, at this stage, very rigid. That any attempt to force the hand through this structure, at this stage, either for the purpose of wholly detaching the placenta, or of turning the

child, must be made at the risk of injuring the womb; and that the dragging the child through the os when in this condition—even when it has not been necessary to pass the hand into the uterus—is a proceeding affording slender chance of life to the child, and dangerous to the mother.

2. That the entire detachment of the placenta is not necessary to ensure the arrest of the hæmorrhage; and that, therefore, it is not necessary either to wholly detach the placenta before the birth of the child, or, in cases uncomplicated with cross-presentation, to proceed to forced delivery, with a view to wholly detach the placenta after the birth of the child.

3. That, since the dilatation of the cervical portion of the womb must take place in order to give passage to the child, and since, during the earlier stages of this necessary dilatation, hæmorrhage is liable to occur, it is desirable to expedite this stage of labour as much as possible.

4. That, in cases where labour appears imminent, with considerable flooding, whilst the os internum uteri is still closed, the arrest of the flooding and the expansion of the os may be favoured by plugging the vagina, and especially the cervix, and by the use of ergot.

5. That, since a cross-presentation, or other unfavourable position of the child at the os internum uteri, is apt to impede or destroy the regular contractions of the uterus which are necessary to the arrest of flooding, it is mostly desirable, in cases complicated with unfavourable positions of the child, to deliver as soon as the condition of the os uteri will permit.

6. That, in some cases, the simple use of means to excite contraction of the uterus, such as ergot, rupturing the membranes, the administration of a purgative, or the employment of galvanism, will suffice to arrest the hæmorrhage.

7. That, in some cases where it is observed that the os uteri has moderately expanded—namely, to the size of a crown piece, or less—the placenta being felt to be detached from the cervical zone, and the hæmorrhage having ceased, it is not necessary to interfere with the course of the labour, now become normal.

8. That, at the critical period, when the total detachment of the placenta or forcible delivery is a dangerous or impractic-

able operation, the introduction of the index finger through
the os, and the artificial separation of that portion of the
placenta which lies within the lower or orificial zone of the
uterus, is a safe and feasible operation.

9. That the artificial detachment of all that portion of the
placenta which adheres within the cervical zone of the uterus
will at once liberate the os internum uteri from those adhesions
which impede its equable dilatation; and, by facilitating the
regular contraction of this segment of the uterus, favour the
arrest of hæmorrhage, and convert a labour complicated with
placental presentation into a natural labour.

10. That the immature uterus, partly paralyzed by loss of
blood, cannot always be trusted to assume the vigorous action
necessary to effect delivery; that it is, therefore, often desirable
to aid by dilating the cervix artificially; that this can be done
safely and quickly by the caoutchouc water-dilators.

The management of the case after the removal of the
placenta falls within the subject of *post-partum* hæmorrhage,
and will be discussed in the next lecture.

C.— *The so-called "Accidental Hæmorrhage."*

The placenta growing within the normal regions, that is,
in the fundal or equatorial zone, hæmorrhage may break out
at any time. The hæmorrhages which occur during the first six
months of pregnancy have been discussed under the head of
abortion; but several of the causes which lead to abortion may
act during the latter three months of pregnancy. The hæmorrhages of the latter three months have been divided by the
elder Rigby into "Unavoidable" and "Accidental," in order
to contrast those depending upon growth of the placenta over
the cervix uteri from the other cases. The distinction is arbitrary and unphilosophical; but it has been generally adopted
in this country because of its convenience. Convenience, however, is here, as is too often the case, purchased at the cost of
truth. I have shown that the word "unavoidable" is much
too absolute. The word "accidental" is even more unfortunate.
It may be doubted whether, with placenta prævia, hæmorrhage
be more unavoidable than it is in the so-called accidental cases.
And why should a hæmorrhage arising from spontaneous detach-

ment of placenta, brought about by disease or emotion, of which we often know nothing—that is, nothing more than we know of the existence of placenta prævia—until its consequence, hæmorrhage, has broken out, be specially characterized as " accidental " ?

If, then, we retain the word, we must be careful to use it simply as an arbitrary formula to exclude hæmorrhage from placenta prævia.

There is, however, sound clinical reason for studying the hæmorrhages that occur during the latter three months of pregnancy, distinguishing those which depend upon placenta prævia, apart from the hæmorrhages of early pregnancy. As we approach the term of gestation, the relations of the placenta to the uterus differ essentially. The adhesion becomes less intimate ; slighter causes determine its premature detachment; and one circumstance, that which lends special significance and gravity to the hæmorrhages of this period, the liability to partial or complete retention or concealment of the extravasated blood in the uterus, is almost unknown in the first half of gestation. This important practical distinction we undoubtedly owe to Dr. Rigby's definitions. This pathological distinction and the outflowing clinical indications are scarcely apprehended by French and German authors.

Hæmorrhage in the latter months depends essentially upon detachment of placenta, so that blood flows from the ruptured and bared utero-placental vessels. This general proposition is no less true of the hæmorrhages which arise when the placenta grows to the fundus than it is where the placenta grows to the lower zone.

The immediate causes of detachment of the placenta are—1. Contractions of the uterus, which disturb and break the relations of surface between uterus and placenta ; 2. Undue, especially sudden, determination of blood to the uterus and placenta ; 3. External violence.

1. Towards the end of gestation, the muscular fibre of the uterus becomes rapidly developed, and contractility becomes more and more pronounced. Hence it is that detachment of the placenta is more frequent at this period. Causes that heretofore would have been harmless, may now excite *active contraction.* Some degree of contraction analogous to the peristaltic motion

of the intestines may commonly be felt by the hand, especially if it be applied cold, on the abdomen. If this contraction be excessive or sudden, the placenta may be partially loosened. The smallest extravasation between uterus and placenta will excite further contraction; the separation is increased, and more blood is effused. This disposition to separation is much facilitated by any morbid condition of the placenta, such as fatty degeneration, calcareous degeneration, fibrinous masses, which impair the uniformity of its structure. A morbid placenta will not follow and adapt itself to the varying movements and superficial area of the uterus, so easily as a healthy one.* A dead fœtus may lead to detachment in two ways. First, through retrogression of tissue the attachment of the placenta is loosened; secondly, the contraction of the uterus is excited as by the presence of a foreign body.

Gendrin gives the following explanation of the mechanism of detachment of the placenta. The muscular structure of the uterus is disposed in two layers, an external and an internal. The relations of these two layers with the vascular layer account for the influence they exercise in the production of hæmorrhage. When spasmodic contractions are excited, the intra-uterine vascular plexus being pressed irregularly by these muscular contractions, blood must flow in some points of the placental disc; hence, a partial congestion, which may cause a rupture of the weak venous ramifications. These contractions, by causing circumscribed puckerings on segments of the uterine globe, necessarily drag upon the placental connections, and may cause their rupture.

2. The second cause may act independently of the first, but it will act with especial force when the first exists. *Emotion* will cause contraction of the uterus; it is also a powerful agent in directing a sudden flow of blood to the uterus. The sudden tension of the vessels, aided or not by contraction of the muscular wall, is relieved by extravasation of blood between uterus and placenta. It has been observed (Kiwisch) that detachment does not always at once ensue upon the nervous shock, and that flooding may not set in until after some hours, even days, have passed. I would explain such cases by sup-

* *See* the Author's memoirs on "Fatty Degeneration of the Placenta," "Med.-Chir. Trans." 1851, 1853.

posing that a slight extravasation takes place at the time, which, acting as an irritant to the uterus, occasions extended contraction a little later. The utero-placental vessels are of extreme delicacy; they form the weakest point of the circulating system. Sometimes extravasation takes place in the placenta itself: placental apoplexy; this will probably lead to detachment. A thin, watery, degraded blood is a frequent factor. Hæmorrhage, with detachment of placenta, is also frequent in small-pox, scarlatina, and typhoid, in acute atrophy of the liver, in leucocythemia.

3. *Violence* in many forms may produce similar results. Direct violence, usually assigned as the most common cause, acts in an obvious manner. A blow, even if not bearing immediately upon that part of the uterus to which the placenta is attached, may, by re-percussion, or *contrè-coup*, throwing the uterine wall into agitation, or making it contract, cast off the placenta. The movements of the fœtus may excite sufficient contraction. In some cases I have suspected that coitus has been the cause. Detachment of the placenta has been known to follow severe vomiting, straining at stool, or coughing, lifting heavy weights, standing at hard work at the wash-tub, and so on. These may act not only by the violent succussion, but also by producing uterine hyperæmia. But the frequency of violence as a cause is, I believe, much exaggerated. Duparcque suggests that compression of the uterus by the abdominal muscles, as under the strain of vomiting, may detach the placenta. But probably the emotional contraction of the uterus participates in this effect.

Predisposing Causes.—This premature separation of placenta rarely occurs in the young and robust. It is most common in women about forty years of age, who have borne many children, whose constitutions are worn by sickness and poverty, and whose tissues are therefore badly nourished, wanting in tone, tending to atrophy or degeneration; in short, in the same class of persons who are most liable to rupture of the uterus. In one case I found fatty degeneration of the heart.

Certain diseases dispose to hæmorrhage; such are variola, albuminuria (Blot), leucocythemia,* acute atrophy of the liver. Should one of these affect a pregnant woman—and the three

* Paterson, " Edin. Med. Journ.," 1870.

last are especially liable to arise in connection with pregnancy —the risk of extravasation at the weakest point of the vascular system, namely, the utero-placental vessels, will be great.

In these predisposing causes lies one source of the great danger attending these cases. A second is, that the blood is often wholly or partially "*concealed.*" The detachment taking place at a distance from the cervix, the blood accumulates in the cavity of the uterus; we lose, therefore, the ordinary evidence of flooding. Thirdly, this sudden irruption of blood into the cavity of the uterus stretches the uterine fibre, produc- ing shock, perhaps collapse. The peritoneal coat has even been known to be torn by this sudden stretching. This is illustrated by the fact observed by Professor Simpson, that forcible injec- tion of water or air into the uterus, for the purpose of inducing labour, has caused similar laceration. Indeed, fatal mischief is often done before the physician is summoned, or has the oppor- tunity of acting.

It must not, however, be concluded that all cases are of this extreme severity. *There is an order of cases in which the symp- toms are comparatively slight.* There is no mark of tension of the uterus, and but slight shock or pain. In this order it will be remarked that *the blood escapes externally;* and it is to this circumstance that the mildness of the symptoms is due. These are the cases in which simple rupture of the membranes, or even expectancy, is sufficient treatment. There being little shock or exhaustion, the uterus quickly re-asserts its power. I believe they truly are chiefly cases in which the placenta encroaches upon the cervical zone. This I have often demon- strated by showing that the rent in the membranes was near the margin of the placenta. This fact must be accepted as circumscribing the number of cases of "accidental" hæmor- rhage. Many cases described under this designation are, in reality, cases of partial placenta prævia. It must at the same time be borne in mind that the same causes produce prema- ture separation of the placenta, whatever be the seat of its attachment.

The Diagnosis and Symptoms.—We must seek for other symptoms than hæmorrhage to guide our diagnosis. The most characteristic are—first, *acute pain* in some part, gene- rally the fundus, of the uterus, increased on pressure; second, *collapse ;* third, *great distension of the fundus of the uterus ;* fourth, *loss of regularity of form of uterus ;* it protrudes more than naturally into the epigastric region, and communicates a doughy feel, the form of the fœtus being lost at this part. These signs are the result of the stretching of the uterus by the accumulation of blood in a circumscribed part. The histories of cases show that the detachment almost always begins in the middle of the placenta, and proceeds towards the margin under the pressure of blood accumulating. A cavity is formed for the reception of the blood, partly by inward compression of the placenta, which tends to be separated, and partly by the bulging outwards of the uterine wall. The placenta examined after expulsion, instead of being convex on its maternal surface, is cup-shaped. Dr. Oldham described * a typical case, in which the placenta retained its adhesion all round the margin only, a large mass of blood being imprisoned in the hollow formed between uterus and placenta. The placenta is in Guy's Museum. A fifth result of the sudden shock and uterine injury is commonly *absence of all true labour-pains.*

It is interesting to note the analogy between these cases and rupture of the uterus. The symptoms often bear a striking resemblance. In both cases there is sudden injury to the uterus. Accidental hæmorrhage may, in some cases, be regarded as an alternative of rupture of the uterus. We have seen that emotion, by producing sudden concentric pressure, may cause rupture. This catastrophe may be averted if the ovum or its attachments give way.

There are indeed the general signs of loss of blood : fainting, blanching, agitation, perhaps deafness or blindness. The skin is cold and clammy, the pulse feeble, dicrotous, or almost extinguished ; the features are pinched ; the whole aspect indicates suffering and depression. The intensity of these symptoms is greater than can be accounted for by the amount of blood which escapes externally. Dr. Roper calls my attention to a characteristic indication, which observation enables me to confirm. In these cases the clot or crassamentum is retained. Under compression the serum is squeezed out and escapes. Hence, when we see a thin watery blood, we may

* "Guy's Reports," 1856.

suspect retention of clot from "accidental" or "concealed" hæmorrhage. When the child is born, placenta and black clots come away with a rush.

Prognosis.—Cases of this nature, occurring as they mostly do under conditions of debility, if not of disease, where there is little inherent power of resistance or of recuperation, must always be looked upon with anxiety. Death may occur in a few hours, even before delivery; and sometimes the additional shock of delivery induces fatal prostration; and sometimes, further hæmorrhage following the birth of the child, extinguishes what little strength and hope remained. Safety often depends upon early recognition of the nature of the case.

Treatment.—The first thing to be done is to *rupture the membranes*. This, by letting off the liquor amnii, takes off the strain upon the uterine fibre, allows the walls to resume their natural condition, and provokes labour. This done, it is desirable to watch and give opportunity to rally. In some cases, where prostration is not marked, this is enough. Nature will do the rest. To proceed quickly to forcible delivery, might prove fatal by adding to the shock. Ergot, I think, is not very useful. If there is great depression, the drug is either not absorbed, and is therefore inert; or, if absorbed, it is injurious by adding to the prostration. *Stimulants* internally, *warmth* to the extremities, and *friction*, are useful in promoting reaction. This effected, the uterus may be able to contract, and labour may go on spontaneously. If not, the next thing is to *dilate the cervix gradually* by means of the water-dilators. Formerly, it was necessary to incur the danger of letting the patient sink from exhaustion, unless we were prepared to encounter the greater danger of forcing the hand through the cervix and turning the child. Collins says, "I know of no operation more truly dangerous both to mother and child than the artificial dilatation of the os uteri and turning the child." He relates a case in which laceration of the uterus was thus caused. If the hydrostatic bags are used, there is no operation more safe. When there is sufficient dilatation, you may deliver by the forceps if the head present; by bi-polar turning if any other part present. The ruling principle should be, to proceed with as little precipitation and force as possible, economizing the woman's strength. With this view it is sometimes desirable to deliver by craniotomy; and especially if the child is dead.

When the child is delivered, the placenta comes away with a mass of clotted and fluid blood ; and often the prostration is increased, death sometimes quickly following. As soon as the placenta is extracted, I strongly advise to *inject or to swab the uterus with perchloride of iron,* instead of trusting to cold or kneading. The great depression contra-indicates the resort to any agents that depend for their efficacy upon a reserve of power in the system. The paralyzed, injured uterine fibre predisposes to fresh hæmorrhage. It is all-important to secure the patient against further loss by the most prompt and trustworthy means.

In these cases the child is almost always born dead. It perishes of asphyxia, arising from the mother's loss of blood and collapse, and the partial or complete detachment of the placenta.

LECTURE XXVIII.

HÆMORRHAGE AFTER THE BIRTH OF THE CHILD—CASES IN
WHICH THE PLACENTA IS RETAINED—CONSEQUENCES OF
RETENTION OF PLACENTA—THE PROPHYLAXIS OF HÆMORRHAGE
—CASES AFTER THE REMOVAL OF THE PLACENTA—SECONDARY
HÆMORRHAGE — CAUSES OF RETENTION OF PLACENTA—
MODE OF CONDUCTING THIRD STAGE OF LABOUR—ENCYSTED
OR INCARCERATED PLACENTA—APPARENT ENCYSTMENT FROM
GROWTH IN ONE HORN OF A TWO-HORNED UTERUS—HOUR-
GLASS CONTRACTION—MEANS OF EFFECTING DETACHMENT AND
EXPULSION OF PLACENTA—ADHESION OF THE PLACENTA—
CAUSES—TREATMENT—PLACENTA SUCCENTURIATA—PLACENTA
DUPLEX—PLACENTA VELAMENTOSA.

WE have now to discuss the nature and management of the
hæmorrhages which occur during and after labour—that is, after
the birth of the child. These may be usefully divided into

a. Cases in which the placenta is retained.

b. Cases in which bleeding persists, or occurs after the removal of the placenta.

c. Cases in which bleeding persists, or occurs some days after
labour: the so-called " secondary puerperal hæmorrhage."

a. Hæmorrhage in which the Placenta is retained.

In the most healthy labour, the supreme effort of expelling
the child is commonly attended by so much pain and expenditure of nervous force, that a period of rest, the result of
temporary exhaustion, follows. Probably the placenta is in
great measure detached during the final contractions which
expel the child. Soon sufficient nerve-force is reproduced; the

uterus contracts again, and completes the detachment of the placenta. If the placenta has been adherent to the fundal or equatorial zones of the uterus, its connection is preserved entire until these final expulsive contractions take place, when it is cast *en masse*. In the case of fundal attachment, the separation begins at the centre of the placenta and extends to the circumference.

But if a portion of the placenta has dipped down within the orificial zone, this part may have been detached during the expulsion of the child, and hæmorrhage will continue afterwards; or, which is as frequent, the part of the placenta which had grown to the upper part of the orificial zone had not been detached during the second stage of labour, and thus, when the uterus contracts, it is only the fundal and equatorial regions of the organ which contract so uniformly as to throw off the placenta; the orificial region not contracting to the same extent, the placenta correspondent to it remains adherent, and again hæmorrhage results. In either of these cases, separation begins at the margin.

If, from any other cause than the foregoing, the nervous force necessary for complete uterine contraction be wanting, so that the uterus is affected unequally, even if the placenta were normally attached, this unequal contraction will cause partial detachment of the placenta and hæmorrhage.

It is not to be forgotten that weakness during expulsion of the child commonly continues into the stage of placental expulsion. This disturbance is the more extensive, the greater the degree of weakness of the pains, the longer the labour, and the more rapid the artificial delivery. Hence the axiom: in delivery by forceps proceed deliberately, so as to give the uterus opportunity to act and help. Help the uterus, do not supplant it. In all labours, let the uterus expel the body and legs of the child.

Or, again, if after the placenta has been wholly detached, it remain in the uterus, and inertia come on, hæmorrhage will also follow.

Dr. Graily Hewitt * has drawn attention to a remarkable cause of hæmorrhage, namely, the presence of peritoneal adhesions on the surface of the uterus, the effect of which is to

* " Obstetrical Transactions," vol. xi.

prevent the regular contraction and descent of the uterus. This cause is rare, but it should not be lost sight of. It is especially a case for the topical use of perchloride of iron.

Not only does the presence of the placenta excite irregular spasmodic action, but, so long as it remains, the full, equable contraction necessary to completely close the uterine sinuses cannot take place. It follows, as a corollary, that the removal of the placenta is the first great end to be attained as a security against hæmorrhage.

How is this best effected? There is error both in precipitation and in delay. If, immediately after the birth of the child, you begin to pull upon the cord, you irritate the uterus at the moment of temporary exhaustion, you make that temporary exhaustion persistent, you induce irregular spasmodic contraction. This will rarely have any other effect than to cause partial detachment, prolonged retention of the placenta, and hæmorrhage.

Hunter and Denman encouraged the practice of leaving the extrusion of the placenta to the natural efforts, even for several hours. This practice had already been tried in Holland under the authority of Ruysch, and abandoned. Hæmorrhage and puerperal fever so frequently followed, that it was abandoned here also. But this sort of moral handcuffing for an arbitrary time can only be applicable to persons who cannot be trusted to observe and interpret accurately the condition of the patient. The uterus will often, indeed, detach the placenta and extrude it into the vagina; but there it will lie for an indefinite time, for the vagina has rarely the power to expel it. What useful purpose can it answer to leave it there? The prevalent practice in this country seems to me the most reasonable. It is to watch for contraction of the uterus, which is ascertained by the hand, and by the consciousness of expulsive pain by the patient. If the uterus is felt, hard, and of the size of a child's head behind the symphysis, and if, on using the cord for a clue, you can feel the insertion of the cord without passing your hand into the uterus, you know the placenta is cast and partially extruded into the vagina. Once beyond the action of the uterus, the spongy mass of the placenta will fill the vagina, adapting itself to the cavity of the pelvis. The vagina, recently distended, has little contractile power, and in it the placenta will be retained. The indication clearly is to complete what the uterus and vagina cannot do, by gently removing it. This you may do sometimes in five minutes; and it is rarely desirable to wait more than ten or fifteen minutes.

The consequences of retention of the placenta are these:—

1. Commonly, hæmorrhage and spasmodic pain. The alternate contractions and relaxations characteristic of uterine action are repeated.

2. Sometimes, no hæmorrhage, but expulsion after an indefinite number of hours or days.

3. Decomposition of blood in the cavity of the uterus, and imprisonment of the products, by the placenta blocking up the orifice, constituting physometra. In this condition the uterus sometimes enlarges, becoming tympanitic. The most horribly offensive discharges escape.

4. Septicæmia, from the absorption of the foul products.

5. Inflammation of the uterus and peritoneum, possibly from escape of foul products by the Fallopian tubes into the peritoneal cavity.

6. Disappearance of the placenta by disintegration, liquefaction, or absorption. I very much doubt whether absorption of the placenta can be established on good evidence. I feel disposed to regard it in the same light as Velpeau regarded " Vagitus uterinus." Since men of credit affirm that they have seen it, I believe it; but if I had seen it myself I should doubt.

The Prophylaxis of Hæmorrhage.

The prophylaxis may even begin during pregnancy. Experience has proved that certain women are prone to bleeding in labour. This proclivity may depend upon one of two conditions especially, namely: an altered state of the blood; or defective muscular tone combined with low nervous power. With albuminuria it has been remarked (Blot) that the disposition to hæmorrhage is greatly increased. In anæmia from any cause, and whenever there is evidence of the blood being reduced in quality, care should be taken to restore it to its due standard. Iron under conditions of this kind is often useful. Under the action of this remedy, abortion even may in many cases be averted. For abortion is often the immediate consequence of hæmorrhage, and therefore, remotely of the conditions that predispose to hæmorrhage. There is an interesting memoir upon this subject in the *Obstetrical Transactions* (vol. xvi.) by Dr. Bassett.

Coming to the time of actual labour, we again sometimes find indications in the character of the labour of the liability to flooding. Where the uterus is largely distended; where the parts of the fœtus are easily felt, showing that the walls are thin; where there is generally want of tone, with flaccid muscles, and hyperæsthesia, excessive anxiety of mind, approaching to terror, it may be expected that the uterus will act feebly, and that hæmorrhage will occur. Dr. Ewing Whittle has given the following description of the signs that threaten hæmorrhage (*British Medical Journal*, 1875): "The pains are strong and quick; they do not culminate into a strong pain and subside again, but they are sharp, quick, and cease almost suddenly; and the intervals between the pains are long in proportion to the length of the pains. After the child is born a relaxation follows: one or two sharp pains expel the placenta with a gush of blood, and the uterus again relaxes, continuing the same tendency which existed before the delivery of the child." When the uterus acts in this manner, Dr. Whittle gives as soon as the os uteri is well dilated a dose of ergot. Indeed the use of ergot at this time has long been relied upon by many practitioners of eminence.

The principle is to study the natural action of the uterus; to take note of any deviation from the course of healthy action; and to endeavour to restore healthy action. Irregular feeble contractions sometimes depend upon loaded bowels. A dose of castor-oil, with ten minims of laudanum or an enema, will in such a case often quickly be followed by efficient pains. Emotion may disturb or quite suspend uterine action. In such a case, the induction of anæsthesia, or a draught compound of thirty minims of Hoffman's anodyne with fifteen minims of the sedative liquor of opium will act with surprising benefit.

Continuing to observe the same principle, we must be careful to respect the great law of labour, that the uterus should as far as possible be allowed to do its own work. Since the time of Smellie, all good authorities have recognized the importance of *avoiding quick delivery*. When you hurry a labour by pulling upon the child, you disturb the sequence and correlation of the factors of labours. The uterus resents this meddling, acts irregularly and imperfectly, and hæmorrhage is the consequence. Another form of interference allied to precipitate manipulations, to be avoided, is the practice too often urged by nurses and officious friends, of bearing-down, of straining violently, fixing the chest, by pulling upon a towel fastened to the bed, in order "to help the pains." Voluntary effort should wait upon the reflex stimulus ; it should not be called into untimely action, as it is when it is made to anticipate the spontaneous reflex contraction of the uterus.

The practice so general, of removing the placenta by pulling upon the cord, has fostered a serious practical error as to the mode in which the placenta is cast off and presents at the os uteri. For example, Baudelocque and Schultze represent the placenta as coming away turned inside out, the fœtal surface presenting externally, whilst the membranes cover the maternal surface. But Caseaux, Lemser, and Matthews Duncan have taken pains to show that this is not the disposition when things are allowed to take their natural course. The placenta presents by its edge at the os ; and if expelled or withdrawn, so as not to disturb the relation of the parts, it comes away edgewise, the membranes covering the fœtal surface and the cord, as during pregnancy. If you compel the uterus, by compressing it, to throw out the placenta ; or if, when you have to take it away, you follow the plan which I was taught of pressing the fingers upon the subtance of the placenta just around the insertion of the cord, and then making the whole mass slide down, you will almost invariably find the membranes preserving their natural relation. By observing this rule the risk of hæmorrhage is very much reduced.

This may be laid down as an axiom in obstetrics : *by the proper management of labour, including the delivery of the placenta, you greatly secure the patient against hæmorrhage and many other dangers.*

In what does this management consist ? To some extent this question is already answered.

The rule laid down by the late Dr. Joseph Clarke, of Dublin, and endorsed by Collins and Beatty, is the one I recommend for your adoption :—When the head and trunk of the child are expelled, follow down the fundus uteri in its contraction by your hand on the abdomen, until the fœtus be entirely expelled, and continue this pressure for some time afterwards, to keep the uterus in a contracted state. When you are satisfied of this, apply the binder. I always apply the binder during the labour; the support thus given to the uterus is invaluable; it tends to preserve the due relation between the axis of the uterus and that of the brim, thus obviating the most serious objection made to the English obstetric position on the left side.

This plan of causing the uterus to contract and expel the placenta by manual compression has within the last few years been introduced into Germany, as a discovery, by Dr. Credé, without a suspicion, apparently, that it has long been a familiar practice in this country. It is insisted upon with detail by Hardy and McClintock.*

The great point is, to let or make the uterus cast the placenta by its own efforts; not to pull the placenta away. The "expression" or squeezing out the placenta should not then be always resorted to. It is "meddlesome" and injurious to practise it immediately after the delivery of the child. A few minutes must be allowed for the uterus and the nervous system to rally from the supreme effort and shock of the second stage. There is commonly a temporary inertia from exhaustion. If, during this primary inertia you squeeze the uterus too resolutely, you might even, as occurred to Schnorr (*Monatsschr. f. Geburtsk.*, 1867) cause inversion of the uterus. Timely applied, "expression" aids and is aided by natural contractile efforts. In this lies security.

Where there is special reason to fear hæmorrhage, as from great laxity of muscular fibre, a mobile emotional or nervous temperament, or a history of liability to flooding, you should be on your guard to *command the uterus* from the moment of the expulsion of the child. Place the patient on her back. This gives more power to the operator, and places the uterus in more favourable relation to the axis of the pelvis. Place the palms of both hands on the fundus of the uterus; compress the fundus steadily downwards, and from side to side between the hands, and thus prevent it from relaxing. The pressure

* "Practical Midwifery," 1848.

should be exerted by preference during an effort at uterine contraction. Another person should feel if the placenta is detached, and when it is so, remove it. The pressure should be maintained some time after the removal. By this plan, I have known delicate women, who in previous labours seemed likely to bleed to death, saved from all loss and make excellent recoveries. It is in this class of cases that a full dose of ergot, given just before or during the expulsion of the child, acts with peculiar advantage.

A retained placenta may become "*encysted*," or "*incarcerated*," that is, locked up in the fundus of the uterus by contraction of the part below. A form of this retention is familiarly spoken of as "*hour-glass construction.*" It is described and figured as a ring-like contraction of the middle part of the uterus, dividing it into two spaces. By others, more accurately, it is described as due to contraction of the os internum uteri. It is agreed by most authors of experience, that hour-glass contraction is very rare. The varieties of irregular contraction of the uterus might be deduced *à priori* from consideration of the arrangement of the muscular fibres. If all the muscular bundles of the uterus contract harmoniously together, there will be the much-desired normal uniform contraction, closing the cavity, and necessarily expelling anything that may happen to be in it. But it is a matter of observation, that occasionally parts of the uterus contract, whilst other parts remain passive. Now, which are the parts most disposed to contract, and which to remain passive? Naturally, those parts in which muscle is most abundantly provided will have the greater power of contracting. Two parts are especially rich in this respect, that is, each corner around the entrance of the Fallopian tubes. At the lower part of the body of the uterus, the bundles of fibres from either side assume a transverse or circular direction.

Now, there are two parts which are specially exposed to conditions that weaken their contractile power, producing even temporary paralysis. These are, the placental site and the cervix. The attachment of the placenta involves a great development of vascular structure; and however it be explained, the fact is certain, that the placental site is very liable to inertia. The paralysis of the cervix is accounted for by the great distension and bruising to which it is exposed during the passage of the child. After delivery it is constantly felt to be quite flaccid, as if it had lost all power of recovering its former shape and tone. Levret points out the necessity of distinguishing from the contractible neck of the womb, that kind of elongation resembling the truncated end of a large intestine which is sometimes met with in the vagina after labour. This is the lower part of the neck. The placental site and the cervix then are the two great sources of hæmorrhage.

If the placental site be at the fundus, extending into the area of Ruysch's muscle on either side, the fundus generally will be liable to paralysis; and the part below being excited to action, there may arise hour-glass contraction, *i.e.*, contraction of the lower part of the body of the uterus.

If the placental site be in one angle of the uterus, occupying the area of one of Ruysch's muscles, the central part of that area will be liable to paralysis, and the circular bundles on the margin of that area, being excited to action, will close in upon the placenta, forming a sac. Clinical observation confirms this. If you feel through the relaxed abdominal wall a spasmodically-acting uterus, you will often find it of irregular shape, and you will perceive a marked prominence at one side of the fundus, caused by the contraction of one circular muscle. Again, if you pass your hand into a uterus contracting irregularly, you will come to the constriction, the seat of which you will be able exactly to determine, between the two hands, to be where I have described it.

It is generally admitted that these forms of irregular contraction are most frequently induced by injudicious meddling, by precipitate artificial delivery, by making too early attempts to bring away the placenta, by pulling on the cord, thus irritating, teazing the uterus. You must remember the precept: Make the uterus cast out the placenta; do not drag it out.

Some cases of *apparent* encysted placenta may be due to the placenta having grown in one horn of a partially duplex uterus (*See Kussmaul,* p. 187).

A method of exciting the uterus to contract and throw off the placenta is *the injection of cold water into the vein of the umbilical cord.* This was proposed by Mojon.* By this

* " Annali Universali di Medicina," 1826.

means cold is applied directly to the placental site, so that the very part of the uterus with which the placenta is connected is stimulated to contract. I have no personal experience of this method; but Scanzoni speaks highly of it.

If the uterus cannot be made to contract, or if pressure fail to cast off the placenta, it will be necessary to *pass the hand into the uterus to remove the placenta*. For this purpose the patient may lie either on her back or left side. Support the uterus firmly, pressing the fundus backwards and downwards towards the pelvis with one hand; at the same time pass your other hand, guided by the cord, into the cavity of the uterus. Feel for the lower edge of the placenta, and with the fingers flattened out between the placenta and uterine wall, insinuate them by a light waving movement upwards, so as to peel off the placenta from the uterine surface; all the while be careful to support the uterus by the hand outside; you gain from its consentaneous action more command over the internal manœuvre, and wonderfully facilitate the accomplishment of your object. When the placenta is entirely separate, seize it well, endeavouring, by outward pressure, to induce the uterus to expel the placenta and hand together. When the placenta is extruded, proceed in like manner to clear the uterine cavity perfectly of clots. Apply the binder firmly, aided by a compress, if necessary.

The whole organ is sometimes affected, assuming the character of tetanus, the contractions are so rigid and persistent. This is especially apt to occur after ergot. To overcome them, you must depend upon steady continuous pressure with the hand or the water-bag; opium, ether, chloral, or chloroform will often assist. Pass the hand in a conical shape into the constriction, carefully pushing the fundus uteri down upon the inside hand by the other hand on the abdomen. You must trust to time to tire out the spasm, not to force. Your hand may very likely be cramped, but you must persevere, or you will only have irritated the uterus, and be obliged to begin again. When you have succeeded in passing the constriction, grasp the placenta, remove it, and keep up steady pressure upon the uterus externally. When the placenta is removed, the harmony of action of the uterus will be restored.

The object is to restore the due relation of contractile energy in the different parts of the uterus. The greater contractile energy should be exerted by the fundus. We must then seek—1, means to relax the crampy constriction of the lower segment; and, 2, means to evoke the supremacy of contraction of the fundus and body.

The cases we have been considering comprise those in which the *placenta is retained simply from want of uterine energy* to detach and expel it. These are by far the most common cases. But *the placenta may be retained by morbid adhesion to the uterus.* Cases of this kind are comparatively rare; they are more troublesome and dangerous than the first, and require more decided treatment.

True adhesion of the placenta commonly depends upon a diseased condition of the decidual element. The most frequent is some form of inflammation with thickening, hyperplasia. This, in all probability, began in the mucous membrane before pregnancy. It is liable to aggravation when the mucous membrane is developed into decidua. Sometimes there is distinct fibrinous deposit on the uterine surface of the placenta; sometimes the decidua is studded with calcareous patches. The maternal origin of the forms of diseased placenta leading to adhesion is proved often by the history of previous endometritis or other disease, and by the frequent recurrence of adherent placenta in successive pregnancies.

I have adverted to this subject under "Abortion." (See p. 468.) Further information will be found in memoirs by myself,[*] Fromont,[†] Hüter,[‡] and Hegar.[§]

Diagnosis of Placental Adhesion.—You may suspect morbid adhesion, if there have been unusual difficulty in removing the placenta in previous labours; if, during the third stage, the uterus contract at intervals firmly, each contraction being accompanied by blood, and yet, on following up the cord, you feel the placenta still *in utero;* if, on pulling on the cord, two fingers being pressed into the placenta at the root, you feel the placenta and the uterus descend in one mass, a sense of dragging pain being elicited; if, during a pain the uterine tumour do

---

[*] "Brit. and For. Med.-Chir. Rev.," 1854.
[†] "Mém. sur la Rétention du Placenta." Bruges, 1857.
[‡] "Die Mutterküchenreste, Monatsschr. f. Geburtsk.," 1857.
[§] "Die Pathol. und Therap. der Placentarretention. Berlin, 1862.

not present a globular form, but be more prominent than usual at the place of placental attachment.

The *removal of a morbidly adherent placenta* must be effected in the same manner as that just described for retained placenta; but you must be prepared to encounter more difficulty. The peeling process must be effected very gently and steadily, keeping carefully in the same plane during your progress, being very careful to avoid digging your finger-nails into the substance of the uterus. In some cases the structures of the uterus and of the placenta are so intimately connected, seeming, in fact, to be continuous parts of one organization, that you cannot tell where placenta ends and uterus begins. In endeavouring to detach the placenta, portions tear away, leaving irregular portions projecting on the surface of the uterus. In trying to take away these adherent portions you must proceed with the utmost caution. The connected uterine tissue is, perhaps, morbidly soft and lacerable. It is very easy to push a finger into it, to the extent of producing fatal mischief. A very serious practical question now arises. To what extent must you persevere in trying to pick off the firmly-adherent portions of placenta? If you leave any portions, hæmorrhage, immediate or secondary, is very likely to follow; in decomposing and breaking-up, septicæmia is likely to follow; and there is, besides, the liability to metritis. If a fatal result ensue, and a portion of placenta be found after death in the uterus, it is but too probable that blame will be cast upon the medical practitioner. The nurse and all the anility of the neighbourhood will be sure to cry out, "Mrs. A. died because Dr. Z. did not take away all the after-birth." The position is a very painful one. The true rule to observe is, simply to do your best; make reasonable effort to remove what adheres. It is safer for the woman to do too little than too much. You cannot repair grave injury to the uterus. To save your own reputation, you must fully explain the nature of the case at the time. You may lessen the risk of hæmorrhage and septicæmia by injecting perchloride of iron and permanganate of potash. In a few days the process of disintegration may loosen the attachment of the placental masses, and they may come away easily. The safest way, if it can be done, of removing these "placental polypi," is to pass a wire-écraseur over them. As this instrument can only shave the uterine surface smoothly, you are secure against injuring the uterus. I have practised this in several cases with perfect success.

As a warning against attempting too much, and as ammunition to repel an unjust charge of having done too little, remember the following passage from Dr. Ramsbotham, the truth of which I can attest from my own experience:—"Instances are sometimes met with in which a portion of the placenta is so closely cemented to the uterine surface that it cannot by any means be detached; nay, I have opened more than one body where a part was left adherent to the uterus, and where, on making a longitudinal section of the organs, and examining the cut edges, I could not determine the boundary-line between the uterus and the placenta, so intimate a union had taken place between them." Smellie, Morgagni, Portal, Simpson, Capuron, relate similar instances; and there is an instructive case reported by Dr. R. T. Corbett, in the *Edinburgh Monthly Journal*, 1850.

A very *soft placenta*, especially if it be *thin*, of *large superficies*, so as to be diffused over a considerable portion of the surface of the uterus, is a frequent cause of adhesion. The contracting uterus does not easily throw off such a placenta, that is, completely. To be cast easily, a placenta must have a certain degree of firmness, and not be too large. Perhaps a large part may be expelled or withdrawn, and appear to be all; but a portion of a cotyledon remains behind; bleeding and irregular action of the uterus are kept up, until the hand is introduced, and the offending substance removed.

A very rare—but on that account the more likely to be overlooked—event is the leaving behind a lobe of a *placenta succenturiata*. I have seen some singular examples of this. At a distance from the main body of the placenta, perhaps three or four inches or more from the margin, a mass of chorion-villi will be developed into true placental structure, and connected with the main body only by a few vessels. It resembles a lobe or cotyledon which has grown far away from the rest by itself. Such an accidental or supernumerary placenta may very easily remain attached after the main body, which is complete in itself, has been removed. These placentæ succenturiatæ rarely exceed in size that of a single cotyledon, *i.e.*, they measure about two or three inches in diameter. But I was once called by a midwife of the Royal Maternity Charity to a case of a different kind, at first very puzzling. The child was born, the cord tied, and the placenta, apparently entire, removed, when there followed a second placenta. Both were of nearly similar size and form. The first and natural conclusion of the midwife was that there was a second child still *in utero;* but she could not feel it, so sent for me. I passed my hand into the uterus, ascertained that it was empty, and made it contract. The second placental mass was developed on the same chorion as the chief placenta; vessels ran from it across the intervening bald space of chorion to join the single umbilical cord which sprang from the chief placenta. Dr. Hall Davis exhibited a similar double placenta to the Obstetrical Society.

Another form of placenta associated with hæmorrhage is where the *cord is velamentous.* In this case, the umbilical vessels, instead of meeting on the surface of the placenta to form the cord, run for some distance along the membranes, uniting perhaps several inches beyond the margin to form the cord. This part of the membranes containing the umbilical vessels spread out, may be placed over the cervix uteri. These vessels must be torn during the passage of the child. The hæmorrhage thus resulting comes from the placenta, and of course endangers the child. Dr. V. Hüter describes this formation of the placenta fully.* Two cases of this kind are related by Caseaux. Joerg, quoted by Hegar, describes a case in which bundles of vessels were found over the greater part of the chorion, but no proper parenchymatous placenta; thus resembling the *diffused placenta* of the pig. Most of the unusual forms of placental development and of arrangement of the vessels in the placenta are beautifully illustrated in Dr. Jos. Hyrtl's splendid work, *Die Blutgefässe der menschlichen Nachgeburt*, Wien, 1870.

* "Monatssch. für Geburtskunde," 1866.

HÆMORRHAGE AFTER THE REMOVAL OF THE PLACENTA — THREE
FORMS: IMMEDIATE, PAULO-POST, AND "SECONDARY" OR
REMOTE—TWO SOURCES: THE PLACENTAL SITE; THE CERVIX
UTERI — THE NATURAL LOSS OF BLOOD — NATURAL AGENTS
IN ARRESTING HÆMORRHAGE—SYMPTOMS, DIAGNOSIS, AND
PROGNOSIS OF HÆMORRHAGE FROM INERTIA, FROM TUMOURS
OR POLYPUS, FROM RETROFLEXION—ARTIFICIAL MEANS OF
ARRESTING HÆMORRHAGE—MEANS DESIGNED TO CAUSE UTE-
RINE CONTRACTION: PASSING THE HAND INTO THE UTERUS,
ERGOT, TURPENTINE, COLD, KNEADING THE UTERUS, COM-
PRESSION OF THE AORTA, COMPRESSION OF THE UTERUS,
BINDER AND COMPRESS, PLUGGING THE UTERUS—INDICATIONS
HOW FAR TO TRUST THE FOREGOING AGENTS—THE DANGERS
ATTENDING THEM—MEANS DESIGNED TO CLOSE THE BLEED-
ING VESSELS: STYPTICS, THE PERCHLORIDE OF IRON—HISTORY:
DANGERS OF, DISCUSSED — SHOCK — SEPTICÆMIA — MODE OF
APPLYING THE IRON—RESTORATIVE MEANS: OPIUM, CORDIALS,
SALINES, REST, IODINE.

THE general or systemic conditions which lead to "Accidental
Hæmorrhage" also predispose to *post-partum* hæmorrhage.

Some of the conditions which lead to hæmorrhage before the
removal of the placenta may also persist and keep up hæmorrhage after its removal. Of these the most formidable is
inertia. When the uterus is perfectly relaxed, it may be said
that the flood-gates are opened, and that the blood issues in
torrents. In a few minutes life may be extinguished. In short,
the accumulation of blood in the uterus, especially if coagulated,

reproduces very closely the same symptoms as those which attend retention of the placenta from inertia.

We may usefully distinguish the hæmorrhages which occur after the removal of the placenta into three forms, differentiated by the time of their appearance. The *first* may be called *primary* or *immediate*. The hæmorrhage follows immediately upon the removal of the placenta; the uterus failing to contract in any degree. The *second* may be called *paulo-post*, in order to distinguish it from the remote hæmorrhage to which the term *secondary* is more especially assigned, and which is the *third* form. The hæmorrhage breaks out after apparently satisfactory contraction of the uterus. An hour or more after the delivery of the child, after the bandage has been applied, when all things seem to be secure, when congratulations have been exchanged, the woman complains of pain, feels faint, says something is flowing from her. You examine, and find the uterus has relaxed, is perhaps irregular in shape, tender to the touch.

Hæmorrhage after the removal of the placenta may arise from *two* sources. The *first* is *from the gaping vessels on the placental site.*

The *second source is from lesion of the cervix* or other part of the uterine structure. In the case of severe rupture of the uterus, this source of hæmorrhage is obvious enough; but minor lacerations of the cervix, especially after forcible delivery, although far more common, are seldom recognized. Contraction of the uterus is less effective in arresting hæmorrhage of this kind than that from the placental site; indeed, it persists when the uterus is well contracted. I have no doubt that laceration of the cervix is the true explanation of some of those cases of hæmorrhage which Gooch described as due to an over-distended circulation, driving blood through the contracted uterus.

If, then, we find oozing or trickling of bright blood going on after labour, and with a well-contracted uterus, we may suspect this injury to be the cause.

It may be useful, at starting, to acquire as accurate an idea as possible of what may be considered the *natural loss of blood.* This standard is very difficult to fix by quantity. Women vary greatly in this respect. Some lose very freely, without appearing to be any the worse. Whereas others cannot bear the loss of even a moderate amount without exhibiting alarming prostration. When the uterus contracts normally, its substance is compressed, so that the blood in its vessels is squeezed out, much as we squeeze water out of a sponge. The quantity of blood so held in the uterus at the moment of separation of the placenta may be regarded as superfluous *quoâd* the wants of the system. It may amount to one pound; but it is often less, and occasionally more. A further quantity drains slowly, constituting the lochia, for several days.

What are *the means which Nature employs to arrest uterine hæmorrhage?*

1. The first, and the most efficient, is *active contraction of the muscular wall of the uterus.* This constricts, with the force of a ligature, the mouths of the arteries and veins on the placental site. So long as firm contraction holds, no blood can escape. To obtain this firm contraction is the great end of the obstetric practitioner. Active contraction settles into what may be called *passive or tonic contraction,* by which the volume of the uterus is permanently reduced. This seems due to a kind of elastic contraction, likened by Leroux to a spring ("ressort"). When this tonic contraction is effected, the patient is secure against a *return* of hæmorrhage.

2. The *uterine arteries have a certain retractile property.* Shrinking inwards, their mouths become narrowed, and the formation of thrombi is favoured.

3. The veins or sinuses of the uterus running obliquely or in strata in the uterine wall, and opening obliquely on the surface, are most favourably disposed for closure by the approximation of their walls, and the *valve-like arrangement* where the sinuses pass from one stratum to another. Even moderate tonic contraction of the uterus will so close the uterine sinuses as to stop bleeding, provided the circulation is not unduly excited.

4. If the stream of blood through the uterine vessels be stopped for a short time, and diverted into the systemic circulation, so that there is temporary rest in the uterus, the opportunity is given for *the formation of clots, thrombi, in the vessels.* Under great losses of blood, the property of coagulating is increased. Andral found the highest proportion of fibrin, 10·2 per 1,000, after a fourth bleeding. Many women are thus rescued, to all appearance from imminent death, after

the most profuse and uncontrollable floodings. Under syncope, or a state approaching to it, the heart beats so feebly that the circulation is almost suspended ; there is suspension of circulation in the uterus ; and if ever so slight tonic contraction of the uterus goes on, the vessels get plugged by coagula. The probability of this event should encourage us never to despair ✗ of a case of hæmorrhage.

The *symptoms, diagnosis, and prognosis of hæmorrhage from atony of the uterus.* The effects of bleeding are—1. To modify the balance between the circulating and respiratory systems. 2. To promote the influx of fluids from all parts of the body into the venous system. 3. To promote the tendency to the separation of fibrin. 4. Syncope. 5. Convulsion of the muscles from the removal of nervous control, muscular irritability being retained. 6. Fall of animal temperature. The first warning generally is the complaint of the patient that she feels blood flowing from her. Never disregard this. Examine the linen and the parts immediately. You will often see a thin stream of florid blood trickling down across the nates. This may seem insignificant in amount; but there should be none; and this "thin red line" is too often the indication of a greater loss, ✗ which is filling and stretching the uterus.

You feel the uterus, and find it has risen above the symphysis, perhaps above the umbilicus, that it is flaccid, or presents irregular hard prominences which shift their position. On compressing its fundus firmly, perhaps blood and clots will be forced out of the vulva. If the uterus is not brought to contract by the usual means, you pass your hand into the cavity, and feel that it is full of blood partly clotted; you feel the enlarged cavity ; you feel the flaccid flabby walls. When the inertia is complete, it is sometimes difficult, by external manipulation, to make out the uterus at all. You miss the hard globe; and this negative sign is all. When the uterus has reached its full measure of distension, spasmodic contraction is sometimes excited, and a furious rush of blood is poured out. Often again, emotion, the dread of flooding, determines blood to the uterus, and a large quantity of blood is poured forth in a gush. Alternate contractions and relaxations, the uterus getting smaller, then larger, a pain attending, and tenderness on grasping, are certain signs of hæmorrhage from atony. These are the *local signs.*

The *general signs* are scarcely less marked. Bleeding often
goes on very insidiously, the woman not complaining. She
may even feel at ease. But this calm may be illusory; she
feels that her eyes darken; that there are strange noises,
singing in the ears; that she "is sinking through the bed,"
that she is unable to move her limbs without difficulty. She
may feel an uncontrollable desire to get up or sit up. This
is due to that disturbance of relation between the circulatory
and respiratory systems already indicated. The pulse is feeble.
Hence every woman should be keenly watched for some hours
after labour. Undoubtedly life has ebbed away under unsuspected bleeding from exhaustion or syncope.   *like a thin*
Blood may issue in a considerable stream, or by gushes.
Or it may simply ooze out in a thin stream. In this latter
case, the discharge is often more watery. It indicates retention of clot in utero. When hæmorrhage is copious enough
to affect the system, a feeling of faintness, sometimes passing
into actual syncope, comes on; irrepressible anxiety, a sense
of fear, depression, are early symptoms; a degree of shock,
of collapse, is conspicuous; the face is pale and cold; the
whole surface is cold; the pulse is almost or quite imper-
ceptible; the heart-beat is feeble and frequent; there is an
indescribable sense of oppression on the chest; the patient
calls out for air, will have the windows open, insists upon
sitting up, sometimes would even get out of bed; the respirations rise to 36, 40, or more in the minute; the
breathing is noisy and laborious; she tosses her arms about,
says she is sinking through the bed, is more or less delirious;
her perception of external objects is dulled, or her appreciation
of them is distorted; partial blindness, double vision, sometimes
complete amaurosis set in, the pupils dilate, the iris seems paralyzed; she ceases to recognize at times the people about her;
she complains of intense headache, noises in the ears, sometimes
is manifestly deaf; she can hardly swallow, unless the fluid
given be poured into the back of the mouth. Brandy, beef-tea,
medicines, lie inert in the stomach, until rejected by vomiting. So great is the loss of nervous force, that every organ,
every tissue seems paralyzed; the uterus refuses to act under any
stimulant; perhaps the sphincters relax, fæces and urine being
voided. There is general muscular paralysis. She rejects help ·

by word or sign entreats to be let alone; she would willingly
die undisturbed.

From these symptoms, desperate as they look, the patient
may recover. If the bleeding stop for awhile, slowly there is
gathered up a little nerve-force; life that seemed ready to flit,
holds its seat, with feeble and doubtful grasp, it is true, but
gradually strengthening, if no fresh loss or accident occur.

But if these signs are followed by marked collapse, contracting
features, by short gasps or sobbing inspirations, which indicate
that the chest-walls unable to expand make but an imperfect
attempt to take in air, then quickly collapse, by convulsions,
the case is indeed desperate. All power to respond to any
remedy is exhausted. To persist in applying remedies now,
except by transfusion, is to harass the last moments of the
patient in vain. At this stage, or even earlier, the slightest
operation may be fatal.

The favourable signs are, returning warmth and moisture of
the surface, disposition to swallow, steady pulsation at the wrist,
evidence of contraction of the uterus, a more tranquil respiration, a feeling of hopefulness and courage, a clearer perception
of surrounding circumstances, a more accurate and steadier
judgment.

Internal hæmorrhage is promoted by any cause that obstructs
the escape of the blood externally. Thus, obliquity, or bagging of the uterus to the side, very likely to occur when the
patient lies on one side, or even tending to the prone position
with the pelvis raised, will form a depending sac in which
blood will readily accumulate. Another condition promoting
internal hæmorrhage is retroflexion of the uterus. This also is
not very uncommon after labour with inertia. It was noticed
by Burns. I have seen it several times myself, the hæmorrhage ceasing when the fundus was lifted into its proper position. Retroflexion is even more common as a cause of secondary
hæmorrhage. The best way of restoring the uterus in the
primary cases is to pass a hand into the cavity, and by it to
lift the fundus forwards.

Hæmorrhage *post-partum* sometimes depends upon a *fibroid
tumour* embedded in the wall of the uterus, or a *fibroid polypus*
projecting into the cavity. No complication can well be more
serious. The tumour disposes to hæmorrhage in two ways. In

the first place, by its density and form, it destroys the uniformity of thickness and density of the uterine wall. This
impairs the power of the uterus to contract equally, and to
maintain its contraction. Secondly, by its independent vitality,
it attracts blood in abnormal quantity to the uterus, and besides
acting as a foreign body, it irritates the uterus, exciting to
irregular spasmodic action. The ordinary means of stimulating contraction are apt either to act badly or to fail. Kneading is especially dangerous, from the liability to bruise and
injure the tumour, and even to lacerate the uterine tissue in
which it is embedded. Some amount of injury of this kind
will probably have been endured under the force of labour.
These cases, then, require the most gentle manipulation. The
safest and most effectual plan is, the placenta being removed, at
once to apply the perchloride of iron. This will arrest the flooding, even where it is difficult to induce the uterus to contract.

The fibroid polypus also requires special management. This
form of tumour, too, is commonly influenced by the developmental stimulus in the pregnant uterus; it enlarges considerably; when the child and placenta are expelled, it is liable to
be extruded from the uterus, and if of large size, it may even
be projected outside the vulva. Hæmorrhage almost constantly
follows, partly, perhaps, from uterine paralysis, partly from the
attraction of blood. Another pressing danger, besides hæmorrhage, calls for decisive action. The polypus, like the
intramural fibroid tumour, is copiously infiltrated with fluid
and new tissue; it has suffered contusion, probably, during
the passage of the child; it is fatally disposed to a low
form of inflammation, tending to necrosis; this is a cause of
constitutional infection replete with danger. The double question
is before us, how to suppress the hæmorrhage, and how to deal
with the polypus, so as to avoid the too probable ulterior mischief? The most immediate difficulty is the hæmorrhage.
Excite contraction by cold, by friction on the uterus, by ergot,
if you can. But lose not much time upon these uncertain
remedies. Inject perchloride of iron before much blood has
been lost. The polypus itself should, I think, be removed by
wire-écraseur without delay. If the pedicle is thick it is better
to use the galvanic wire cautery. It can scarcely be doubted
that the necessity of supporting a large parasitic growth of this

kind must, as the least evil, impede the healthy involution of the uterus.

Burns describes *retroflexion* as a cause. " I believe," he says, " it is more likely to take place if the placenta be still retained. Uterine hæmorrhage may be a consequence; but, if not, attention may at first be directed to the case by retention of urine and bearing-down. The hand is to be introduced into the uterus, and the position rectified." Boivin describes a case in point. Contraction took place, hand and placenta were expelled, after raising the fundus to its proper position.

One or two *general rules* should be observed, whatever the particular method we may choose to rely upon to restrain the hæmorrhage.

1. *Place the Patient on her Back.*—The value of this rule is very great. Gravitation helps to let the uterus sink into the pelvis, instead of bagging over, as when the patient is on her side; the face is open to observation, to access of air, to administration of stimulants and food; the chest-walls can expand for better respiration; the uterus and aorta are under more easy control by the hand of the physician.

2. *Pass the Catheter.*—A full bladder diverts the nervous force from the uterus, inducing irregular contractions, and impedes manipulation on the abdomen.

The indications in practice are drawn from our knowledge of the sources and natural modes of arrest of hæmorrhage, and observation of the symptoms.

In hæmorrhage after the removal of the placenta, we have, *first*, a class of remedies whose power is limited by one imperative condition. These *remedies act by exciting contraction of the uterine muscular fibre*. To effect this, there must exist a certain degree of nervous force, able to respond to centric or peripheral irritation. All the remedies commonly relied upon postulate this condition. *Ergot, compression of the uterus, cold*, all depend upon their power of inducing uterine contraction. If the nervous exhaustion be so great that irritability is lost, these agents are useless or injurious. The patient may indeed rally after syncope, but it may be truly said that at this point the art of the physician fails.

It fails unless, *secondly*, he has the courage to call to his aid another hæmostatic power, which acts even when to evoke con-

traction is impossible. When this is gone he may still stop
bleeding by the direct application of powerful *styptics* to the
bleeding surface. The most useful of these agents is the *perchloride or persulphate of iron*, the obstetric application of which
I shall presently describe.

Passing the Hand into the Uterus to Clear out the Cavity.—
Although this proceeding is deprecated by many, and looked
upon by most men as hazardous, I am satisfied that it is the
first thing to be done whenever there is hæmorrhage with an
enlarged uterus, and a suspicion that clots or other substances
are retained in the cavity. If the enlarged uterus refuse to
contract on external pressure, so as to expel its contents completely, I think there is no rule in obstetrics more imperative
than to pass in the hand. If the patient be lying on her back,
you can support the uterus externally by one hand whilst you
introduce the other, without admitting much air into the cavity;
you gain certain knowledge as to what is in the uterus; if there
is nothing, the hand will act as a stimulus to contraction; if
there are placental remains or clots, you have the opportunity
of removing at once one of the most certain causes of flooding,
primary and secondary, and of averting a very frequent cause
of puerperal disease. In many cases this operation is followed
by immediate success, little else being wanted. Often have I,
when called in to a bad case of puerperal fever, wished that
I could be fully assured there was nothing in the uterus. I
perfectly agree with Collins and others, who insist upon the
importance of this proceeding. I do not hesitate to repeat it
two or three times if the uterus fills again; and in this latter
case, the hand is ready to carry up the uterine-tube for the
injection of perchloride of iron. It is needless to insist that the
operation should be performed gently. The shirt-sleeve should
be rolled up to the shoulder; the back of the hand and the arm
be well oiled; the hand should be directed in the axes of the
pelvis; the other hand must support the uterus externally.

Ergot, we know, possesses the property of causing uterine
contraction. But how often does it fail to arrest uterine hæmorrhage? It is best given before the expulsion of the child. The
subcutaneous injection of *ergotine* is extolled by some. The
caution I would urge in respect to ergot is this: if the first
dose be not quickly followed by contraction, trust it no longer;

do not lose all-precious time in repeating it. If it fail to act at once, it is probably because the nervous power is too much exhausted to respond to the excitation. I am sure I have seen the administration of ergot, when the system was much prostrated by loss of blood, cause further depression, doing harm instead of good ; and if it does no harm, it is inert. In states of great depression, no absorption goes on in the stomach. Ergot, brandy, beef-tea, all alike simply load the stomach until rejected by vomiting.

Vomiting is, indeed, often beneficial. It seems to check flooding. On this indication some practitioners give an emetic dose of *ipecacuanha*. *Turpentine*, if the stomach will tolerate it, is a very efficient hæmostatic.

Cold is very much relied upon. How does it act ? Producing a kind of shock, it excites reflex action, inducing contraction of the uterine muscles. The essential condition, therefore, of its action is, that there be sufficient nervous power to respond to the excitation. If this power be deficient, then the shock of cold only adds to the general depression ; no contraction follows, or it is only momentary, the uterus quickly relaxing again. Nor does the harm cease here. The continuous application of ice and cold water will often cause congestion of the internal organs, and even lead to pleurisy, broncho-pneumonia, or peritonitis. I have seen cases of these forms of puerperal fever which I have no doubt were due to the patients being deluged in water, and being left for an hour or more chilled in wet clothes, because extreme prostration made it dangerous to change. Velpeau bears testimony to the same effect.

The rule, then, in the application of cold should be, not to trust to it, if it be not quickly successful in evoking uterine contraction. The best form of using it, if the case is slight, if the nervous energy is good, is to apply the cold hand, or a lump of ice, or a plate taken out of ice-water, to the abdomen or back of the neck. Taking a draught of cold water into the stomach will sometimes excite uterine action.

The douche method, practised by Gooch and extolled by Collins, of pouring a stream of cold water from a height upon the abdomen, is the most certain to evoke contraction, if any degree of contractility remain. It is open to the objection that it swamps the bed, and exposes the patient to the danger of subsequent chill and inflammation. A very efficient plan, and free from the foregoing objection, is to flap the abdomen smartly with the corner of a wet towel. A good plan is to inject cold water into the rectum. Cold is more effectual if applied internally. Levret was, I believe, the first who used ice in this way. Perfect says Levret "hit upon a very odd and ingenious expedient; he introduced a piece of ice into the uterus, which being struck with a sudden chill, immediately contracted and put a stop to the hæmorrhage." Of late years it has been a frequent practice to inject cold water into the uterus. I know it is often effectual. It has the advantage of washing out clots, as well as of exciting contraction; but I am not sure that it is free from danger. Levret's plan of placing a lump of ice in the uterus is the safest and best mode of applying cold. Ice also has the power, if continued long enough, of condensing the tissues, and thus of diminishing the calibre of the vessels. But I must repeat the warning, that unless there exist in the patient the power of reaction, cold will do harm instead of good. Unless it answer quickly, give it up. The necessary effect of great loss of blood is to reduce animal temperature. Cold can only increase this injurious effect. At this stage, warmth, not cold, is often necessary, as was well observed by Crosse.

Kneading the uterus, or compressing it firmly with the hand, is a means of exciting the uterus to contract much trusted to, and often useful. Even gentle friction or compression will often cause the uterus to contract and expel the placenta. A similar force applied when the uterus remains flaccid, or contracting spasmodically, will commonly induce firm, equable contraction. In states of greater exhaustion, with profuse flooding, even firm grasping is apt to fail. Whilst a strong hand is powerfully compressing the uterus, the hæmorrhage is certainly checked. But who can long keep up the requisite grasp? As soon as the tired hand relaxes, the hæmorrhage returns. Another hand takes the place of the first, and is in its turn exhausted; and so on, until the condition of the operator is almost as pitiable as that of the patient.

In resorting to this manœuvre, it is all-important to economize your own strength. You may do this greatly by placing the patient on her back near the side of the bed, so that you can stand well over her, and aid direct manual compression by the weight of your body. This manœuvre is often rewarded by success. If the hæmorrhage can be restrained for a few minutes, the strength will rally, and tonic contraction may return.

On the other hand, when there is deficient power, the uterus relaxes again and again; each renewal of flooding makes it more difficult to excite permanent contraction. Clots form again in the cavity, excite spasmodic action, and compel repeated introduction of the hand to remove them, for squeezing out will not answer well unless it be backed up by firm uterine contraction. Alternate relaxation and contraction of the uterus acts thus: expanding, it sucks up more blood from the aorta and vena cava; contracting, this will escape again on the uterine surface; and so the uterus goes on relaxing and contracting, drawing blood from the system and discharging it into the uterine cavity. Thus, the uterus acts like a Higginson's syringe pumping blood out of the body.

Not only is kneading uncertain, most painful to the patient, and exhausting to the physician, but it entails a special danger. This severe handling of the uterus, attended by bruising of the tissues, is liable to cause metritis. I have seen cases of metritic puerperal fever which I could only assign to this cause.

The compression of the abdominal aorta is a plan that deserves attention. It is practised in two ways. Ploucquet, who was the first to insist upon the method, compressed the aorta by the hand in the uterus. Baudelocque and Ulsamer compressed it through the flaccid abdominal walls. This is the plan most generally followed. It is recommended by Chailly, Caseaux, Jacquemier, and others. By it you may arrest the flooding, gaining a respite and time for the preparation of other means. It is performed in the following manner:—The patient lies on her back near the edge of the bed, the thighs drawn up. The physician stands at the right side; he presses the three middle fingers of his left hand gently into the abdominal wall near the umbilicus, curving them so as to fall obliquely upon the aorta, compressing it equally with all three fingers. The aorta is thus fixed upon the left side of the spinal column, avoiding the vena cava. The right hand can be used to aid the left, by supporting it. Compression kept up for a minute will often control the bleeding, giving time for the reproduction of nervous power.

and the contraction of the uterus. Faye relies mainly upon this method; Kiwisch objects to it, that compression on the aorta will simply compel the blood to go round by the spermatic arteries, and that the compression of the vena cava being unavoidable, the blood gets to the uterus by the numerous anastomoses of the pelvic circulation. The good resulting, he says, is due to the compression of the uterus. Boër and Hohl take similar objections. I have occasionally derived advantage from it; and look upon it as a momentary resource.

A form of pressure sometimes preferable to outward grasping of the uterus, is one recommended by Dr. G. Hamilton (*Edin. Med. Journ.*, 1861). The fingers of one hand introduced into the vagina are placed *under* the uterus, which the relaxed state of the parts usually allows, then, with the other hand upon the uterus externally, the organ is firmly compressed between the two hands. The cavity is closed by the anterior and posterior surfaces being flattened together. This manœuvre involves less violence, and is, for the time at least of its application, quite as effectual as external kneading.

Another form of applying pressure is by means of *the binder*, with or without a compress. I am a decided advocate for the binder after labour. That it is a source of comfort to the patient, almost every woman testifies; that it offers some security against hæmorrhage, by exciting contraction of the uterus and preventing accumulation of blood in the uterine cavity, scarcely any practitioner of experience doubts. It has another advantage: in severe flooding from inertia, if the woman be lying on her side or prone, the bulky flaccid uterus falls towards the diaphragm and forwards, encountering no support or resistance from the distended and paralyzed abdominal walls; this falling forwards, creating a vacuum, draws air into the vagina and cavity of the uterus, supplying the conditions for decomposition of clots, and thus favouring septicæmia. If the hand be introduced into the uterus, as it is likely to be in these emergencies, to remove placental remains or clots, air may often be *felt and heard* rushing in. Another urgent reason is this: just as, after tapping an ovarian cyst, the sudden release of the abdominal vessels from pressure, by diminishing the tension under which the heart acts, disposes to syncope. A well-applied bandage is a security against these possibly fatal accidents.

But it should not be applied until the uterus is contracted. Its use is rather to maintain contraction than to produce it.

Plugging has been practised in various ways. Paul of Ægina placed a sponge, soaked in vinegar, in the vagina and uterus. In later times, Leroux (1810) was a warm advocate of this method. D. D. Davis also advocated it. Rouget (1810) advised the introduction of a sheep's or pig's bladder into the uterus, and then distending it. Diday (1850) used a vulcanized caoutchouc bladder in this way. If there is nervous force enough to respond to the irritation provoked by the presence of these foreign bodies, the plan may be useful. But as excitants to contraction they are inferior to other means ; and, as plugs, I am sure they cannot be trusted.

Now, let us assume that the means which act by inducing uterine contraction have failed. Some men, indeed, affirm that they never fail to induce contraction. This is simply evidence that their experience is limited. No one largely engaged in consulting practice has ever made such a statement. The hæmorrhage goes on. Must we abandon the patient to a too probable death ? Is there no other condition for stopping the stream of blood from the open vessels than contraction ? There are styptics powerful enough. Why should we not use them ? If they will act precisely when other means fail, is it not unreasonable to reject their aid ?

This is the place to discuss the application of the styptic salts of iron. Until after the publication of this work, this remedy was not mentioned in English or French text-books ; it is still feared by many practitioners as a dangerous innovation. These men should ask themselves whether hæmorrhage to extreme degrees is not more certainly dangerous ? To withhold this remedy, then, from a woman bleeding to death, because it *may* do some immediate or ulterior harm, is at once illogical and wrong. The first pressing duty is to save the woman from dying. The case is, that, other means being exhausted, she will die unless local styptics be applied. Where, then, is the force of the objection that these styptics may work ulterior harm ? They cannot harm the dead. Is not this a legitimate *reductio ad absurdum ?*

As I am mainly responsible for the introduction of iron styptics into obstetric practice, at least in this country, it is desirable that I should give a more connected account of the matter than has hitherto been done.

The styptic virtue of the perchloride and other salts of iron has long been known. These salts have been applied to bleeding surfaces by swabs by many surgeons. The idea of applying styptics to the bleeding uterus is also not new; but until recent times there prevailed an unreasonable dread of touching the inner surface of the uterus with styptics. Yet from time to time they have been used with a success that might have encouraged us to look upon them with more confidence. Perfect says he stuffed the vagina with tow and oxycrate (vinegar and water). Hoffman succeeded in stopping a profuse hæmorrhage by passing pledgets of lint dipped in a solution of " colcothar of vitriol "—a mixture of sulphate and oxide of iron—as high into the vagina as possible.

Smellie says : " Others order ligatures for compressing the returning veins at the hams, arms, and neck to retain as much blood as possible in the extremities and head. Besides these applications, the vagina may be filled with tow or rags dipped in oxycrate, etc., in which a little alum, or saccharum saturni, hath been dissolved ; nay, some practitioners inject proof spirits warmed, or, soaking them up in a rag or sponge, introduce or squeeze them in the uterus, in order to constringe the vessels."

It appears from Hohl that the injection of perchloride of iron was first used by D'Outrepont ; but reference to the subject in those of his works which I have read has escaped me. It is strongly praised by Kiwisch,* who used a solution of two drachms of the muriate of iron in eight ounces of water. Professor Faye, of Christiania (*Obstetrical Journal*, 1874), says he has used it since D'Outrepont's time in his Maternity Hospital.

The first idea of using this means in uterine hæmorrhage was suggested to my mind on reading that perchloride of iron had been injected into an aneurismal sac. If the salt could thus be thrown almost into the stream of the circulation with safety, *à fortiori* the open mouths of the sinuses on a free surface might be bathed with it ; and surely the often irresistible and desperate course of uterine hæmorrhage was motive enough to try even a doubtful remedy. In the Lettsomian Lectures on

* " Beiträge zur Geburtskunde," 1846.

Placenta Prævia, delivered by me in 1857,* I recommended this practice. In a lecture on " The Obstetric Bag," published in the *Lancet*, in 1862, I again recommended that perchloride of iron should be carried for the purpose of arresting hæmorrhage. Again, at the Obstetrical Society's meeting, in February, 1865, I made the following remarks, which were soon afterwards reported in the medical journals.† The subject under discussion was puerperal fever. " As a means of preventing the loss of blood—as hæmorrhage undoubtedly predisposed to puerperal fever—I have found nothing of equal efficacy to the injection of a solution of perchloride of iron into the uterus after clearing out the cavity of placental remains and clots. I have used this plan for several years, and in a large number of cases after labour and abortion, and have always had reason to congratulate myself upon the result. The perchloride had the further advantage of being antiseptic. It instantly coagulated the blood in the mouths of the uterine vessels." As many of the cases referred to had occurred in consultation, of course my practice had been observed by many professional brethren. Dr. Mendenhall, of Cincinnati, published (*Cincinnati Lancet*, 1860), a successful case of injection of persulphate of iron. Scanzoni, in his systematic work on " Obstetrics" (edition 1867), recommends it to be used in the most desperate cases, and quotes D'Outrepont and Kiwisch as advising it. Since the publication of the first edition of this work, in 1869, the practice there recommended has been widely adopted; and it may fairly be said that an overwhelming mass of evidence has established its value. This evidence is now far too·copious to admit of even comprehensive analysis. I shall therefore limit the discussion to the objections that have been urged, and to some of the cases relied upon in support of those objections. It is incumbent, however, upon me to state the accidents that have attended the use of the remedy.

Reasoning from general knowledge, it may be apprehended that the injection of perchloride of iron into the uterus is not free from danger.

May not some of the fluid penetrate into the circulation, and

* " Lancet," 1857, vol. ii., and " The Physiology and Treatment of Placenta Prævia," 1858.

† " Obstetrical Transactions," 1866, vol. vii.

cause thrombi in the blood-vessels or heart? This risk is met by the property the iron possesses of instantly coagulating the blood it comes in contact with at the mouths of the vessels. The thrombi there formed protect the circulation beyond. Thrombi indeed always or nearly so form in the mouths of the uterine vessels under ordinary circumstances. These artificially-produced thrombi break down in a few days, and under the contraction and involution of the uterus, get extruded into the cavity. For some days the *débris*, small coagula formed by free blood, and serous oozing, come away. The discharge is black; it is apt to stain linen; and the nurse should be warned of this, lest she should take alarm at the unusual appearance. The discharge sometimes becomes a little offensive. This may be corrected by washing the vagina with Condy's fluid, weak carbolic acid, or weak solution of iodine. But the perchloride of iron is a valuable antiseptic; and I believe this property is useful as a preservative against the septicæmia to which women who have gone through a labour with placenta prævia are so prone.

Several disastrous accidents have happened from the injection of minute quantities of a solution of iron into nævi. Mr. R. B. Carter (*Med. Times and Gazette,* 1864) refers to a case where immediate death followed the injection of five minims into a nævus on the nose of an infant eleven weeks old. Another case is recorded by Mr. N. Crisp, of Swallowfield. Examination showed that the point of the syringe had penetrated the transverse facial vein, and that the blood in the right cavity of the heart had been immediately coagulated. Dr. Aveling has informed me of another case. Santesson (*Journ. für Kinderkrankh.* 1868) records one. He injected a few drops of the tincture of oxymuriate of iron of the Swedish Pharmacopœia into a nævus on the face of a child eight weeks old. Death occurred in a few minutes. The veins adjoining the nævus were found empty. The jugulars contained no blood in the upper part of the neck; but in the lower part, near the thoracic cavity, the blood was mostly coagulated. These clots becoming firmer and firmer downwards through the subclavian vein and vena cava superior, reached into the right auricle, and even into the right ventricle of the heart, which was distended with clotted blood. Mr. Kesteven relates a case (*Lancet,* 1874) in which a child died suddenly, after he had injected five minims into a nævus on the head. There was no autopsy; but Mr. Kesteven thinks it was not the result of embolism. He attributes it to "spasm of the glottis, from mental emotion." These cases suggest that there is some special danger attending the injection of perchloride of iron into nævi. It will be observed that an injecting force is used; that in one case the point of the syringe actually penetrated a vein; and that in two cases the blood in the venous cavities of the heart was coagulated. Did the perchloride reach the heart, and there cause coagulation? This is doubtful. The process, more probably, was as follows:—The moment the perchloride touched the blood in the facial vein it created a small clot, or thrombus. This thrombus was carried along the veins, and there served as a nucleus for further ✗ coagulation.

May not air be carried into the uterine sinuses, and thence into the heart? Possibly. The only case known to me in which symptoms causing a suspicion of this accident arose, was one of abortion, in which an injection was made by means of a clumsy india-rubber bottle. Symptoms like those described as following on air entering the circulation, and death ensued. In the face of these catastrophes, we cannot regard the application of the perchloride to the uterus without some misgiving. Why should not the styptic be carried in like manner along the sinuses into the iliac veins and vena cava inferior, and so to the heart? But there are reassuring considerations. The small, rigid uterus, with an imperfectly dilated os, as in abortion, is different from the large, flaccid uterus, with widely expanded os, ⋎ after labour. In abortion, I have already insisted that it is better to swab than to inject. The veins about the face and neck present peculiar facilities to the entry of injected air or fluid. Most of the recorded accidents in this region have occurred during operations where the veins come within the range of the aspiration or suction-force of the chest, no injection being used. Poiseuille denies that there is a suction-power in the abdominal veins. In this he is possibly wrong. But certainly the conditions are unlike those of the chest-walls.

May not some of the injected fluid run along the Fallopian tubes, and escape into the peritoneal cavity? Three cases are ⊹ known to me in which this accident happened. The first is related by Dr. V. Haselberg, and reported in full by me in the *British and Foreign Medical and Chirurgical Review*, 1870. The second was reported to me by Dr. Herman. It occurred at the London Hospital. The third is reported in *Joulin's Journal*, 1873. All these subjects were in the non-pregnant state. The two first cases are related in detail in the second edition of this work, and in my work on the "Diseases of Women." In the latter work I have discussed the conditions under which injected fluids run along the Fallopian tubes in the non-pregnant woman. In the lecture on "Induction of Labour" I have referred to cases of sudden death from injecting water into the gravid womb. These facts must be considered in relation to the subject.

Is there any other styptic equally efficacious and more safe than the salts of iron? The agent that comes most prominently into competition is *iodine*. In 1857 there appeared in the *North American Medico-Chirurgical Review* a communication by M. Dupierris, containing the histories of three cases in which iodine was used successfully to control *post-partum* hæmorrhage. The injection consisted of half an ounce of tincture of iodine. In 1871 (*Bull. de Thérap.*) Dupierris collected twenty-four cases, in which iodine had been resorted to. In all the result was successful, no accident occurring. Dr. Trask informs us that strong injections of iodine have been lately used in New York. But the amount of evidence is still too small to justify a decided opinion. And Dr. Trask's estimate of the dangers of perchloride of iron is based upon statements which, although publicly challenged, have never been substantiated.

I have always held it as an open question, whether perchloride of iron was the best styptic for intra-uterine application. That the contact of this substance with the inner surface of the uterus is not seldom attended by acute severe pain is certain. And although this may be affirmed of many other substances, and even of cold water, or a clot of blood, I am satisfied from observation that the perchloride of iron is especially apt to produce this pain. It is probable that the pain is in some measure due to the great excess of free acid in the perchloride. To obviate this objection, Piazza (see *Braun on the use of "Acid-free Iron-solutions in Gynæcology,"* *Wiener med. Wchnschr*, 1867) suggests neutralization by carbonate of soda, which forms common salt, and he says increases the styptic action. The persulphate of iron has been preferred by many practitioners in America. I have tried it. I believe it is not less efficacious than the perchloride, and it may prove to be less irritating. A form of iron that deserves further trial than I have yet been able to give it is the liquor ferri chloroxydi, recommended to me by Mr. Squire. It is almost neutral.

Too great injecting force may be a cause of pain. To obviate this, only the gentlest propelling force should be used. Associated with the pain occasionally caused by the iron-injection arises the question of *shock*. The truth of the following propositions, as expressed by Legroux, has been affirmed by many of the highest authorities. " 1st, *acute anæmia* (that is, succeeding to repeated hæmorrhages) lays the subjects under the imminence of instant, unexpected death, and calls for all the solicitude of the physician; 2nd, that the most trifling obstetrical manœuvres may add to this debility a fatal perturbation." Now, this is exactly the condition which exists in those cases where hæmorrhage urges to give prompt assistance. I have seen the simple introduction of the hand into the uterus followed by almost sudden death. I have seen the same thing from the injection of cold water into the uterus. I have seen it from the injection of perchloride of iron. Sometimes death occurs more slowly, as in a case which I saw with two medical friends in Camden Town. The patient had been delivered of her eleventh child. Hæmorrhage set in four hours afterwards. Dr. A. introduced his hand, and removed clot and a bit of placenta. On the third day he injected a warm solution of permanganate of potash. At the time the pulse was 120, the respiration ordinary. Immediately afterwards the pulse rose to 160 or more, and the respirations to 60, attended by great pain in the abdomen. He feared instant death, under the shock. She rallied somewhat, but died two days later.

I met Dr. Arthur Farre, on the case of a lady of exaggerated nervous susceptibility. Ergot had been given, and the utmost pains taken to prevent hæmorrhage. Notwithstanding, some hæmorrhage occurred. Dr. F. injected perchloride, feeling that in so delicate a woman even a moderate loss might prove fatal. She remained very prostrate. The injection was repeated; and the hæmorrhage was arrested. She was so low, that resorting to transfusion was anxiously discussed. The proposition was abandoned, under the fear that the most trifling disturbance might extinguish her. Under absolute rest she slowly recovered.

The case which has made the most impression upon me is that described by Dr. Bantock (*Obstetr. Trans.*, 1873). There had been considerable loss from "accidental hæmorrhage," occurring a fortnight before term. There was marked prostration from anæmia. I delivered by forceps. Acting on the experience that in these cases atony of the uterus is especially to be feared, I injected some perchloride of iron. It was attended by severe pain in the hypogastrium. A fluid drachm of nepenthe was given, but the pain increased in severity, and the patient died seven or eight hours afterwards. I am disposed to concur in Dr. Bantock's opinion, that the iron-injection was the immediate cause of death. The case is an illustration of the influence of shock in persons already depressed by hæmorrhage and labour.

One of the most accomplished practitioners in the provinces sent me the following history :—A lady whom he had attended before pregnancy for pulmonary hæmorrhage and albuminuria, who, to use his expression, " at her best had but a slender hold on life," having pale lips and a faint aortic bruit, came under his care in her fifth labour. She was delivered of twins. "Enormous hæmorrhage" ensued. She eventually recovered. In her next labour, the uterus was carefully supported throughout the labour. She had several doses of ergot. The uterus was steadily compressed until after the placenta was removed. But grasping it expelled a pint of blood and clot at one gush ; the flow continued. The patient became pulseless, blanched, unconscious. A stream of ice-cold water was injected into the uterus ; contraction ensued, she rallied, the binder was applied. About three hours after the labour flooding reappeared. The patient was again blanched ; unconscious, breathing heavily, in cold sweat, with little or no pulse. There was no external flow. The uterus contained about four ounces of firm coagula ; the vagina was full of blood and soft coagula. All this Dr. X. at once expelled ; but *"this time no contraction followed the irritation of the internal surface of the uterus."* Death seemed imminent. A little more than an ounce of liquor ferri was mixed with eight ounces of water. "There were still tails and shreds of coagula lining the uterus, which were rubbed off with a last hope of inducing contraction; but instead, warm blood was felt running along the hollow of the hand." The syringe-tube was passed up to the fundus uteri, air having been previously expelled, and the fluid was injected. Some blackened coagula came away, but the latter part of the injection came out uncoloured by blood. The hæmorrhage ceased from that moment. The patient cried out, " Oh, my legs, oh, my back; don't press me so; let me die in peace. I am getting blind, I can't see." She threw her arms about. She died in an hour. I have quoted this account in some detail, because it presents a vivid picture of the effects of hæmorrhage drawn by a trustworthy observer and skilful practitioner. Did the injection cause death? Possibly the shock attending the operation ✗ might have turned the trembling scale against life. But the symptoms that followed the injection are those of hæmorrhage. At any rate, we here see an example of ergot, compression and cold-water injection failing to prevent or to stop hæmorrhage. We see a uterus refusing to contract under irritation; we see hæmorrhage stopped under this inexcitable condition by a small injection of perchloride of iron. Had the styptic been used at an earlier stage, before extreme exhaustion, the result might have been more favourable.

It has been urged that the perchloride of iron may give rise to *septicæmia*. The evidence upon this point is not conclusive. In Dr. Routh's case (*Obstetr. Trans.*, 1873) of *post-partum* hæmorrhage, I injected a solution of equal parts of the weak tincture of steel and water. " On the third or fourth day puerperal fever set in," and ended fatally. There was no such continuity of symptoms as might be expected, had the septicæmia been started by the injection. The history differed in no respect from a multitude of cases of septicæmia, in which no injections had been used. And hæmorrhage is a condition which powerfully predisposes to septicæmia.

Another case cited at the discussion at the Obstetrical Society is open to still graver doubt. A solution of one in four of the tincture was used. The hæmorrhage was restrained. After three days pain set in, the lochia became arrested, and the patient died from puerperal peritonitis and other grave complications five weeks after delivery. In this case it transpired that the uterus had not been cleared out before injecting. A large clot was found which had been condensed by the injection. The subsequent decomposition of this probably set up the inflammation and septicæmia.

Dr. Heywood Smith's case (*Obstetr. Trans.*, 1873) cannot be admitted as telling against the judicious use of perchloride. The subject complained of severe pain in the hypogastrium on the third day, no injection having yet been made. On the tenth day hæmorrhage occurred, and continuing on the eleventh, an injection of one part of strong liquor ferri to eight of water was made. On the sixteenth, bleeding continuing, the injection was repeated; and again on the "eighteenth day with iron one in four; and again on the twentieth with equal parts of the solution of iron and water. On the twenty-first day a strong solution of iron was injected into the uterus with an intra-uterine syringe holding about two drachms. This produced severe pain, but completely stopped the hæmorrhage." On the twenty-third day the patient was delirious, and the discharge offensive and brown. On the twenty-eighth day she died. Autopsy showed a depression stained black, near the junction of the upper third with the lower two-thirds of the uterus. Near the centre of this depression an artery hung out more than an eighth of an inch, and close to it was "a rounded mass of placenta the size of a small filbert." "The anterior and posterior surfaces of the uterus were marked with black streaks." The case is in several respects remarkable. Injections of increasing strength were used, ending with the concentrated liquor, which is highly caustic. Although continuous bleeding lasted for several days, no attempt was made to dilate the cervix, to explore the cavity of the uterus for the cause of the hæmorrhage. Had this been done, the bit of placenta might have been removed, and in all probability the artery which it is presumed yielded the blood would have retracted. Had some perchloride been applied to the spot immediately after the removal of the bit of placenta, the woman's life might have been saved.

Faye, who has used the iron-injections freely, says: "The theory about blood-poisoning and *septicæmia* is a mere phantom" (*Obstetr. Journ.*, 1874).

I am not prepared to express so absolute a negative. But there are several considerations which, if fairly weighed, would moderate the confidence with which the affirmative is held by some men whose personal knowledge of the subject is very limited. 1st. I have probably resorted to iron-injections in severe post-partum hæmorrhages as frequently as any one in this country; and I declare that I have seen no case in which septicæmia could reasonably be traced to the practice; and I am not aware that any one who has had considerable experience, looks upon the remote risk of septicæmia as a sufficient reason for withholding the use of iron when death is threatened by hæmorrhage. 2nd. The cases reported, even assuming them to be correctly interpreted, are very few in number. It is doubtful whether they exceed or equal the proportion in which septicæmia occurs in cases of severe hæmorrhage in which iron-injections had not been used, or in those in which ergot, cold water or kneading the uterus had been trusted to. 3rdly. It must not be forgotten that hæmorrhage predisposes to septicæmia; and that quite independently of all treatment. The consequent relaxed state of the uterine tissue favours the collection of small clots in the uterus; the thrombi that form in the sinuses not being readily squeezed out into the uterine cavity and expelled linger *in situ*, undergo decomposition; whilst the increased activity of the absorptive process, brought about by the empty state of the circulatory apparatus, and the need of restorative nutrition, promotes the entry of septic matters into the blood. 4thly. Iron is a powerful antiseptic. For a while, at least, blood coagulated by it will be less liable to enter into decomposition than blood spontaneously coagulated. 5thly. The conditions that favour septicæmia are to a great extent under control. No one who throws iron into a uterus full of blood, is entitled to say that the iron caused septicæmia. The plain duty is to clear the uterus first. No one who throws repeated injections of iron, ending with a caustic solution, into the uterus whilst placenta adheres, is entitled to say that septicæmia was caused by the iron. The placenta should first be removed. But when these things have been done, small clots may still remain; and if contraction of the uterus do not follow, may give rise to septicæmia. Thus, Dr. Hicks said (*Obstetr. Trans.*, 1873), "The only case in which he had seen any trouble was one of severe flooding after twins. The injection was used with complete success.

Twenty-four hours after, pains arose, and it was found that the uterus contained hard blackened coagula which it could not expel. These were broken up and washed out, and the patient did well."

But, even on the hypothesis that shock and septicæmia are real dangers, do those who insist upon them to the extent of rejecting the perchloride of iron in extreme danger from hæmorrhage hold a very logical position? Certainly, women who are allowed to die outright of hæmorrhage will escape shock and septicæmia. And, if shock and septicæmia are dangers that must in any case be avoided, we should equally be called upon not to pass the hand into the uterus, not to inject cold water, not to give ergot. In short, we should, by this reasoning, be compelled to give up all treatment, for shock, and septicæmia and death have followed every mode of treatment.

It may be affirmed, as a general fact, that wherever the practice has been largely used, the greatest confidence in its efficacy and safety has been established; and that the condemnation of it is absolutely in proportion to the smallness of the experience of those who condemn. Thus it is largely adopted in London, in the provinces, in our Colonies, in the United States, in Dublin, in Christiania, in Vienna. In Edinburgh, the experience is limited. In a discussion at the Edinburgh Obstetrical Society (December, 1874), the prevailing opinion was one of suspense or opposition. It was doubted whether the remedy was necessary, since hæmorrhage could always be averted or arrested by other means. A most distinguished member stated that "in the Maternity Hospital there, he believed a death from post-partum hæmorrhage had never occurred." There is sometimes a difficulty in assigning a death to its main cause, where several causes have acted concurrently or consecutively. But cases of fatal hæmorrhage cannot be unknown in Edinburgh. I am informed in a letter from a former pupil and resident-assistant at St. Thomas's Hospital, written during his residence in Edinburgh (February, 1874), that "Professor Simpson's assistant, up to within four weeks ago, invariably told the clinical class that it was always possible to stop post-partum hæmorrhage by compression, etc., and that it was the fault of the practitioner if death occurred. But during the last month, and within a few days, he has had in his own practice two deaths arising in this way, and in spite of every means usually advocated. As a last resort he injected the iron; but not having it by him, valuable time being lost, the injection was used too late. The hæmorrhage, however, immediately stopped in both cases, directly the remedy was used; and this practitioner states that he shall be more careful in the future to have the iron by him. In neither case does he attribute death to the injection, but is confident that had it been used earlier, life would have been saved. The first case was absolutely dying when the injection was made, and died shortly after; the other one, although gasping, lived on until the next evening."

Let us now endeavour to lay down precise rules for using the iron-styptics.

In the case of hæmorrhage from abortion, it is proper, first, to dilate the cervix well with laminaria-tents; secondly, to pass a finger into the uterus to remove all placental *débris*; thirdly, if hæmorrhage still persist, to apply the styptic on a swab of sponge or cotton-wool attached to a piece of whalebone, or by the tubes to be described in the next paragraph. Injection is rarely necessary.

In the case of hæmorrhage from the non-pregnant uterus, the same method of dilatation and swabbing will commonly be the best. Where injection is thought necessary, a tube should be used provided with *very small holes at the extremity, directed backwards*, so that the fluid will emerge with moderated force in fine streams running downwards towards the os uteri. It is better also that the patient should lie on her back during the operation. I have lately used silver or vulcanite tubes having large slits at the uterine end, in which pieces of sponge, soaked with the iron-solution, are packed. When this is *in situ* the sponge is compressed by a piston-rod, and the styptic oozes out. I have these tubes of different sizes, so as to meet cases where the uterus is small, as in cases of abortion, and even after ordinary labour. By this means all injecting force is obviated.

I believe also that greater security is obtained by using a strong solution, not less than one part of perchloride of iron to ten of water. The constringing effect is greater, and thus there is a greater tendency to corrugate the inner surface of the uterus and to close the Fallopian tubes.

To show the effect of a concentrated solution, the following case is cited. Dr. Tissier (*Gaz. des Hôpitaux*, 1869), in a case of profuse flooding, non-puerperal, introduced into the vagina a plug of charpie dipped in strong perchloride of iron. The plug was removed in forty-eight hours. On the seventeenth day a piece of mucous membrane was discharged. The patient had a slow convalescence; and great contraction of the vagina followed. The perchloride acted as a powerful caustic, producing a slough of the mucous membrane. The case is valuable as an example of the greatest local injury that caustic perchloride of iron can cause. Bouchacourt relates a similar case (*Lyon Médical*, 1875). Atthill has used successfully the solid perchloride.

The case of the uterus after labour at term differs essentially from that of the non-pregnant uterus; and it appears to me that the difference is in favour of increased safety. The free opening of the cervix, permitting a hand to pass, is a security against retention of injected fluid. The ragged state of the lining membrane from decidual shreds and clots must serve as a great protection against the entry of injected fluid into the tubes or uterine tissues. The muscular contraction which is also generally provoked, is a further protecting condition.

The mode of using the perchloride of iron in post-partum flooding.—An important preliminary step is to clear the uterus of placental remains and clots, so that the fluid, when injected, may come into immediate contact with the walls of the cavity. At one time it was my habit to wash out the uterus with iced water first. I now prefer not to lose time in doing this. Besides, at the period when the perchloride is especially indicated, the exhaustion is generally so great that the injection of iced water is ill borne. I think that, under the circumstances, the injection of iced water is scarcely less hazardous than the iron injection, and without equal prospective advantage.

Place the woman on her back. Get an assistant to press firmly with a hand on either side of the uterus, whilst you inject.

You have the Higginson's syringe adapted with an uterine tube eight or nine inches long. Into a deep basin or shallow jug, pour a mixture of four ounces of the liquor ferri perchloridi fortior of the British Pharmacopœia and twelve ounces of water, or dissolve half an ounce of solid perchloride or persulphate of iron in ten ounces of water. The suction-tube of the syringe should reach to the bottom of the vessel. Pump through the delivery-tube two or three times to expel air, and to insure the filling of the apparatus with the fluid before passing the uterine tube into the uterus. This, guided by the fingers of the left hand in the os uteri, should be passed quite up to the fundus. Then inject slowly and gently. Cease injecting as soon as the effect of the styptic is noted. A few strokes is often enough.

The hæmostatic effect of the iron is produced in three ways : first, there is its direct action in coagulating the blood in the mouths of the vessels; secondly, it acts as a powerful astringent on the inner membrane of the uterus, strongly corrugating the surface, and thus constringing the mouths of the vessels; thirdly, it often provokes some amount of contractile action of the muscular wall. This contraction, if accompanied by constriction closing the os uteri, may prevent the escape of the styptic fluid. The concentric contraction of the uterine globe upon its fluid contents will then tend to drive the fluid along the Fallopian tubes. It is therefore imperative to observe the rule to pass one or two fingers through the cervix along with the tube. The fingers in this position further act as sentinels, informing you of the operation of the styptic in checking the hæmorrhage and in corrugating the inner surface of the uterus.

It is rare to find any renewal of hæmorrhage after the injection has been made as described. If there should be any return the injection may be repeated, first taking care that there is no clot in the uterus. If the discharge prove offensive, especially if the temperature and pulse rise, if there be uterine colic, or shivering, wash out the uterus with Condy's fluid, or a weak solution of iodine.

The injection of the perchloride to arrest hæmorrhage after labour has been followed :—1st. In many cases in which flooding persisted, in spite of the use of ordinary means, i.e., ergot, compression, cold-water injections, and after refusal of the uterus to respond to any excitation, by instant arrest of the bleeding and steady recovery. In a large proportion of these cases the patients have borne children subsequently.

2ndly. In several cases where profuse bleeding had taken place, it was also instantly arrested; phlegmasia dolens appeared. All the cases known to me recovered well in the end. One had had phlegmasia dolens after a previous labour, with profuse flooding, the injection not having been resorted to.

3rdly. In several cases where the patients were already *in extremis*, from loss of blood, the injection stopped further bleeding; but death ensued. In these cases I was unable to see that the injection in any way contributed to the fatal issue. The patients were moribund; the remedy was used too late.

The commentary which I would offer upon these three classes of cases is, that, of the first class, some women would clearly have died but for the remedy; whilst others were clearly spared an amount of blood which secured them against the severe secondary effects of flooding. In making this statement, I make full allowance for the difficulty that not seldom occurs in satisfying oneself that some of the women who recovered would not equally have recovered if they had been let alone.

That the supervention of phlegmasia dolens may possibly, in some cases, have been due to the extension of the thrombi formed in the mouths of the uterine sinuses by the action of the perchloride; but that, on the other hand, we cannot regard this thrombotic process as altogether injurious. It is undoubtedly often a conservative process, cutting off from the general circulation noxious matter generated on the surface of the uterus. And, again, it must be borne in mind that phlegmasia dolens is by no means uncommon after severe flooding, where the iron injection is not used; and this is especially true of placenta prævia. Finally, some of these patients would probably have died. It was surely better to run the hazard of phlegmasia dolens.

That, in the last class of cases, those in which death ensued after the injection, the cause being collapse from loss of blood, recovery was not to be expected from anything less than the supply of that which was lost. Here the remedy is transfusion. The practical lesson to be deduced from these unfortunate cases is, not to defer resort to the injection too long. It will, at almost any stage, stop the bleeding, but it cannot restore the blood which has been lost. It cannot recall the life that is ebbing away. Two remarkable illustrations of the combined use of styptic injections and transfusion are related in the section on "Transfusion." Although, therefore, the injection of perchloride of iron will not seldom rescue women from death by flooding after other means have failed, it is important to use it betimes, before the stage of utter collapse, whilst there is still inherent in the system the power of reaction.

This end will be secured if the warning I have given as to the extent to which ergot, compression, and cold should be trusted, be regarded. If these agents be abandoned the moment we observe that the expected response in uterine contraction is not given, we may then count upon the best results from injecting perchloride of iron. One more emphatic reason I would plead in favour of early resort to this treatment, is, that it is not enough to save bare life; we must save all the blood we can. For this reason, and having acquired confidence in the efficacy and safety of this crowning remedy, I never now lose time by persisting in other means that may fail, and which, having failed, even partially, leave the patient the worse by the quantity of blood lost.

We have, then, three stages of hæmorrhage to deal with :

First, there is hæmorrhage with active contractility of the uterus. Here the diastaltic function may be relied upon; excitants of contraction find their application.

Secondly, there is a stage beyond the first, when contractility is seriously impaired, or even lost. Here excitants of contraction are useless. Our reliance must be upon the direct application of styptics to the bleeding surface.

Thirdly, there is the stage beyond the two first, when not only contractility, but all vital force is spent—when no remedy holds out a hope unless it be transfusion; and when even this will probably be too late.

In these cases, and pending the preparation for transfusion, we should put in practice the method of restoration proposed by Nélaton in collapse from chloroform, namely, raising the body so that the head depending may receive by gravitation what blood there is left in the vessels.

A few words upon the *restorative treatment*. The great physiological principle of *rest* must govern all treatment. The system drained of blood, and therefore barely nourished enough to keep the ordinary functions going on the most reduced scale compatible with life, can ill tolerate any unnecessary call. The first effort must be directed to produce reaction from the stage of collapse. This is especially necessary if the patient have been much prostrated by the persistent use of cold. And to prostration from this cause is added that from the rapid cooling influence of accelerated respiration.

Since cold is so universally trusted to, and is pushed so far, it requires some firmness to insist upon resort to its opposite, warmth. But unless you would see your patient sink under collapse, you must restore the circulation. To this end, hot bottles should be applied to the feet; wet, cold linen removed, if possible, without disturbing the patient; if not, warm dry clothes must be wrapped around her; gentle friction of the hands and feet may be used. Cordials and stimulants must be given in small quantities, *warm*. It is the common practice to give everything cold, under the fear that warmth in any form is likely to cause a renewal of the bleeding. But we are now dealing with a state of vital depression greatly caused by cold. You will not cure by making the patient colder. Cordials are, no doubt, increased in efficacy if taken warm. What are the best cordials? The best of all, undoubtedly, is cognac. Give it at first in small sips, diluted with equal parts of hot water. A good mull of claret and cinnamon or nutmeg is often useful. Watchfulness is required lest over-anxiety ply the weak, all but paralyzed stomach too freely. The power of imbibition is very feeble; if you pour in any quantity, it simply accumulates, loads the stomach, until it is forced to reject the whole by vomiting.

It is at this stage, or when the system is rallying, that *opium* is so valuable. Opium is, in my opinion, decidedly contra-indicated during the flooding. It then tends to relax the uterus. But when the object is to support the system, to allay nervous irritability, there is no remedy like opium. It may be given in doses of thirty or forty drops of laudanum or Battley's solution, every two or three hours. In confirmation, I cannot help quoting the conclusion of the late Dr. Beatty: "That opium is capable of inducing hæmorrhage when given soon after delivery, I once learned to my cost. Opium given pre-maturely must do mischief. Ergot is our sheet-anchor at first, when we want to arrest the hæmorrhage; and opium afterwards, when we wish to restore exhausted vital action." If the stomach will not bear it, it may be given in an enema. It may be worth while to try the subcutaneous injection of $\frac{1}{8}$ to $\frac{1}{4}$ grain of mor-phia; but this must be done cautiously.

When a little strength has been gathered up, you must think of nutrition. "Little and often" is a good maxim. What you give, if it be absorbed, is soon used up; but if you give much it is lost by vomiting; therefore, the supplies must be frequent and moderate. When the stomach will not tolerate even small quantities, the difficulty may be tided over by enemata, consisting of six ounces of beef-tea or milk, an ounce of wine or brandy, and ten or twenty drops of laudanum. A state of atonic dyspepsia lasts for some time; and during this the stomach must be most sedulously studied.

When reaction has set in, the heart beats painfully, the temples throb, agonizing headache follows. Then salines, of which the best is the fresh-made acetate of ammonia, are most serviceable. The opium can be combined with it. Nothing tranquillizes the hæmorrhagic excitation like salines; and it is more than probable that salines supply a pabulum the circulation wants. Salines continue to be useful for many days during what has been aptly called the "hæmorrhagic fever." Ice will now be useful; it quenches thirst, and acts as a sedative. Do not be in a hurry to give iron. If given prematurely, it parches the mucous membranes, increases headache and fever. In short, it cannot be borne at this stage; it comes in usefully at a more advanced period of convalescence. Five or ten-drop doses of tincture of digitalis, or a minim of hydrocyanic acid, added to the saline, are often of great service in calming the irritability of the heart.

In cases of great exhaustion from hæmorrhage, do not suffer the patient to rise on any pretext; linen can be changed without quitting the recumbent posture; and it is better to use the catheter two or three times a day than to incur the risk of sitting up. In fine, all depends upon constant attention to the smallest details, upon good, noiseless, gentle nursing. Guard your patient against all excitement, even from the effort of talking. Bar the hall-door against officious, sympathizing friends. After labour with flooding, there is great danger of defective contraction of the uterus. This state disposes to secondary hæmorrhage, and lays the circulation open to the invasion of septic matter. To counteract this, give three times a day a mixture of quinine two grains, dilute sulphuric acid ten minims, liquor of ergot twenty minims, or some equivalent medicine. Fordyce Barker insists much upon this. I can attest the value of the practice.

LECTURE XXX.

Secondary Puerperal Hæmorrhage.

WHEN the immediate perils of childbirth seem to have been weathered; when the patient has been deemed safe from hæmorrhage; at any rate, at a period more or less remote from labour, varying from a few hours to two or three weeks, this terrible complication may yet make its appearance. But if we have taken due care in the conduct of the labour, and especially in the management of the placenta and any primary hæmorrhage, we shall rarely experience the mortification of seeing secondary hæmorrhage. This will be evident if we examine the principal ascertained causes of this catastrophe. They are :

A. *Local.*—1. A portion of placenta or membranes may have remained *in utero*.

2. Clots of blood may have formed and been retained.

3. Laceration or abrasion of the cervix, vagina, or perinæum, or a vesico-vaginal or recto-vaginal fistula.

4. Hæmatocele or thrombus of the cervix, vagina, vulva, or perinæum.

5. Chronic hypertrophy, congestion or ulceration of the cervix uteri.

6. General relaxation of the uterine tissues.

7. Fibroid tumours and polypi.

8. Inversion of the uterus.

9. Retroflexion or retroversion of the uterus.

10. Pelvic cellulitis or peritonitis fixing the uterus.

B. *Constitutional or remote* conditions causing disturbance of the vascular system.

1. Emotions.

2. Sexual intercourse.

3. Returning ovarian action, manifested commonly at the end of one month from labour.

3A. Ovarian action, favoured by failure to suckle, and imperfect involution of the uterus.

4. Heart disease, including imperfect involution.

5. Liver disease, Bright's disease.

6. Leucocythæmia.

7. General debility of tissue, malnutrition of nervous system, and irritable heart from anæmia.

A. 1. *Retention of Placenta.*—This is perhaps the most common cause. It is that which I have most frequently met with, it is that which may be most surely avoided. Since it has become a recognized practice not to allow the placenta to remain more than an hour, it must be very rare to find secondary hæmorrhage occurring from retention of the entire placenta, or even the greater part of it. But I have been called in to such cases. In these the flooding has usually come on within twenty-four hours of labour. The explanation has been that rigidity of the cervix uteri rendered it impossible to get the placenta away. I have removed the whole placenta several days, even a week after labour. In all these cases there was not only flooding—indeed, flooding was not always the urgent symptom—but there were unmistakable marks of septicæmia: a pallid straw-coloured complexion, with hectic flush; rigors; quick pulse, running over 100, to 120 or 130; often foul breath, suggesting the odour of decomposing lochia; thirst; distressing perspirations alternating with hot skin; vomiting; general prostration. The local signs were: some distension of the abdomen; tenderness on pressure over the uterus, which could be felt rising perhaps as high as the umbilicus; on examination by vagina, the uterus was felt enlarged; the cervix more or less open, at least admitting one or two fingers, and on withdrawing the finger, it was found smeared with muco-purulent sanguineous fluid generally of peculiarly offensive odour; and, sometimes, retention of urine.

These symptoms are enough to justify, even to demand, exploration of the cavity of the uterus. The tenderness of the uterus and abdomen and the partial closure of the cervix render the introduction of the hand, or even a finger, a very distressing operation. You may sometimes acquire information to guide further proceedings by passing in the uterine sound. This will give you the exact measure of the cavity; and you may be able by it to make out the presence of a foreign body in the cavity. If the sound goes four inches or more beyond the os, you may infer that the uterus is of abnormal size, and that it contains something solid. If you perceive any fluid issuing from it, or feel any substance in it, you must make up your mind to pass your finger fairly into the cavity, so as to explore fully, and to bring away whatever is in it. At an early period, that is, within two or three days after labour, you may squeeze out placenta or clot, by grasping the uterus firmly above the symphysis.

It will be next to impossible to remove an entire placenta without passing your hand into the uterus; and this is an operation of no small difficulty. If the cervix be so contracted as not to admit more than a finger, it will be wise to introduce a fagot of three or four large laminaria-tents, and to leave the patient for a few hours to let them expand. The next step may be to introduce the medium or largest sized caoutchouc bag, and distend with water until there is room to pass three or four fingers. When you can do this, the patient being anæsthetized, by steady pressure, supporting and pushing the fundus down upon the inside hand by the hand outside, you will succeed in getting your left hand into the cavity. This done, take care to grasp well the whole mass of the placenta, and withdraw it. When you have got it out, you will be glad to get rid of it as quickly as possible. There is, perhaps, no stench more offensive than that of a placenta that has decomposed *in utero*.

This being so, what must be the condition of the uterus itself? It clearly wants disinfecting. It should be washed out with chlorine water, Condy's fluid, weak solution of carbolic acid, creasote, weak iodic solution, or perchloride of iron; and if none of these are at hand, with tepid water. The uterine tube should be carried fairly into the uterus. The operation should be repeated two or three times a day for some days. An excellent plan is to irrigate the parts by means of such an apparatus as that contrived by Dr. Rasch. The fluid is carried into the uterus by a syphon, and runs out by a discharge-tube, which empties itself into a vessel by the side of the bed.

More common is the retention of a portion of placenta. This may be either adherent or loose. It is more likely to have adhered at the time of the removal of the bulk of the placenta. In this case the uterus remains large; there is constantly tenderness on pressure; often there is an offensive discharge attending or alternating with hæmorrhage. Comparing a considerable number of observations, I find that where portions of placenta are retained the secondary hæmorrhage commonly appears at the end of seven or eight days. At about this period the progress of involution brings the walls of the uterus into closer contact with the foreign body in its cavity, which then becomes a more active source of irritation.

The same proceedings must be adopted as in the case of retention of the whole placenta. But there may be great difficulty in separating all the adherent portions from the surface of the uterus. You must take pains to break off all that will readily come, carefully avoiding digging your nails into the substance of the uterus. It often answers, as in the case of early abortion, to break up any adherent masses if you cannot detach them. It is far better to be content with this than to err by excess of pertinacity. If a portion project at all prominently in a polypoid form, I strongly advise removing it by the wire-écraseur, which does it cleanly, and without injuring the uterus. Having done your best in removing *débris*, wash out with disinfecting fluid.

You may generally infer that the uterus is free from placental remains if, at the end of ten days, 1, there is no bleeding; 2, the os uteri closes; 3, the uterine sound goes not more than three inches.

2. A *blood-clot* may form in the uterine cavity from imperfect contraction after labour; it grows by receiving accretions on

its surface; and being compressed by the uterus, it becomes condensed into a mass having the shape of the cavity. This acts as a foreign body. It excites spasmodic action; it attracts blood to the uterus; and the congestion is relieved by oozing of blood and serum from the uterine wall. Some of this escapes externally, but not always in sufficient quantity to arrest attention as "flooding." The napkins are sodden with a serous fluid streaked with blood, and often somewhat offensive to the smell. This serous discharge is indicative of retained clot.

Sometimes, when the uterus remains flaccid, the clot is not moulded and compressed, but more blood collects, and forms a number of dark, loose clots, mixed with fluid blood.

In either case the uterus must be emptied. It is advisable, first, to try to empty the uterus by squeezing. If this is not satisfactory the fingers or hand must be passed into the cavity. Here, again, irrigation by disinfectants is advisable.

I ought to mention, for the benefit of those who are led by authority rather than by reason, that the immortal Harvey used the sound for diagnosis, and intra-uterine injections as a remedy, in cases of this kind. His knowledge of uterine pathology, and his practice, were indeed vastly superior to those of the great bulk of our modern English physicians. He was superior in freedom from prejudice, and therefore in diagnostic and therapeutical success.

3. *Laceration or abrasion of the cervix* is more frequent than is commonly suspected. Hæmorrhage from this cause is sometimes protracted and serious in quantity. We arrive at the diagnosis by a method of exclusion. If we find the body of the uterus firmly contracted, of size corresponding to the date after delivery—a point accurately determinable by bi-manual examination and the introduction of the uterine sound—our attention is fixed upon the state of the cervix. A rent here may then be felt, and even seen by the speculum. The blood is usually florid. The treatment is to apply a solution of perchloride of iron to the wound by a swab, carried through the speculum, or on pledgets of lint, in the form of a compress or plug. Dr. Roper was called to a patient of the Royal Maternity Charity eight hours after delivery. The pains had been very violent. The placenta had been expelled. There had been much hæmorrhage; but it had stopped. The uterus was *well*

contracted. On the sixth day very profuse flooding occurred; life seemed in jeopardy. Dr. R. found a large ragged laceration extending the whole length of the cervix uteri on the left side as far as the reflection of the vagina. The uterus was well contracted. Dr. R. passed dry cotton-wool into the gaping cervix, and filled the vagina too. Next day I saw the woman with Dr. R., and swabbed the rent with perchloride of iron. Fifteen days after delivery there was another severe flooding. Dry cotton-wool was again used. Two months after delivery a wide red cicatrix remained. The healing was perfect, without deformity.

Laceration of the Vagina or Perinæum.—Any part of the vagina may be torn during labour. Laceration of the upper part is, commonly, an extension of rupture of the lower part of the uterus; and symptoms indicating the gravity of the injury will be apparent at the time of labour. Laceration of the middle part of the vagina is rare; but laceration of the lower part, merging in rent of the perinæum, is common. This may become a source of secondary bleeding. Digital and visual examination will reveal it. If not discovered for some days, the better course will be to stop the bleeding by compresses steeped in perchloride of iron, if necessary, and to leave restoration to future consideration.

A little bleeding may also continue from vesico-vaginal or recto-vaginal rents or fistulæ; but the quantity is rarely enough to cause anxiety as hæmorrhage.

4. *Thrombus.*—A collection of blood may form before, during, or after labour, in the sub-mucous tissue of the cervix uteri, some part of the vagina, or of the perinæum or vulva. Thrombus of the cervix uteri is apt to occur in connection with hypertrophic elongation of that part. The part most exposed to the formation of blood-tumour is the vulva, especially the labia majora. The folds of skin and mucous membrane which constitute the labia, enclose a copious arrangement of veins, arteries, cellular filaments. The vessels anastomose freely. Rupture at any part gives free vent to blood, which finds ample space to accumulate. In pregnancy the veins are frequently varicose, always excessively full of blood, with a disposition to stagnation; they easily rupture, sometimes under external violence, sometimes spontaneously. Most frequently, one labium only is

affected. Thrombus may occur at any period of pregnancy;
is more frequent as the term of gestation approaches; and most
common during labour. The blood-tumour mostly forms when
the head is about to clear the vulva. The rupture of the
vessels is thus the result of the excessive distension to which
they are subjected by the pressure upon the soft parts above.
I have known a large thrombus dissecting the vagina away
from the rectum following the administration of ergot. Occasionally the tumour is developed rapidly before the head comes
down to the outlet, forming a mechanical obstacle to the completion of labour. The distension is at times so rapid and great
that the walls burst, and considerable hæmorrhage of an arterial
character takes place.* But more often, although the rupture
of the vessels may occur before the passage of the child, the
tumour is developed gradually after its birth. The passage
of the head, carrying before it the inner layer of the labium,
increases the lesion of the vessels. There is a glacier-like movement of the mucous membrane upon the subjacent tissues.
These *post-partum* thrombi are especially dangerous, because
they are more liable to be overlooked. They may burst; their
walls may undergo mortification.

Acute pain generally marks the beginning of thrombus, due
probably (Caseaux) to the rupture of the vessels. The effusion
may be limited to the loose tissue of the vulva; but it may be
very extended. Thus Caseaux relates a case in which he traced
the blood dissecting up the peritoneum all up the iliac fossa,
where it formed a large coagulated mass; it extended, still
behind the peritoneum, up the left and posterior part of the
abdomen, as high as the right hypochondrium, bathing all the
cellular tissue surrounding the kidney, and even to the attachments of the diaphragm.

If the tumour burst, the hæmorrhage may be so great and
rapid as to prove quickly fatal. If it fail to burst, it may
become so large as to close the vagina, and lead to retention of
lochia (Lachapelle); it may also cause retention of urine, or
fæces.

The *diagnosis* is not always easy. When the tumour has
acquired a moderate size, it presents a shining aspect of purple

* *See* an excellent lecture on this subject by Fordyce Barker, " Medical
Record," 1870; also, this author's " Puerperal Diseases," 1874.

or bluish-black colour. It tends to occlude the entry to the vagina, whilst the finger, passed above it, defines its extent and relations. Its rapid formation with pain, the hardness of the tumour if the blood has coagulated, and its fluctuation if the blood remains fluid, are characteristic.

The *prognosis* is serious. Deneux collected sixty-two cases, of which twenty-two were fatal. In twenty-one of these the child also was lost. The cause of death is usually loss of blood, internal or external. Gangrene or suppuration may prove fatal at a later period. Other cases have ended by resolution, suppuration, bursting, peritonitis, or septicæmia.

Hugenberger * summarizes the issues of the puerperal effusions of blood into the cellular tissue, observed by him— 1. Perinæal hæmatoma before labour, consecutive abscess, perforation of rectum, recovery. 2. Labial hæmatoma, apparently arising during labour; suppuration of the burst blood-gathering, pyæmia, death. 3. Labial hæmatoma before labour; suppuration of the burst gathering during childbed; fatal metroperitonitis and pyæmia. 4. Labial hæmatoma, bursting, recovery. 5. Labial hæmatoma, incision, recovery. 6. Labial hæmatoma, incision, recovery. 7. Labial hæmatoma, incision, recovery. 8. Peri-vaginal after labour, bursting, recovery. 9. Peri-vaginal after labour, bursting, recovery. 10. Peri-uterine during labour; violent labour-pains, fatal hæmorrhage. 11. Peri-uterine after labour, with cross-birth and turning; bursting of the sac, and death by bleeding into the abdominal cavity.

The two following cases illustrate some of the features of hæmatocele or thrombus:

R. M. Charity, 31st Oct., 1859. Primipara: ordinary labour. After it, midwife observed a mass protruding externally, which she took to be bladder and vagina prolapsed. I saw her two hours after labour. There was a soft tumour, the size of a child's head, projecting from pubes to anus. In front it seemed tense, shining, translucent; fluctuating; its sides presented similar characters; the circumference, the base, was continuous with the skin of labia and thighs. Posteriorly was an inflexion of the tumour, having an anterior lip much ecchymosed, and posterior lip forming a sharp crescent, the whole

* "St. Petersburg Medical Zeitung," 1865.

much resembling the os uteri after labour. Pursuing examination up this orifice I found that the swelling was caused by the enormous distension of the labia vulvæ, especially the left, by blood and serum. On the internal aspect of left labium, about one inch from orifice, was a jagged hole which communicated with the sac formed by the effusion in the labium. Through this some sanious matter escaped on pressing the swelling. It was a rent made in the mucous membrane by the head in the birth. Fomentations used. During the two succeeding days the tumour much diminished. She did well.

R. M. Charity, June, 1863. *Hypertrophy; procidentia; hæmatoma of cervix.*—Sent for by midwife in great alarm, she thinking the whole uterus had come out after child and placenta. I found patient prostrate, cold, agitated; a large mass lay forth beyond the vulva of dark colour, like coagulated blood, resembling in bulk and colour the placenta. I next thought it was the inverted uterus. Some hæmorrhage from it. The mass was soft, its covering easily lacerating on slight pressure; it was found to consist of two lobes, and between them was an opening admitting two or three fingers some distance. This was the os uteri; the two lobes were the lips of the cervix uteri, enormously enlarged by infiltration with serum and blood. On pressing the mass to reduce it, the mucous membrane easily tore, and blood oozed out. To guard against this, I covered the whole with a napkin; then by careful and gradual compression it was returned within the vagina. A good perineal compress was applied to prevent descent. Patient did well.

This was a most aggravated example of the contusions and injury which the cervix undergoes during labour. It was hypertrophied and elongated.

It is probable that a varicose condition of the cervix, similar to that which so frequently exists in the labia vulvæ, may be the cause of thrombus of the cervix.

Dr. Murray relates (*Obst. Journ.*, 1873) a case in which severe hæmorrhage occurred in two successive labours before the passage of the head. The source appeared to be the bursting of a mass of varicose veins—a form of hæmatocele—just within the cervical zone. On one occasion hæmorrhage recurred a fortnight after labour.

The *treatment* will vary according to the time of the appearance of the tumour. If it form before the descent of the child, we may first endeavour to restrain the effusion by ice and pressure; if this is not successful, whether the tumour by its bulk impede labour or not, I believe it is better to open it by the lancet, and to deliver by forceps. Thus the danger of further injury under the attrition of the head is lessened. If the tumour have burst, and hæmorrhage be at all profuse, the first effort should be to deliver by forceps, if the head present. Then we can employ means to arrest the bleeding. F. Barker advises to enlarge the opening by incision, to clear out the clots, and to compress the cavity with lint soaked in solution of iron. When the hæmorrhage is fairly arrested, care is required to avoid sloughing and septicæmia from decomposition of the clots. The styptic plug should be replaced by dressings with carbolic acid oil, and syringing with a solution of carbolic acid, or permanganate of potash or iodine. It is well to remember that hæmorrhage may arise from laceration of the vulva apart from thrombus (*see* p. 349).

5. *Chronic hypertrophy, congestion or ulceration* of the cervix, will be detected by touch and by speculum. The parts may be touched lightly with nitrate of silver every third day. Astringent lotions are also useful.

6. *General relaxation of the uterine tissues* is mostly associated with systemic debility and malnutrition. Constitutional treatment is here especially serviceable. Iron, strychnine, phosphoric acid, cinchona, ergot are particularly valuable.

7. The complication of *fibroid tumours and polypi* has been previously discussed (*see* p. 285). If first discovered some days after labour, the treatment is still the same as that recommended when found at the time of labour. In the case of fibroid tumours, we must first restrain hæmorrhage by the topical application of perchloride of iron; sometimes the tumours can be removed by enucleation. Polypi should be removed by the wire-écraseur.

8. *Inversion* has been already discussed (*see* p. 376.)

9. *Retroflexion of the uterus* is, in my experience, a frequent cause of secondary hæmorrhage. The displacement no doubt occurs soon after the labour, the heavy fundus falling backwards, whilst the tissues are still in a relaxed state. In some cases, I ascertained that there had been retroflexion in the non-

pregnant state. It is highly probable that there has frequently been antecedent retroflexion, and also, that occurring for the first time after a labour, the condition becomes permanent. Ed. Martin says retroflexion and anteflexion of the labour are caused by non-involution of the placental-site. Thus, if the anterior wall be the placental-site, it will remain larger than the posterior wall, and so throw the fundus backwards. When it occurs, the free return of blood from the body of the uterus is impeded by the flexion at the neck; the involution of the fundus and body is arrested, these parts become congested and relaxed, the body is found bulky, and the pressure upon the pelvic organs no doubt also favours local hyperæmia.

The diagnosis is made out by the finger in the vagina feeling the rounded mass of the fundus of the uterus behind the os uteri, bulging downwards and forwards the posterior and upper part of the vagina; by the finger in the rectum, which determines the rounded mass of the fundus even more accurately; by the finger in the vagina passing up in front of the os uteri to meet the hand pressed down from above the symphysis, revealing the absence of the uterus between them; and still more absolutely, by the uterine sound, the point of which must be turned back to enter the body of the uterus.

The treatment is, first, to restore the uterus. This may be done by the sound at once, aided or not by the pressure of the finger in the vagina or rectum below on the fundus. It may be maintained in position by an air pessary placed in the fundus of the vagina, or better still, by a large Hodge's pessary.

If bleeding continue after restoration, the interior of the uterus should be swabbed with perchloride of iron. The constringing effect of this application, by lessening the bulk of the fundus, tends still further to correct the retroflexion. At a later stage, the insertion of a five-grain stick of sulphate of zinc inside the uterus every fourth day will much conduce to bring about a healthy condition.

10. *Perimetritis fixing the Uterus.*—When this condition arises, involution is impeded; the uterus is liable to engorgement, the circulation being difficult; and as there is commonly some shedding of epithelium from the cervical cavity and os externum, bleedings are frequent. Rest and quinine best promote absorption of effused fibrin. The bared surface may be

touched occasionally with nitrate of silver or nitric acid, if the bleeding is troublesome.

B. The management of secondary hæmorrhage depending upon the constitutional or remote causes must obviously consist in avoiding or lessening the influence of those causes, and in pursuing the treatment indicated for the cure or mitigation of the diseases disposing to uterine hæmorrhage.

Some women are so excitable that the mere application of the child to the breast will cause hæmorrhage. Here we have an example of the influence of breast-irritation acting unfavourably. The normal influence of suckling promotes uterine contraction; and thus may be regarded as a provision against hæmorrhage.

We may observe the influence of ovarian irritation in the not infrequent occurrence of hæmorrhage exactly a <u>month</u> after labour. This is more especially the case where suckling has not been instituted. The ovaries then more readily assert their power, and, resuming active work, excite the menstrual flow, which easily exceeds the normal quantity, assuming the proportions of hæmorrhage.

The enumeration of the conditions which predispose to, or cause hæmorrhage (*see* the Table, p. 563) will, in many cases, suggest the indications, prophylactic or curative, of treatment.

LECTURE XXXI.

TRANSFUSION.

WHEN the patient has sunk to the last extremity from loss of blood, the hope that springs eternal in the human breast under severest trials, is still justified by the remarkable recoveries that have followed the operation of transfusion.

This hope will be strengthened by the reflection that the recoveries—resuscitations they may fairly be called—have occurred under the most desperate circumstances. The operation appears terrible to the bystanders; the conditions of success are not yet established; there are practical difficulties attending it; but there seems no reason to doubt that every difficulty may be overcome, and that we shall not long have to regret that the operation occupies an equivocal position amongst the resources of obstetric medicine.

Under extreme prostration from hæmorrhage the stomach either rejects nutriment, or the faculty of conversion and absorption is lost. Nor can the bowel be trusted to. In this conjuncture, assimilation being either too slow or altogether at a standstill, what remains but to throw aliment directly into the circulation? It is the last hope.

Two great questions to settle are, first, the form in which the renovating fluid shall be prepared; secondly, the apparatus to be used. It is obviously essential that the fluid be in a condition to answer the great end of rallying the powers of the system, that it enter as an harmonious constituent into the body of the patient. It is also essential that the apparatus be simple, and of convenient manipulation.

Three kinds of fluid have been used for the purpose: 1st,

whole blood, as it flows from the vein of the giver; 2nd, *defi-brinated blood;* 3rdly, fluids artificially prepared, chiefly *saline solutions* to which alcohol may be added, or *milk.*

The most animated contest has been raised between the advocates of whole blood on the one side, and of defibrinated blood on the other. But in truth, exclusive merit does not rest with either. Both have been used with success, and accidental circumstances attending a particular case may fairly determine which is to be preferred.

1. We will first discuss the question of *whole blood. Primâ facie,* it seems most natural to conclude that pure blood would be the best. Many cases are on record in which the trans-fusion of pure blood has been attended with complete success. The objection to it is its tendency to coagulate. This property has often baffled the operator, by plugging the apparatus or the vein near the point of injection; and sometimes by the blood clotting in the heart. The blood of many animals begins to coagulate almost immediately on escaping from the vein. Human blood sometimes begins to coagulate in a minute, and within a few minutes, four or five, there is no security against this event. Can coagulation be in any way averted? Oré shows that <u>cold</u>, and preventing contact of air, retard coagu-lation. Hewson showed that various salts prevent coagulation. Dr. Richardson (*Cause of Coagulation of the Blood*) proved that ammonia prevented fibrillation, and he used it for this purpose in transfusion.

We may, then, preserve the fluidity of the blood by keeping it from contact with the air; or by the addition of various salts or ammonia. To obviate contact with air, we must adopt the immediate or vein-to-vein system. This is the plan adopted by Oré and Aveling. Aveling's apparatus is simple. It may be described as a small Higginson's syringe without valves, and armed at each end with a silver canula to enter the vessels. It is, in truth, a continuous tube with a dilatation in the middle. This dilated part holds two drachms. It may be used to propel the blood, if flowing slowly. By pinching the tube on either side of the expanded portion, valves are dispensed with. If a bit of glass tube be interpolated, the operator can see if the blood is flowing. In animals the apparatus answers perfectly. It has been doubted whether a sufficient supply could be drawn by it from the human giver. But numerous experiments of Richardson and Oré answer this objection. Air may be got rid of by first pumping water through the instrument, or better still, a saline solution. But although avoiding contact with air retards coagulation it does not give security against it.

Some form of injecting apparatus has been most generally used. The blood has been collected as it flows from the vein of the giver in a reservoir, generally a funnel communicating directly with the pipe of the syringe, so that the blood is propelled partly by gravitation. An excellent instrument of this kind is Mr. Higginson's, who by it saved ten lives out of fifteen, using whole blood (see *Liverpool Med.-Chir. Journ.*, 1857, and *Obst. Journ.*, 1873). In this apparatus there is no piston; gravitation is aided by squeezing the barrel, which is made of vulcanized india-rubber. The apparatus of Dr. Little, figured and described in the *London Hospital Reports*, 1866, has answered in practice. Dr. Richardson, who prefers mediate transfusion, uses a glass syringe holding about eight ounces, to which is attached an elastic delivery-tube. The syringe has a piston with a jointed rod. Where the rod is united to the piston, this is perforated, so that on pulling up the rod fluid will run through, getting below. On pushing the rod down, this opening is closed, and the fluid is propelled. A little water is first put in the syringe. The blood is received direct from the giver into the syringe, the jointed rod and perforated piston admitting of this without any difficulty. Coagulation is prevented by adding about three drops of ammonia to every ounce of blood. This enables the operator to proceed leisurely and without fear of obstruction.

2. *Defibrinated blood* is by some preferred, on the grounds that it obviates the inconveniences and danger of coagulation; that the fibrin is really useless as a renovating ingredient; and that the use of defibrinated blood renders it unnecessary to bring the giver into the sick room. All the preparations can be made in an adjoining room.

In using defibrinated blood, the blood is first received into a basin (not less than eight ounces should be collected); then it is stirred round rapidly with a clean silver fork or a glass rod, to let the fibrine cohere and be removed; it should then be filtered through fine muslin, to stop any small coagula or air-bubbles which might be dispersed in the fluid.

The fluid thus prepared may be poured into Wagstaffe's or Little's or McDonnell's apparatus, or pumped directly from the basin by Aveling's syringe. If human blood cannot be procured, a lamb may be used. This has been successfully done (*see* Hasse, *Allg. Wiener Med. Zeitung*, 1873). To supply an answer to a vulgar dread that with the blood of animals some noxious vital principle may be imparted, it ought to be enough

Fig. 124.

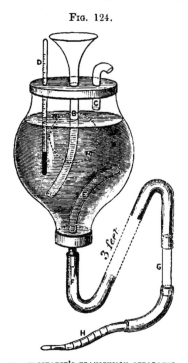

MR. WAGSTAFFE'S TRANSFUSION APPARATUS.

A, a glass cylinder holding about half a pint, closed above by a cork, the under surface of which is coated with gutta percha. B, tube, funnel-shaped, externally to receive the fluid; it dips to near the bottom of the cylinder, and to one side. C, a small tube, allowing of escape of air from the cylinder during filling. D, thermometer. E, outflow tube. F, india rubber tube, 3 ft. long. G, glass tube. H, nozzle for insertion into the vein of patient. It has its hole about half an inch from the extreme point, so as to allow of its being partially withdrawn to clear the hole without removing the nozzle.

to remember that man lives upon the flesh and blood of animals; and that it cannot matter whether lamb's blood be taken first into the stomach or directly into the veins. Moreover, F. Galton put the doctrine of pangenesis to the test by breeding from rabbits of a pure variety, into whose circulation blood

P P

taken from other varieties had previously been largely injected. The results were absolutely opposed to the idea of pangenesis; for the animals continued to maintain in their offspring the purity of the breed. Blood is mere nutriment. It is not the seat of the original developmental or formative force.

Mr. Wagstaffe (*Obst. Journ.*, 1874) has constructed an apparatus upon the following principles : " 1. That suction rather than artificial pressure should regulate the admission of the fluid (into the patient). 2. That great precaution is advisable to prevent the admission of air with the fluid. 3. To prevent the admission of solid sedimentary or other foreign matter (or, in the case of blood, to hinder the formation of coagulum). 4. That the amount of artificial pressure used should be capable of ready estimation. 5. That the temperature of the liquid should be indicated and maintained. 6. That the rate of flow should be known, and therefore the proper action of the apparatus determinable." These conditions all seem to be fulfilled in his apparatus.

When respiration is going on, the natural suction-power is considerable. It is a moving force in the circulation ; and for the time being the fluid supplied and the vessel which connects it with the patient's vein form part of the circulation. *A priori*, it is desirable to let the operation proceed as far as possible on this principle. In the course of one of Mr. Wagstaffe's cases, it was observed that " during inspiration the flow was rapid ; when she was crying out, there was some regurgitation of fluid." Mr. Wagstaffe used milk as the renovating fluid.

An apparatus, the value of which in transfusing defibrinated blood has been proved by successful experience, is that of Dr. McDonnell (*Trans. of Dubl. Obst. Soc.*, 1871). It consists of a strong glass pipette, capable of holding about seven ounces, to the lower end of which is attached an elastic tube, and to the end of this the nozzle for insertion into the patient's vein. It acts primarily by gravitation ; but, this being insufficient, propulsion is gained by strongly blowing into the pipette. The apparatus is extremely simple ; but I would suggest that it might be improved by being made double the size. It would thus gain in hydraulic pressure.

3. *Saline Fluids.*—But it will not rarely happen that to procure blood is impossible, or entails too much loss of time. In

such cases I would strenuously urge the expediency of giving the sinking woman the chance of recovery promised by saline
✗ alcoholic injections or milk. If such injections will restore a patient dying from a disease such as cholera, which to extreme draining of the circulating fluid adds a dire poisonous influence, it may fairly be expected that they will act not less effectually where there is nothing but the loss of blood to contend with. Dr. Hodder, of Toronto (*Practitioner*, 1873), in two patients moribund from cholera injected fourteen and twenty-eight ✗ ounces of pure milk. Both recovered.

An argument in favour of saline fluids lies in the fact that the drained circulatory system labours not alone under the loss of the vivifying element, but also under the purely mechanical difficulty of vacuity. The heart and arteries have nothing to contract upon. Now the simple injection of ten or twelve ounces of fluid restores, to a certain extent, the dynamic condition of the circulation. And it is proved by Little's experience that saline solution is well tolerated. The case of Dr. Henderson cited (*see* p. 584) shows the value of saline injection, and also the importance of supplementing it by the addition of defibrinated blood. Probably further observations will prove that the best fluid is a mixture of saline solution and defibrinated blood. I have referred to the dynamic relations of venous injections in connection with the history of a case in which I resorted to the mixed saline and blood. (*Lancet*, 1874).

The method of saline injections has the incontestable advantages of being always available at short notice; of avoiding risk of failure from coagulation; of enabling us to use simple forms of apparatus; and of being a more simple proceeding.

If, then, it be determined to employ saline injections, we have first *to prepare the fluid*. This may be either Mr. Little's fluid, adding two drachms of pure alcohol to the pint, or a similar solution, omitting the chloride of potassium. Three or four pints of this may be prepared in a basin, and kept at a temperature of 98° to 100°. Mr. Little's solution is composed of—

Chloride of sodium . . .	60 grains.
Chloride of potassium .	6 ,,
Phosphate of soda . .	3 ,,
Carbonate of soda	20 ,,
Distilled water. .	20 ounces.

If these ingredients cannot be procured, a suitable saline may be improvised by dissolving a teaspoonful of common salt and half the quantity of carbonate of soda in a pint of water, at the temperature of 98° to 100° Fahr., or a solution of phosphate of soda alone, or water with a few drops of ammonia in it. The solution should be of about equal specific gravity to that of blood. It should be filtered.

The apparatus may be Mr. Little's, figured in the *London Hospital Reports*, 1866.

With this apparatus, the whole operation can be performed by one person. The vessel holds forty ounces; a lamp underneath and a thermometer in it regulate the temperature; from the tap near the bottom proceeds a thick india-rubber tube four feet long, with a silver nozzle at the end. When this instrument was placed at the bedside, about on a level with the patient's head, and the nozzle inserted into a vein at the bend of the elbow, its contents flowed into the vein in about ten minutes. The instrument might be made to contain eighty ounces. In Mr. Little's more successful cases, eighty ounces were introduced at a time, generally in between twenty minutes and half an hour. These were cases of cholera.

Such an apparatus cannot conveniently enter into the obstetric bag, but it might very well be a part of the armamentarium of lying-in hospitals.

But a small Higginson's syringe armed with an appropriate nozzle, as used by Dr. Woodman and Mr. Heckford, should be carried for emergencies of this kind; or Aveling's little instrument might serve for the saline injections as well as for immediate transfusion. Wagstaffe's apparatus would answer perfectly for defibrinated blood, or for saline fluids.

Dr. Woodman informs me that he examined the blood-globules in a patient who had undergone saline injection; he found them quite unaltered. Pure water, we know, will affect the globules.

Whichever the method preferred, it is well to obviate the fatal tendency arising from the rapid cooling of the body, by keeping warm bottles to the feet and chest; and to raise the body above the level of the head, as advised by Nélaton in cases of asphyxia from chloroform.

The Operation.—The instruments required in addition to the injecting apparatus are—1. An ordinary bleeding-lancet, or what is better, a fine-pointed small knife mounted on a handle. 2. A pair of dissecting forceps. 3. A silver probe.

Failing a special apparatus *ad hoc*, you need not therefore abandon the operation. Almost any injecting apparatus may be made to answer. Thus I have found an aspirator-syringe, adapted to work as a stomach-pump, do perfectly. One of the aspirator-trocars was made to do duty for a canula by filing off the sharp point before inserting it into the vein.

The proceedings will of course be modified by the method selected. If we determine on the use of defibrinated blood or saline solution, two distinct operations have to be performed. First, there is the renovating fluid to prepare ; secondly, there is the injection itself to be done.

Let us first prepare the fluid and the apparatus. This should be done in an adjoining room. The best subject we can get to give the blood is bled in the ordinary way, about ten or twelve ounces of blood being received in a clean basin. This is whipped with a silver fork or a glass rod to separate the fibrin. It is then filtered through muslin, to free it from small coagula and air-bubbles. To the filtrate may be added an equal quantity of Little's saline solution ; and this addition is especially important when the quantity of blood at our disposal is small. The apparatus is then charged with the fluid, and worked so that we are sure there is no hitch or defect, and that air is expelled from it. The fluid is kept at about 98° Fahr.

We then prepare for the injection. Expose a vein at the bend of the patient's elbow. To do this, pinch up a fold of skin ; transfix it transversely to the course of the vein ; and, if necessary, dissect down gently to the vein, so as fairly to bare it for a quarter of an inch. Seize the vein with forceps ; open it by a longitudinal slit. It may be useful to pass a fine probe under it to secure it.

Now, having first allowed some of the renovating fluid to flow through the tube and canula, insert the canula in the vein, directing its point towards the heart.

The fluid may now be allowed to flow into the patient's system. This should be done very gently and slowly. If McDonnell's or Wagstaffe's apparatus be used, the natural suction-force, aided perhaps by a moderate and regulated gravitation-force, will carry the fluid on. If a syringe, as Higginson's, Aveling's, or Richardson's be used, more especially the two latter, a little propelling-force may be required. And here great delicacy is essential. Propulsion by squeezing the barrel, or by driving a piston, is inferior in smoothness to the force of gravitation. Sudden jerks or too rapid propulsion are to be avoided.

Watch the effect upon the patient. When ten or twelve ounces of fluid have been taken in, the operation is usually completed.

The canula is then removed, whilst you compress the wounded vein with your thumb. Over the spot, apply a small compress of lint or rags, and bind up the arm *secundum artem*, that is, according to an almost lost art, which many of you may never have seen, and may never see. A broad piece of tape is placed by its middle over the compress, the two heads are carried, one above the elbow-joint, the other below; they are then brought forward in the opposite direction, so as to form a figure of 8 round the elbow; they are then crossed just over the compress and tied.

If we determine on direct or vein-to-vein transfusion, you still first get your apparatus in order. Aveling's apparatus is the most convenient for the purpose. It should be charged with saline fluid, and so worked as to expel all air. Next prepare the patient. Expose and open a vein as already described. Commonly a little thin serous blood escapes. Let an assistant compress the vein with his thumb whilst you open the vein of the giver. Into this insert the canula in the distal direction, that is, towards the giver's hand. Keeping it *in sitû*, insert the other canula in the patient's vein, directing its point towards the heart. All being arranged, the blood may be allowed to flow. Sometimes the natural suction-force of the patient, aided by the natural driving-force of the giver's heart, will give movement enough. But if the current flags, movement is imparted by squeezing the barrel or ball of the instrument. To send the blood in, you close the tube by compressing it with finger and thumb between the barrel and the giver; then, squeezing the barrel, the contents must go on into the patient's vein. This done, to replenish the barrel, you let go the giver's side of the tube and close the patient's side; the barrel expanding, draws in blood from the giver. Thus charged again, you repeat the proceeding described as often as necessary. The barrel holds about two drachms, so that the charges being counted, you know very nearly the quantity injected.

Before connecting the veins of giver and receiver, I think it better to inject two or three ounces of the saline fluid.

In this mode of proceeding we have in the saline fluid an additional security against fibrillation; at the same time, additional security against the entry of air; and, further, what is almost certainly an advantage, a quantity of saline fluid injected as well as blood. A reservoir containing saline fluid might be so adapted to the tube as to give a further supply to mix with the blood as desired; or we might throw into the barrel or expanded portion of the tube a drop or two of ammonia by means of a hair-trocar or a subcutaneous injecting syringe, from time to time, so as to ensure liquidity of the blood.

Whole blood may also be transfused by Higginson's apparatus. In using this also, it would, I think, be advisable to secure against coagulation by putting some saline or a little ammonia in the apparatus.

A very small quantity of blood will sometimes suffice. Patients have rallied and recovered in a marked manner after the injection of six or eight ounces of blood, whole or defibrinated.

The Indications for the Operation.—The quantity of blood lost is no criterion. We must look to the state of the system. Absence of pulse, sinking of the features, gasping for air, jactitation, unwillingness to be disturbed, arrest of absorption, are signs of threatening dissolution. It should be our endeavour to act before these are far advanced. We shall come by-and-by, I have little doubt, to regard transfusion and venous injection as much less formidable operations than they now seem; we shall then take courage to act earlier, before the case is desperate, and the results will be proportionally better.

As to Re-injection.—If the patient flag again, showing symptoms of sinking, the operation should be repeated.

Transfusion has been performed before delivery in cases of placenta prævia, where the patient was too prostrated to be able to survive delivery unless previously rallied. Many patients perish under the attempt to deliver them when in a state of extreme depression. If rallied first, they may bear artificial delivery, and recover.

In cases of hæmorrhage during and after labour, it has been apprehended that the new blood transfused might again escape on the uterine surface. This source of failure may be overcome by the local application of perchloride of iron. The two methods of arrest by styptics, and of replenishing by transfusion, should be thoughtfully studied together. In most cases, the timely use of the styptic will render resort to transfusion unnecessary. But in some we should be prepared to follow up the styptic by transfusion.

This suggestion has been successfully carried out since the publication of the second edition of this work, by my friends, Drs. Henderson, Johnston, and Little, of Shanghai. The case is one of the most interesting in the records of obstetric medicine. It is related in detail in the *Lancet* (1874). Mrs. A. had had five children. She was very anæmic from protracted lactation and long residence in the East. The labour being slow, Dr. H. ruptured the membranes, gave ergot, and the nurse followed the fundus uteri down during the expulsion of the child. The placenta not being cast, the patient was put upon her back, when the first flooding took place. Dr. H. had to detach the placenta, which he found closely adherent to the fundus. The womb was then felt to contract, but it soon relaxed, and flooding recurred. The situation was "sufficiently alarming." Ergot was again given. The uterus continued to contract feebly and relax alternately. After a hurried consultation it was decided, in the first instance, to try the effect of external and internal cold. The long elastic mount of Higginson's syringe was passed fairly to the fundus, and iced water was freely injected into the cavity. On withdrawing the syringe, a piece of ice was placed in the uterus, and other pieces in the vagina. By these means and by compression, the bleeding was at last arrested, the uterus contracting fairly. She was then left, Drs. H. and J. "congratulating themselves on the successful issue of the treatment by cold. But the bleeding returned." The uterus was found full of blood, and oozing was going on. The patient was blanched; her surface and extremities cold; the breathing sighing and interrupted; the pulse feeble and intermitting." After immediately emptying the uterus, twelve ounces of the iron-solution, prepared according to Dr. Barnes's directions, was injected. As " he withdrew the pipes the uterus contracted, *and the hæmorrhage ceased, never again to recur.* So far reason for satisfaction; but Mrs. A.'s appearance was ghastly in the extreme, and I feared that Dr. Barnes's powerful remedy had been used too late to save life." It was then decided to resort to transfusion. Mr. Little assisted. The operation was delayed for some time, as she appeared to rally a little under the use of some soup prepared out of raw meat and hydrochloric acid. But she became worse, and " appeared to be sinking rapidly." Little's fluid was accordingly used. " The change in the patient's condition which followed the operation was no less rapid than wonderful; the marked improvement in breathing, pulse, and general appearance, leading us to form a much more favourable opinion as to the ultimate result. This was at 4 p.m. At 6 p.m. her condition was, if possible, worse than before. Twelve ounces of blood were taken from the husband, defibrinated and strained. The apparatus (Little's) being adjusted, some saline fluid was first injected, then some of the defibrinated blood was added to the saline fluid, when it was ascertained that the first fluid was satisfactorily entering the circulation." The result of this last transfusion was favourable beyond the most sanguine anticipations. " The pulse once more filled and steadied; the breathing became full and regular; and by slow degrees warmth returned to the extremities." She made a perfect recovery.

One of Dr. Ringland's cases is scarcely less interesting as illustrating the propositions I have so much insisted upon, that, although styptic injection may stop hæmorrhage, death may follow from the loss sustained; and that the hope of rescue lies in transfusion. The subject, in her two previous labours, had flooded; " her life had, moreover, been on two occasions placed in imminent jeopardy by hæmorrhage connected with miscarriage."

In her third labour, which was under Dr. Ringland's care, the secundines were expelled entire within ten minutes after the birth of the child. In consequence of her hæmorrhagic diathesis, Dr. R. gave two doses of ergot, and retained a grasp of the uterus for an unusually long time. This pressure was continued even after the pad and binder were applied. After

some time, notwithstanding, severe hæmorrhage set in. The
uterus responded to the grasp of the cold hand, but almost
immediately relaxed, and the hæmorrhage returned with renewed violence. Brandy and ergot were given with temporary
benefit. Injection of cold water into the uterus, and the introduction of the hand in its cavity, were had recourse to, but
with no better result, contraction following the application, but
relaxation instantly succeeding. Dr. George Johnson assenting,
a solution of perchloride of iron was injected into the uterus.
Subsequent to its employment, not one drop of blood was lost.
She had brandy, cold beef-tea, ammonia; but still the pulse
indicated excessive prostration, being easily quite extinguished.
She became very restless, sighed frequently, and respiration
became very laboured. It was felt that transfusion was the
only hope. Dr. McDonnell came. During the necessary preparations the patient's state became materially worse. About
twelve ounces of defibrinated blood was injected. Marked
improvement was immediately manifest. She made a perfect
recovery. Dr. R. gave an anodyne soon after. This lady
having come to reside in London, was by Dr. Ringland recommended to my care in her next labour. Impressed by her
history, the utmost pains were taken before and throughout
her labour to counteract her tendency to hæmorrhage. She had
ergot. Aided by the same nurse whom she had brought from
Ireland—and a more efficient assistant I never had—the uterus
was kept under constant control. The placenta was expelled,
not taken away. The uterus contracted. It was firmly compressed by pad and binder; and we thought she would this time
escape all serious loss. But the old diathesis asserted itself.
The uterus relaxed; rather free hæmorrhage occurred. I felt it
was not a case to dally with. Dr. Ringland's experience proved
that, although one might obtain contraction, it would only be
temporary. I therefore at once injected about eight ounces of a
solution of iron. This stopped the hæmorrhage. There was no
return. The puerperal state was perfectly physiological. She is
now in good health.

These cases of Dr. Henderson and Dr. Ringland form the
best possible commentary upon the principles advocated in this
work. Had they not proceeded to transfusion, the women
would in all probability have perished, and it might have been

concluded that the iron-styptic which stopped the bleeding, had been the cause of death.

I would further urge the expediency *of extending the application of transfusion of blood, or of the injection of saline fluid, to other conditions of great emergency as well as to hæmorrhage.* In three conditions, the indication is especially strong. I mean puerperal convulsions, extreme prostration from uncontrollable vomiting in pregnancy, and puerperal fever. In convulsions with albuminuria, the blood is undoubtedly poisoned; most men of experience in this disease still recognize that venesection is beneficial; that the benefit thus derived is partly due to the abstraction of a portion of the poison, it seems reasonable to believe; and there seems to be ground for hope that the injection of healthy blood or saline fluids, to replace the poisoned blood abstracted, may be useful in diluting the poison still circulating and in rallying the system from the prostration that threatens to be fatal. I believe one successful case of this kind has been published.

With regard to the application in cases of extreme exhaustion from obstinate vomiting, I have even less hesitation. Here there is undoubtedly not only a high degree of anæmia, but the little blood circulating is degraded by admixture with the products of diseased action. The source of nutrition is stopped up; there is no other door open for the admission of fresh supplies than through the veins.

Dr. Tyler Smith narrated* the case of a woman seemingly dying of puerperal fever, in which he injected a solution of one part of liquor ammoniæ to three of water, to the extent of half a drachm, into one of the veins of the forearm. Recovery ensued. Here, as in other cases where strong ammonia has been injected, considerable inflammation arose at the point of injection. The practice here was suggested by that of Dr. Halford, of Melbourne, who injects ammonia as a remedy in snake-bites. I am much inclined to think that the injection of a saline fluid, after the formula of Dr. Little, to the extent of six or eight ounces, will be found more practicable and useful.

After-treatment.—When transfusion has been performed, great care is still required to economize and to increase the little store of heat and strength the system may retain. Take care that

* " Obstetrical Transactions," 1870.

warm dry linen surround the patient's body; apply hot bottles to the feet, legs, and sides of the body; when opportunity offers, throw an enema of a pint of warm gruel or beef-tea, containing an ounce or two of brandy and twenty minims of laudanum into the rectum. When the woman can swallow, give small quantities of hot brandy and water.

To encourage resort to transfusion, even in extreme collapse, the experiments of Brown-Séquard may be cited. He made the following observations on decapitated crimimals. Decapitation took place at 8 a.m. Eleven hours later all trace of irritability had disappeared. Injection was then made of blood drawn from the veins of the subject into the radial artery. It went in vermillion, and returned dark as in life. In ten minutes irritability had returned; movements in the muscles could be excited.

Is it not strange that in these days, when blood is looked upon as so precious, when venesection is almost gone out of use, when the dread of losing a pint of blood by an operation is too often an obstacle to relief, that restoring this vital fluid to a patient dying for the want of it, should be still struggling for a settled place in obstetric medicine?

I think it ought to be an accepted aphorism in medicine, *that no one should be permitted to die of hæmorrhage.* It may be long before this aphorism will be realized in practice. But we should never rest until the means of preventing dangerous losses of blood, and of restoring those who are in danger from such losses, shall be brought to perfection.

INDEX OF AUTHORITIES.

INDEX OF SUBJECTS.

DISEASES OF WOMEN.

——◆◦◆——

THE SURGERY, SURGICAL PATHOLOGY, and Surgical Anatomy of the Female Pelvic Organs, in a Series of Coloured Plates taken from Nature: with Commentaries, Notes, and Cases by HENRY SAVAGE, M.D. Lond., F.R.C.S., Consulting Physician to the Samaritan Free Hospital. Third Edition. Revised and greatly extended, with an additional Plate, and 36 Engravings and Special Illustrations of the Operations of Vesico-vaginal Fistula, Ovariotomy, and Perineal Operations. 4to, 35s.

A PRACTICAL TREATISE ON UTERINE DISEASES. By J. HENRY BENNET, M.D. Fourth Edition, 8vo, 12s.

LECTURES ON THE DISEASES OF WOMEN. By CHARLES WEST, M.D., F.R.C.P. Third Edition, 8vo, 16s.

DISEASES OF THE OVARIES: their Diagnosis and Treatment. By T. SPENCER WELLS, F.R.C.S., Surgeon to the Queen's Household and to the Samaritan Hospital. 8vo, with about 150 Engravings, 21s.

OVARIOTOMY. By THOMAS BRYANT, F.R.C.S., Surgeon to Guy's Hospital. 8vo, 4s. 6d.

SCHROEDER'S MANUAL OF MIDWIFERY, including the Pathology of Pregnancy and the Puerperal State. Translated into English by CHARLES H. CARTER, B.A., M.D. 8vo, with Engravings, 12s. 6d.

UTERINE DISORDERS: their Constitutional Influence and Treatment. By HENRY G. WRIGHT, M.D., M.R.C.P. 8vo, 7s. 6d.

ON UTERINE AND OVARIAN INFLAMMATION, and on the Physiology and Diseases of Menstruation. By E. J. TILT, M.D., M.R.C.P. Third Edition, 8vo, 12s.

By the same Author,

A HANDBOOK OF UTERINE THERAPEUTICS and of Diseases of Women. Third Edition, post 8vo, 10s.

THE COMPLETE HANDBOOK OF OBSTETRIC SURGERY; or, Short Rules of Practice in every Emergency, from the Simplest to the most Formidable Operations connected with the Science of Obstetricy. By CHARLES CLAY, M.D., late Senior Surgeon and Lecturer on Midwifery, St. Mary's Hospital, Manchester. Third Edition, Enlarged, with 91 Engravings, fcap. 8vo, 6s. 6d.

A MANUAL FOR HOSPITAL NURSES AND OTHERS ENGAGED IN ATTENDING ON THE SICK. By EDWARD J. DOMVILLE, L.R.C.P., M.R.C.S., Devon and Exeter Hospital. Second Edition, Revised and Enlarged, crown 8vo, 2s. 6d.

J. & A. CHURCHILL, NEW BURLINGTON STREET.

A CLINICAL HISTORY OF THE MEDICAL AND SUR-
GICAL DISEASES OF WOMEN. By ROBERT BARNES, M.D., F.R.C.P.,
Obstetric Physician to St. George's Hospital. With 169 Engravings. 8vo,
28s.

OBSTETRIC APHORISMS for the Use of Students commencing
Midwifery Practice. By J. G. SWAYNE, M.D., Physician-Accoucheur to
the Bristol General Hospital. Sixth Edition, fcap. 8vo, with Engravings,
3s. 6d.

OBSTETRIC MEDICINE AND SURGERY (The Principles
and Practice of). By F. H. RAMSBOTHAM, M.D., F.R.C.P. Fifth Edition,
with 120 Plates on Steel and Wood, 8vo, 22s.

THE PUERPERAL DISEASES: Clinical Lectures by FOR-
DYCE BARKER, M.D., Obstetric Physician to Bellevue Hospital, Fellow of
the New York Academy of Medicine, Honorary Fellow of the Obstetrical
Societies of London and Edinburgh. 8vo, 15s.

J. & A. CHURCHILL, NEW BURLINGTON STREET.

THE

OBSTETRICAL JOURNAL

OF

GREAT BRITAIN AND IRELAND,

INCLUDING

MIDWIFERY AND THE DISEASES OF WOMEN AND CHILDREN.

EDITED BY J. H. AVELING, M.D.

This is the only periodical published in the United Kingdom which is entirely
devoted to the above important branch of Medicine. The information it contains is necessarily of interest to the greater part of the Profession—not to
Specialists only, but also to General Practitioners, whose daily work consists so
largely of Midwifery and the Treatment of the Diseases of Women and
Children. The necessity of such help as this Journal affords is evidenced by
the Association of Medical Men into "Obstetrical Societies" for mutual aid and
counsel; such Societies exist in each division of the kingdom—that of London
alone including nearly 700 members.

Each number contains Original Communications (illustrated, when necessary
by lithographic plates and woodcuts)—Reports of Hospital Practice—Abstracts
of Societies' Proceedings—Obstetric, Gynecic, and Pediatric Summary—News—
and a Leading Article.

All the back numbers (from April, 1873) can be supplied at 1s. 6d. each.
Annual subscriptions, including free delivery or postage to any part of the
United Kingdom, 20s.; to Foreign Countries within the Postal Union, 21s.;
other Foreign Countries and the Colonies, 24s.

J. & A. CHURCHILL, NEW BURLINGTON STREET.

London, New Burlington Street,
October, 1875.

SELECTION

FROM

MESSRS J. & A. CHURCHILL'S

General Catalogue

COMPRISING

ALL RECENT WORKS PUBLISHED BY THEM

ON THE

ART AND SCIENCE

OF

MEDICINE

INDEX

THE PRACTICE OF SURGERY:

a Manual by THOMAS BRYANT, F.R.C.S., Surgeon to Guy's Hospital. Crown 8vo, with 507 Engravings, 21s.   [1872]

THE PRINCIPLES AND PRACTICE OF SURGERY

by WILLIAM PIRRIE, F.R.S.E., Professor of Surgery in the University of Aberdeen. Third Edition, 8vo, with 490 Engravings, 28s.   [1873]

A SYSTEM OF PRACTICAL SURGERY

by Sir WILLIAM FERGUSSON, Bart., F.R.C.S., F.R.S., Serjeant-Surgeon to the Queen. Fifth Edition, 8vo, with 463 Engravings, 21s.   [1870]

OPERATIVE SURGERY

by C. F. MAUNDER, F.R.C.S., Surgeon to the London Hospital, formerly Demonstrator of Anatomy at Guy's Hospital. Second Edition, post 8vo, with 164 Engravings, 6s.   [1873]

THE SURGEON'S VADE-MECUM

by ROBERT DRUITT. Tenth Edition, fcap. 8vo, with numerous Engravings, 12s. 6d.   [1870]

THE SCIENCE AND PRACTICE OF SURGERY:

a complete System and Textbook by F. J. GANT, F.R.C.S., Surgeon to the Royal Free Hospital. 8vo, with 470 Engravings, 24s.   [1871]

OUTLINES OF SURGERY AND SURGICAL PATHOLOGY

including the Diagnosis and Treatment of Obscure and Urgent Cases, and the Surgical Anatomy of some Important Structures and Regions, by F. LE GROS CLARK, F.R.S., Consulting Surgeon to St. Thomas's Hospital. Second Edition, Revised and Expanded by the Author, assisted by W. W. WAGSTAFFE, F.R.C.S., Assistant-Surgeon to, and Joint-Lecturer on Anatomy at, St. Thomas's Hospital. 8vo, 10s. 6d.   [1872]

CLINICAL AND PATHOLOGICAL OBSERVATIONS IN INDIA

by J. FAYRER, C.S.I., M.D., F.R.C.P. Lond., F.R.S.E., Honorary Physician to the Queen. 8vo, with Engravings, 20s.   [1873]

SURGICAL EMERGENCIES

together with the Emergencies attendant on Parturition and the Treatment of Poisoning: a Manual for the use of General Practitioners, by WILLIAM P. SWAIN, F.R.C.S., Surgeon to the Royal Albert Hospital, Devonport. Fcap. 8vo, with 82 Engravings, 6s.   [1874]

ILLUSTRATIONS OF CLINICAL SURGERY,

consisting of Plates, Photographs, Woodcuts, Diagrams, &c., illustrating Surgical Diseases, Symptoms and Accidents; also Operations and other methods of Treatment. By JONATHAN HUTCHINSON, F.R.C.S., Senior Surgeon to the London Hospital. In Quarterly Fasciculi, 6s. 6d. each. [1875]

FRACTURES OF THE LOWER END OF THE RADIUS,

Fractures of the Clavicle, and on the Reduction of the Recent Inward Dislocations of the Shoulder Joint. By ALEXANDER GORDON, M.D., Professor of Surgery in Queen's College, Belfast. With Engravings, 8vo, 5s. [1875]

MINOR SURGERY AND BANDAGING

a Manual for the Use of House-Surgeons, Dressers, and Junior Practitioners, by CHRISTOPHER HEATH, F.R.C.S., Surgeon to University College Hospital. Fifth Edition, fcap 8vo, with 86 Engravings, 5s. 6d. [1875]

BY THE SAME AUTHOR,

INJURIES AND DISEASES OF THE JAWS:

JACKSONIAN PRIZE ESSAY. Second Edition, 8vo, with 164 Engravings, 12s. [1872]

DICTIONARY OF PRACTICAL SURGERY

and Encyclopædia of Surgical Science, by SAMUEL COOPER. New Edition, brought down to the present Time by SAMUEL A. LANE, Consulting Surgeon to St. Mary's and to the Lock Hospitals; assisted by various Eminent Surgeons. 2 vols. 8vo, 50s. [1861 and 1872]

THE FEMALE PELVIC ORGANS

(the Surgery, Surgical Pathology, and Surgical Anatomy of), in a Series of Coloured Plates taken from Nature: with Commentaries, Notes, and Cases, by HENRY SAVAGE, M.D. Lond., F.R.C.S., Consulting Physician to the Samaritan Free Hospital. Second Edition, 4to, £1 11s. 6d. [1870]

FRACTURES OF THE LIMBS

(Treatment of) by J. SAMPSON GAMGEE, Surgeon to the Queen's Hospital, Birmingham. 8vo, with Plates, 10s. 6d. [1871]

DISEASES AND INJURIES OF THE EAR

by W. B. DALBY, F.R.C.S., M.B., Aural Surgeon and Lecturer on Aural Surgery at St. George's Hospital. Crown 8vo, with 21 Engravings, 6s. 6d. [1873]

AURAL CATARRH;
or, the Commonest Forms of Deafness, and their Cure, by PETER ALLEN, M.D., F.R.C.S.E., late Aural Surgeon to St. Mary's Hospital. Second Edition, crown 8vo, with Engravings, 8s. 6d.   [1874]

PRINCIPLES OF SURGICAL DIAGNOSIS
especially in Relation to Shock and Visceral Lesions, Lectures delivered at the Royal College of Surgeons by F. LE GROS CLARK, F.R.C.S., Consulting Surgeon to St. Thomas's Hospital. 8vo, 10s. 6d.   [1870]

CLUBFOOT:
its Causes, Pathology, and Treatment; being the Jacksonian Prize Essay by WM. ADAMS, F.R.C.S., Surgeon to the Great Northern Hospital. Second Edition, 8vo, with 106 Engravings and 6 Lithographic Plates, 15s.   [1873]

INJURIES AND DISEASES OF THE KNEE-JOINT
and their Treatment by Amputation and Excision Contrasted: Jacksonian Prize Essay by W. P. SWAIN, F.R.C.S., Surgeon to the Royal Albert Hospital, Devonport. 8vo, with 36 Engravings, 9s.   [1869]

DEFORMITIES OF THE HUMAN BODY:
a System of Orthopædic Surgery, by BERNARD E. BRODHURST, F.R.C.S., Surgeon to the Royal Orthopædic Hospital. 8vo, with Engravings, 10s. 6d.   [1871]

OPERATIVE SURGERY OF THE FOOT AND ANKLE
by HENRY HANCOCK, F.R.C.S., Consulting Surgeon to Charing Cross Hospital. 8vo, with Engravings, 15s.   [1873]

THE TREATMENT OF SURGICAL INFLAMMATIONS
by a New Method, which greatly shortens their Duration, by FURNEAUX JORDAN, F.R.C.S., Professor of Surgery in Queen's College, Birmingham. 8vo, with Plates, 7s. 6d.   [1870]

BY THE SAME AUTHOR,
SURGICAL INQUIRIES
With numerous Lithographic Plates. 8vo, 5s.   [1873]

INTERNAL ANEURISM:
Successful Treatment of, by Consolidation of the Contents of the Sac. By T. JOLIFFE TUFNELL, F.R.C.S.I., President of the Royal College of Surgeons in Ireland. With Coloured Plates. Second Edition, royal 8vo, 5s.   [1875]

HERNIAL AND OTHER TUMOURS
of the Groin and its Neighbourhood, with some Practical Remarks on the Radical Cure of Ruptures, by C. HOLTHOUSE, F.R.C.S., Surgeon to the Westminster Hospital. 8vo, 6s. 6d.   [1870]

THE SURGERY OF THE RECTUM:
Lettsomian Lectures by HENRY SMITH, F.R.C.S., Surgeon to King's
College Hospital. Third Edition, fcap 8vo, 3s. 6d. [1871]

FISTULA, HÆMORRHOIDS, PAINFUL ULCER,
Stricture, Prolapsus, and other Diseases of the Rectum: their Diagnosis
and Treatment, by WM. ALLINGHAM, F.R.C.S., Surgeon to St. Mark's
Hospital for Fistula, &c., late Surgeon to the Great Northern Hospital.
Second Edition, 8vo, 7s. [1872]

THE URINE AND ITS DERANGEMENTS
with the Application of Physiological Chemistry to the Diagnosis and
Treatment of Constitutional as well as Local Diseases. Lectures
by GEORGE HARLEY, M.D., F.R.S., F.R.C.P., formerly Professor in
University College. Post 8vo, 9s. [1872]

STRICTURE OF THE URETHRA
and Urinary Fistulæ; their Pathology and Treatment: Jacksonian
Prize Essay by Sir HENRY THOMPSON, F.R.C.S., Surgeon-Extraordi-
nary to the King of the Belgians. Third Edition, 8vo, with Plates,
10s. [1869]

BY THE SAME AUTHOR,
PRACTICAL LITHOTOMY AND LITHOTRITY;
or, An Inquiry into the best Modes of removing Stone from the
Bladder. Second Edition, 8vo, with numerous Engravings. 10s. [1871]

ALSO,
DISEASES OF THE URINARY ORGANS
(Clinical Lectures). Third Edition, crown 8vo, with Engravings, 6s. [1872]

ALSO,
DISEASES OF THE PROSTATE:
their Pathology and Treatment. Fourth Edition, 8vo, with numerous
Plates, 10s. [1873]

STRICTURE OF THE URETHRA
(the Immediate Treatment of), by BARNARD HOLT, F.R.C.S.,
Consulting Surgeon to the Westminster Hospital. Third Edition,
8vo, 6s. [1868]

KIDNEY DISEASES, URINARY DEPOSITS
and Calculous Disorders by LIONEL S. BEALE, M.B., F.R.S., F.R.C.P.,
Physician to King's College Hospital. Third Edition, 8vo, with
70 Plates, 25s. [1868]

THE IRRITABLE BLADDER:
its Causes and Treatment, by F. J. GANT, F.R.C.S., Surgeon to the
Royal Free Hospital. Third Edition, crown 8vo, with Engravings,
6s. [1872]

RENAL DISEASES:

a Clinical Guide to their Diagnosis and Treatment by W. R. BASHAM, M.D., F.R.C.P., Senior Physician to the Westminster Hospital. Post 8vo, 7s. [1870]

BY THE SAME AUTHOR,

THE DIAGNOSIS OF DISEASES OF THE KIDNEYS

(Aids to). 8vo, with 10 Plates, 5s. [1872]

MICROSCOPIC STRUCTURE OF URINARY CALCULI

by H. V. CARTER, M.D., Surgeon-Major, H.M.'s Bombay Army. 8vo, with 4 Plates, 5s. [1873]

THE REPRODUCTIVE ORGANS

in Childhood, Youth, Adult Age, and Advanced Life (Functions and Disorders of), considered in their Physiological, Social, and Moral Relations, by WILLIAM ACTON, M.R.C.S. Sixth Edition, 8vo, 12s. [1875]

URINARY AND REPRODUCTIVE ORGANS

(Functional Diseases of) by D. CAMPBELL BLACK, M.D., L.R.C.S. Edin. Second Edition. 8vo, 10s. 6d. [1875]

PRACTICAL PATHOLOGY:

containing Lectures on Suppurative Fever, Diseases of the Veins, Hæmorrhoidal Tumours, Diseases of the Rectum, Syphilis, Gonorrheal Ophthalmia, &c., by HENRY LEE, F.R.C.S., Surgeon to St. George's Hospital. Third Edition, in 2 vols. 8vo, 10s. each. [1870]

BY THE SAME AUTHOR,

LECTURES ON SYPHILIS,

and on some forms of Local Disease, affecting principally the Organs of Generation. With Engravings, 8vo, 10s. [1875]

GENITO-URINARY ORGANS, INCLUDING SYPHILIS

A Practical Treatise on their Surgical Diseases, designed as a Manual for Students and Practitioners, by W. H. VAN BUREN, M.D., Professor of the Principles of Surgery in Bellevue Hospital Medical College, New York, and E. L. KEYES, M.D., Professor of Dermatology in Bellevue Hospital Medical College, New York. Royal 8vo, with 140 Engravings, 21s. [1874]

SYPHILIS

A Treatise by WALTER J. COULSON, F.R.C.S., Surgeon to the Lock Hospital. 8vo, 10s. [1869]

BY THE SAME AUTHOR,

STONE IN THE BLADDER:

Its Prevention, Early Symptoms, and Treatment by Lithotrity. 8vo, 6s. [1868]

SYPHILITIC NERVOUS AFFECTIONS

(Clinical Aspects of) by THOMAS BUZZARD, M.D., F.R.C.P. Lond., Physician to the National Hospital for Paralysis and Epilepsy. Post 8vo, 5s. [1874]

SYPHILITIC OSTEITIS AND PERIOSTITIS

Lectures by JOHN HAMILTON, F.R.C.S.I., Surgeon to the Richmond Hospital and to Swift's Hospital for Lunatics, Dublin. 8vo, with Plates, 6s. 6d. [1874]

THE CIRCULATION OF THE BLOOD

(Forces which carry on) by ANDREW BUCHANAN, M.D., Professor of Physiology in the University of Glasgow. Second Edition, 8vo, with Engravings, 5s. [1874]

PRINCIPLES OF HUMAN PHYSIOLOGY

by W. B. CARPENTER, M.D., F.R.S. Seventh Edition by Mr. HENRY POWER. 8vo, with nearly 300 Illustrations, 28s. [1869]

HANDBOOK FOR THE PHYSIOLOGICAL LABORATORY

by E. KLEIN, M.D., F.R.S., Assistant Professor in the Pathological Laboratory of the Brown Institution, London; J. BURDON-SANDERSON, M.D., F.R.S., Professor of Practical Physiology in University College, London; MICHAEL FOSTER, M.D., F.R.S., Prælector of Physiology in Trinity College, Cambridge; and T. LAUDER BRUNTON, M.D., D.Sc., Lecturer on Materia Medica at St. Bartholomew's Hospital; edited by J. BURDON-SANDERSON. 8vo, with 123 Plates, 24s. [1873]

HISTOLOGY AND HISTO-CHEMISTRY OF MAN

A Treatise on the Elements of Composition and Structure of the Human Body, by HEINRICH FREY, Professor of Medicine in Zurich. Translated from the Fourth German Edition by ARTHUR E. J. BARKER, Assistant-Surgeon to University College Hospital. And Revised by the Author. 8vo, with 608 Engravings, 21s. [1874]

PRACTICAL HISTOLOGY

(Outlines of) by WILLIAM RUTHERFORD, M.D., Professor of the Institutes of Medicine in the University of Edinburgh. With Engravings. Crown 8vo, interleaved, 3s. [1875]

THE MARRIAGE OF NEAR KIN

Considered with respect to the Laws of Nations, Results of Experience, and the Teachings of Biology, by ALFRED H. HUTH. 8vo, 14s. [1875]

STUDENTS' GUIDE TO HUMAN OSTEOLOGY

By WILLIAM WARWICK WAGSTAFFE, F.R.C.S., Assistant-Surgeon and Lecturer on Anatomy, St. Thomas's Hospital. With 23 Plates and 66 Engravings. Fcap. 8vo, 10s. 6d. [1875]

§

HUMAN OSTEOLOGY:

with Plates, showing the Attachments of the Muscles, by LUTHER HOLDEN, F.R.C.S., Surgeon to St. Bartholomew's Hospital. Fourth Edition, 8vo, 16s. [1869]

BY THE SAME AUTHOR,

THE DISSECTION OF THE HUMAN BODY

(A Manual). Third Edition, 8vo, with Engravings, 16s. [1868]

MEDICAL ANATOMY

by FRANCIS SIBSON, M.D., F.R.C.P., F.R.S., Consulting Physician to St. Mary's Hospital. Imp. folio, with 21 coloured Plates, cloth, 42s.; half-morocco, 50s. [Completed in 1869]

THE ANATOMIST'S VADE-MECUM:

a System of Human Anatomy by ERASMUS WILSON, F.R.C.S., F.R.S. Ninth Edition, by Dr. G. BUCHANAN, Professor of Anatomy in Anderson's University, Glasgow. Crown 8vo, with 371 Engravings, 14s. [1873]

PRACTICAL ANATOMY:

a Manual of Dissections by CHRISTOPHER HEATH, F.R.C.S., Surgeon to University College Hospital. Third Edition, fcap 8vo, with 226 Engravings, 12s. 6d. [1874]

PATHOLOGICAL ANATOMY

Lectures by SAMUEL WILKS, M.D., F.R.S., Physician to, and Lecturer on Medicine at, Guy's Hospital; and WALTER MOXON, M.D., F.R.C.P., Physician to, and Lecturer on Materia Medica at, Guy's Hospital. Second Edition, 8vo, with Plates, 18s. [1875]

PATHOLOGICAL ANATOMY

A Manual by C. HANDFIELD JONES, M.B., F.R.S., Physician to St. Mary's Hospital, and EDWARD H. SIEVEKING, M.D., F.R.C.P., Physician to St. Mary's Hospital. Edited by J. F. PAYNE, M.D., F.R.C.P., Assistant Physician and late Demonstrator of Morbid Anatomy at St. Thomas's Hospital. Second Edition, crown 8vo, with nearly 200 Engravings, 16s. [1875]

DIAGRAMS OF THE NERVES OF THE HUMAN BODY

Exhibiting their Origin, Divisions, and Connexions, with their Distribution, by WILLIAM HENRY FLOWER, F.R.S., Conservator of the Museum of the Royal College of Surgeons. Second Edition, roy. 4to, 12s. [1872]

STUDENT'S GUIDE TO SURGICAL ANATOMY:

a Text-book for the Pass Examination, by E. BELLAMY, F.R.C.S., Senior Assistant-Surgeon and Lecturer on Anatomy at Charing Cross Hospital. Fcap 8vo, with 50 Engravings, 6s. 6d. [1873]

THE STUDENT'S GUIDE TO MEDICAL DIAGNOSIS

by SAMUEL FENWICK, M.D., F.R.C.P., Assistant Physician to the London Hospital. Third Edition, fcap 8vo, with 87 Engravings, 6s. 6d. [1873]

A MANUAL OF MEDICAL DIAGNOSIS

by A. W. BARCLAY, M.D., F.R.C.P., Physician to, and Lecturer on Medicine at, St. George's Hospital. Third Edition, fcap 8vo, 10s. 6d. [1870]

THE MEDICAL REMEMBRANCER;

or, Book of Emergencies. By E. SHAW, M.R.C.S. Fifth Edition by JONATHAN HUTCHINSON, F.R.C.S., Senior Surgeon to the London Hospital. 32mo, 2s. 6d. [1867]

THE ANATOMICAL REMEMBRANCER;

or, Complete Pocket Anatomist. Seventh Edition, 32mo, 3s. 6d. [1972]

PRACTICAL THERAPEUTICS

A Manual by E. J. WARING, M.D., F.R.C.P. Lond. Third Edition, fcap 8vo, 12s. 6d. [1871]

ANTAGONISM OF MEDICINES

(Researches into the) being the Report of the Edinburgh Committee of the British Medical Association. By J. HUGHES BENNETT, M.D. Post 8vo, 3s. 6d. [1875]

HOOPER'S PHYSICIAN'S VADE-MECUM;

or, Manual of the Principles and Practice of Physic, Ninth Edition by W. A. GUY, M.B., F.R.S., and JOHN HARLEY, M.D., F.R.C.P. Fcap 8vo, with Engravings, 12s. 6d. [1874]

CLINICAL MEDICINE

Lectures and Essays by BALTHAZAR FOSTER, M.D., F.R.C.P. Lond., Professor of Medicine in Queen's College, Birmingham. 8vo, 10s. 6d. [1874]

DISCOURSES ON PRACTICAL PHYSIC

by B. W. RICHARDSON, M.D., F.R.C.P., F.R.S. 8vo, 5s. [1871]

MATERIA MEDICA

A Manual by J. F. ROYLE, M.D., F.R.S., and JOHN HARLEY, M.D. Sixth Edition, crown 8vo, with numerous Engravings.

A DICTIONARY OF MATERIA MEDICA

and Therapeutics by ADOLPHE WAHLTUCH, M.D. 8vo, 15s. [1868]

MATERIA MEDICA AND THERAPEUTICS:

(Vegetable Kingdom), by CHARLES D. F. PHILLIPS, M.D., F.R.C.S.E. 8vo, 15s. [1871]

THE STUDENT'S GUIDE TO MATERIA MEDICA

by JOHN C. THOROWGOOD, M.D. Lond., Physician to the City of London Hospital for Diseases of the Chest. Fcap 8vo, with Engravings, 6s. 6d. [1874]

THE DISEASES OF CHILDREN

A Practical Manual, with a Formulary, by EDWARD ELLIS, M.D., Physician to the Victoria Hospital for Children. Second Edition, crown 8vo, 7s. [1873]

THE WASTING DISEASES OF CHILDREN

by EUSTACE SMITH, M.D. Lond., Physician to the King of the Belgians, Physician to the East London Hospital for Children. Second Edition, post 8vo, 7s. 6d. [1870]

THE DISEASES OF CHILDREN

Essays by WILLIAM HENRY DAY, M.D., Physician to the Samaritan Hospital for Diseases of Women and Children. Fcap 8vo, 5s. [1873]

COMPENDIUM OF CHILDREN'S DISEASES

A Handbook for Practitioners and Students, by JOHANN STEINER, M.D., Professor of the Diseases of Children in the University of Prague. Translated from the Second German Edition by LAWSON TAIT. F.R.C.S., Surgeon to the Birmingham Hospital for Women. 8vo, 12s. 6d. [1874]

PUERPERAL DISEASES

Clinical Lectures by FORDYCE BARKER, M.D., Obstetric Physician to Bellevue Hospital, New York. 8vo, 15s. [1874]

OBSTETRIC OPERATIONS,

including the Treatment of Hæmorrhage, and forming a Guide to the Management of Difficult Labour; Lectures by ROBERT BARNES, M.D., F.R.C.P., Obstetric Physician to, and Lecturer on Midwifery at, St. George's Hospital. Second Edition, 8vo, with 113 Engravings, 15s. [1871]

BY THE SAME AUTHOR,

MEDICAL AND SURGICAL DISEASES OF WOMEN

(a Clinical History). 8vo, with 169 Engravings, 28s. [1873]

OBSTETRIC SURGERY

A Complete Handbook, giving Short Rules of Practice in every Emergency, from the Simplest to the most Formidable Operations connected with the Science of Obstetricy, by CHARLES CLAY, Ext.L.R.C.P. Lond., L.R.C.S.E., late Senior Surgeon and Lecturer on Midwifery, St. Mary's Hospital, Manchester. Fcap 8vo, with 91 Engravings, 6s. 6d. [1874]

OBSTETRIC MEDICINE AND SURGERY

(Principles and Practice of) by F. H. RAMSBOTHAM, M.D., F.R.C.P.
Fifth Edition, 8vo, with 120 Plates, 22s.   [1867]

OBSTETRIC APHORISMS

for the Use of Students commencing Midwifery Practice by J. G.
SWAYNE, M.D., Physician-Accoucheur to the Bristol General Hospital. Fifth Edition, fcap 8vo, with Engravings, 3s. 6d.   [1871]

SCHROEDER'S MANUAL OF MIDWIFERY,

including the Pathology of Pregnancy and the Puerperal State.
Translated by CHARLES H. CARTER, B.A., M.D. 8vo, with Engravings, 12s. 6d.   [1873]

A HANDBOOK OF UTERINE THERAPEUTICS

and of Diseases of Women by E. J. TILT, M.D., M.R.C.P. Third
Edition, post 8vo, 10s.   [1868]

BY THE SAME AUTHOR,

THE CHANGE OF LIFE

in Health and Disease: a Practical Treatise on the Nervous and other
Affections incidental to Women at the Decline of Life. Third Edition,
8vo, 10s. 6d.   [1870]

DISEASES OF THE OVARIES :

their Diagnosis and Treatment, by T. SPENCER WELLS, F.R.C.S.,
Surgeon to the Queen's Household and to the Samaritan Hospital.
8vo, with about 150 Engravings, 21s.   [1872]

HANDBOOK FOR NURSES FOR THE SICK

by Miss VEITCH. Crown 8vo, 2s. 6d.   [1870]

A MANUAL FOR HOSPITAL NURSES

and others engaged in Attending on the Sick by EDWARD J. DOM-
VILLE, L.R.C.P., M.R.C.S. Second Edition, crown 8vo, 2s. 6d. [1875]

LECTURES ON NURSING

by WILLIAM ROBERT SMITH, L.R.C.S.E, Resident Surgeon, Royal
Hants County Hospital, Winchester. With 26 Engravings. Post
8vo, 6s.   [1875]

ENGLISH MIDWIVES:

their History and Prospects, by J. H. AVELING, M.D., Physician to
the Chelsea Hospital for Women, Examiner of Midwives for the
Obstetrical Society of London. Crown 8vo, 5s.   [1872]

A COMPENDIUM OF DOMESTIC MEDICINE

and Companion to the Medicine Chest; intended as a Source of Easy Reference for Clergymen, and for Families residing at a Distance from Professional Assistance, by JOHN SAVORY, M.S.A. Eighth Edition, 12mo, 5s.   [1871]

THE WIFE'S DOMAIN

The Young Couple—The Mother—The Nurse—The Nursling, by PHILOTHALOS. Second Edition, post 8vo, 3s. 6d.   [1874]

WINTER COUGH

(Catarrh, Bronchitis, Emphysema, Asthma), Lectures by HORACE DOBELL, M.D., Consulting Physician to the Royal Hospital for Diseases of the Chest. Third Edition, with Coloured Plates, 8vo, 10s. 6d.   [1875]

BY THE SAME AUTHOR,

THE TRUE FIRST STAGE OF CONSUMPTION

(Lectures). Crown 8vo, 3s. 6d.   [1867]

DISEASES OF THE CHEST:

Contributions to their Clinical History, Pathology, and Treatment, by A. T. H. WATERS, M.D., F.R.C.P., Physician to the Liverpool Royal Infirmary. Second Edition, 8vo, with Plates, 15s.   [1873]

PHTHISIS AND THE STETHOSCOPE;

or, the Physical Signs of Consumption, by R. P. COTTON, M.D., F.R.C.P., Senior Physician to the Hospital for Consumption, Brompton. Fourth Edition, fcap 8vo, 3s. 6d.   [1869]

DISEASES OF THE HEART

and of the Lungs in Connexion therewith—Notes and Observations by THOMAS SHAPTER, M.D., F.R.C.P. Lond., Senior Physician to the Devon and Exeter Hospital. 8vo, 7s. 6d.   [1874]

DISEASES OF THE HEART

Their Pathology, Diagnosis, Prognosis, and Treatment (a Manual), by ROBERT H. SEMPLE, M.D., Physician to the Hospital for Diseases of the Throat. 8vo, 8s. 6d.   [1875]

DISEASES OF THE HEART AND AORTA

By THOMAS HAYDEN, F.K.Q.C.P. Irel., Physician to the Mater Misericordiæ Hospital, Dublin. With 80 Engravings. 8vo, 25s.   [1875]

VALVULAR DISEASE OF THE HEART

(some of its causes and effects). Croonian Lectures for 1865. By THOMAS B. PEACOCK, M.D., F.R.C.P., Physician to St. Thomas's Hospital. With Engravings, 8vo, 5s.   [1865]

THE ACTION AND SOUNDS OF THE HEART

Researches by GEORGE PATON, M.D., author of numerous papers published in the British and American Medical Journals. Re-issue, with Appendix, 8vo, 3s. 6d. [1874]

NOTES ON ASTHMA;

its Forms and Treatment, by JOHN C. THOROWGOOD, M.D. Lond., F.R.C.P., Physician to the Hospital for Diseases of the Chest, Victoria Park. Second Edition, crown 8vo, 4s. 6d. [1873]

GROWTHS IN THE LARYNX,

with Reports and an Analysis of 100 consecutive Cases treated since the Invention of the Laryngoscope by MORELL MACKENZIE, M.D. Lond., M.R.C.P., Physician to the Hospital for Diseases of the Throat. 8vo, with Coloured Plates, 12s. 6d. [1871]

IRRITATIVE DYSPEPSIA

and its Important Connection with Irritative Congestion of the Windpipe and with the Origin and Progress of Consumption by C. B. GARRETT, M.D. Crown 8vo, 2s. 6d. [1868]

MINERAL SPRINGS OF HARROGATE

By Dr. KENNION. Revised and enlarged by ADAM BEALEY, M.A., M.D. Cantab., F.R.C.P. Lond. Seventh Thousand. Crown 8vo, 1s. [1875]

SKETCH OF CANNES AND ITS CLIMATE

by TH. DE VALCOURT, M.D. Paris, Physician at Cannes. Second Edition, with Photographic View and 6 Meteorological Charts. Crown 8vo, 2s. 6d. [1873]

WINTER AND SPRING

on the Shores of the Mediterranean; or, the Genoese Rivieras, Italy, Spain, Greece, the Archipelago, Constantinople, Corsica, Sardinia, Sicily, Corfu, Malta, Tunis, Algeria, Smyrna, Asia Minor, with Biarritz and Arcachon, as Winter Climates. By HENRY BENNET, M.D. Fifth Edition, post 8vo, with numerous Plates, Maps, and Engravings, 12s. 6d. [1874]

BY THE SAME AUTHOR,

TREATMENT OF PULMONARY CONSUMPTION

by Hygiene, Climate, and Medicine. Second Edition, 8vo, 5s. [1871]

EGYPT AS A HEALTH RESORT;

with Medical and other Hints for Travellers in Syria, by A. DUNBAR WALKER M.D. Fcap 8vo, 3s. 6d. [1873]

FAMILY MEDICINE FOR INDIA

A Manual, by WILLIAM J. MOORE, M.D., Surgeon-Major H.M. Indian Medical Service. Published under the Authority of the Government of India. Post 8vo, with 57 Engravings, 8s. 6d. [1874]

DISEASES OF TROPICAL CLIMATES

and their Treatment: with Hints for the Preservation of Health in the Tropics, by JAMES A. HORTON, M.D., Surgeon-Major, Army Medical Department. Post 8vo, 12s. 6d. [1874]

HEALTH IN INDIA FOR BRITISH WOMEN

and on the Prevention of Disease in Tropical Climates by EDWARD J. TILT, M.D., Consulting Physician-Accoucheur to the Farringdon General Dispensary. Fourth Edition, crown 8vo, 5s. [1875]

BAZAAR MEDICINES OF INDIA

and Common Medical Plants: Remarks on their Uses, with Full Index of Diseases, indicating their Treatment by these and other Agents procurable throughout India, &c., by EDWARD J. WARING, M.D., F.R.C.P. Lond., Retired Surgeon H.M. Indian Army. Third Edition. Fcap 8vo, 5s. [1875]

SOME AFFECTIONS OF THE LIVER

and Intestinal Canal; with Remarks on Ague and its Sequelæ, Scurvy, Purpura, &c., by STEPHEN H. WARD, M.D. Lond., F.R.C.P., Physician to the Seamen's Hospital, Greenwich. 8vo, 7s. [1872]

DISEASES OF THE LIVER:

Lettsomian Lectures for 1872 by S. O. HABERSHON, M.D., F.R.C.P., Senior Physician to Guy's Hospital. Post 8vo, 3s. 6d. [1872]

THE STOMACH AND DUODENUM

Their Morbid States and their Relations to the Diseases of other Organs, by SAMUEL FENWICK, M.D., F.R.C.P., Assistant-Physician to the London Hospital. 8vo, with 10 Plates, 12s. [1868]

CONSTIPATED BOWELS:

the Various Causes and the Different Means of Cure, by S. B. BIRCH, M.D., M.R.C.P. Third Edition, post 8vo, 3s. 6d. [1868]

FOOD AND DIETETICS

Physiologically and Therapeutically Considered. Second Edition, 8vo, 15s. [1875]

THE INDIGESTIONS;

or, Diseases of the Digestive Organs Functionally Treated, by T. K. CHAMBERS, M.D., F.R.C.P., Lecturer on Medicine at St. Mary's Hospital. Second Edition, 8vo, 10s. 6d. [1867]

IMPERFECT DIGESTION:

its Causes and Treatment by ARTHUR LEARED, M.D., F.R.C.P., Senior Physician to the Great Northern Hospital. Fifth Edition, fcap 8vo, 4s. 6d. [1870]

THE ISSUE OF A SPIRIT RATION

during the Ashanti Campaign of 1874; with two Appendices containing Experiments to show the Relative Effects of Rum, Meat Extract and Coffee during Marching, and the Use of Oatmeal Drink during Labour, by EDMUND A. PARKES, M.D., F.R.S., Professor of Hygiene to the Army Medical School, Netley. 8vo, 2s. 6d. [1875]

MEGRIM, SICK-HEADACHE,

and some Allied Disorders: a Contribution to the Pathology of Nerve-Storms, by EDWARD LIVEING, M.D. Cantab., Hon. Fellow of King's College, London. 8vo, with Coloured Plate, 15s. [1873]

IRRITABILITY:

Popular and Practical Sketches of Common Morbid States and Conditions bordering on Disease; with Hints for Management, Alleviation, and Cure, by JAMES MORRIS, M.D. Lond. Crown 8vo, 4s. 6d. [1868]

FUNCTIONAL NERVOUS DISORDERS

Studies by C. HANDFIELD JONES, M.B., F.R.C.P., F.R.S., Physician to St. Mary's Hospital. Second Edition, 8vo, 18s. [1870]

NEURALGIA AND KINDRED DISEASES

of the Nervous System: their Nature, Causes, and Treatment, with a series of Cases, by JOHN CHAPMAN, M.D., M.R.C.P. 8vo, 14s. [1873]

THE SYMPATHETIC SYSTEM OF NERVES

and their Functions as a Physiological Basis for a Rational System of Therapeutics by EDWARD MERYON, M.D., F.R.C.P., Physician to the Hospital for Diseases of the Nervous System. 8vo, 3s. 6d. [1872]

GOUT, RHEUMATISM

and the Allied Affections; a Treatise by P. HOOD, M.D. Crown 8vo, 10s. 6d. [1871]

RHEUMATIC GOUT,

or Chronic Rheumatic Arthritis of all the Joints; a Treatise by ROBERT ADAMS, M.D., M.R.I.A., Surgeon to H.M. the Queen in Ireland, Regius Professor of Surgery in the University of Dublin. Second Edition, 8vo, with Atlas of Plates, 21s. [1872]

TEMPERATURE OBSERVATIONS

containing (1) Temperature Variations in the Diseases of Children, (2) Puerperal Temperatures, (3) Infantile Temperatures in Health and Disease, by WM. SQUIRE, M.R.C.P. Lond. 8vo, 5s.   [1871]

MYCETOMA ;

or, the Fungus Disease of India, by H. VANDYKE CARTER, M.D., Surgeon-Major H.M. Indian Army.   4to, with 11 Coloured Plates, 42s.
[1874]

THE ORIGIN OF CANCER

considered with Reference to the Treatment of the Disease by CAMPBELL DE MORGAN, F.R.S., F.R.C.S., Surgeon to the Middlesex Hospital. Crown 8vo, 3s. 6d.   [1872]

CANCER:

its varieties, their Histology and Diagnosis, by HENRY ARNOTT, F.R.C.S., Assistant-Surgeon to, and Lecturer on Morbid Anatomy at, St. Thomas's Hospital.   8vo, with 5 Plates and 22 Engravings, 5s. 6d.
[1872]

CANCEROUS AND OTHER INTRA-THORACIC GROWTHS:

their Natural History and Diagnosis, by J. RISDON BENNETT, M.D., F.R.C.P., Member of the General Medical Council. Post 8vo, with Plates, 8s.   [1872]

CERTAIN FORMS OF CANCER

with a New and successful Mode of Treating it, to which is prefixed a Practical and Systematic Description of all the varieties of this Disease, by ALEX. MARSDEN, M.D., F.R.C.S.E., Consulting Surgeon to the Royal Free Hospital, and Senior Surgeon to the Cancer Hospital. Second Edition, with Coloured Plates, 8vo, 8s. 6d.   [1873]

DISEASES OF THE SKIN:

a System of Cutaneous Medicine by ERASMUS WILSON, F.R.C.S., F.R.S. Sixth Edition, 8vo, 18s., with Coloured Plates, 36s.   [1867]

BY THE SAME AUTHOR,

LECTURES ON EKZEMA

and Ekzematous Affections: with an Introduction on the General Pathology of the Skin, and an Appendix of Essays and Cases. 8vo, 10s. 6d.   [1870]

ALSO,

LECTURES ON DERMATOLOGY

delivered at the Royal College of Surgeons, 1870, 6s. ; 1871-3, 10s. 6d., 1874-5, 10s. 6d.

ATLAS OF SKIN DISEASES:

a series of Illustrations, with Descriptive Text and Notes upon Treatment. By TILBURY FOX, M.D., F.R.C.P., Physician to the Department for Skin Diseases in University College Hospital. In monthly parts, each containing Four Coloured Plates, 6s. 6d.   [1875]

ECZEMA

by McCALL ANDERSON, M.D., Professor of Clinical Medicine in the University of Glasgow. Third Edition, 8vo, with Engravings, 7s. 6d.   [1874]

BY THE SAME AUTHOR,

PARASITIC AFFECTIONS OF THE SKIN

Second Edition, 8vo, with Engravings, 7s. 6d.   [1868]

PSORIASIS OR LEPRA

by GEORGE GASKOIN, M.R.C.S., Surgeon to the British Hospital for Diseases of the Skin. 8vo, 5s.   [1875]

DISEASES OF THE SKIN

in Twenty-four Letters on the Principles and Practice of Cutaneous Medicine, by HENRY EVANS CAUTY, Surgeon to the Liverpool Dispensary for Diseases of the Skin, 8vo, 12s. 6d.   [1874]

FOURTEEN COLOURED PHOTOGRAPHS OF LEPROSY

as met with in the Straits Settlements, with Explanatory Notes by A. F. ANDERSON, M.D., Acting Colonial Surgeon, Singapore. 4to, 31s. 6d.   [1872]

WORMS:

a Series of Lectures delivered at the Middlesex Hospital on Practical Helminthology by T. SPENCER COBBOLD, M.D., F.R.S. Post 8vo, 5s.   [1872]

OXYGEN:

its Action, Use, and Value in the Treatment of Various Diseases otherwise Incurable or very Intractable, by S. B. BIRCH, M.D., M.R.C.P. Second Edition, post 8vo, 3s. 6d.   [1868]

THE MEDICAL ADVISER IN LIFE ASSURANCE

by EDWARD HENRY SIEVEKING, M.D., F.R.C.P., Physician to St. Mary's and the Lock Hospitals; Physician-Extraordinary to the Queen; Physician-in-Ordinary to the Prince of Wales, &c. Crown 8vo, 6s.   [1874]

THE LAWS AFFECTING MEDICAL MEN

a Manual by ROBERT G. GLENN, LL.B., Barrister-at-Law; with a Chapter on Medical Etiquette by Dr. A. CARPENTER. 8vo, 14s.   [1871]

MEDICAL JURISPRUDENCE

(Principles and Practice of) by Alfred S. Taylor, M.D., F.R.C.P., F.R.S. Second Edition, 2 vols., 8vo, with 189 Engravings, £1 11s. 6d.
[1873]

BY THE SAME AUTHOR,

A MANUAL OF MEDICAL JURISPRUDENCE

Ninth Edition. Crown 8vo, with Engravings. 14s.   [1874]

ALSO,

POISONS

in Relation to Medical Jurisprudence and Medicine. Third Edition, crown 8vo, with 104 Engravings, 16s.   [1875]

A TOXICOLOGICAL CHART,

exhibiting at one View the Symptoms, Treatment, and mode of Detecting the various Poisons—Mineral, Vegetable, and Animal: with Concise Directions for the Treatment of Suspended Animation, by William Stowe, M.R.C.S.E. Thirteenth Edition, 2s.; on roller, 5s.   [1872]

MADNESS

in its Medical, Legal, and Social Aspects, Lectures by Edgar Sheppard, M.D., M.R.C.P., Professor of Psychological Medicine in King's College; one of the Medical Superintendents of the Colney Hatch Lunatic Asylum. 8vo, 6s. 6d.   [1873]

HANDBOOK OF LAW AND LUNACY;

or, the Medical Practitioner's Complete Guide in all Matters relating to Lunacy Practice, by J. T. Sabben, M.D., and J. H. Balfour Browne, Barrister-at-Law. 8vo, 5s.   [1872]

CEREBRIA

and other Diseases of the Brain by Charles Elam, M.D., F.R.C.P., Assistant-Physician to the National Hospital for Paralysis and Epilepsy. 8vo, 6s.   [1872]

INFLUENCE OF THE MIND UPON THE BODY

in Health and Disease, Illustrations designed to elucidate the Action of the Imagination, by Daniel Hack Tuke, M.D., M.R.C.P. 8vo, 14s.   [1872]

OBSCURE DISEASES OF THE BRAIN AND MIND

by Forbes Winslow, M.D., D.C.L. Oxon. Fourth Edition, post 8vo, 10s. 6d.   [1868]

PSYCHOLOGICAL MEDICINE:

a Manual, containing the Lunacy Laws, the Nosology, Ætiology, Statistics, Description, Diagnosis, Pathology (including Morbid Histology), and Treatment of Insanity, by J. C. BUCKNILL, M.D., F.R.S., and D. H. TUKE, M.D. Third Edition, 8vo, with 10 Plates and 34 Engravings, 25s. [1873]

A MANUAL OF PRACTICAL HYGIENE

by E. A. PARKES, M.D., F.R.C.P., F.R.S., Professor of Hygiene in the Army Medical School. Fourth Edition, 8vo, with Plates and Engravings, 16s. [1873]

A HANDBOOK OF HYGIENE

for the Use of Sanitary Authorities and Health Officers by GEORGE WILSON, M.D. Edin., Medical Officer of Health for the Warwick Union of Sanitary Authorities. Second Edition, crown 8vo, with Engravings, 8s. 6d. [1873]

HANDBOOK OF MEDICAL ELECTRICITY

by HERBERT TIBBITS, M.D., M.R.C.P.E., Medical Superintendent of the National Hospital for the Paralysed and Epileptic. 8vo, with 64 Engravings, 6s. [1873]

CLINICAL USES OF ELECTRICITY

Lectures delivered at University College Hospital by J. RUSSELL REYNOLDS, M.D. Lond., F.R.C.P., F.R.S., Professor of Medicine in University College. Second Edition, post 8vo, 3s. 6d. [1873]

MEDICO-ELECTRIC APPARATUS

and How to Use it; or, a Practical Description of every Form of Medico-Electric Apparatus in Modern Use, with Plain Directions for Mounting, Charging, and Working, by T. P. SALT. 8vo, with 31 Engravings, 2s. 6d. [1875]

ATLAS OF OPHTHALMOSCOPY:

representing the Normal and Pathological Conditions of the Fundus Oculi as seen with the Ophthalmoscope : composed of 12 Chromolithographic Plates (containing 59 Figures), accompanied by an Explanatory Text by R. LIEBREICH, Ophthalmic Surgeon to St. Thomas's Hospital. Translated into English by H. ROSBOROUGH SWANZY, M.B. Dub. Second Edition, 4to, £1 10s. [1870]

DISEASES OF THE EYE

a Manual by C. MACNAMARA, Surgeon to the Calcutta Ophthalmic Hospital Second Edition, fcap 8vo, with Coloured Plates, 12s. 6d. [1872]

AUTOBIOGRAPHICAL RECOLLECTIONS

of the Medical Profession, being personal reminiscences of many distinguished Medical Men during the last forty years, by J. FERNANDEZ CLARKE, M.R.C.S., for many years on the Editorial Staff of the 'Lancet.' Post 8vo, 10s. 6d.   [1874]

A DICTIONARY OF MEDICAL SCIENCE

containing a concise explanation of the various subjects and terms of Anatomy, Physiology, Pathology, Hygiene, Therapeutics, Medical Chemistry, Pharmacology, Pharmacy, Surgery, Obstetrics, Medical Jurisprudence and Dentistry ; Notices of Climate and Mineral Waters ; formulæ for Officinal, Empirical, and Dietetic Preparations ; with the Accentuation and Etymology of the terms and the French and other Synonyms, by ROBLEY DUNGLISON, M.D., LL.D. New Edition, by RICHARD J. DUNGLISON, M.D. Royal 8vo, 28s.   [1874]

A MEDICAL VOCABULARY;

being an Explanation of all Terms and Phrases used in the various Departments of Medical Science and Practice, giving their derivation, meaning, application, and pronunciation, by ROBERT G. MAYNE, M.D., LL.D. Fourth Edition, fcap 8vo, 10s.   [1875]

OPHTHALMIC MEDICINE AND SURGERY

a Manual by T. WHARTON JONES, F.R.S., Professor of Ophthalmic Medicine and Surgery in University College. Third Edition, fcap 8vo, with 9 Coloured Plates and 173 Engravings, 12s. 6d.   [1865]

DISEASES OF THE EYE

A Treatise by J. SOELBERG WELLS, F.R.C.S., Ophthalmic Surgeon to King's College Hospital and Surgeon to the Royal London Ophthalmic Hospital. Third Edition, 8vo, with Coloured Plates and Engravings, 25s.   [1873]

BY THE SAME AUTHOR,

LONG, SHORT, AND WEAK SIGHT,

and their Treatment by the Scientific use of Spectacles. Fourth Edition, 8vo, 6s.   [1873]

DISEASES OF THE EYE

A Practical Treatise by HAYNES WALTON, F.R.C.S., Surgeon to St. Mary's Hospital and in charge of its Ophthalmological Department. Third Edition, 8vo, with 3 Plates and nearly 300 Engravings, 25s.

[1875]

DISEASES OF THE EYE

Illustrations of, with an Account of their Symptoms, Pathology, and Treatment, by HENRY POWER, F.R.C.S., M.B. Lond., Ophthalmic Surgeon to St. Bartholomew's Hospital. 8vo, with 12 Coloured Plates, 20s. [1867]

A SYSTEM OF DENTAL SURGERY

by JOHN TOMES, F.R.S., and CHARLES S. TOMES, M.A., Lecturer on Dental Anatomy and Physiology, and Assistant Dental Surgeon to the Dental Hospital of London. Second Edition, fcap 8vo, with 268 Engravings, 14s. [1873]

A MANUAL OF DENTAL MECHANICS

with an Account of the Materials and Appliances used in Mechanical Dentistry, by OAKLEY COLES, L.D.S., R.C.S., Surgeon-Dentist to the Hospital for Diseases of the Throat. Crown 8vo, with 140 Engravings, 7s. 6d. [1873]

HANDBOOK OF DENTAL ANATOMY

and Surgery for the use of Students and Practitioners by JOHN SMITH, M.D., F.R.S. Edin., Surgeon-Dentist to the Queen in Scotland. Second Edition, fcap 8vo, 4s. 6d. [1871]

EPIDEMIOLOGY;

or, the Remote Cause of Epidemic Diseases in the Animal and in the Vegetable Creation, by JOHN PARKIN, M.D.; F.R.C.S. Part I, 8vo, 5s. [1873]

GERMINAL MATTER AND THE CONTACT THEORY:

an Essay on the Morbid Poisons by JAMES MORRIS, M.D. Lond. Second Edition, crown 8vo, 4s. 6d. [1867]

DISEASE GERMS;

and on the Treatment of the Feverish State, by LIONEL S. BEALE, M.B., F.R.C.P., F.R.S., Physician to King's College Hospital. Second Edition, crown 8vo, with 28 Plates, 12s. 6d. [1872]

THE GRAFT THEORY OF DISEASE

being an Application of Mr. DARWIN'S Hypothesis of Pangenesis to the Explanation of the Phenomena of the Zymotic Diseases, by JAMES ROSS, M.D. 8vo, 10s. [1872]

ZYMOTIC DISEASES:

their Correlation and Causation by A. WOLFF, F.R.C.S. Post 8vo, 5s. [1872]